AP® Calculus BC
Prep Plus
2020 & 2021

PUBLISHING

New York

ACKNOWLEDGEMENTS

Lead Editor
Brandon Deason, MD

AP Expert
Adam Robison, MS, MA

Special thanks to our faculty writers and editors on this edition:
Jonathan Habermacher, Jack Hayes

Special thanks to the following for their contributions to this text: Laura Aitcheson, Matthew Callan, Michael Cook, M. Dominic Eggert, Tim Eich, Dan Frey, Adam Grey, Allison Harm, Jonathan Habermacher, Katy Haynicz-Smith, Justin Heftel, Rebecca Knauer, Jessica Knoch, Liz Laub, Terry McMullen, Jenn Moore, Kristin Murner, Christopher Murphy, Oscar Velazquez, Shayna Webb-Dray, Lee Weiss, Dan Wittich, Michael Wolff, Jessica Yee, and Nina Zhang.

AP® is a registered trademark of the College Board, which was not involved in the production of, and does not endorse, this product.

This publication is designed to provide accurate information in regard to the subject matter covered as of its publication date, with the understanding that knowledge and best practice constantly evolve. The publisher is not engaged in rendering medical, legal, accounting, or other professional service. If medical or legal advice or other expert assistance is required, the services of a competent professional should be sought. This publication is not intended for use in clinical practice or the delivery of medical care. To the fullest extent of the law, neither the Publisher nor the Editors assume any liability for any injury and/or damage to persons or property arising out of or related to any use of the material contained in this book.

TABLE OF CONTENTS

Table of Contents

PART 4: INTEGRALS AND THE FUNDAMENTAL THEOREM OF CALCULUS

Table of Contents

PART 1

Getting Started

CHAPTER 1

What You Need to Know About the AP Calculus BC Exam

Congratulations—you have chosen Kaplan to help you get a top score on your AP Calculus BC exam. Kaplan understands your goals and what you're up against: conquering a tough exam while participating in everything else that high school has to offer.

You expect realistic practice, authoritative advice, and accurate, up-to-the-minute information on the exam. And that's exactly what you'll find in this book. To help you reach your goals, we have conducted extensive research and have incorporated insights from an AP Expert who has 15 years of experience with AP Calculus.

ABOUT THE AP EXPERT

Adam Robison has earned a BA in Mathematics from SUNY Geneseo, an MS in Mathematics Learning and Teaching from Drexel University, and an MA in Teaching from St. Cloud State University. Adam has been a teacher at Naples High School for fifteen years and also serves as an adjunct instructor in the Mathematics Department at Finger Lakes Community College.

As a member of the New York State Master Teachers Program, he is privileged to collaborate with some of the most dedicated and motivated teachers in the state, including a Professional Learning Team focused specifically on AP Calculus. Adam has contributed AP Calculus review questions for Barron's Educational Series. To promote mathematics in his area, Adam hosts the AMC, is a coach for the M3 challenge, and assists with the local robotics team to explore the application of mathematics.

ABOUT THE AP CALCULUS EXAM

There are actually two AP Calculus tests offered to students, a Calculus AB exam and a Calculus BC exam. The BC exam covers some more advanced topics; it's not necessarily a harder test, but its scope is broader. Most students opt to take the AB exam because they find it somewhat easier, but the BC exam is worth more credit at many universities.

Your decision to take the AP Calculus BC exam involves many factors, but, in essence, it boils down to a question of choosing between three hours of time spent on the AP Calculus exam and a significant amount of time spent in a college classroom. Depending on the college, a score of 3, 4, or 5 on the AP Calculus exam can allow you to leap over the introductory courses and jump right into more advanced classes. These advanced classes are usually smaller, more specifically focused, more intellectually stimulating, and simply more interesting than a basic course. If you are concerned solely about fulfilling your mathematics requirement so you can get on with your study of pre-Columbian art, Elizabethan music, or some other area apart from calculus, the AP exam can help you there, too. If you do well on the AP Calculus exam, you may never have to take a math class again, depending on the requirements of the college you choose.

ABOUT THIS BOOK

If you're holding this book, chances are you are already gearing up for the AP Calculus exam. Your teacher has spent the year cramming your head full of the math know-how you will need to have at your disposal. But there is more to the AP Calculus exam than math know-how. You have to be able to work around the challenges and pitfalls of the test—and there are many—if you want your score to reflect your abilities. Studying math and preparing for the AP Calculus exam are not the same thing. Rereading your textbook is helpful, but it's not enough.

That's where this book comes in. This guide offers much more than a review of basic content. We'll show you how to put your knowledge to brilliant use on the AP exam through structured practice and efficient review of the areas you need to work on most. We'll explain the ins and outs of the exam structure and question formats so you won't experience any surprises. We'll even give you test-taking strategies that successful students use to earn high scores.

Are you ready for your adventure in the study and mastery of everything AP Calculus BC? Good luck!

EXAM STRUCTURE

The AP Calculus BC exam consists of two sections, each with two parts, as outlined in the table below.

Section	Part	Question Count	Allotted Time
Section I: Multiple-Choice	Part A: No Calculator	30	60 minutes
	Part B: Calculator	15	45 minutes
	TOTAL	45	105 minutes
Section II: Free-Response	Part A: Calculator	2	30 minutes
	Part B: No Calculator	4	60 minutes
	TOTAL	6	90 minutes

If you do the math, your pacing during the test should look roughly like this:

- Multiple-choice with no calculator: 2 minutes per question

- Multiple-choice with calculator: 3 minutes per question

- Free-response: 15 minutes per question

✔ **AP Expert Note**

You almost certainly will use most—probably *all*—of the 90 minutes allotted for Section II. Although these free-response problems are long and are made up of multiple parts, they don't usually cover an obscure topic. Instead, they take a fairly basic calculus concept and ask you a *bunch* of questions about it. It's a lot of calculus work, but it's fundamental calculus work.

EXAM SCORING

Once you complete your AP exam, it will be sent to the College Board for grading. Student answer sheets for the multiple-choice section (Section I, Part A) are scored by machine. Scores are based on the number of questions answered correctly. No points are deducted for wrong answers, and no points are awarded for unanswered questions.

The free-response section (Section II) is evaluated and scored by hand by trained AP readers. Rubrics based on each specific free-response prompt are released on the AP central website after the exams are administered.

The score from the multiple-choice section of the exam counts for 50% of your total exam score. The other 50% is the combined score from the six free-response questions (which each count for 8.3 percent of your score).

After your total scores from Sections I and II are calculated, your results are converted to a scaled score from 1 to 5. The range of points for each scaled score varies depending on the difficulty of the exam in a particular year, but the significance of each value is constant from year to year. According to the College Board, AP scores should be interpreted as follows:

    5 = Extremely well qualified

    4 = Very well qualified

    3 = Qualified

    2 = Possibly qualified

    1 = No recommendation

Colleges will generally not award course credit for any score below a 3, with more selective schools requiring a 4 or 5. Note that some schools will not award college credit regardless of your score. Be sure to research schools that you plan to apply to so you can determine the score you need to aim for on the AP exam.

REGISTRATION AND FEES

To register for the exam, contact your school guidance counselor or AP coordinator. If your school does not administer the AP exam, contact the College Board for a listing of schools that do.

There is a fee for taking AP exams. The current cost can be found at the official exam website listed below. For students with acute financial need, the College Board offers a fee reduction equal to about one third of the cost of the exam. In addition, most states offer exam subsidies to cover all or part of the remaining cost for eligible students. To learn about other sources of financial aid, contact your AP Coordinator.

For more information on all things AP, contact the Advanced Placement Program:

Phone: (888) 225-5427 or (212) 632-1780

Email: apstudents@info.collegeboard.org

Website: https://apstudent.collegeboard.org/home

CHAPTER 2

How to Get the Score You Need

HOW TO GET THE MOST OUT OF THIS BOOK

Kaplan's *AP Calculus BC Prep Plus* contains precisely what you'll need to get the score you want in the time you have to study. The unique format of this book allows you to customize your prep experience to make the most of your time.

Start by going to kaptest.com/moreonline to register your book and get a glimpse of the additional online resources available to you.

BOOK FEATURES

Specific Strategies

This chapter features both general test-taking strategies and strategies tailored specifically to the AP Calculus BC exam. You'll learn about the types of questions you'll see on the official exam and how to best approach them to achieve a top score. This book also provides some calculator basics (Chapter 3). In addition, the final two chapters are devoted specifically to questions that can't be solved without a calculator (Chapter 19) and free-response questions (Chapter 20).

Customizable Study Plans

We recognize that every student is a unique individual, and there is no single recipe for success that works for everyone. To give you the best chance to succeed, we have developed three customizable study plans. Each offers guidance on how to make the most of your available study time. With the instructions offered in the following section of this chapter, you'll be able to select and customize the study plan that is right for you, maximizing your chances of earning the score you need.

A Review of ALL the Relevant Subjects Based on Specific Test-like Questions

The best test-taking strategies can't help you get a good score if you don't know the difference between a limit and a derivative. At its core, the AP Calculus exam covers a wide range of calculus topics, and it's necessary for you to know that material if you are to do well on the test. However, chances are good that you're already familiar with these subjects, so an exhaustive review is not needed. In fact, it would be a waste of your time. No one wants that, so we've tailored our review section to focus on how the relevant topics typically appear on the exam and what you need to know to answer the questions correctly. You'll also see High-Yield icons for certain topics. These icons identify topics that *ALWAYS* appear on the test.

Pre- and Post-Chapter Quizzes

Each chapter begins and ends with a mini-quiz—one to test what you already know about the given topic ("Test What You Already Know") and one to test what you learned by working through the practice sets in the chapter ("Test What You Learned"). You can use the results of these quizzes to guide your studying and chart your progress in mastering the Learning Objectives.

Full-Length Practice Exams

In addition to all of the exam-like practice questions featured in the chapter quizzes, we have provided six full-length practice exams. These full-length exams mimic the multiple-choice and free-response questions on the real AP exam. Taking a full-length practice exam gives you an idea of what it's like to answer exam-like questions for about three hours. Granted, that's not exactly a fun experience, but it is a helpful one. And the best part is that it doesn't count; mistakes you make on our practice exams are mistakes you won't make on your real exam.

After taking each practice exam, you'll score your multiple-choice and free-response sections using the answers and explanations. Then, you'll navigate to the scoring section in your online resources and input your raw scores to see what your overall score would be with a similar performance on the official exam.

Online Quizzes

While this book contains hundreds of exam-like multiple-choice questions, you may still find yourself wanting additional practice on particular topics. That's what the online quizzes are for! Your online resources contain additional quizzes for content chapters 4 throught 14. Go to kaptest.com/moreonline to find them all.

CHOOSING THE BEST STUDY PLAN FOR YOU

There's a lot of material to review before the AP exam, so it's essential to have a solid game plan that optimizes your available study time. The tear-out sheet in the front of the book consists of three separable bookmarks, each of which covers a specific, customizable study plan. You can use one of these bookmarks both to hold your place in the book and to keep track of your progress in completing one of these study plans. But how do you choose the study plan that's right for you?

Fortunately, all you need to know to make this decision is how much time you have to prep. If you have two months or more with plenty of time to study, then we recommend using the Two Months

Plan. If you only have about a month, or if you have more than a month but your time will be split among competing priorities, you should probably choose the One Month Plan. Finally, if you have less than a month to prep, your best bet is the Two Weeks Plan.

Regardless of your chosen plan, you have flexibility in how you follow the instructions. You can stick to the order and timing that the plan recommends or tailor those recommendations to fit your particular study schedule. For example, if you have six weeks before your exam, you could use the One Month Plan but spread out the recommended activities for Week 1 across the first two weeks of your studying.

After you've made your selection, tear out the perforated study plan page, separate the bookmark that contains your choice of plan, and use it to keep track of both your place in the book and your progress in the plan. You can further customize any of the study plans by skipping over chapters or sections that you've already mastered or by adjusting the recommended time to better suit your schedule. Don't forget to also use the guidelines in the Rapid Review chapters to further customize how you study.

GENERAL TEST-TAKING STRATEGIES

Pacing

Because many tests are timed, proper pacing allows someone to attempt as many questions as possible in the time allotted. Poor pacing causes students to spend too much time on some questions while leaving others untouched because they run out of time before getting a chance to attempt every problem.

Two-Pass System

Using the two-pass system is one way to help pace yourself better on a test. The key idea is that you don't simply start with the first question and trudge onward from there. Instead, you start at the beginning, but take a first pass through the test answering all the questions that are easy for you. If you encounter a tough problem, you spend only a small amount of time on it and then move on in search of easier questions that might come after that problem. This way, you don't get bogged down on a tough problem when you could be earning points answering later problems that you *do* know. On your second pass, go back through the section and attempt all the tougher problems that you passed over the first time. You should be able to spend a little more time on them, and this extra time might help you answer the problem. Even if you don't reach an answer, you might be able to employ techniques such as process of elimination to cross out some answer choices and then take a guess.

To make this strategy work for you, you must be organized and careful. You cannot just flip haphazardly through the test, hoping to find problems you can solve. You must go through the test in an orderly way, analyzing problems quickly and deciding how you will handle them. You will also need to mark the questions that you choose to save for later and make sure that you *really do* come back and answer them. For each question you skip, be careful to skip that number on the answer grid as well until you go back and answer it.

Process of Elimination

On every multiple-choice test, the correct answer is given to you. The only difficulty lies in spotting the correct answer hidden among incorrect choices. Even so, the multiple-choice format means you don't have to pluck the answer out of the air. Instead, if you can eliminate the answer choices you know are incorrect and only one choice remains, that must be the correct answer.

Patterns and Trends

The key word here is the *standardized* in "standardized testing." Standardized tests don't change greatly from year to year. Sure, the particular questions won't be the same and different topics will be covered from one administration to the next, but there will also be a lot of overlap from year to year. That's the nature of standardized testing: If the test changed dramatically each time it came out, it would be useless as a tool for comparison. Because of this, certain patterns can be uncovered about any standardized test. Learning about these trends and patterns can help students do better on tests they have not taken previously.

The Right Approach

Having the right mindset plays a large part in how well people do on a test. Those who are nervous about the exam and hesitant to make guesses often fare much worse than students with an aggressive, confident attitude. Students who start with the first question and struggle methodically forward through each problem don't score as well as students who deal with the easy questions before tackling the harder ones. People who take a test cold have more problems than those who take the time to learn about the test beforehand. In the end, factors like these are what separate people who are good test takers from those who struggle even when they know the material.

These points are valid for every standardized test, but they are quite broad. The next section of this chapter will discuss how these general ideas can be modified to apply specifically to the AP Calculus exam. These test-specific strategies—combined with the factual information covered in this book's review—are the one-two punch that will help you succeed on this specific test.

HOW TO APPROACH THE MULTIPLE-CHOICE QUESTIONS

The multiple-choice questions on the AP Calculus exam count for 50% of your total score. The multiple-choice section consists of two parts: Part A contains 30 multiple-choice questions for which you are not allowed to use your graphing calculator, and Part B contains 15 multiple-choice questions for which you may (and, in fact, will most likely need to) use your calculator.

Although you might not like multiple-choice questions, there's no denying the fact that it's easier to guess on a multiple-choice question than it is to guess the correct answer to an open-ended question. Contrast this with Section II of the AP Calculus exam, which accounts for the other 50% of your score. In that section, if you don't know how to work a problem, you have to write down what you *do* know and hope to earn at least part of the available points.

Every multiple-choice question on the AP Calculus exam can be described as a "stand-alone" question. A stand-alone question covers a specific topic and is not part of a set; the question that follows it covers a different topic. Here's a typical question:

The cost of producing x units of a certain item is $c(x) = 2000 + 8.6x + 0.5x^2$. What is the instantaneous rate of change of c with respect to x when $x = 300$?

(A) 297.2   (B) 300.0   (C) 308.6   (D) 313.6

In this question, you get some information, and then you're expected to answer the question. Where this question occurs in the test makes no difference, because there's no patterned order of difficulty in which questions are presented on the AP Calculus exam. Tough questions are scattered between easy and moderately difficult questions.

The stand-alone questions look like a bunch of disconnected calculus questions one after the other and that's just what they are. Because they aren't connected to each other, there's no reason you have to answer these questions in sequential order. The two-pass system, discussed next, should be used here. You can tweak the general idea of the two-pass system and apply it specifically to the AP Calculus exam.

The Two-Pass System on the AP Calculus Exam

If you wanted to, you could take all the AP Calculus questions and arrange them in a spectrum ranging from "fastest to answer" to "slowest to answer." For example, questions involving a graph are often much faster to answer, as long as you can interpret the visual data correctly.

Picking out questions with graphs is not an especially critical way to analyze the exam questions, but some students do no more than that. You should realize that the more advanced your pacing system is, the more time you might have at the end of Section I to answer the questions that you find difficult. To further refine your two-pass approach before Test Day, draw up two lists of exam topics. Label one list "Calculus Concepts I Enjoy and Know About" and label the other list "Calculus Concepts That Are Not My Strong Points."

When you get ready to begin the multiple-choice section, keep these two lists in mind. On your first pass through the section, answer all the questions that deal with concepts you like and know a lot about. If a question covers a subject that's not one of your strong points, skip it and come back on your second pass. The overarching goal is to use the time available to answer the maximum number of questions correctly.

This refinement of the basic two-pass system should give you a clear idea about how to approach the multiple-choice section of the AP Calculus exam. Now that you've got an idea of the correct approach, let's talk a bit about the correct mindset for test-taking.

Comprehensive, Not Sneaky

Some tests are sneakier than others. A sneaky test has questions that are written in convoluted ways; they're designed to trip you up mentally and manipulate your score by using a host of other little tricks. Students taking a sneaky test often have the proper facts, but get many questions wrong because of traps in the questions themselves.

The AP Calculus exam is *not* a sneaky test. It aims to see how much calculus knowledge you have. To do this, it asks a wide range of questions from an even wider range of calculus topics. The exam tries to cover as many different calculus facts as it can, which is why the questions jump from topic to topic. The test makers work hard to design the test so that it is comprehensive, which means that students who only know one or two calculus topics will soon find themselves struggling.

Understanding these facts about how the test is designed can help you to answer its questions. The AP Calculus exam is comprehensive, not sneaky; it makes questions hard by asking about hard subjects, not by using rhetorical tricks to create hard questions.

> ✔ **AP Expert Note**
>
> **Trust your instincts when guessing. If you think you know the right answer, chances are that you dimly remember the topic being discussed in your AP course. The test is about knowledge, not traps, so trusting your instincts will help more often than not.**

You don't have enough time to think deeply about every tough question, so trusting your instincts can keep you from getting bogged down and wasting precious time on a problem. Some of your educated guesses are likely to be incorrect, but again, the point is not to get a perfect score. A perfect score would certainly be nice, but most people are going to lose at least some points on the AP exam (and may still get a 5). Your basic goal should be to get as good a score as you can; surviving hard questions by going with your gut feelings can help you to achieve this aim.

On other questions, though, you might have no inkling of what the correct answer should be. In that case, turn to the following key idea.

Think "Good Math!"

The AP Calculus exam rewards good mathematicians. It covers fundamental topics and expects you to use logical thinking and good mathematical techniques to answer questions. What the test doesn't want is sloppy math or sloppy thinking. It doesn't want answers that are factually incorrect, too extreme to be true, or irrelevant to the topic.

Still, bad math answers invariably appear, because it's a multiple-choice test that includes three incorrect answer choices along with the one correct answer. So if you don't know how to answer a question, look at the answer choices and think "Good Math." This may lead you to find some poor answer choices that can be eliminated. Here's an example:

Which of the following gives the derivative of the function $f(x) = x^2$ at the point $(2, 4)$?

(A) $\displaystyle\lim_{h \to 0} \frac{(x+2)^2 - x^2}{4}$

(B) $\displaystyle\lim_{h \to \infty} \frac{(2+h)^2 - 2^2}{h}$

(C) $\displaystyle\lim_{h \to 0} \frac{(2+h)^2 - 2^2}{h}$

(D) $\displaystyle\lim_{h \to 0} \frac{(4+h)^2 - 4^2}{h}$

✔ **AP Expert Note**

For each multiple-choice question, no points are deducted for wrong answers, but no points are awarded for unanswered questions. Therefore, you should answer every question, even if you have to guess.

Students who recall the definition of a derivative based on limits will have no problem determining that **(C)** is the correct answer. However, if you don't know how to answer this question, you can use "Comprehensive, not Sneaky" and "Good Math" to give yourself a chance at guessing the right answer. The limit definition involves h approaching 0 (because it's getting very small), so you can eliminate (B) right away (because h is approaching ∞). Choice (A) can also be eliminated, because there is nowhere to plug in 0 for h (all the terms have x's in them, but no h's). Finally, think back to Algebra I: The first coordinate in an ordered pair is x, and this function is all about x^2, so you should see a 2^2 in the definition, not a 4^2. This means you can eliminate (D), leaving **(C)** as the only remaining choice.

HOW TO APPROACH THE FREE-RESPONSE QUESTIONS

Section II of the AP Calculus exam is worth 50% of your grade. The section consists of two parts, just like the multiple-choice section. Part A contains two free-response questions for which you are allowed (and in fact, will most likely need) to use your graphing calculator. Part B contains four free-response questions for which you may not use your calculator.

The Good News ...

The good news is that the topics covered in these questions are usually fairly common calculus topics. You won't see trick questions asking about an obscure subject. However, the fact that a topic is well known doesn't mean the problem will be simple or easy. All free-response questions are divided into parts—that is, they are stuffed with smaller questions. You won't usually get one broad question like, "Is a Riemann sum ever really happy?" Instead, you'll get an initial setup followed by questions labeled (a), (b), (c), and so on.

✔ **AP Expert Note**

Sometimes these parts are related to each other, and sometimes they are independent.

Be Organized and Use Only the Space You Need

Expect to write at least one paragraph (or provide multistep equations) for each lettered part of a free-response question, and to clearly label the part you are answering (e.g., [a] or [part a]). Use proper notation and always include units where applicable. Don't try to fill up every inch of the space provided. If you only use half the space, but have a complete solution, that's fine. The test makers provide extra space for students who write big or for corrections. You don't want to add extra information (that may contain a mistake) if you already have the correct answer written on the page.

How You Earn Points: Scoring Rubrics

For the free-response questions, you receive points for responding properly to each subquestion prompt. The more points you score, the better off you are on that question. Going into the details about how points are scored would make your head spin, but in general, the AP Calculus readers have a rubric that works as a blueprint for a good answer. Every subsection of a question has one to four key ideas attached to it. If you write about one of those ideas, you earn yourself a point. Readers always use the same rubric for a question and all questions are evaluated using a rubric. Any reader who reads your exam should, in theory, award you the same number of points on a given question. Readers check and cross-check each other to ensure that each answer is evaluated in the same way.

There's a limit to how many points you can earn on a single subquestion and there are other complex rules guiding the grading of the AP exams, but it boils down to this: Writing smart things about each question will earn you points on that question.

> ✔ **AP Expert Note**
>
> **Review the sample Scoring Rubrics provided. Reading through the rubrics will give you an even better idea of how to earn maximum points on Test Day.**

Don't rush or be unnecessarily terse. You have about 15 minutes for each free-response problem. Use that time to be as precise as you can be for each part of the question. If the answer to one part depends on the answer to the previous part, but you don't think you got the first part right, you should still forge ahead using your answer. In most cases, the grader will award you points for the second part as long as your math is valid relative to that particular part of the question.

Don't Panic

If you get a question on a subject you don't know well, things might look grim for that problem. Still, take heart. Quite often, you'll be able to earn at least some points on every question because there will be some subquestion or segment in each question that you know. Remember, the goal is not perfection. If you can ace four of the questions and slug your way to partial credit on the other two, you will put yourself in position to get a good score on the entire section. That's the big picture, so don't lose sight of it just because you don't know the answer to every subquestion of every question.

Calculator Questions

Finally, because you are allowed to use a calculator on Part A of Section II, you should *expect* to use it. There's no point in permitting a calculator for part of the exam if you aren't going to actually need it. This isn't 100 percent certain, of course, because nothing on an unknown test ever is, but it is a good guess. As a general rule, you can expect to have to:

- Find the zeros of a complicated function (for example, to determine where a particle is at rest)

- Evaluate a messy derivative at a specified value (for example, to find an instantaneous rate of change)

- Calculate a definite integral (for example, to find the volume of a solid of revolution)

Be sure to work through the questions in Chapter 19 to review how to complete each of the tasks described above using a graphing calculator.

Practice Using These Tips

Be sure to use all the strategies discussed in this chapter when taking the practice exams. Trying out the strategies there will help you become comfortable with them, and you should be able to put them to good use on the real exam.

Of course, all the strategies in the world can't save you if you don't know anything about calculus. Chapters 4–14 of this book will help you review the primary concepts and facts that you can expect to encounter on the AP Calculus BC exam.

COUNTDOWN TO THE EXAM

This book contains detailed review, guidance, and practice for you to utilize in the weeks leading up to your AP exam. In the final few days before your exam, we recommend the following steps.

Three Days Before the Exam

Take a full-length practice exam under timed conditions. Use the techniques and strategies you've learned in this book. Approach the exam strategically, actively, and confidently. (Note that you should *not* take a full-length practice exam with fewer than 48 hours left before your real exam. Doing so will probably exhaust you and hurt your score.)

Two Days Before the Exam

Go over the results of your latest practice exam. Don't worry too much about your score or whether you got a specific question right or wrong. Instead, examine your overall performance on the different topics, choose a few of the topics where you struggled the most, and brush up on them one final time.

Know exactly where you're going to take the official exam, how you're getting there, and how long it takes to get there. It's probably a good idea to visit your testing center sometime before the day of your exam so that you know what to expect: what the rooms are like, how the desks are set up, and so on.

The Night Before the Exam

Do not study! You cannot cram for a test as extensive as the AP Exam. Worse, pulling an all-nighter will simply deplete your stamina ahead of the exam. If you feel you must review some AP material, only do so for a little while and stick to broad review (such as the Essential Content sections of this book). The best, most effective way to prepare for the AP Exam at this point is to rest the night beforehand.

Get together an "AP Exam Kit" containing the following items:

- A few No. 2 pencils (Pencils with slightly dull points fill the ovals better; mechanical pencils are NOT permitted.)

- A few pens with black or dark blue ink (for the free-response questions)

- Erasers

- A watch (as long as it doesn't have Internet access, have an alarm, or make noise)

- Your 6-digit school code (Home-schooled students will be provided with their state's or country's home-school code at the time of the exam.)

- Photo ID card

- Your AP Student Pack

- If applicable, your Student Accommodation Letter verifying that you have been approved for a testing accommodation such as braille or large-type exams

Make sure that you don't bring anything that is *not* allowed in the exam room. You can find a complete list at the College Board's website (https://apstudent.collegeboard.org/home). Your school may have additional restrictions, so make sure you get this information from your school's AP coordinator prior to the exam.

Again, try to relax. Read a good book, take a hot shower, watch something you enjoy. Go to bed early and get a good night's sleep.

The Morning of the Exam

Wake up early, leaving yourself plenty of time to get ready without rushing. Dress in layers so that you can adjust to the temperature of the testing room. Eat a solid breakfast: something substantial, but nothing too heavy or greasy. Don't drink a lot of coffee, especially if you're not used to it; bathroom breaks cut into your time, and too much caffeine is a bad idea. Read something as you eat breakfast, such as a newspaper or a magazine; you shouldn't let the exam be the first thing you read that day.

Leave extra early so that you can ensure you are on time to the testing location. Allow yourself extra time for any traffic, mass transit delays, and/or detours.

During the Exam

Breathe. Don't get shaken up. If you find your confidence slipping, remind yourself how well you've prepared. You know the structure of the exam; you know the material covered on it; you've had practice with every question type.

If something goes really wrong, do not panic! If you accidentally misgrid your answer page or put the answers in the wrong section, raise your hand and tell the proctor. He or she may be able to arrange for you to regrid your exam after it's over, when it won't cost you any time.

After the Exam

You might walk out of the AP exam thinking that you blew it. This is a normal reaction. Lots of people—even the highest scorers—feel that way. You tend to remember the questions that stumped you, not the ones that you knew. Keep in mind that almost nobody gets everything correct. You can still score a 4 or 5 even if you get some multiple-choice questions incorrect or miss several points on a free-response question.

We're positive that you will have performed well and scored your best on the exam because you followed the Kaplan strategies outlined in this chapter and reviewed all the content provided in the rest of this book. Be confident and celebrate the fact that, after many hours of hard work and preparation, you will have just completed the AP Calculus BC exam!

CHAPTER 3

Calculator Basics

WHEN YOU CAN USE YOUR GRAPHING CALCULATOR

Calculators are good—that's what professional organizations for math teachers have decided. This is lucky for you, because it means you can use a calculator on the AP exam. In fact, on some parts of the exam, you are *expected* to use a calculator—a *graphing calculator*, to be specific. Graphing calculators have become a standard, essential part of AP Calculus. You should be using yours on a regular basis to become comfortable with it.

You'll need your calculator for Section I, Part B (which consists of 15 multiple-choice questions) and Section II, Part A (which consists of 2 free-response questions). Let's do the math here. This means that for 34 out of the 51 questions on the test (more than 60 percent of the exam), you *cannot* use a calculator. It's not a crutch or a cure-all, even if you have a calculator that takes derivatives, computes integrals, and does your laundry. The bottom line is that you need to know calculus and know it well to succeed on the AP exam.

WHAT YOU CAN EXPECT TO DO WITH YOUR CALCULATOR

As a general rule, you can expect to:

- Adjust the viewing window to zoom in on a particular part of a function's graph

- Find the zeros of a complicated function (for example, to determine where a particle is at rest)

- Evaluate a messy derivative at a specified value (for example, to find an instantaneous rate of change)

- Calculate a definite integral (for example, to find the volume of a solid of revolution)

Be sure to work through the questions in Chapter 19 to review how to complete each of the tasks described above using a graphing calculator.

You can bring up to two graphing calculators into the exam with you, along with whatever programs they contain. You can use whatever shortcuts you've programmed into your calculator. To maintain the integrity of the test, you can't take any test information out in your calculator.

CHOOSING THE RIGHT CALCULATOR FOR YOU

There is a huge variety of graphing calculators available on the market, and their capabilities have increased as technology has advanced. Some of the newer models are much more powerful than their older, first-generation siblings. It does not really matter which particular brand or type of calculator you have, though you should make certain the calculator meets your needs. What is truly important is that you know how to *use* your calculator.

The AP Calculus committee works hard to develop questions that can be solved with any appropriate calculator, but it can't develop tests that are fair to all students if the range of calculator capabilities is too large. To level the playing field, your calculator must meet a minimum performance level to ensure that you can solve all of the problems on the test. Some calculators with more advanced, complicated features are not allowed because the AP Calculus Exam Committee feels that they would give students who use them an unfair advantage. The trick, then, is to find a calculator that can do everything you need it to do, but not so much that it will be barred from the exam.

The people who write the test assume that your calculator can:

- Plot a graph in the appropriate viewing window
- Find the zeros of a function (a technique often used to solve equations)
- Calculate the value of the derivative of a function at a point
- Calculate the value of a definite integral numerically

If you can perform these functions with your calculator, then you have enough computational power to solve any problem on the AP exam.

A non-graphing scientific calculator simply isn't enough on the exam—they don't give you enough computing power. You can't use tools that provide too much computing power either, which means no computers—like laptops—and no mini-computers—including pocket organizers, pen-input devices like tablets, or anything with a QWERTY keyboard, such as a TI-92 Plus. The AP Course Description has a list of approved calculators; read it and make sure yours is on the list. Then make sure that you *really* know your calculator. Your calculator is your friend, so treat it like one. Use it often. Take it with you. Learn its secrets. Let it help you find answers in class and on tests.

Does a Fancier Calculator Help You More?

The authors of the AP exam try to write questions that present an equal challenge to students, regardless of what sort of calculators they may have. They try, but they don't always succeed, and it is definitely to your advantage to have the most advanced calculator you can.

For example, suppose you are asked to find the value of c that satisfies the mean value theorem for $f(x) = x \ln x$ on the closed interval [2, 4].

Using a less advanced calculator, you need to compute the derivative of f by hand; i.e., you need to use the product rule to compute $f'(x) = \ln x + 1$. Then you need to solve the equation $\frac{4 \cdot \ln 4 - 2 \cdot \ln 2}{4 - 2} = \ln c + 1$. You need a calculator to solve this equation, and at this point there are lots of ways to proceed.

More advanced calculators can handle symbolic manipulation, compute the derivative for you, and solve the entire problem in one line of code. For example, to solve this problem using a TI-89, on the home screen type $y = x(\ln x)$. Press ENTER, then type F2: *Solve* $(d(y(x), x, c) = (y(4) - y(2))/(4 - 2), c$. Press ENTER and your answer appears. With an advanced calculator, you don't need to do any computations yourself.

Distinctions Among Different Types of Calculators

Lower-level calculators cannot perform symbolic manipulations. That is, they cannot multiply or factor polynomials, differentiate or integrate symbolically, or take limits. They evaluate a polynomial at a point; they can evaluate the derivative of a function at the point, but they can't tell you the derivative function. They can compute a numerical approximation of a definite integral, but they can't tell you the exact value. For example, a lower-level graphing calculator will compute

$\int_1^3 \frac{1}{\sqrt{x}}\, dx$ as 1.4641016 but it will not give the exact answer $2\sqrt{3} - 2$. It also cannot compute anti-derivatives such as $\int \frac{1}{\sqrt{x}}\, dx = 2\sqrt{x}$.

The TI-83 Plus is an example of a lower-level graphing calculator. A more powerful graphing calculator can perform symbolic manipulations. It can multiply and factor polynomials, differentiate and integrate symbolically, and take limits of functions. It shows the answer symbolically as an equation, when appropriate. It can show a precise answer with square roots and fractions. The TI-89 is an example of a more powerful model.

> ✔ **AP Expert Note**
>
> Be sure to use the calculator that you've been using through the school year. The TI-83/84 models have comparable keystrokes, but they are significantly different from the TI-89 calculators. So, it would not be wise to switch to an unfamiliar calculator right before the exam.

There are lots of other calculators that fall into each of these categories. Yours, whatever it is, is just fine as long as you know how to use it. Many calculators use similar keystrokes. The important thing is to become familiar with *your* calculator. If you want to perform an operation and don't know how to do it, refer to your owner's manual. If you've lost yours (have you cleaned your room lately?), most calculator companies have a free copy of the instruction manual on their websites.

USING CALCULATORS WITH THIS BOOK

In subsequent chapters, we'll help you to get familiar with the ways you'll need to use your calculator on the AP exam. For calculator-specific questions, we'll give general guidelines for solving problems with any graphing calculator and detailed instructions for solving problems with a TI-83 Plus. The TI-83 is an example of a basic, no-frills, graphing calculator.

What we won't do is tell you how to get started with your graphing calculator. You should refer to your owner's manual for that basic information. For example, we'll assume that you know how to enter a function into the calculator (e.g., using the $\boxed{Y=}$ button on a TI-83) and that you know how to set the viewing window for a graph.

Throughout the book, in the chapter Practice Sets, questions that require a calculator will have a calculator icon to the left of the question number. For example:

The growth of a population P is modeled by the differential equation $\dfrac{dP}{dt} = 0.713P$. If the population is 2 at $t = 0$, what is the population at $t = 5$?

    (A) 5.565   (B) 20.401   (C) 50.810   (D) 70.679

> ✔ **AP Expert Note**
>
> **The correct answer to the question above is (D), by the way, and your calculator can do all the heavy lifting here, with just a bit of conceptual understanding on your part to get started.**

There's also an entire chapter (Chapter 19) devoted to questions that require the use of a calculator. Finally, in the Practice Exams, the sections are divided into No Calculator and Calculator, just as they are on the real exam. All in all, you should get plenty of practice using your calculator.

PART 2

Limits

CHAPTER 4

Basic Limits

LEARNING OBJECTIVES

4.1 Evaluate limits graphically, including one-sided limits

4.2 Evaluate limits algebraically

4.3 Evaluate limits of composite functions from graphs and tables

4.4 Determine when a limit does not exist

TEST WHAT YOU ALREADY KNOW

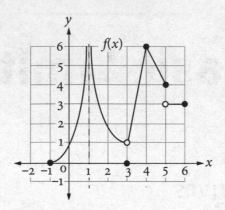

1. The graph of the function f is shown above. Which of the following statements is false?

(A) $\lim\limits_{x \to 1} f(x) = \infty$

(B) $\lim\limits_{x \to 3} f(x) = 1$

(C) $\lim\limits_{x \to 4} f(x) = 6$

(D) $\lim\limits_{x \to 5} f(x) = 4$

2. $\lim\limits_{x \to 3} \dfrac{\dfrac{1}{x-1} - \dfrac{1}{2}}{x-3} =$

(A) $-\dfrac{1}{2}$　　　(B) $-\dfrac{1}{4}$　　　(C) 0　　　(D) $\dfrac{1}{2}$

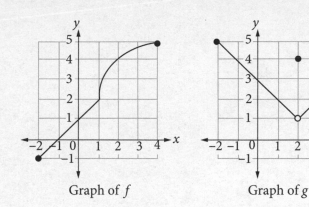

Graph of f Graph of g

3. The graphs of the functions f and g are shown above. What is the value of $\lim\limits_{x \to 2} f\big(g(x)\big)$?

(A) 2   (B) 3   (C) 4   (D) 5

4. $\lim\limits_{x \to 2} \dfrac{|x-2|}{x-2} =$

(A) -1 (B) 0   (C) 1   (D) nonexistent

Answers to this quiz can be found at the end of this chapter.

4.1 EVALUATING LIMITS GRAPHICALLY

To answer a question like this:

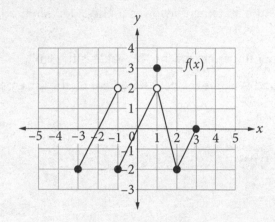

Consider the graph of the function *f* shown above. Which of the following statements is true?

(A) $\lim\limits_{x \to -3^+} f(x) = 0$

(B) $\lim\limits_{x \to -1^-} f(x) = 2$

(C) $\lim\limits_{x \to -1^+} f(x)$ does not exist.

(D) $\lim\limits_{x \to 1} f(x) = 3$

You need to know this:

Conceptually:

- Limits describe what happens as *x* *approaches* (gets close to) a number. A limit DOESN'T describe, and doesn't care about, what happens when *x* actually *equals* that number.

Notation:

- If the value of a function *f* gets close to a number *L* as *x* gets close to *c* from both sides, then you write $\lim\limits_{x \to c} f(x) = L$.

- If *f* approaches a number *L* as *x* approaches *c* from the right (i.e., if you only consider the function at values of *x* that are greater than *c*), then you write $\lim\limits_{x \to c^+} f(x) = L$.

- Similarly, if *f* approaches a number *L* as *x* approaches *c* from the left (i.e., if you only consider the function at values of *x* that are less than *c*), then you write $\lim\limits_{x \to c^-} f(x) = L$.

Limits

You need to do this:

- Evaluate each statement, one at a time.

- For each limit, lightly draw a dashed vertical line at *c*. Then trace along the graph near *c*. Trace from only one side if the statement involves a one-sided limit; otherwise, trace from both sides.

- What *y*-value does your finger approach? This value is the limit.

- Erase your markings and evaluate the next statement on the list, crossing out false statements as you go.

Answer and Explanation:

B

(A)

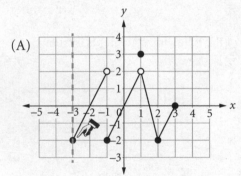

$$\lim_{x \to -3^+} f(x) = -2 \therefore \text{(A) is false.}$$

(B)

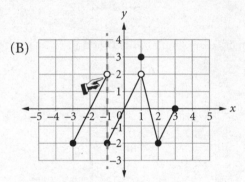

$$\lim_{x \to -1^-} f(x) = 2 \therefore \textbf{(B) is true.}$$

(C)

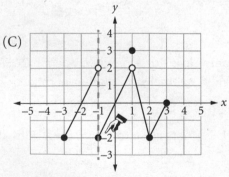

$$\lim_{x \to -1^+} f(x) = -2 \therefore \text{(C) is false.}$$

(D)

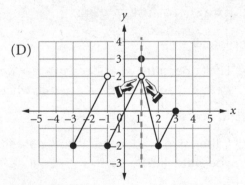

$$\lim_{x \to 1} f(x) = 2 \therefore \text{(D) is false.}$$

(B) is the only true statement and is therefore correct.

PRACTICE SET

Use the graph of *f* shown below to answer questions 1 and 2.

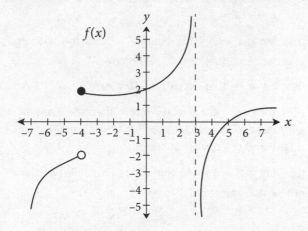

1. $\lim\limits_{x \to -4} f(x) =$

   (A) -2

   (B) 0

   (C) 2

   (D) nonexistent

2. $\lim\limits_{x \to 3^+} f(x) =$

   (A) $-\infty$

   (B) -5

   (C) 5

   (D) ∞

Use the graph of *g* shown below to answer questions 3 and 4.

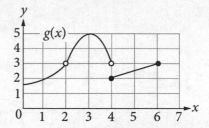

3. Which of the following statements about *g* is false?

   (A) $\lim\limits_{x \to 2} g(x) = 3$

   (B) $\lim\limits_{x \to 4^+} g(x) = 2$

   (C) $\lim\limits_{x \to 2} g(x)$ does not exist.

   (D) $\lim\limits_{x \to 4} g(x)$ does not exist.

4. Suppose $\lim\limits_{x \to k^-} g(x) = 3$. The value of *k* could be which of the following?

   (A) 2 only

   (B) 6 only

   (C) 2 and 6 only

   (D) 2, 4, and 6

ANSWERS AND EXPLANATIONS

Limits

1. D

Trace along the function from each side of $x = -4$:

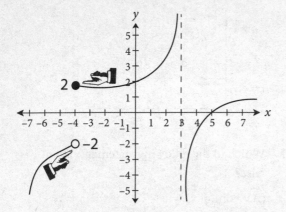

$\lim\limits_{x \to -4^-} f(x) = -2$ while $\lim\limits_{x \to -4^+} f(x) = 2$. Because the one-sided limits are not the same, $\lim\limits_{x \to -4} f(x)$ does not exist, making **(D)** the correct answer.

2. A

By definition, *limit* means *as x gets very close to*. Pay careful attention to the notation—you are interested in the limit as *x* gets very close to 3 from the right-hand side, so trace along the function just to the right of 3.

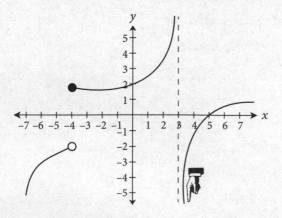

The function decreases without bound (falls *down* along the asymptote), so the limit is $-\infty$, matching **(A)**.

3. C

Evaluate each statement, one at a time, until you find one that is false.

(A): $\lim\limits_{x \to 2^-} g(x) = 3$ and $\lim\limits_{x \to 2^+} g(x) = 3$, so the limit as *x* approaches 2 does equal 3. Eliminate (A).

(B): $\lim\limits_{x \to 4^+} g(x)$ does equal 2, so eliminate (B).

(C): From (A), you already know that the limit as *x* approaches 2 equals 3, so the limit does exist. Thus, **(C)** is false and the correct answer.

(D): $\lim\limits_{x \to 4^-} g(x) = 3$ while $\lim\limits_{x \to 4^+} g(x) = 2$, so $\lim\limits_{x \to 4} g(x)$ does not exist. Eliminate (D).

4. D

Pay careful attention to the notation in the question—you are only interested in the limit from the left. Trace along the function, just to the left of each number of interest (2, 4, and 6).

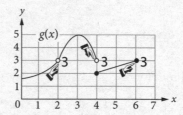

For each of these values of *x*, the limit of the function from the left is indeed 3, so **(D)** is correct.

4.2 EVALUATING LIMITS ALGEBRAICALLY

To answer a question like this:

$$\lim_{x \to 0} \frac{\dfrac{1}{x+4} - \dfrac{1}{4}}{x} =$$

(A) $-\dfrac{1}{16}$ (B) $-\dfrac{1}{4}$ (C) 0   (D) nonexistent

You need to know this:

Conceptually:

- When you are asked to compute $\lim_{x \to c} f(x)$, the first thing you should try is to plug the value c into the function. If you get an answer that's a real number, you're done.

- If you don't get a real-number answer, there are three main simplification strategies: factoring, combining terms in complex fractions, and multiplying by the conjugate (usually of a radical term in either the numerator or the denominator).

Operations with limits:

- The limit as x approaches any number (or even infinity) of a constant function is that constant, or $\lim_{x \to c} a = a$.

- The limit of a sum is the sum of the limits, or $\lim_{x \to c} (a + b) = \lim_{x \to c} a + \lim_{x \to c} b$.

- The same is true for other operations (subtraction, multiplication, division, and square rooting).

> ✔ **AP Expert Note**
>
> There is no need to memorize the rules for performing operations with limits. Standard operations behave exactly as you would expect them to.

Limits

You need to do this:

- Substitute 0 for each x in the expression. If the result is not a real number, simplify the expression.

- Decide which of the three strategies you need to use.

- Carefully perform the algebraic manipulation.

- Re-evaluate the limit.

Answer and Explanation:

A

Plugging 0 in for x yields $\frac{0}{0}$, so try simplifying the expression by writing the numerator as a single term. Then you can multiply by the reciprocal of x (which is $\frac{1}{x}$):

$$\lim_{x \to 0} \frac{\frac{1}{x+4} - \frac{1}{4}}{x} = \lim_{x \to 0} \frac{\frac{4}{4(x+4)} - \frac{1(x+4)}{4(x+4)}}{x}$$

$$= \lim_{x \to 0} \frac{\frac{4-(x+4)}{4(x+4)}}{\frac{x}{1}}$$

$$= \lim_{x \to 0} \frac{4-x-4}{4(x+4)} \times \frac{1}{x}$$

$$= \lim_{x \to 0} \frac{\cancel{4}-x-\cancel{4}}{4x(x+4)}$$

$$= \lim_{x \to 0} \frac{-\cancel{x}}{4\cancel{x}(x+4)}$$

$$= \lim_{x \to 0} \frac{-1}{4(x+4)}$$

Now, re-evaluate the limit by plugging in 0 again:

$$\lim_{x \to 0} \frac{-1}{4(x+4)} = \lim_{x \to 0} \frac{-1}{4(0+4)} = -\frac{1}{16}$$

(A) is correct.

✔ **AP Expert Note**

You'll learn in a later chapter that you could also use L'Hopital's rule to answer this question, but that only works when the limit equals $\frac{0}{0}$ or $\frac{\infty}{\infty}$.

PRACTICE SET

1. $\lim\limits_{x \to -1} \cos(\pi x) =$

   (A) -1

   (B) 0

   (C) 1

   (D) π

2. $\lim\limits_{x \to -1} \dfrac{2x^2 - x - 3}{x^2 - 2x - 3} =$

   (A) -2

   (B) 0

   (C) 1.25

   (D) nonexistent

3. If $a \neq 0$, then $\lim\limits_{x \to a} \dfrac{a^2 - x^2}{3x^4 - 3a^4} =$

   (A) $-\dfrac{1}{6a^2}$

   (B) $-\dfrac{1}{2a^2}$

   (C) 0

   (D) $\dfrac{1}{3a^4}$

4. $\lim\limits_{x \to 1} \dfrac{\sqrt{x} - 1}{x - 1} =$

   (A) 0

   (B) $\dfrac{1}{2}$

   (C) 1

   (D) nonexistent

Limits

ANSWERS AND EXPLANATIONS

1. A

Not all limit questions require algebraic manipulation, so your first step should always be to plug in the number of interest. Here, plugging in -1 for x gives the following:

$$\lim_{x \to -1} \cos(\pi x) = \lim_{x \to -1} \cos\big(\pi(-1)\big)$$
$$= \lim_{x \to -1} \cos(-\pi)$$
$$= -1$$

Thus, **(A)** is correct.

2. C

Plugging -1 in for x yields $\frac{0}{0}$, so simplify the expression by factoring the numerator and denominator, then re-evaluate the limit. Factoring tip: Because $x = -1$ makes both polynomials equal 0, it must be true that $(x + 1)$ is a factor of each.

$$\lim_{x \to -1} \frac{2x^2 - x - 3}{x^2 - 2x - 3} = \lim_{x \to -1} \frac{(2x - 3)\,\cancel{(x+1)}}{(x - 3)\,\cancel{(x+1)}}$$
$$= \lim_{x \to -1} \frac{2x - 3}{x - 3}$$
$$= \frac{2(-1) - 3}{-1 - 3} = \frac{-5}{-4} = \frac{5}{4} = 1.25$$

That's **(C)**.

3. A

Plugging a in for x yields $\frac{0}{0}$, so simplify the expression by factoring the numerator and denominator, then re-evaluate the limit. Factoring tip: When you see lots of squared terms, think *difference of squares*:

$$\lim_{x \to a} \frac{a^2 - x^2}{3x^4 - 3a^4} = \lim_{x \to a} \frac{-\big(x^2 - a^2\big)}{3\big(x^4 - a^4\big)}$$
$$= \lim_{x \to a} \frac{-\big(\cancel{x^2 - a^2}\big)}{3\big(\cancel{x^2 - a^2}\big)\big(x^2 + a^2\big)}$$
$$= \lim_{x \to a} \frac{-1}{3\big(x^2 + a^2\big)}$$
$$= \frac{-1}{3\big(a^2 + a^2\big)}$$
$$= -\frac{1}{3\big(2a^2\big)} = -\frac{1}{6a^2}$$

Therefore, **(A)** is correct.

4. B

Plugging 1 in for x yields $\frac{0}{0}$, so simplify the expression by multiplying the numerator and denominator by the conjugate of the numerator, which is $\sqrt{x} + 1$:

$$\lim_{x \to 1} \frac{\sqrt{x} - 1}{x - 1} = \lim_{x \to 1} \frac{\sqrt{x} - 1}{x - 1} \cdot \frac{\sqrt{x} + 1}{\sqrt{x} + 1}$$
$$= \lim_{x \to 1} \frac{x + \sqrt{x} - \sqrt{x} - 1}{(x - 1)\big(\sqrt{x} + 1\big)}$$
$$= \lim_{x \to 1} \frac{\cancel{x - 1}}{\cancel{(x - 1)}\big(\sqrt{x} + 1\big)}$$
$$= \lim_{x \to 1} \frac{1}{\sqrt{x} + 1} = \frac{1}{\big(\sqrt{1} + 1\big)} = \frac{1}{2}$$

That's **(B)**.

4.3 LIMITS OF COMPOSITE FUNCTIONS

To answer a question like this:

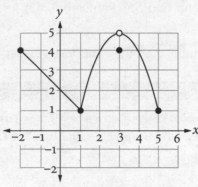

Graph of *f*

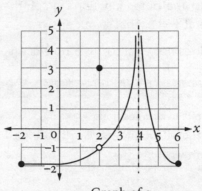

Graph of *g*

The graphs of the functions *f* and *g* are shown above. What is the value of $\lim_{x \to 2} (f \circ g)(x)$?

(A) 1   (B) 3   (C) 4   (D) nonexistent

You need to know this:

Rule: $\lim_{x \to a} f(g(x)) = f\left(\lim_{x \to a} g(x)\right) = f(b)$

The rule above is valid provided all of the following are true:

- $\lim_{x \to a} g(x) = b$

- $\lim_{x \to b} f(x)$ exists

- $\lim_{x \to b} f(x) = f(b)$, assuming $f(b)$ is defined

Translation:

- In plain English, this means take the desired limit of *g(x)*—and then you're done with the limit part—then plug the result into *f(x)*.

- Graphically, this means find the *y*-value that the *g* function gets *very close to* (from both sides) when *x* = *a*, think of the result as *b*, and then find the *y*-value of the *f* function when *x* = *b*.

- If the limit of *g(x)* doesn't exist, then the limit of the composite function doesn't exist. Similarly, if *f(b)* is not defined, then the limit of the composite function doesn't exist.

You need to do this:

- To find the limit, trace the inner function of the composition on its graph (here, g) from both sides of the number that x is approaching (here, 2). Call the result b.

- Move to the outer function's graph (here, f) and find $f(b)$. Remember, $f(b)$ just means the y-value at $x = b$. That's your answer.

- If $f(b)$ does not exist (i.e., is a hole or an asymptote), then the limit doesn't exist.

> ✔ **AP Expert Note**
>
> Taking limits of compositions works much the same way as evaluating compositions—start with the inner function. If the question is written using the notation $\lim_{x \to a}(f \circ g)(x)$, rewrite it as $\lim_{x \to a} f(g(x))$ so you'll know where to start.

Answer and Explanation:

B

First, rewrite the composition as $\lim_{x \to 2} f(g(x))$. Then apply the rule:

$$\lim_{x \to 2} f(g(x)) = f\left(\lim_{x \to 2} g(x)\right)$$

This tells you to start by finding $\lim_{x \to 2} g(x)$ using the graph of g. Then use the graph of f to find f of the result.

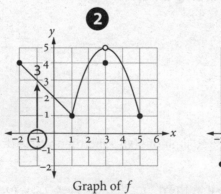
Graph of f

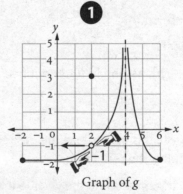
Graph of g

As you can see from the graph labeled 1, $\lim_{x \to 2} g(x) = -1$, (not 3, which is the functional value of g at $x = 2$). Based on the graph labeled 2, $f(-1) = 3$, so **(B)** is correct.

PRACTICE SET

x	-2	-1	0	1	2	3
$f(x)$	3	4	5	6	5	4

1. Several ordered pairs for the function f are given in the table above. If $\lim\limits_{x \to 3} g(x) = 1$, then $\lim\limits_{x \to 3} f(g(x)) =$

   (A) -2

   (B) 3

   (C) 4

   (D) 6

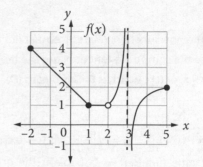

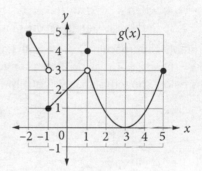

3. The graphs of two functions f and g are shown above. Which of the following does not exist?

   I. $\lim\limits_{x \to -1} (f \circ g)(x)$

   II. $\lim\limits_{x \to 1} (f \circ g)(x)$

   III. $\lim\limits_{x \to 3} (f \circ g)(x)$

   (A) I only

   (B) II only

   (C) I and II only

   (D) I, II, and III

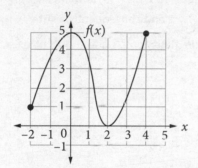

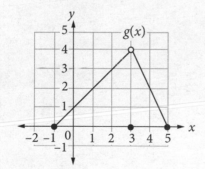

2. Given the graphs of f and g above, which of the following could be the value of a if $\lim f(g(x)) = 0$?

   (A) 2 only

   (B) 1 and 4 only

   (C) -1 and 5 only

   (D) -1, 3, and 5

ANSWERS AND EXPLANATIONS

1. D

Start with the inner function: You're given that $\lim_{x \to 3} g(x) = 1$. To evaluate the composite function, use the table to find $f\left(\lim_{x \to 3} g(x)\right)$ x, or $f(1)$, which is 6. This means **(D)** is correct.

2. B

Rewrite $\lim_{x \to a} f(g(x)) = 0$ as $f\left(\lim_{x \to a} g(x)\right) = 0$. This translates to $f(\text{some number}) = 0$. Look at the graph of f: The y-value is 0 when $x = 2$, so $f(2) = 0$. This means $\lim_{x \to a} g(x) = 2$, so find the x-value(s) on the graph of g for which the graph is approaching 2 (from both sides). Those values are $x = 1$ and $x = 4$, making **(B)** the correct answer. The graphs below show this analysis visually.

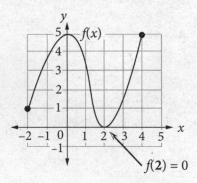

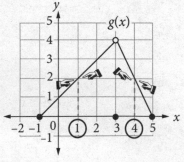

3. C

In each case, think of $f \circ g$ as $f(g(x))$, which tells you to start with g. Evaluate the given limit (of g), then find f of the result (if possible).

Statement I: Look at the graph of g: $\lim_{x \to -1} g(x)$ does not exist because the limit from the left of -1 is 3, while the limit from the right of -1 is 1. Thus, $f\left(\lim_{x \to -1} g(x)\right)$ cannot be evaluated and consequently, does not exist.

Statement II: Look at the graph of g: $\lim_{x \to 1} g(x) = 3$. Now look at the graph of f: There is an asymptote at $x = 3$, so $f(3)$ does not exist. Thus, $f\left(\lim_{x \to 1} g(x)\right)$ does not exist.

Statement III: Look at the graph of g: $\lim_{x \to 3} g(x) = 0$. Now look at the graph of f to find that $f(0) = 2$. Thus, $f\left(\lim_{x \to 3} g(x)\right) = 2$ and consequently, exists.

The limits in statements I and II do not exist, so **(C)** is correct.

4.4 LIMITS THAT DON'T EXIST

To answer a question like this:

If $f(x) = \dfrac{1}{2 + e^{\frac{1}{x}}}$, then $\displaystyle\lim_{x \to 0} f(x) =$

(A) $-\infty$ (B) 0 (C) $\dfrac{1}{2}$ (D) nonexistent

You need to know this:

- The $\displaystyle\lim_{x \to a} f(x)$ exists if and only if $\displaystyle\lim_{x \to a^-} f(x) = \lim_{x \to a^+} f(x)$. In other words, the limit does not exist (DNE), or is nonexistent, if the two one-sided limits are not the same value.

- Technically, a limit that equals positive or negative infinity doesn't exist because it's not a specific value. However, saying a limit is equal to $\pm\infty$ is commonly used to describe a function that increases (or decreases) without bound.

- Some useful "calculations":

$\dfrac{1}{\infty} = 0$	$\dfrac{1}{\text{tiny positive number}} = \infty$	For $c > 1$, $c^{\infty} \to \infty$
$\dfrac{1}{-\infty} \to 0$	$\dfrac{1}{\text{tiny negative number}} = -\infty$	For $c > 1$, $c^{-\infty} = 0$

> ✔ **AP Expert Note**
>
> Technically, you can't plug positive or negative infinity into an expression. Rather, this notation is used to represent plugging a huge positive or negative number in for the variable.

Anticipating limits that don't exist:

- Because the graphs of $y = \sin x$ and $y = \cos x$ oscillate back and forth between -1 and 1, the limit as x approaches positive or negative infinity for either of these doesn't exist. However, when the trig function is in the numerator or denominator of a rational function, you need to analyze further.

- Aside from trig functions, in *most* cases, a limit that doesn't exist involves a rational expression or a piecewise function.

You need to do this:

- As always, start by plugging in the number. If the result is a real number, then you're done.

- If the result is $\dfrac{\text{a number}}{0}$, then you must find both one-sided limits and compare the results.

- If the result is $\dfrac{0}{0}$, consider using L'Hopital's rule. If that doesn't work, you must find both one-sided limits and compare the results.

- To find the one-sided limits, reason them out logically or apply brute force—that is, plug in a tiny number just to the left and another just to the right and see what happens. For example, if the limit approaches 0, try -0.01 and 0.01; you don't need to do complicated computations—just determine if the values are the same.

Answer and Explanation:

D

Plugging in 0 yields $\dfrac{1}{2 + e^{\frac{1}{0}}}$, which can't be evaluated, so you must find the one-sided limits and compare the results. Don't forget—e is just a number $(2.718\ldots)$, so it behaves as any other constant.

As $x \to 0^-$, $\dfrac{1}{x} \to \dfrac{1}{\text{tiny negative number}} = -\infty$. Therefore,

$e^{\frac{1}{x}} \to e^{-\infty} = 0$, which means $f(x) \to \dfrac{1}{2 + 0} = \dfrac{1}{2}$.

As $x \to 0^+$, $\dfrac{1}{x} \to \dfrac{1}{\text{tiny positive number}} = \infty$.

Therefore, $e^{\frac{1}{x}} \to e^{\infty} = \infty$, which means $f(x) \to \dfrac{1}{2 + \infty} = \dfrac{1}{\infty} = 0$.

Because the one-sided limits are not the same, $\lim\limits_{x \to 0} f(x)$ does not exist, which is **(D)**.

PRACTICE SET

$$f(x) = \frac{x}{5-x}$$

1. For the function f defined above, which of the following statements is true?

   (A) $\lim_{x \to 0} f(x)$ does not exist.

   (B) $\lim_{x \to 5} f(x)$ does not exist.

   (C) $\lim_{x \to 5^-} f(x) = 0$

   (D) $\lim_{x \to 5^+} f(x) = 0$

$$f(x) = \begin{cases} -x, & x < -1 \\ x+1, & x \geq -1 \end{cases}$$

2. A function f is defined above. What is $\lim_{x \to -1} f(x)$?

   (A) -1

   (B) 0

   (C) 1

   (D) nonexistent

3. If $f(x) = 4 \cos(x + \pi)$, then $\lim_{x \to -\infty} f(x) =$

   (A) $-\infty$

   (B) -4

   (C) 4

   (D) nonexistent

4. If $f(x) = \dfrac{3}{3 + 10^{\frac{1}{x}}}$, then $\lim_{x \to 0} f(x) =$

   (A) 0

   (B) 1

   (C) ∞

   (D) nonexistent

ANSWERS AND EXPLANATIONS

1. B

Evaluate each statement, one at a time, until you find one that's true.

(A): $\lim\limits_{x \to 0} f(x) = \dfrac{0}{5-0} = \dfrac{0}{5} = 0$. Eliminate (A).

(B): $\lim\limits_{x \to 5} f(x) = \dfrac{5}{5-5} = \dfrac{5}{0}$. Find the one-sided limits, so skip this choice for now.

(C): $\lim\limits_{x \to 5^-} f(x) = \dfrac{4.99}{5-4.99} = \dfrac{4.99}{\text{tiny positive number}} = \infty$. Eliminate (C).

(D): $\lim\limits_{x \to 5^+} f(x) = \dfrac{5.01}{5-5.01} = \dfrac{5.01}{\text{tiny negative number}} = -\infty$. Eliminate (D).

Because the results of (C) and (D) are not the same, **(B)** must be correct.

2. D

To evaluate the limit where a piecewise function "breaks" (here, at $x = -1$), evaluate the limit from the left of the break ($x < -1$) and the limit from the right of the break ($x \geq -1$) and see what happens:

To the left of -1, the function is defined by $-x$, therefore:

$$\lim\limits_{x \to -1^-} f(x) = \lim\limits_{x \to -1^-} (-x) = -(-1) = 1$$

To the right of -1, the function is defined by $x + 1$, therefore:

$$\lim\limits_{x \to -1^+} f(x) = \lim\limits_{x \to -1^+} (x + 1) = -1 + 1 = 0$$

Because the limit from the left (1) is not equal to the limit from the right (0), $\lim\limits_{x \to -1} f(x)$ does not exist, which is **(D)**.

3. D

Because the graphs of $y = \sin x$ and $y = \cos x$ oscillate back and forth between –1 and 1, the limit as x approaches positive or negative infinity for either of these typically doesn't exist. This is true even when there is a change to the amplitude and/or period of the function, or when there is a phase shift. This particular trig function is shifted to the left π units and has an amplitude of 4, but it still oscillates back and forth, so $\lim\limits_{x \to -\infty} f(x)$ does not exist. Thus, **(D)** is correct.

4. D

Because plugging in 0 doesn't yield an answer that is a real number (you can't evaluate $10^{\frac{1}{0}}$), you must check the limit from the left of 0 and the limit from the right of 0.

As $x \to 0^-$, $\dfrac{1}{x} \to -\infty$. Therefore, $10^{\frac{1}{x}} \to 10^{-\infty} = 0$, which means $f(x) \to \dfrac{3}{3+0} = 1$.

As $x \to 0^+$, $\dfrac{1}{x} \to \infty$. Therefore, $10^{\frac{1}{x}} \to 10^{\infty} = \infty$, which means $f(x) \to \dfrac{3}{3+\infty} = \dfrac{3}{\infty} = 0$.

Because the one-sided limits are not the same, $\lim\limits_{x \to 0} f(x)$ does not exist, which is **(D)**.

RAPID REVIEW

If you take away only 7 things from this chapter:

1. Limits describe what happens as *x approaches* (gets close to) a number. A limit DOESN'T describe, and doesn't care about, what happens when *x* actually *equals* that number.

2. When you are asked to compute $\lim\limits_{x \to c} f(x)$, the first thing you should try is to plug the value *c* into the function.

3. If you don't get a real-number answer, there are three main simplification strategies: factoring, combining terms in complex fractions, and multiplying by the conjugate (usually of a radical term).

4. The limit as *x* approaches any number (or even infinity) of a constant function is that constant, or $\lim\limits_{x \to c} a = a$.

5. To find the limit of a composite function, use the rule $\lim\limits_{x \to a} f(g(x)) = f\left(\lim\limits_{x \to a} g(x)\right) = f(b)$. In plain English, this means take the desired limit of *g*(*x*)—and then you're done with the limit part—then plug the result into *f*(*x*).

6. The $\lim\limits_{x \to a} f(x)$ exists if and only if $\lim\limits_{x \to a^-} f(x) = \lim\limits_{x \to a^+} f(x)$. In other words, the limit does not exist if the one-sided limits are not the same value.

7. Because the graphs of $y = \sin x$ and $y = \cos x$ oscillate back and forth between -1 and 1, the limit as *x* approaches positive or negative infinity for either of these typically doesn't exist.

Limits

TEST WHAT YOU LEARNED

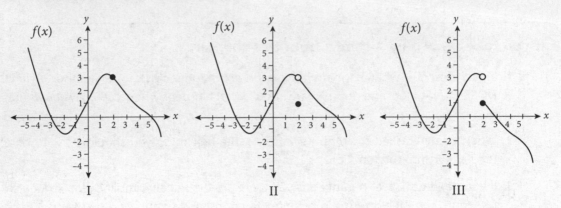

I   II   III

1. Each graph shown above represents a function f. For which graph(s) does $\lim\limits_{x \to 2} f(x) = 3$?

    (A) I only

    (B) I and II only

    (C) II and III only

    (D) I, II, and III

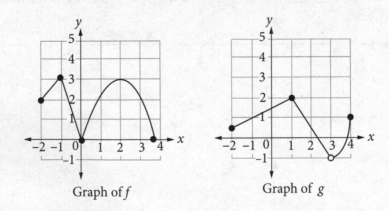

Graph of f Graph of g

2. The graphs of two functions f and g are shown above. If $\lim\limits_{x \to k} f(g(x)) = 3$, what is the value of k?

    (A) −1   (B) 1   (C) 3   (D) 1 or 3

3. The limit as $x \to 0$ of which of the following functions exists?

    (A) $f(x) = \dfrac{1}{x}$

    (B) $f(x) = \dfrac{|x|}{x}$

    (C) $f(x) = \dfrac{x^2 - x}{x}$

    (D) $f(x) = \begin{cases} |x + 2| & \text{if } x < 0 \\ (x + 2)^2 & \text{if } x \geq 0 \end{cases}$

---

4. $\displaystyle \lim_{x \to -2} \sqrt{\dfrac{x + 2}{4x + 8}} =$

    (A) -4 (B) $-\dfrac{1}{4}$ (C) $\dfrac{1}{4}$ (D) $\dfrac{1}{2}$

Answer Key

Test What You Already Know	Test What You Learned
1. **D** **Learning Objective:** 4.1	1. **B** **Learning Objective:** 4.1
2. **B** **Learning Objective:** 4.2	2. **D** **Learning Objective:** 4.3
3. **A** **Learning Objective:** 4.3	3. **C** **Learning Objective:** 4.4
4. **D** **Learning Objective:** 4.4	4. **D** **Learning Objective:** 4.2

 REFLECTION

Test What You Already Know score: _____

Test What You Learned score: _____

Use this section to evaluate your progress. After working through the pre-quiz, check off the boxes in the "Pre" column to indicate which Learning Objectives you feel confident about. Then, after completing the chapter, including the post-quiz, do the same to the boxes in the "Post" column. Keep working on unchecked Objectives until you're confident about them all!

Pre	Post		
☐	☐	**4.1**	Evaluate limits graphically, including one-sided limits
☐	☐	**4.2**	Evaluate limits algebraically
☐	☐	**4.3**	Evaluate limits of composite functions from graphs and tables
☐	☐	**4.4**	Determine when a limit does not exist

 FOR MORE PRACTICE

Complete more practice online at kaptest.com. Haven't registered your book yet? Go to kaptest.com/booksonline to begin.

ANSWERS AND EXPLANATIONS

Test What You Already Know

1. D Learning Objective: 4.1

The limit of a function exists at $x = a$ if both one-sided limits approach the same value. The limit does *not* have to equal the actual value of the function at that point. For this function, the limit from the left and from the right of $x = 1$ is $+\infty$ (because the graph increases without bound on both sides of the vertical asymptote), so the limit equals ∞ and you can eliminate (A). The limit from the left and from the right of $x = 3$ is 1 (even though $f(3)$ is 0), so eliminate (B). The limit from the left and from the right of $x = 4$ is 6, so eliminate (C). This means **(D)** must be correct. (The limit from the left of $x = 5$ is 4, while the limit from the right is 3, so the limit does not exist.)

2. B Learning Objective: 4.2

Plugging 3 in for x yields $\frac{0}{0}$, so try simplifying the expression by writing the numerator as a single term. Chances are that something will cancel out nicely.

$$\lim_{x \to 3} \frac{\frac{1}{x-1} - \frac{1}{2}}{x-3} = \lim_{x \to 3} \frac{\frac{2}{2(x-1)} - \frac{1(x-1)}{2(x-1)}}{x-3}$$

$$= \lim_{x \to 3} \frac{\frac{2-x+1}{2(x-1)}}{\frac{(x-3)}{1}}$$

$$= \lim_{x \to 3} \frac{3-x}{2(x-1)} \cdot \frac{1}{x-3}$$

$$= \lim_{x \to 3} \frac{-(x-3)}{2(x-1)} \cdot \frac{1}{x-3}$$

$$= \lim_{x \to 3} \frac{-1}{2(x-1)} = \frac{-1}{2(3-1)} = -\frac{1}{4}$$

Hence, **(B)** is correct. Note that you could also use L'Hôpital's rule here. If you rewrite the first term in the numerator of the complex fraction as $(x-1)^{-1}$, taking the derivative is fairly quick:

$$\lim_{x \to 3} \frac{\frac{1}{x-1} - \frac{1}{2}}{x-3} = \lim_{x \to 3} \frac{(x-1)^{-1} - \frac{1}{2}}{x-3}$$

$$= \lim_{x \to 3} \frac{-1(x-1)^{-2}}{1}$$

$$= \lim_{x \to 3} \frac{-1}{(x-1)^2} = \frac{-1}{(3-1)^2} = -\frac{1}{4}$$

3. A Learning Objective: 4.3

Knowing the appropriate limit theorem is the key to answering this question. The theorem states:

$$\lim_{x \to a} f(g(x)) = f\left(\lim_{x \to a} g(x)\right)$$

According to the graph of g, as x gets close to 2, the function gets close (from the left and from the right) to 1 (not 4, which happens to be $g(2)$). According to the graph of f, the value of $f(1)$ is 2, which is **(A)**.

4. D Learning Objective: 4.4

When finding the limit of a function that involves an absolute value, you could rewrite the function as a piecewise function and evaluate the one-sided limits based on the value of x where the function breaks. However, it's easier to simply choose a convenient value that is just barely to the left (and another barely to the right) of the number that you're interested in, and plug them into the function. Here, choose 1.9 and 2.1 and see what happens:

$$\lim_{x \to 2^-} \frac{|x-2|}{x-2} \approx \frac{|1.9-2|}{1.9-2} = \frac{0.1}{-0.1} = -1$$

$$\lim_{x \to 2^+} \frac{|x-2|}{x-2} \approx \frac{|2.1-2|}{2.1-2} = \frac{0.1}{0.1} = 1$$

The limit from the left (-1) does not equal the limit from the right (1), so the limit as $x \to 2$ does not exist. Therefore, **(D)** is correct.

Test What You Learned

1. B Learning Objective: 4.1

Examine each graph to see what happens as x approaches 2 (from both the left and the right). In graph I, the function approaches 3 from both sides, so the limit equals 3. The same is true for graph II, even though the functional value at $x = 2$ happens to be 1. However, in graph III, the limit from the left is 3, while the limit from the right is 1, so the limit as x approaches 2 for this function does not exist. Thus, **(B)** is correct.

2. D Learning Objective: 4.3

To take the limit of $f(g(x))$ as x approaches k, take the limit as x gets very close to k of $g(x)$—and then you're done with the limit part—then plug the result into $f(x)$. Check each answer choice until you find one that yields 3. Because (D) includes multiple values, check all the answer choices.

$$(A): f\left(\lim_{x \to -1} g(x)\right) = f(1) \approx 2.3$$
$$(B): f\left(\lim_{x \to 1} g(x)\right) = f(2) \approx 3$$
$$(C): f\left(\lim_{x \to 3} g(x)\right) = f(-1) \approx 3$$

Both (B) and (C) yield 3, so **(D)** is correct.

3. C Learning Objective: 4.4

For each of the functions, the limit as $x \to 0$ exists if the limit from the left of 0 is equal to the limit from the right. However, your first thought should be to simply plug 0 into each function. (A), (B), and (C) all yield undefined values, so that doesn't help much. However, plugging 0 in for x in (D) yields 2 for the top piece and 4 for the bottom piece, which means the limit doesn't exist. Eliminate (D).

For (A) and (B), try plugging in a value just to the left of 0 and another just to the right, such as −0.1 and 0.1. In

(A), the first gives $-\infty$ while the second gives $+\infty$, so the limit doesn't exist; the same is true for (B). This means **(C)** must be correct. To check, simplify the expression algebraically and then plug 0 in again:

$$\lim_{x \to 0} \frac{x^2 - x}{x} = \lim_{x \to 0} \frac{\cancel{x}(x-1)}{\cancel{x}}$$
$$= \lim_{x \to 0}(x-1) = 0 - 1 = -1$$

The limit does indeed exist, so **(C)** is correct.

4. D Learning Objective: 4.2

Plugging −2 in for x yields $\frac{0}{0}$, so try simplifying the expression by factoring the denominator of the fraction:

$$\lim_{x \to -2} \sqrt{\frac{x+2}{4x+8}} = \lim_{x \to -2} \sqrt{\frac{x+2}{4(x+2)}}$$
$$= \lim_{x \to -2} \sqrt{\frac{1}{4}}$$
$$= \lim_{x \to -2} \frac{1}{2}$$

The limit as x approaches any value of a constant function is that constant, so the limit you're looking for is $\frac{1}{2}$. That's **(D)**.

CHAPTER 5

More Advanced Limits

LEARNING OBJECTIVES

5.1   Determine when a limit is infinite

5.2   Evaluate limits at infinity

5.3   Identify horizontal and vertical asymptotes of a function

5.4   Evaluate limits involving trig functions

TEST WHAT YOU ALREADY KNOW

1. $\displaystyle\lim_{x \to 1} \frac{2x}{(x-1)^2} =$

    (A) $-\infty$ (B) 0   (C) 2   (D) ∞

2. $\displaystyle\lim_{x \to \infty} \frac{(2x+1)(x-3)}{5-x^2} =$

    (A) -2 (B) $-\dfrac{3}{5}$ (C) 1   (D) nonexistent

3. Which statement is true about the graph of $y = \dfrac{2x^2 + 13}{x^3 - 1}$?

   (A) The line $x = -1$ is a vertical asymptote.

   (B) The line $y = 0$ is a horizontal asymptote.

   (C) The line $y = 3$ is a horizontal asymptote.

   (D) The graph has no vertical or horizontal asymptotes.

---

4. $\displaystyle\lim_{x \to 0} \frac{4 \sin x}{x^2 - x} =$

   (A) -4 (B) $-\dfrac{1}{4}$ (C) 1 (D) ∞

Answers to this quiz can be found at the end of this chapter.

5.1 INFINITE LIMITS

To answer a question like this:

$$\lim_{x \to 3} \frac{x}{x^2 - 6x + 9} =$$

(A) $-\infty$ (B) 0 (C) $\frac{1}{3}$ (D) ∞

You need to know this:

- The $\lim\limits_{x \to a} f(x) = \infty$ if $f(x)$ gets arbitrarily large (increases without bound) for all x sufficiently close to $x = a$, from both sides, without actually letting $x = a$.

- The $\lim\limits_{x \to a} f(x) = -\infty$ if $f(x)$ gets arbitrarily small (decreases without bound) for all x sufficiently close to $x = a$, from both sides, without actually letting $x = a$.

- Graphically, when a function has a vertical asymptote, the limit on either—or both—sides of the asymptote goes to positive or negative infinity.

✔ AP Expert Note

When we say that the limit of a function is positive or negative infinity, this does not mean that the limit exists. We use the notation $\lim\limits_{x \to c} f(x) = \infty$ to describe a particular way that the limit does not exist. The limit does not exist *because* the function's values are approaching positive or negative infinity. In this case, you should write: The limit $= \infty$ or $-\infty$; do not write the limit DNE.

Anticipating limits that equal positive or negative infinity:

- When you plug the value of a into a limit and get "$\dfrac{\text{something}}{0}$" (as long as "something" $\neq 0$), it means that infinity is coming into play. The limit will either be infinity, negative infinity, or will not exist (because the limit from the left of a and the limit from the right of a have opposite signs).

- Logarithmic functions

- Tangent functions

General rules of thumb to follow:

- For a simplified rational function, if the denominator equals 0 when $x = a$ and the denominator is a quantity raised to an even power, such as x^2, $(x + 5)^2$, $(3x - 1)^4$, etc., the limit probably equals infinity.

- For a simplified rational function, if the denominator equals 0 when $x = a$ and the denominator is a quantity raised to an odd power, such as x, $(x + 5)^3$, $(3x - 1)^5$, etc., the limit most likely does not exist (DNE).

- For logarithmic functions and tangent functions, knowing what the graph looks like (and what happens on each side of the asymptotes) is the key.

You need to do this:

- As with all limit questions, start by plugging the value of a into the function. If the result is a plain number, you're done.

- If the result is 0 over 0, use L'Hôpital's rule. (This is covered in chapter 7.)

- If the result is a number over 0, simplify the function as much as possible (particularly the denominator of a rational function).

- If the result is still a number over 0, check the limit from the left and from the right of a by plugging in a number just to the left of a and another just to the right of a. (You don't need exact values, just an idea of where the y-values are headed.)

Answer and Explanation:

D

Substituting 3 into the expression yields:

$$\lim_{x \to 3} \frac{x}{x^2 - 6x + 9} = \frac{3}{9 - 18 + 9} = \frac{3}{0}$$

Factor the denominator and then check the limit from the left and the limit from the right by plugging in a number just to the left of 3 (such as 2.99) and another just to the right of 3 (such as 3.01). Remember—you don't need to carry out all the calculations; just determine whether the y-values are getting arbitrarily large or arbitrarily small.

$$\lim_{x \to 3} \frac{x}{x^2 - 6x + 9} = \lim_{x \to 3} \frac{x}{(x - 3)^2}$$

From the left: $\dfrac{2.99}{(3 - 2.99)^2} = \dfrac{2.99}{(0.01)^2} = \dfrac{2.99}{\text{tiny positive \#}} \to \infty$

From the right: $\dfrac{3.01}{(3 - 3.01)^2} = \dfrac{3.01}{(-0.01)^2} = \dfrac{3.01}{\text{tiny positive \#}} \to \infty$

Because the limit from the left and the limit from the right both approach ∞, **(D)** is correct.

PRACTICE SET

1. $\lim\limits_{x \to \frac{3\pi}{2}^+} \tan x =$

   (A) $-\infty$

   (B) 0

   (C) 1

   (D) ∞

2. $\lim\limits_{x \to 0^+} \dfrac{\cos x}{x} =$

   (A) $-\infty$

   (B) -1

   (C) 1

   (D) ∞

3. $\lim\limits_{x \to 0} \dfrac{2x - 10}{x^3 - 5x^2} =$

   (A) $-\infty$

   (B) 0

   (C) $\dfrac{1}{5}$

   (D) ∞

4. $\lim\limits_{x \to -1} \dfrac{5e^x}{xe^x + e^x} =$

   (A) $\dfrac{1}{e}$

   (B) $\dfrac{5}{e}$

   (C) ∞

   (D) nonexistent

ANSWERS AND EXPLANATIONS

1. A

Occasionally, you'll be fortunate enough to get a limit question in the calculator section of the exam. When this happens, use your calculator to your advantage. Here, the function is $y = \tan x$, which is a good candidate for an infinite limit, so graph the function (in a reasonable viewing window) and examine its behavior near the value of interest. For the tangent function, interesting things (asymptotes and zeros) occur at multiples of π and $\frac{\pi}{2}$, so set your viewing window to 0 to 2π, with the Xscl (x scale) set to $\frac{\pi}{2}$. Then take a look at what happens as x approaches $\frac{3\pi}{2}$ from the right-hand side.

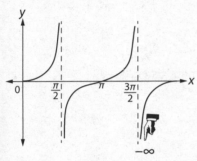

The function decreases without bound, so **(A)** is correct.

2. D

Plugging in 0 yields $\frac{1}{0}$, but there is nothing to simplify here, so reason this one out logically. As x approaches 0 from the right, the numerator approaches 1 (because cos (0) = 1) and the denominator approaches zero but through positive values. Thus, the fraction goes to 1 over an increasingly small but positive number, which is an increasingly large positive number. Thus, the limit grows to infinity, which is **(D)**.

3. D

Substituting 0 into the expression yields $\frac{-10}{0}$, so simplify the numerator and the denominator and re-evaluate:

$$\lim_{x \to 0} \frac{2x - 10}{x^3 - 5x^2} = \lim_{x \to 0} \frac{2(x - 5)}{x^2(x - 5)} = \lim_{x \to 0} \frac{2}{x^2}$$

The limit still equals a number over 0, so check the limit from the left and the limit from the right by plugging in a number just to the left of 0 (such as −0.01) and another just to the right of 0 (such as 0.01).

From the left: $\dfrac{2}{(-0.01)^2} = \dfrac{2}{\text{tiny positive \#}} \to \infty$

From the right: $\dfrac{2}{(0.01)^2} = \dfrac{2}{\text{tiny positive \#}} \to \infty$

Because the limit from the left and the limit from the right both approach ∞, **(D)** is correct.

Note that you could also use the general rule of thumb: After being simplified, the denominator is raised to an even power, so the limit is equal to infinity.

4. D

Substituting −1 into the expression yields:

$$\lim_{x \to -1} \frac{5e^x}{xe^x + e^x} = \frac{5e^{-1}}{-1e^{-1} + e^{-1}} = \frac{\frac{5}{e}}{0}$$

The value of $\frac{5}{e}$ is just some number, so simplify the fraction and re-evaluate:

$$\lim_{x \to -1} \frac{5e^x}{xe^x + e^x} = \lim_{x \to -1} \frac{5e^x}{e^x(x + 1)} = \lim_{x \to -1} \frac{5}{x + 1}$$

The limit still equals a number over 0, so check the limit from the left and the limit from the right by plugging in a number just to the left of −1 (such as –1.01) and another just to the right of −1 (such as −0.99).

From the left: $\dfrac{5}{-1.01 + 1} = \dfrac{5}{\text{tiny negative \#}} \to -\infty$

From the right: $\dfrac{5}{-0.99 + 1} = \dfrac{5}{\text{tiny positive \#}} \to \infty$

Because the limit from the left is not the same as the limit from the right, the limit does not exist, or is nonexistent, which is **(D)**.

Note that you could also use the general rule of thumb: After the expression is simplified, the denominator is raised to an odd power, so the limit does not exist.

5.2 LIMITS AT INFINITY

To answer a question like this:

$$\lim_{x \to -\infty} \frac{\sqrt{25x^6 - 10x^3 + 50}}{x^3 + 25} =$$

(A) −25   (B) −5   (C) 1   (D) 5

You need to know this:

Conceptually:

- As x approaches positive or negative infinity, the limit of a function depends on the dominant term (or terms).

- The *dominant term* is the term that grows the fastest (in either a positive or negative direction, depending on the circumstances).

- In general, as $x \to \infty$, polynomials of higher degree grow faster than polynomials of lower degree, and exponential functions grow faster than polynomials.

- The limit (including when x is approaching ∞) of a constant function is that constant.

- Periodic functions, such as $\sin x$, do not have limits as x approaches infinity because their graphs oscillate. That is, $\lim_{x \to \infty} \sin x$ does not exist (DNE).

Rules for limits of *rational* functions as $x \to \pm\infty$:

- If the degree of the numerator is less than the degree of the denominator, the limit equals 0.

- If the degree of the numerator is equal to the degree of the denominator, the limit equals the ratio of the leading coefficients.

- If the degree of the numerator is greater than the degree of the denominator, the limit goes to $\pm\infty$.

Limits

You need to do this:

- This is a rational function, so identify the dominant term in the numerator and in the denominator.

- Disregard the negligible terms (the terms that are dominated and thus don't affect the overall limit).

- Re-evaluate the limit. This step often requires logical thinking and some basic knowledge of limits, especially when terms include square roots.

- Think about the nature of extremely large positive and negative numbers. Some general rules follow in the table below.

Rules for ∞	Rules for $-\infty$
∞	$-\infty$
$\dfrac{c}{\infty} = 0$	$\dfrac{c}{-\infty} = 0$
For $c > 1$, $c^{\infty} = \infty$	For $c > 1$, $c^{-\infty} = 0$
For $c > 0$, $\infty^{c} = \infty$	For $c > 0$ and even, $(-\infty)^{c} = \infty$ For $c > 0$ and odd, $(-\infty)^{c} = -\infty$
For $c < 0$, $\infty^{c} = \dfrac{1}{\infty} = 0$	For $c < 0$, $(-\infty)^{c} = \pm\dfrac{1}{\infty} = 0$

Answer and Explanation:

B

Because x approaches $-\infty$, you can disregard the negligible terms (the terms that are completely dominated by another term).

$$\lim_{x \to -\infty} \frac{\sqrt{25x^6 - 10x^3 + 50}}{x^3 + 25} \to \lim_{x \to -\infty} \frac{\sqrt{25x^6}}{x^3} = \lim_{x \to -\infty} \frac{5|x^3|}{x^3}$$

The variable part in the numerator and denominator is *almost* the same, but not quite, so don't cancel. Think of it this way:

$$\lim_{x \to -\infty} \frac{5|x^3|}{x^3} = \lim_{x \to -\infty} 5\left(\frac{|x^3|}{x^3}\right)$$

As x approaches $-\infty$, the circled part is equivalent to -1 because the quantities in the numerator and denominator have the same value, but opposite signs. Thus, the limit equals $5 \times (-1) = -5$, which is **(B)**.

PRACTICE SET

1. $\displaystyle\lim_{x \to \infty} \frac{x^3 - 3x^2 + 2x - 1}{2^x - 1} =$

    (A) -1

    (B) 0

    (C) 1

    (D) ∞

2. $\displaystyle\lim_{x \to \infty} \frac{\cos x}{x} =$

    (A) 0

    (B) 1

    (C) π

    (D) ∞

3. $\displaystyle\lim_{x \to -\infty} \frac{e^x + 2x^3}{6 - 3x^3} =$

    (A) $-\infty$

    (B) $-\dfrac{2}{3}$

    (C) $\dfrac{2}{3}$

    (D) ∞

4. $\displaystyle\lim_{x \to \infty} \left(\sqrt{x + 400} - \sqrt{x} \right) =$

    (A) 0

    (B) 20

    (C) 400

    (D) ∞

ANSWERS AND EXPLANATIONS

1. B

Because the limit is approaching infinity, you can disregard all terms other than the dominant term in the numerator and in the denominator. That's x^3 in the numerator and 2^x in the denominator.

$$\lim_{x \to \infty} \frac{x^3 - 3x^2 + 2x - 1}{2^x - 1} \to \lim_{x \to \infty} \frac{x^3}{2^x}$$

Now reason logically: Exponential functions grow faster than regular power functions, so the denominator grows to infinity faster than the numerator, and the fraction approaches zero. **(B)** is correct.

2. A

The numerator and denominator here each have only one term, so reason this one out logically. Think about how the numerator behaves as x gets larger and larger, and how the denominator behaves under the same condition. Because the numerator oscillates (bounces) back and forth between ± 1, and the denominator grows to infinity, the fraction gets closer to 0 as x gets larger, making **(A)** the correct answer.

3. B

Because x is approaching negative infinity, e^x approaches zero, so $2x^3$ is the dominant term in the numerator while $-3x^3$ dominates in the denominator. Disregard the other terms and re-evaluate the limit:

$$\lim_{x \to -\infty} \frac{e^x + 2x^3}{6 - 3x^3} \to \lim_{x \to -\infty} \frac{2x^3}{-3x^3} = \lim_{x \to -\infty} \left(-\frac{2}{3}\right)$$

The limit of a constant function is always that constant, so the correct answer is $-\frac{2}{3}$, or **(B)**.

4. A

Intuitively, you can treat this limit just as you do limits of rational functions: consider only the dominant term in each piece of the equation. As x gets very large (say a million), the 400 is so small in comparison that you can disregard it. The limit then becomes $\lim_{x \to \infty} \left(\sqrt{x} - \sqrt{x}\right) = \lim_{x \to \infty} 0$, which is 0. **(A)** is correct.

Alternatively (but a lot more work), you could multiply the original expression by its conjugate, simplify, and then re-evaluate the limit. If you go this route, save yourself some time by recalling that the middle two terms always drop out when you multiply by a conjugate (which means you can ignore the OI in FOIL when you multiply).

$$\lim_{x \to \infty} \frac{\left(\sqrt{x + 400} - \sqrt{x}\right)}{1} \left(\frac{\left(\sqrt{x + 400} + \sqrt{x}\right)}{\left(\sqrt{x + 400} + \sqrt{x}\right)}\right)$$

$$= \lim_{x \to \infty} \frac{x + 400 - x}{\sqrt{x + 400} + \sqrt{x}}$$

$$= \lim_{x \to \infty} \frac{400}{\sqrt{x + 400} + \sqrt{x}}$$

$$= \frac{400}{\text{huge number}}$$

Even after all this algebraic manipulation, you still have to think logically: The numerator is fixed (400) while the denominator keeps getting larger and larger, so the fraction approaches 0, which is **(A)**.

5.3 HORIZONTAL AND VERTICAL ASYMPTOTES

High-Yield

To answer a question like this:

Consider the function: $f(x) = \dfrac{x^2 - 5x + 6}{x^2 - 4}$. Which of the following statements is true?

    I. $f(x)$ has a vertical asymptote of $x = 2$.

    II. $f(x)$ has a vertical asymptote of $x = -2$.

    III. $f(x)$ has a horizontal asymptote of $y = 1$.

(A) II only

(B) I and III only

(C) II and III only

(D) I, II and III

You need to know this:

Horizontal asymptotes:

- The horizontal line $y = c$ is an asymptote if $\lim\limits_{x \to \infty} f(x) = c$ and/or $\lim\limits_{x \to -\infty} f(x) = c$.

- A function can, in principle, have different horizontal asymptotes as x approaches infinity and as x approaches negative infinity. However, in reality such functions are complicated constructions and almost never appear on the AP exam.

Vertical asymptotes:

- A function can have zero, one, or many vertical asymptotes.

- The candidates for vertical asymptotes of a function are the values of x that make the denominator equal 0. Many people mistakenly assume that a function has a vertical asymptote at every point where the denominator is zero. This is *not* true. Just because a function with a denominator is undefined at a point where the denominator equals zero, it does not necessarily have a vertical asymptote there; it may only have a hole (which occurs when the numerator and denominator share a common factor that can be canceled).

- Some functions without a denominator can also have vertical asymptotes. The function $f(x) = \tan x$ is one example—although not the best example, because we can rewrite $f(x) = \tan x$ as a function *with* a denominator, $f(x) = \dfrac{\sin x}{\cos x}$.

- Logarithmic functions have vertical asymptotes. For example, $f(x) = \ln x$ has a vertical asymptote at $x = 0$, and $f(x) = \ln (x - 5)$ has a vertical asymptote at $x = 5$.

You need to do this:

- In questions like these, treat options I, II and III as true/false statements. Draw a line through any statement(s) that you determine is false.

- To check for vertical asymptotes, factor the numerator and denominator (if possible). After simplifying the expression (canceling common factors), plug in the value that x is approaching. If the result is 0 in the denominator, you've found a vertical asymptote.

- To check for horizontal asymptotes, compute the limit as x approaches infinity or negative infinity.

Answer and Explanation:

C

Statement I: If you substitute $x = 2$ into the function, $f(x) \to \dfrac{0}{0}$, which indicates a probable hole at $x = 2$, and not an asymptote. To confirm this, factor and simplify:

$$\lim_{x \to 2} \frac{x^2 - 5x + 6}{x^2 - 4} = \lim_{x \to 2} \frac{(x-3)\cancel{(x-2)}}{(x+2)\cancel{(x-2)}} = \lim_{x \to 2} \frac{x-3}{x+2} = -\frac{1}{4}$$

Because the factor $(x - 2)$ cancels, there is a hole, not a vertical asymptote, at $x = 2$, so statement I is false. This means you can eliminate (B) and (D).

Statement II: If you substitute $x = -2$ into the function, $f(x) \to \dfrac{20}{0}$, which indicates a vertical asymptote at $x = -2$. (You can also plug -2 into the simplified function because you already did the work when considering statement I.) This already tells you that **(C)** is correct. If you're not sure, check statement III.

Statement III: To check for horizontal asymptotes, consider the limit as x goes to infinity or negative infinity. To do this, identify the dominant term in the numerator and in the denominator and evaluate the limit using only those two terms.

$$\lim_{x \to \infty} \frac{\cancel{x^2} - 5x + 6}{\cancel{x^2} - 4} \to \lim_{x \to \infty} \frac{x^2}{x^2} = \lim_{x \to \infty} 1 = 1$$

This indicates that $y = 1$ is a horizontal asymptote. Therefore, II and III are both true, confirming that **(C)** is the correct answer.

PRACTICE SET

1. If $y = a$ is a horizontal asymptote of the function $y = f(x)$, which of the following statements must be true?

   (A) $\lim\limits_{x \to \infty} f(x) = a$

   (B) $\lim\limits_{x \to a} f(x) = \infty$

   (C) $f(a) = 0$

   (D) None of the above

2. Which of the following functions has a vertical asymptote at $x = 4$?

   (A) $f(x) = \dfrac{x+5}{x^2-4}$

   (B) $f(x) = \dfrac{x^2-16}{x-4}$

   (C) $f(x) = \dfrac{4x}{x+1}$

   (D) $f(x) = \dfrac{x+6}{x^2-7x+12}$

3. Let f be the function defined by

   $f(x) = \dfrac{(4x+7)(1-3x)}{(2x+5)^2}$. Which of the following is a horizontal asymptote to the graph of f?

   (A) $y = -6$

   (B) $y = -3$

   (C) $y = -0$

   (D) The graph has no horizontal asymptote.

4. Which of the following functions has both a vertical and horizontal asymptote?

   (A) $f(x) = \dfrac{1}{1+e^{-x}}$

   (B) $f(x) = \dfrac{x}{x^2+2}$

   (C) $f(x) = \dfrac{x}{x^2-2}$

   (D) $f(x) = \dfrac{x^2+2}{x}$

ANSWERS AND EXPLANATIONS

1.   D

If $y = a$ is a horizontal asymptote of the function $y = f(x)$, then EITHER $\lim\limits_{x \to \infty} f(x) = a$ OR $\lim\limits_{x \to -\infty} f(x) = a$. Because either of these is sufficient for a horizontal asymptote at $y = a$, (A) might be true, but does not have to be true. (B) is one of the conditions for a vertical asymptote because a limit that equals ∞ indicates a function that increases without bound, not one that has a horizontal asymptote. (C) is discussing the behavior of the function specifically at $x = a$, and because horizontal asymptotes are concerned with the function's behavior as it approaches either positive or negative infinity, it doesn't matter what is happening at $x = a$, so again, this might be true but does not have to be. This means **(D)** must be correct.

2.   D

You can eliminate (A) immediately; the only candidates for the vertical asymptotes of this function are the points where the denominator equals zero, which are $x = \pm 2$, not $x = 4$.

You can also eliminate answer (C), because the only candidate for a vertical asymptote of this function is $x = -1$.

For the function in (B), $x = 4$ *is* a candidate for a vertical asymptote because the function is not defined at this point (the denominator equals 0). To determine whether this point is in fact a vertical asymptote, simplify the function and then compute the limit:

$$\lim_{x \to 4} \frac{x^2 - 16}{x - 4} = \lim_{x \to 4} \frac{(x + 4)\,\cancel{(x - 4)}}{\cancel{x - 4}} = 4 + 4 = 8$$

Because the limit is finite, the function *does not* have a vertical asymptote at $x = 4$.

This leaves **(D)** as the correct answer. To confirm, simplify the function:

$$f(x) = \frac{x + 6}{x^2 - 7x + 12} = \frac{x + 6}{(x - 3)(x - 4)}$$

Because the denominator equals 0 when $x = 4$, and the factor $(x - 4)$ cannot be canceled, this function does have a vertical asymptote at $x = 4$.

3.   B

To check for a horizontal asymptote, compute the limit as x approaches infinity or negative infinity by considering only the dominant terms in the numerator and denominator. Before you can do that, you need to expand both the numerator and the denominator, but not all the way. You don't need to carry out the entire FOIL process, just multiply the highest powers of x together:

$$\lim_{x \to \infty} \frac{(4x + 7)(1 - 3x)}{(2x + 5)^2} \to \lim_{x \to \infty} \frac{(4x)(-3x)}{(2x)^2}$$

$$= \lim_{x \to \infty} \frac{-12x^2}{4x^2} = \lim_{x \to \infty} (-3)$$

The limit of a constant function is that constant, so $y = -3$ is a horizontal asymptote to the graph of f, which is **(B)**.

4.   C

The only candidates for a vertical asymptote are the values of x that make the denominator equal 0. Quickly examine each function, one at a time.

(A): The denominator never equals zero because $e^{-x} = \dfrac{1}{e^x} \neq 0$. Therefore, the function has no vertical asymptotes. Eliminate (A).

(B): Again, the denominator never equals zero (because x^2 is always positive and you're adding 2 more), so the function has no vertical asymptotes. Eliminate (B).

(C): The denominator equals 0 when $x = \pm\sqrt{2}$ (and no factors can be canceled), so the function does have a vertical asymptote (in fact, it has two). To check for a horizontal asymptote, compute the limit as x approaches infinity or negative infinity (by considering only the dominant terms in the numerator and denominator):

$$\lim_{x \to \infty} \frac{x}{x^2 - 2} \to \lim_{x \to \infty} \frac{x}{x^2} = \lim_{x \to \infty} \frac{1}{x} = \frac{1}{\infty} = 0$$

The function has a horizontal asymptote at $y = 0$, making **(C)** the correct answer.

(D): The function has a vertical asymptote at $x = 0$, but does not have a horizontal asymptote.

5.4 LIMITS INVOLVING TRIG FUNCTIONS

To answer a question like this:

$$\lim_{x \to 0} \frac{\sin 3x}{4x} =$$

(A) 0   (B) $\frac{3}{4}$ (C) 1   (D) $\frac{4}{3}$

You need to know this:

- $\lim\limits_{x \to a} \sin x = \sin a$: Just like any other limit, plug the value of a into the function for x.

- $\lim\limits_{x \to 0} \frac{\sin x}{x} = 1$: Some form of this special trig limit appears relatively often on the AP Calc exam, so be sure to memorize it before Test Day.

- $\lim\limits_{x \to 0} \frac{1 - \cos x}{x} = 0$

- The limit as x approaches positive or negative infinity of cosine or sine does not exist (DNE) because their graphs oscillate (bounce back and forth) between -1 and 1.

Memorize the trig values for common benchmark angles:

angle (in radians)	0	$\frac{\pi}{6}$	$\frac{\pi}{4}$	$\frac{\pi}{3}$	$\frac{\pi}{2}$	π	$\frac{3\pi}{2}$	2π
$\sin x$	0	$\frac{1}{2}$	$\frac{\sqrt{2}}{2}$	$\frac{\sqrt{3}}{2}$	1	0	-1	0
$\cos x$	1	$\frac{\sqrt{3}}{2}$	$\frac{\sqrt{2}}{2}$	$\frac{1}{2}$	0	-1	0	1

You need to do this:

- As with all limit questions of the form $\lim_{x \to a} f(x)$, start by plugging the value of a into the function. If the result is a plain number, you're done.

- If the result is not a plain number (e.g., $\frac{a \text{ number}}{0}$ or $\frac{0}{0}$), examine the function and see if anything can be factored, rewritten using a trig identity, or made to resemble one of the two special trig limits.

Answer and Explanation:

B

Plugging in 0 for x yields $\frac{\sin(0)}{0} = \frac{0}{0}$, so some algebraic manipulation is in order. The function looks somewhat similar to the special trig limit that involves $\frac{\sin x}{x}$, so the goal is to get $\frac{\text{sine of something}}{\text{that } exact \text{ same something}}$ using valid mathematical manipulations. Be careful here—you can factor the 4 from the denominator, but not the 3 from the numerator because the 3 is part of the argument of the trig function. The steps are shown below:

$$\lim_{x \to 0} \frac{\sin 3x}{4x} = \frac{1}{4} \lim_{x \to 0} \frac{\sin 3x}{x}$$

$$= \frac{1}{4} \lim_{x \to 0} \frac{\sin 3x}{x} \cdot \frac{3}{3}$$

$$= \frac{3}{4} \boxed{\lim_{x \to 0} \frac{\sin 3x}{3x}}$$

$$= \frac{3}{4} \cdot 1 = \frac{3}{4}$$

Thus, **(B)** is correct.

✔ **AP Expert Note**

You'll practice L'Hôpital's rule in chapter 7, which could be used to answer this question as well.

PRACTICE SET

1. $\lim\limits_{x \to 0} \dfrac{2 \cos x}{\sin x - 2} =$

    (A) -1

    (B) 0

    (C) 1

    (D) 2

$$f(x) = \begin{cases} \cos x & \text{for } x < 0 \\ \dfrac{\sin x}{x} & \text{for } x \geq 0 \end{cases}$$

2. A piecewise-defined function f is given above. Which of the following statements about f is true?

    I. $\lim\limits_{x \to 0^+} f(x) = 1$

    II. $\lim\limits_{x \to 0^-} f(x) = 1$

    III. $\lim\limits_{x \to 0} f(x) = 1$

    (A) I only

    (B) II only

    (C) I and II only

    (D) I, II and III

3. $\lim\limits_{x \to \infty} \dfrac{\sin \frac{1}{x}}{\frac{1}{x}} =$

    (A) -1

    (B) 0

    (C) 1

    (D) ∞

4. $\lim\limits_{z \to 0} \dfrac{\tan 4x}{6x} =$

    (A) 0

    (B) $\dfrac{1}{3}$

    (C) $\dfrac{2}{3}$

    (D) nonexistent

ANSWERS AND EXPLANATIONS

1. A

Don't automatically assume you need to use a special rule when computing the limit of a trig function. As with any limit, you should always start by plugging in the value that x is approaching. Here, the result is:

$$\lim_{x \to 0} \frac{2\cos x}{\sin x - 2} = \frac{2\cos(0)}{\sin(0) - 2} = \frac{2(1)}{0 - 2} = -1$$

(A) is correct.

2. D

Evaluate each statement based on the definition of the function. The limit as x approaches 0 from the right (+) means values of x that are greater than 0, so use the bottom piece (which happens to be one of the special trig limits):

$$\lim_{x \to 0^+} f(x) = \boxed{\lim_{x \to 0^+} \frac{\sin x}{x}} = 1$$

Thus statement I is true.

The limit as x approaches 0 from the left (–) means from values of x that are less than 0, so use the top piece. For this piece, you can simply plug in 0:

$$\lim_{x \to 0^-} f(x) = \lim_{x \to 0^-} \cos x = \cos(0) = 1$$

Thus statement II is true.

Because statements I and II are true, statement III is true by definition, which means **(D)** is the correct answer.

3. C

Even though x approaches infinity here, this limit is simply a twist on the special trig limit $\lim_{x \to 0} \frac{\sin x}{x} = 1$. Notice that if $x \to \infty$, then $\frac{1}{x} \to 0$; therefore if you let $u = \frac{1}{x}$, then:

$$\lim_{x \to \infty} \frac{\sin \frac{1}{x}}{\frac{1}{x}} = \boxed{\lim_{u \to 0} \frac{\sin u}{u}} = 1$$

That's **(C)**.

4. C

Plugging 0 in yields $\frac{0}{0}$, so use rules for trig identities, manipulating limits, and the special trig limit $\lim_{x \to 0} \frac{\sin x}{x} = 1$ to obtain:

$$\lim_{x \to 0} \frac{\tan 4x}{6x} = \lim_{x \to 0} \frac{\sin 4x}{6x \cos 4x}$$

$$= \frac{1}{6} \lim_{x \to 0} \frac{1}{\cos 4x} \cdot \frac{\sin 4x}{x}$$

$$= \frac{1}{6} \lim_{x \to 0} \frac{1}{\cos 4x} \cdot \frac{\sin 4x}{x} \cdot \frac{4}{4}$$

$$= \frac{1}{6} \lim_{x \to 0} \frac{4}{\cos 4x} \cdot \frac{\sin 4x}{4x}$$

$$= \frac{1}{6} \left(\lim_{x \to 0} \frac{4}{\cos 4x} \right) \cdot \boxed{\left(\lim_{x \to 0} \frac{\sin 4x}{4x} \right)}$$

$$= \frac{1}{6} \left(\lim_{x \to 0} \frac{4}{\cos 4x} \right) \cdot (1)$$

$$= \frac{1}{6} \cdot \frac{4}{1} \cdot 1 = \frac{2}{3}$$

(C) is correct.

RAPID REVIEW

If you take away only 6 things from this chapter:

1. Graphically, when a function has a vertical asymptote, the limit on either—or both—sides of the asymptote goes to positive or negative infinity.

2. When finding the limit of a simplified rational function, $\lim\limits_{x \to a} \dfrac{f(x)}{g(x)}$: if plugging in the value of a yields a number over 0, check the limit from the left and from the right of a. If the values are not the same, the limit does not exist.

3. As x approaches positive or negative infinity, the limit of a function depends on the dominant term (or terms). The *dominant term* is the term that grows the fastest.

4. The candidates for vertical asymptotes of a rational function are the values of x that make the denominator equal 0. To check for vertical asymptotes, factor the numerator and denominator (if possible). After simplifying the expression (canceling common factors), find the values of x that make the denominator equal zero. These are your vertical asymptotes. Remember, $y = \ln x$ is not rational, but it has a vertical asymptote at $x = 0$.

5. To check for horizontal asymptotes, compute the limit as x approaches infinity and negative infinity.

6. Some form of the special trig limit $\lim\limits_{x \to 0} \dfrac{\sin x}{x} = 1$ appears relatively often on the AP Calc exam.

Limits

TEST WHAT YOU LEARNED

1. $\lim\limits_{x \to \infty} \dfrac{\sqrt{4x^4 + 3}}{x^2 - 4x + 1} =$

    (A) 2   (B) 3   (C) 4   (D) ∞

2. $\lim\limits_{x \to 0} \dfrac{\sin x}{x + 2x^2} =$

    (A) -2 (B) 0   (C) $\dfrac{1}{2}$ (D) 1

3. $\lim\limits_{x \to 3} \dfrac{2 - x}{(x - 3)^2} =$

    (A) $-\infty$ (B) 0   (C) $\dfrac{2}{3}$ (D) ∞

4. Which statement is true about the graph of $y = \dfrac{3x^2 + 2}{2 - x - 3x^2}$?

(A) The line $y = 0$ is a horizontal asymptote.

(B) The line $x = 1$ is a vertical asymptote.

(C) The line $x = \dfrac{2}{3}$ is a vertical asymptote.

(D) The graph has no horizontal or vertical asymptotes.

Answer Key

Test What You Already Know

1. **D** **Learning Objective:** 5.1

2. **A** **Learning Objective:** 5.2

3. **B** **Learning Objective:** 5.3

4. **A** **Learning Objective:** 5.4

Test What You Learned

1. **A** **Learning Objective:** 5.2

2. **D** **Learning Objective:** 5.4

3. **A** **Learning Objective:** 5.1

4. **C** **Learning Objective:** 5.3

 **REFLECTION**

Test What You Already Know score: _____

Test What You Learned score: _____

Use this section to evaluate your progress. After working through the pre-quiz, check off the boxes in the "Pre" column to indicate which Learning Objectives you feel confident about. Then, after completing the chapter, including the post-quiz, do the same to the boxes in the "Post" column. Keep working on unchecked Objectives until you're confident about them all!

Pre	Post		
☐	☐	**5.1**	Determine when a limit is infinite
☐	☐	**5.2**	Evaluate limits at infinity
☐	☐	**5.3**	Identify horizontal and vertical asymptotes of a function
☐	☐	**5.4**	Evaluate limits involving trig functions

 FOR MORE PRACTICE

Complete more practice online at kaptest.com. Haven't registered your book yet? Go to kaptest.com/booksonline to begin.

ANSWERS AND EXPLANATIONS

Test What You Already Know

1.   D   Learning Objective: 5.1

Substituting 1 into the expression yields $\frac{2}{0}$, so check the limit from the left and the limit from the right by plugging in a number just to the left of 1 and another just to the right of 1. (You don't need exact values, just an idea of where the y-values are headed.)

$$\frac{2(0.99)}{(0.99-1)^2} = \frac{1.98}{(-0.01)^2} = \frac{1.98}{\text{tiny positive \#}} \to \infty$$

$$\frac{2(1.01)}{(1.01-1)^2} = \frac{2.02}{(0.01)^2} = \frac{2.02}{\text{tiny positive \#}} \to \infty$$

Because the limit from the left and the limit from the right both approach ∞, **(D)** is correct.

Note that technically a limit that equals $\pm\infty$ doesn't exist because it's not a specific value. However, saying a limit is equal to $\pm\infty$ is commonly used to describe a function that increases (or decreases) without bound.

2.   A   Learning Objective: 5.2

When evaluating a limit that approaches ∞, the term with the highest exponent is so much larger than all other terms, you can ignore the other terms (they are negligible in comparison). When the expression is rational, consider the term in the numerator with the highest exponent and the term in the denominator with the highest exponent. In this question, before you can do that, you must expand the numerator (using FOIL):

$$\lim_{x\to\infty} \frac{(2x+1)(x-3)}{5-x^2} = \lim_{x\to\infty} \frac{2x^2-5x-3}{5-x^2}$$

Now, disregard the negligible terms:

$$\lim_{x\to\infty} \frac{2x^2 - 5x - 3}{5 - x^2} \to \lim_{x\to\infty} \frac{2x^2}{-x^2} = \lim_{x\to\infty}(-2)$$

The limit of a constant value is that constant value, so the limit here is -2, so **(A)** is correct.

Note that you can disregard negligible terms ONLY when the limit approaches positive or negative infinity (not when x approaches a numerical value).

3.   B   Learning Objective: 5.3

Evaluate each answer choice, one at a time, until you find one that is true. Start with (A): The numerator in the given expression is not factorable, so to check for vertical asymptotes, set the denominator equal to zero and solve:

$$x^3 - 1 = 0$$
$$x^3 = 1$$
$$x = 1$$

You now know $x = 1$ is a vertical asymptote. However, this isn't one of the answer choices. [Note that had the numerator been factorable, you would need to check for removable discontinuities (factors that cancel) before identifying vertical asymptotes.]

To check for horizontal asymptotes, consider the limit of the function as x approaches positive and negative infinity (which allows you to disregard all terms except the term in the numerator with the highest exponent and the term in the denominator with the highest exponent):

$$\lim_{x\to\infty} \frac{2x^2+13}{x^3-1} \to \lim_{x\to\infty} \frac{2x^2}{x^3}$$
$$= \lim_{x\to\infty} \frac{2}{x}$$
$$= \frac{2}{\text{huge number}} = 0$$

This tells you that the curve has a horizontal asymptote at $y = 0$, making **(B)** the correct choice.

Note that if you like to memorize rules, you can also compare the degrees of the numerator and the denominator. The degree of the numerator is 2, and the degree of the denominator is 3. Because the degree of the denominator is larger than the degree of the numerator, the curve has a horizontal asymptote at $y = 0$.

4.   A   Learning Objective: 5.4

Knowing the rule $\lim\limits_{x \to 0} \dfrac{\sin x}{x} = 1$ comes in very handy here. Factor the denominator of the expression and you'll see why:

$$\lim_{x \to 0} \frac{4 \sin x}{x^2 - x} = \lim_{x \to 0} \frac{4 \sin x}{x(x-1)}$$

$$= \lim_{x \to 0} \left(\boxed{\frac{\sin x}{x}} \cdot \frac{4}{x-1} \right)$$

$$= 1 \cdot \lim_{x \to 0} \frac{4}{x-1}$$

$$= 1 \cdot \frac{4}{0-1} = -4$$

That's **(A)**.

Note that you could also use L'Hôpital's rule to answer this question.

Test What You Learned

1.   A   Learning Objective: 5.2

Because x approaches ∞, you can disregard the negligible terms (the terms that are completely dominated by another term). In the numerator, disregard the 3, and in the denominator, disregard $-4x$ and $+1$:

$$\lim_{x \to \infty} \frac{\sqrt{4x^4 + 3}}{x^2 - 4x + 1} \to \lim_{x \to \infty} \frac{\sqrt{4x^4}}{x^2}$$

Now, recall that $\sqrt{x^2} = |x|$, which means $\sqrt{x^4} = |x^2|$. Because x^2 is always positive, you can write this as $\sqrt{x^4} = x^2$ (without the absolute value symbols) and reevaluate the limit:

$$\lim_{x \to \infty} \frac{\sqrt{4x^4}}{x^2} = \lim_{x \to \infty} \frac{2x^2}{x^2} = \lim_{x \to \infty} 2$$

The limit of a constant, regardless of what x is approaching, is that constant, so **(A)** is correct.

2.   D   Learning Objective: 5.4

You need to manipulate the expression because substituting 0 yields:

$$\frac{\sin(0)}{0 + 2(0)^2} = \frac{0}{0}$$

A commonly seen trig limit on the AP Calculus exam is $\lim\limits_{x \to 0} \dfrac{\sin x}{x} = 1$, so whenever you see the limit as x approaches 0 of a rational trig expression, try to rewrite the expression in this form. Here, factor the denominator of the expression and see what happens:

$$\lim_{x \to 0} \frac{\sin x}{x + 2x^2} = \lim_{x \to 0} \frac{\sin x}{x(1 + 2x)}$$

$$= \lim_{x \to 0} \left(\boxed{\frac{\sin x}{x}} \cdot \frac{1}{1 + 2x} \right)$$

$$= 1 \cdot \lim_{x \to 0} \frac{1}{1 + 2x}$$

$$= \frac{1}{1 + 2(0)} = \frac{1}{1} = 1$$

(D) is correct.

Note that you could also use L'Hôpital's rule to answer this question.

3.   A   Learning Objective: 5.1

Substituting 3 into the expression yields $\dfrac{-1}{0}$, so check the limit from the left and the limit from the right by plugging in a number just to the left of 3 and another just to the right of 3. (You don't need exact values, just an idea of where the y-values are headed.)

$$\frac{2 - 2.99}{(2.99 - 3)^2} = \frac{-0.99}{(-0.01)^2} = \frac{-0.99}{\text{tiny positive \#}} \to -\infty$$

$$\frac{2 - 3.01}{(3.01 - 3)^2} = \frac{-1.01}{(0.01)^2} = \frac{-1.01}{\text{tiny positive \#}} \to -\infty$$

Because the limit from the left and the limit from the right both approach $-\infty$, **(A)** is correct.

4. C Learning Objective: 5.3

Evaluate each answer choice, one at a time, until you find one that is true. Start with (A): To find the horizontal asymptote, take the limit as x approaches ∞. Remember, you can disregard the negligible terms:

$$\lim_{x \to \infty} \frac{3x^2 \cancel{+2}}{2 \cancel{-x} - 3x^2} \to \lim_{x \to \infty} \frac{3x^2}{-3x^2} = \lim_{x \to \infty} -1 = -1$$

The horizontal asymptote is $y = -1$, not $y = 0$, so eliminate A and move on to vertical asymptotes. The numerator is not factorable (which means no factors can cancel with any factors in the denominator), so to find the vertical asymptotes, set the denominator equal to 0, factor, and solve for x:

$$0 = 2 - x - 3x^2$$
$$0 = -\left(3x^2 + x - 2\right)$$
$$0 = -(3x - 2)(x + 1)$$

You can see from the second factor that $x = -1$ is a vertical asymptote, but this isn't one of the answer choices, so set the other factor equal to 0 and solve:

$$3x - 2 = 0$$
$$3x = 2$$
$$x = \frac{2}{3}$$

The other vertical asymptote is $x = \frac{2}{3}$, which is **(C)**.

CHAPTER 6

Continuity

LEARNING OBJECTIVES

6.1   Analyze functions to determine continuity

6.2   Identify types of discontinuities

6.3   Determine when piecewise functions are continuous

TEST WHAT YOU ALREADY KNOW

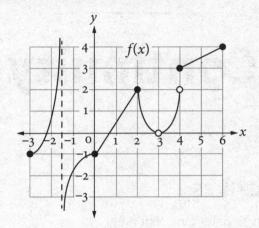

1. The graph of the function f is shown above. For how many values of x in the open interval $(-3, 6)$ is f discontinuous?

   (A) two   (B) three   (C) four   (D) five

2. For the function f defined by $f(x) = \dfrac{x^2 - 13x + 36}{x^2 + 5x - 36}$, which of the following statements is true?

   (A) The function has a removable discontinuity at $x = 4$ only.

   (B) The function has a removable discontinuity at $x = -9$ only.

   (C) The function has a removable discontinuity at $x = 9$ only.

   (D) The function has removable discontinuities at $x = 4$ and $x = -9$.

$$f(x) = \begin{cases} \dfrac{x^2 - 10x + 21}{b(x-3)} & \text{for } x \neq 3 \\ b & \text{for } x = 3 \end{cases}$$

3. Let f be defined as shown above. For what value of b is f continuous at $x = 3$?

    (A) -7

    (B) $\sqrt{3}$

    (C) 3

    (D) There is no such value of b.

Answers to this quiz can be found at the end of this chapter.

6.1 CONTINUITY AND LIMITS

To answer a question like this:

Which of the following functions are continuous for all real numbers x?

    I. $f(x) = x^{\frac{2}{3}}$

    II. $f(x) = \tan(x + \pi)$

    III. $f(x) = e^{x-1}$

   (A) None

   (B) I only

   (C) I and III only

   (D) I, II, and III

You need to know this:

Conceptually:

- A function is *continuous* if it has no holes or breaks. That is, the function is uninterrupted.

- Graphically, a function is continuous over a given interval if you can draw it without lifting up your pencil.

- In general, functions that have vertical asymptotes (e.g., logarithmic functions, tangent functions, and some rational functions) are not continuous for all real numbers.

Technically:

A function $f(x)$ is continuous at the point $x = c$ if all three of the following conditions are met.

- $f(x)$ is defined at $x = c$

- $\lim\limits_{x \to c} f(x)$ exists

- $\lim\limits_{x \to c} f(x) = f(c)$

If a function is not continuous at $x = c$, we say it is discontinuous there, or has a discontinuity at $x = c$. Each of the graphs below has a discontinuity at $x = 1$ because it does not meet one of the conditions above.

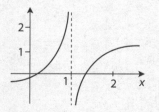

Not defined

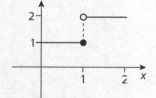

Defined, but limit does not exist

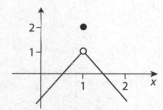

Defined, limit exists, but values of function and limit are different

You need to do this:

- Think about (or sketch) the graph of each of the functions.

- Determine whether the function is undefined for any values of x.

- Eliminate functions that have any discontinuities.

✔ **AP Expert Note**

Learning the basic shapes of certain types of functions can save you time on Test Day.

Answer and Explanation:

C

Quickly analyze each function by thinking about what its graph looks like and/or what values of x may cause the function to be undefined.

Function I: $f(x) = x^{\frac{2}{3}}$ is the same as $f(x) = \sqrt[3]{x^2}$. You can square any number and you can take the cube root of any number (including negative numbers), so this function is defined, and continuous, for all real numbers. Eliminate (A).

Function II: The graph of $f(x) = \tan(x + \pi)$ is a translation of the graph of $f(x) = \tan x$, which has vertical asymptotes at odd multiples of $\frac{\pi}{2}$ and is therefore *not* continuous for all real numbers. Eliminate (D).

Function III: The graph of $f(x) = e^{x+1}$ is a translation of the graph of $f(x) = e^x$, which has a horizontal asymptote (along the x-axis), but no vertical asymptotes, holes, or jump discontinuities, so this function is continuous for all real numbers.

This means **(C)** is correct.

PRACTICE SET

1. Suppose that f is not continuous at $x = 1$, f is defined at $x = 1$, and $\lim_{x \to 1} f(x) = L$, where L is finite. Which of the following could be the graph of f?

(A)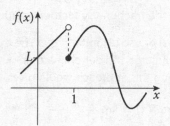

(B)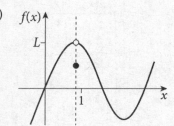

(C)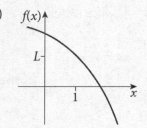

(D)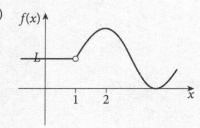

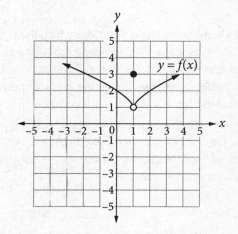

2. Consider the function $y = f(x)$ shown above. Which of the following statements is true?

(A) $\lim_{x \to -} f(x) = 3$

(B) $f(1) = 1$

(C) $\lim_{x \to 1} f(x) = f(1)$

(D) None of the above

3. Let f be the function given by
$f(x) = \dfrac{(x+1)^2 (x-5)}{(x+1)(x-2)}$. For which of the following values of x is f not continuous?

(A) 2 only

(B) 5 only

(C) -1 and 2 only

(D) -1, 2, and 5

4. Let $f(x)$ be a continuous function for all x.
If $f(x) = \dfrac{2x^2 - 2}{x^2 - 3x + 2}$ for $x^2 - 3x + 2 \neq 0$,
then what is $f(1)$?

(A) -4

(B) -2

(C) 0

(D) 4

ANSWERS AND EXPLANATIONS

1. B

Begin by eliminating the graph(s) for which the function *is* continuous at $x = 1$. This will eliminate (C). Next, eliminate graphs for which f is not defined at $x = 1$. This eliminates (D). Of the two remaining graphs, choose the one for which $\lim\limits_{x \to 1^+} f = \lim\limits_{x \to 1^-} f = L$. In other words, select the graph that approaches the same y-value (L) from the left and the right of 1, which is **(B)**. (In (A), the limit from the right of 1 is L, but the limit from the left of 1 is some value that is greater than L.)

2. D

Evaluate each statement, one at a time. Draw a line through false statements as you go.

$\lim\limits_{x \to 1} f(x) = 1$ (not 3), so (A) is incorrect.

There is a solid dot at $(1, 3)$, so $f(1) = 3$, making (B) incorrect.

The previous two statements prove that (C) is incorrect, because $1 \neq 3$.

Thus, **(D)** is the correct answer.

3. C

The function is undefined (and thus not continuous) for values of x that cause the denominator to equal 0, regardless of whether those values indicate a "hole" or an asymptote in the function's graph. Thus, the function is not continuous where $x + 1 = 0$, or $x = -1$, and where $x - 2 = 0$, or $x = 2$. That's **(C)**.

Note that $x = 5$ is the x-intercept of the graph, not a point of discontinuity.

4. A

By definition, in order for $f(x)$ to be continuous at $x = 1$, $\lim\limits_{x \to 1} f(x) = f(1)$.

To find the limit, start by plugging in 1 for x. This yields $\dfrac{0}{0}$, so simplify the fraction (by factoring and canceling common factors) or use L'Hôpital's rule. The expressions in both the numerator and denominator look fairly easy to factor, so using this approach would give:

$$
\begin{aligned}
\lim_{x \to 1} f(x) &= \lim_{x \to 1} \frac{2x^2 - 2}{x^2 - 3x + 2} \\
&= \lim_{x \to 1} \frac{2(x+1)(x-1)}{(x-2)(x-1)} \\
&= \lim_{x \to 1} \frac{2(x+1)}{(x-2)} \\
&= \frac{2(1+1)}{1-2} \\
&= \frac{4}{-1} = -4
\end{aligned}
$$

Thus, $f(1)$ must equal -4, which is **(A)**.

6.2 TYPES OF DISCONTINUITIES

To answer a question like this:

$$f(x) = \frac{x^2 + x + k}{x + 3}$$

Given the function f defined above, for what value of k will the function have a removable discontinuity at $x = -3$?

(A) -6 (B) -3 (C) 3   (D) 6

You need to know this:

There are three types of discontinuities:

- *Removable discontinuities*: Whenever the zero in the denominator of a rational function can be eliminated by factoring and canceling out common factors, the discontinuity is removable (as it can be "removed" from the function). The result is a hole in the graph at that value of *x*.

- *Vertical asymptotes*: Whenever the zero in the denominator of a rational function cannot be eliminated, the discontinuity is a vertical asymptote. This means that the right- and left-hand limits of the function are either positive or negative infinity there.

- *Jump discontinuities*: Whenever the left- and right-hand limits are both finite, but not equal, the function is considered to "jump" at that number. This commonly occurs in piecewise functions.

- Vertical asymptotes and jump discontinuities are often referred to as *nonremovable discontinuities*.

You need to do this:

- Pay careful attention to the wording—here, you're looking for a *removable* discontinuity.

- Think about what scenario results in this type of discontinuity.

- Chances are that factoring is involved.

- Can you cancel any factors? If so, you've found a removable discontinuity.

Answer and Explanation:

A

If the given function is to have a removable discontinuity, then $(x + 3)$ must be canceled from the denominator. This means the numerator must be factorable and one of its factors must be $(x + 3)$.

You could try each of the answer choices, but thinking about how FOIL works could save you some time. Ask yourself what value of k could possibly produce a middle term of $1x$. If $k = -3$ or 3, then a middle term of $1x$ cannot result because the only factors of 3 are 1 and 3, and those numbers differ by 2. Thus, you can eliminate (B) and (C) right away. There are only two remaining choices, so try each.

(A): If $k = -6$, then:

$$f(x) = \frac{x^2 + x - 6}{x + 3} = \frac{(x - 2)\boxed{(x + 3)}}{\boxed{x + 3}}$$

When $k = -6$, the factor $x + 3$ can be canceled, so there is a removable discontinuity at $x = -3$ and **(A)** is correct.

PRACTICE SET

Use the graph of f shown below to answer questions 1 and 2.

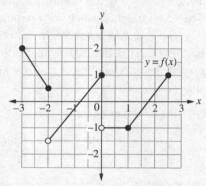

1. The function f has a removable discontinuity at

   (A) $x = -2$ only

   (B) $x = 0$ only

   (C) $x = -2$ and $x = 0$

   (D) $f(x)$ has no removable discontinuities.

2. The function has a jump discontinuity at

   (A) $x = -2$ only

   (B) $x = 0$ only

   (C) $x = 1$ only

   (D) $x = -2$ and $x = 0$

3.

   I $\quad f(x) = \dfrac{\ln (x - 4)}{3x - 12}$

   II. $\quad f(x) = \dfrac{4}{x + 4}$

   III. $\quad f(x) = \dfrac{x^2 - 16}{x + 4}$

   Which of the functions given above has a removable discontinuity at $x = 4$?

   (A) None

   (B) I only

   (C) III only

   (D) I and III only

4. $\quad f(x) = \dfrac{x^2 - (k + 2)x + 6}{x - k}$

   Given the function f defined above, for what value of k will the function have a removable discontinuity at $x = k$?

   (A) $k = -1$

   (B) $k = 1$

   (C) $k = 2$

   (D) $k = 3$

Limits

ANSWERS AND EXPLANATIONS

1. D

The only discontinuities that are removable are holes and holes with a point above or below—this function has neither. The function has 2 jump discontinuities (gaps), at $x = -2$ and $x = 0$, but neither of these is removable. **(D)** is correct.

2. D

A jump discontinuity occurs at a value x if the left- and right-hand limits are finite but not equal. Here, that occurs at $x = -2$ and $x = 0$ because:

$$\lim_{x \to -2^-} f(x) = \frac{1}{2} \text{ while } \lim_{x \to -2^+} f(x) = -\frac{3}{2} \text{ and}$$
$$\lim_{x \to 0^-} f(x) = 1 \text{ while } \lim_{x \to 0^+} f(x) = -1$$

Because the one-sided limits are different for each of these values of x, there is a gap in the graph, which is clearly the case at $x = -2$ and $x = 0$. **(D)** is correct.

3. A

The function(s) will have a removable discontinuity at $x = 4$ as long as a factor of $x - 4$ can be removed from the denominator. Only function **I** has a factor of $x - 4$ in the denominator, so that is the only one you need to check. Be careful—although the denominator in function **I** can be factored as $3(x - 4)$, you cannot cancel this factor with the $x - 4$ in the numerator because that quantity is *part* of the natural logarithm. Thus, none of the functions has a removable discontinuity at $x = 4$, making **(A)** the correct answer.

4. D

Occasionally, brute force is the cleanest approach to a question. The given function will have a removable discontinuity at $x = k$ if, when written in factored form, $x - k$ can be canceled. The easiest way to proceed at this point is to try all the answer choices and see which one produces a removable factor.

When $k = -1$, the function becomes:

$$\frac{x^2 - (-1 + 2)x + 6}{x - (-1)} = \frac{x^2 - x + 6}{x + 1}$$

The numerator cannot be factored, which means $x + 1$ can't be canceled, resulting in a vertical asymptote at $x = -1$, not a removable discontinuity. Eliminate (A).

When $k = 1$, the function becomes:

$$\frac{x^2 - (1 + 2)x + 6}{x - (1)} = \frac{x^2 - 3x + 6}{x - 1}$$

Again, the numerator cannot be factored, so there is a vertical asymptote at $x = 1$ for this function. Eliminate (B).

When $k = 2$, the function becomes:

$$\frac{x^2 - (2 + 2)x + 6}{x - (2)} = \frac{x^2 - 4x + 6}{x - 2}$$

The situation is the same as in (A) and (B), so eliminate (C), which leaves **(D)** as the correct answer.

Just in case you're curious, when $k = 3$, the function becomes:

$$\frac{x^2 - (3 + 2)x + 6}{x - (3)} = \frac{x^2 - 5x + 6}{x - 3}$$

$$= \frac{(x - 3)(x - 2)}{x - 3}$$

When $k = 3$, the factor $x - 3$ can be canceled, so there is a removable discontinuity at $x = 3$ and **(D)** is indeed correct.

6.3 PIECEWISE FUNCTIONS

Limits

High-Yield

To answer a question like this:

$$f(x) = \begin{cases} \sqrt{x-3} & \text{for } 3 \le x < 7 \\ x-5 & \text{for } x \ge 7 \end{cases}$$

Which of the following statements is true about the piecewise function f given above?

(A) The function is discontinuous at $x = 4$.

(B) The function is discontinuous at $x = 7$.

(C) The function is discontinuous at $x = 4$ and at $x = 7$.

(D) The function is continuous for all values of x over the closed interval $[3, 7]$.

You need to know this:

Working with piecewise functions:

- A piecewise function is simply a function that is defined in "pieces," each of which corresponds to a certain interval that makes up the entire function's domain.

- To check the continuity of a piecewise function, check each of the given pieces over the corresponding interval and then check the x-values where the function is broken into pieces.

- Each piece of the function will be continuous at $x = a$ if the limit exists at $x = a$ and that limit is equal to the value of the function at that point.

Continuity of certain types of functions:

- Polynomials are continuous everywhere, so there is no need to check the limits at individual points.

- Exponential functions are continuous everywhere.

- Rational functions are continuous everywhere that their denominators are not equal to zero.

- Square root functions are continuous for all values of x such that the radicand (the quantity inside the square root) is greater than or equal to 0.

- Cube root functions are continuous everywhere.

- Logarithmic functions of the form $\log (ax + b)$ are continuous for all x such that $ax + b > 0$. That is, the quantity inside the parentheses must be positive.

- $y = \sin x$ and $y = \cos x$ are continuous everywhere, while $y = \tan x$ is discontinuous at odd multiples of $\frac{\pi}{2}$.

You need to do this:

- Examine each piece of the function for continuity over the given closed interval, [3, 7].

- Use the answer choices to guide you to the piece or pieces of the function that you're interested in. Here, you're interested in $x = 4$, which falls in the top interval, and in $x = 7$, which is the x-value at which the function breaks (from using the top piece to using the bottom piece).

- When trying to decide whether a function is continuous at an x-value where a function breaks ($x = 7$), use both pieces that involve that x-value. In this case, use the top piece for x-values close to (but smaller than) 7 and the bottom piece for $x = 7$ (or larger than 7).

✔ AP Expert Note

Continuity questions on the AP Calc exam that involve piecewise functions almost always require finding the limit as x approaches the "break point" and using the result to determine whether the functional value at that point is equal to the limit.

Answer and Explanation:

D

For $x = 4$, use only the top piece, because $3 \leq 4 < 7$. The graph of $y = \sqrt{x - 3}$ does not have any points of discontinuity for x-values in its domain (which is $x \geq 3$), so the function is continuous at $x = 4$.

For $x \geq 7$, the function is a polynomial, so it is continuous for all values in its domain (which is all real numbers).

You must pay special attention at $x = 7$; this is the value at which the function breaks. Make sure that $f(x)$ has the same value when x is close to 7 and when x equals 7. Compute each to check:

$$\lim_{x \to 7^-} \sqrt{x - 3} = \sqrt{7 - 3} = \sqrt{4} = 2$$
$$f(7) = 7 - 5 = 2$$

Because the two values are equal, the function is continuous at $x = 7$.

Thus, the function is continuous for all values of x over the closed interval [3, 7], and **(D)** is correct.

PRACTICE SET

1. $g(x) = \begin{cases} \dfrac{x^2 - 25}{x + 5} & \text{for } x \neq -5 \\ k & \text{for } x = -5 \end{cases}$

Let g be the function defined above. If g is continuous for all x, then what is the value of k?

(A) -10

(B) -5

(C) 0

(D) 5

2. $f(x) = \begin{cases} \dfrac{e^x - 1}{x} & \text{for } x \neq 0 \\ k & \text{for } x = 0 \end{cases}$

A function f is defined above. The function is continuous at $x = 0$ if $k =$

(A) 0

(B) 1

(C) $e - 1$

(D) e

3. $f(x) = \begin{cases} \sin \dfrac{\pi}{x} & \text{for } x \leq 4 \\ k\sqrt{\dfrac{x}{2}} & \text{for } x > 4 \end{cases}$

For which value of k is the function defined above continuous at $x = 4$?

(A) $k = -1$

(B) $k = -\dfrac{1}{2}$

(C) $k = \dfrac{1}{2}$

(D) $k = 2$

4. $f(x) = \begin{cases} \dfrac{-2\,x(x + 1)(x - 2)}{x^2 - x - 2} & \text{for } x \neq -1, x \neq 2 \\ 2 & \text{for } x = -1 \\ -4 & \text{for } x = 2 \end{cases}$

The function f defined above is continuous everywhere except at

(A) $x = -1$ only

(B) $x = 2$ only

(C) $x = -1$ and $x = 2$

(D) $f(x)$ is continuous for all real values of x.

Limits

ANSWERS AND EXPLANATIONS

1.   A

For g to be continuous, $g(-5)$ must equal the limit of g as x approaches -5, so evaluate the limit:

$$\lim_{x \to -5} g(x) = \lim_{x \to -5} \frac{x^2 - 25}{x + 5} \Rightarrow \frac{0}{0}$$
$$= \lim_{x \to -5} \frac{(x + 5)(x - 5)}{x + 5}$$
$$= \lim_{x \to -5} (x - 5) = -5 - 5 = -10$$

For g to be continuous, $g(-5) = -10$. Thus, **(A)** is correct.

2.   B

In order for $f(x)$ to be continuous at $x = 0$, $\lim_{x \to 0} f(x) = f(0)$.

Substituting $x = 0$ into $f(x)$ yields $\frac{0}{0}$, so use L'Hôpital's rule to find the limit:

$$\lim_{x \to 0} f(x) = \lim_{x \to 0} \frac{e^x - 1}{x} \Rightarrow \lim_{x \to 0} \frac{e^x}{1} = e^0 = 1.$$

Then, for continuity, $f(0) = k = 1$. Hence, **(B)** is correct.

3.   C

The function $f(x)$ will be continuous at $x = 4$ if the limit exists at $x = 4$ and is equal to the value of the function at that point. At $x = 4$, the function is defined by the formula $\sin \frac{\pi}{x}$, therefore $f(4) = \sin \frac{\pi}{4} = \frac{\sqrt{2}}{2}$. In addition:

$$\lim_{x \to 4^-} f(x) = \lim_{x \to 4^-} \sin \frac{\pi}{4} = \frac{\sqrt{2}}{2}$$

Thus, f will be continuous at $x = 4$ if:

$$\lim_{x \to 4^+} f(x) = \lim_{x \to 4^+} k\sqrt{\frac{x}{2}} = \frac{\sqrt{2}}{2}$$

Using the rules for manipulating limits, compute that:

$$\lim_{x \to 4^+} k\sqrt{\frac{x}{2}} = k\sqrt{\frac{4}{2}} = k\sqrt{2}$$

Therefore, to find k, solve:

$$k\sqrt{2} = \frac{\sqrt{2}}{2}$$
$$k\sqrt{2} \times \frac{1}{\sqrt{2}} = \frac{\sqrt{2}}{2} \times \frac{1}{\sqrt{2}}$$
$$k = \frac{1}{2}$$

That's **(C)**.

4.   D

The top piece of the function is rational, and rational functions are continuous everywhere except where their denominators are equal to zero. Factoring the denominator yields $(x + 1)(x - 2)$, which has zeros at $x = -1$ and $x = 2$; these are the x-values where the function breaks. Thus, you only need to check that the function is continuous at $x = -1$ and $x = 2$. Also, notice that $f(-1) = 2$ and $f(2) = -4$.

In order for the function to be continuous at $x = -1$, the limit must exist when x approaches -1 and must be equal to the value of the function (which is 2) when $x = -1$.

$$\lim_{x \to -1} f(x) = \lim_{x \to -1} \frac{-2x(x + 1)(x - 2)}{(x + 1)(x - 2)}$$
$$= \lim_{x \to -1} (-2x)$$
$$= -2(-1) = 2$$

$\therefore f(x)$ is continuous at $x = -1$

In order for the function to be continuous at $x = 2$, the limit must exist when x approaches 2 and must be equal to the value of the function (which is -4) when $x = 2$.

$$\lim_{x \to 2} f(x) = \lim_{x \to 2} \frac{-2x(x + 1)(x - 2)}{(x + 1)(x - 2)}$$
$$= \lim_{x \to 2} (-2x)$$
$$= -2(2) = -4$$

$\therefore f(x)$ is continuous at $x = 2$

Hence, $f(x)$ is continuous for all real values of x. This means **(D)** is the correct answer.

RAPID REVIEW

If you take away only 3 things from this chapter:

1. A function $f(x)$ is continuous at the point $x = c$ if all three of the following conditions are met:

   - $f(x)$ is defined at $x = c$

   - $\lim\limits_{x \to c} f(x)$ exists

   - $\lim\limits_{x \to c} f(x) = f(c)$

2. There are three types of discontinuities: removable discontinuities, jump discontinuities, and vertical asymptotes. A discontinuity is removable if the zero in the denominator of a rational function can be eliminated by factoring and canceling out common factors. A discontinuity is a vertical asymptote if the zero in the denominator of a rational function cannot be eliminated. A discontinuity jumps if its left- and right-hand limits are finite, but not equal.

3. To check the continuity of a piecewise function, check each of the given pieces over the corresponding interval and then check where the function is broken into pieces. Each piece of the function will be continuous at $x = a$ if the limit exists as x approaches a and is equal to the value of the function at $x = a$.

TEST WHAT YOU LEARNED

1. For the function f defined by $f(x) = \dfrac{x^2 - x - 42}{x^3 + 11x^2 + 30x}$, which of the following statements is true?

   (A) The function has one removable discontinuity and one vertical asymptote.

   (B) The function has one removable discontinuity and two vertical asymptotes.

   (C) The function has two removable discontinuities and one vertical asymptote.

   (D) The function has three vertical asymptotes.

$$g(x) = \begin{cases} -x^2 + 1 & \text{for } x < 3 \\ 2x + k & \text{for } x \geq 3 \end{cases}$$

2. Let g be defined as shown above. If g is continuous for all real numbers, what is the value of k?

   (A) -6 (B) -10 (C) 0 (D) 4

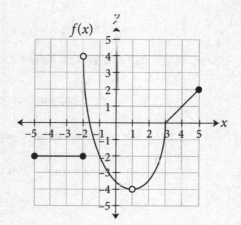

3. Part of the graph of f on the closed interval $[-5, 5]$ is shown above. If f is discontinuous for only one value of x in the open interval $(-5, 5)$, then what must $f(1)$ equal?

(A) -4

(B) 4

(C) $f(1)$ can take on any value.

(D) $f(1)$ cannot be determined.

Answer Key

Test What You Already Know

1. **B** **Learning Objective:** 6.1

2. **A** **Learning Objective:** 6.2

3. **D** **Learning Objective:** 6.3

Test What You Learned

1. **B** **Learning Objective:** 6.2

2. **A** **Learning Objective:** 6.3

3. **A** **Learning Objective:** 6.1

 **REFLECTION**

Test What You Already Know score: _____

Test What You Learned score: _____

Use this section to evaluate your progress. After working through the pre-quiz, check off the boxes in the "Pre" column to indicate which Learning Objectives you feel confident about. Then, after completing the chapter, including the post-quiz, do the same to the boxes in the "Post" column. Keep working on unchecked Objectives until you're confident about them all!

Pre	Post		
☐	☐	**6.1**	Analyze functions to determine continuity
☐	☐	**6.2**	Identify types of discontinuities
☐	☐	**6.3**	Determine when piecewise functions are continuous

 FOR MORE PRACTICE

Complete more practice online at kaptest.com. Haven't registered your book yet? Go to kaptest.com/booksonline to begin.

ANSWERS AND EXPLANATIONS

Test What You Already Know

1. B Learning Objective: 6.1

A function is continuous at a value of x (e.g., $x = 2$) if $\lim_{x \to 2} f(x) = f(2)$. Graphically, this means that a function is continuous if you can draw its graph without lifting up your pencil. Here, you would have to lift your pencil at $x = -1.5$, at $x = 3$, and again at $x = 4$. Thus, the function is not continuous (and therefore discontinuous) for three values of x in the interval $(-3, 6)$, making **(B)** the correct answer.

2. A Learning Objective: 6.2

There are three types of discontinuities: removable discontinuities (holes), jump (or gap) discontinuities, and vertical asymptotes. Based on the answer choices here, you're interested in removable discontinuities (holes). Whenever a zero in the denominator of a rational expression can be eliminated by factoring and canceling out terms, the discontinuity is removable (as it can be "removed" from the function). So, factor the numerator and the denominator:

$$\frac{x^2 - 13x + 36}{x^2 + 5x - 36} = \frac{(x-4)(x-9)}{(x-4)(x+9)} = \frac{\cancel{(x-4)}(x-9)}{\cancel{(x-4)}(x+9)}$$

The factor $(x - 4)$ can be canceled (thus removing the discontinuity), so the correct answer is **(A)**.

Note that at $x = -9$, the discontinuity is a vertical asymptote, and at $x = 9$, the *numerator* equals zero (which indicates an x-intercept, not a discontinuity).

3. D Learning Objective: 6.3

To be continuous at $x = 3$, $\lim_{x \to 3} f(x)$ must equal $f(3)$, which is given as b. Thus:

$$\lim_{x \to 3} \frac{x^2 - 10x + 21}{b(x-3)} = b$$

$$\lim_{x \to 3} \frac{(x-7)\cancel{(x-3)}}{b\cancel{(x-3)}} = b$$

$$\frac{3-7}{b} = b$$

$$-4 = b^2$$

The square of a number cannot be negative, so this is not possible. In other words, there is no such value of b, and **(D)** is correct.

Test What You Learned

1. B Learning Objective: 6.2

There are three types of discontinuities: removable discontinuities, jump (or gap) discontinuities, and vertical asymptotes. Based on the answer choices here, you're interested in removable discontinuities and vertical asymptotes. Whenever a zero in the denominator of an expression can be eliminated by factoring and canceling out terms, the discontinuity is removable (as it can be "removed" from the function) but it is still undefined at that x-value (there is a hole). Whenever a zero in the denominator of an expression cannot be eliminated, it indicates a vertical asymptote, as the simplified fraction is undefined at that x-value. So, factor the numerator and denominator:

$$\frac{x^2 - x - 42}{x^3 + 11x^2 + 30x} = \frac{(x+6)(x-7)}{x(x^2 + 11x + 30)}$$

$$= \frac{(x+6)(x-7)}{x(x+6)(x+5)}$$

$$= \frac{\cancel{(x+6)}(x-7)}{x\cancel{(x+6)}(x+5)}$$

$$= \frac{x-7}{x(x+5)}$$

The factor $(x + 6)$ can be canceled, so you have one removable discontinuity. You still have two factors in the denominator, x and $x + 5$ (resulting in zeros at $x = 0$ and $x = -5$ respectively), which correspond to two vertical asymptotes (at $x = 0$ and $x = -5$), so all together you have one removable discontinuity and two vertical asymptotes, which matches **(B)**.

2. A Learning Objective: 6.3

You are given a piecewise function g and told that it is continuous for all real numbers. Both individual "pieces" of the function are continuous on their own, so all you need to check is where the function breaks, which is at $x = 3$.

To be continuous at $x = 3$, $\lim\limits_{x \to 3^-} f(x)$ must equal $\lim\limits_{x \to 3^+} f(x)$, which also gives $f(3)$, so plug 3 into each piece of the function, set them equal to each other, and solve for k. This gives $-(3)^2 + 1 = 2(3) + k$, or $-8 = 6 + k$. Solving this equation yields $k = -14$, so **(A)** is correct.

3. A Learning Objective: 6.1

The graph shown has a gap at $x = -2$ and a hole at $x = 1$. For there to be only one discontinuity, one of the apparent discontinuities needs to be continuous. It's not possible to make the gap continuous, but if $f(1) = -4$, the discontinuity is eliminated, leaving f with only one discontinuity. Thus, **(A)** is correct.

PART 3

Derivatives

CHAPTER 7

Derivatives—The Basics

LEARNING OBJECTIVES

7.1   Recognize the derivative as the limit of a difference quotient

7.2   Use graphs and tables to estimate derivatives or functional values

7.3   Use the power rule to find the derivative of a polynomial, rational, or radical function

7.4   Use the product and quotient rules to find the derivative of a function

7.5   Apply the chain rule to calculate derivatives of basic composite functions

7.6   Calculate derivatives of logarithmic and exponential functions

7.7   Apply L'Hôpital's rule to evaluate limits of the form $\dfrac{0}{0}$ and $\dfrac{\infty}{\infty}$

7.8   Apply the intermediate value, extreme value, and/or mean value theorems

TEST WHAT YOU ALREADY KNOW

1. $\lim\limits_{h \to 0} \dfrac{(5+h)^2 - 5^2}{h} =$

    (A) 0   (B) 10   (C) 25   (D) nonexistent

2. If $f'(x) > 0$ and $f''(x) > 0$ for all x, which of the following could be a table of values for the function f ?

    (A)

x	$f(x)$
-1	7
0	5
1	4

    (B)

x	$f(x)$
-1	2
0	4
1	6

    (C)

x	$f(x)$
-1	4
0	5
1	6

    (D)

x	$f(x)$
-1	2
0	4
1	7

3. If $f(x) = \dfrac{9}{x} - \dfrac{1}{2}x^4 + 2$, then $f'(3) =$

   (A) -55 (B) -36 (C) 0 (D) 27

---

4. If $y = \dfrac{4x}{x^2 - 3}$, then $\dfrac{dy}{dx} =$

   (A) $\dfrac{4}{(2x)^2}$

   (B) $\dfrac{4}{\left(x^2 - 3\right)^2}$

   (C) $\dfrac{4x^2 + 12}{\left(x^2 - 3\right)^2}$

   (D) $\dfrac{-4x^2 - 12}{\left(x^2 - 3\right)^2}$

5. If $y = \left(\sqrt{x} + 2x^3\right)^6$, then $y' =$

(A) $6\left(\sqrt{x} + 2x^3\right)^5$

(B) $6\left(\dfrac{1}{2\sqrt{x}} + 6x^2\right)^5$

(C) $6\left(\sqrt{x} + 2x^3\right) \cdot \left(\dfrac{1}{2\sqrt{x}} + 6x^2\right)$

(D) $6\left(\sqrt{x} + 2x^3\right)^5 \cdot \left(\dfrac{1}{2\sqrt{x}} + 6x^2\right)$

6. If $f(x) = \ln\left(\dfrac{4x-5}{x+7}\right)$, then $f'(x) =$

(A) $2(2x - 5)(x + 7)$

(B) $\dfrac{4}{4x-5} \cdot \dfrac{1}{x+7}$

(C) $\dfrac{4}{(4x-5)} - \dfrac{1}{(x+7)}$

(D) $4\left(\dfrac{1}{4x-5} + \dfrac{1}{x+7}\right)$

7. $\displaystyle\lim_{x \to \pi} \frac{\cos(2x) - 1}{\sin(4x)} =$

    (A) $-\dfrac{1}{4}$

    (B) 0

    (C) $\dfrac{1}{2}$

    (D) nonexistent

---

8. Let f be a function that is continuous on the closed interval $[0, 5]$ with $f(0) = -2$ and $f(5) = 13$. Which of the following is guaranteed by the intermediate value theorem?

    (A) f attains a minimum on the open interval $(0, 5)$.

    (B) $f'(x) > 0$ for all x in the open interval $(0, 5)$.

    (C) $f(x) = 3$ has at least one solution in the open interval $(0, 5)$.

    (D) $f'(x) = 3$ has at least one solution in the open interval $(0, 5)$.

Answers to this quiz can be found at the end of this chapter.

7.1 THE LIMIT DEFINITION OF THE DERIVATIVE

To answer a question like this:

$$\lim_{h \to 0} \frac{\sqrt{16+h} - 4}{h} =$$

(A) 0   (B) $\frac{1}{8}$ (C) $\frac{1}{4}$ (D) 2

You need to know this:

The Difference Quotient

- The fraction $\dfrac{f(a + h) - f(a)}{h}$ is called the *difference quotient* and represents the average rate of change of $f(x)$ from $x = a$ to $x = a + h$.

- The difference quotient can also be expressed as $\dfrac{f(x) - f(a)}{x - a}$, which you should recognize as the slope formula you learned in algebra.

- Geometrically, this represents the slope of the secant line between two given points.

The Limit Definition of the Derivative

- The limit as h approaches 0 of the difference quotient represents the instantaneous rate of change of f at a given point.

- Geometrically, this represents the slope of the curve at a given point. The tangent to the curve at this point is the line that passes through the point and has that same slope.

- Memorize this limit definition of the derivative before Test Day:

$$f'(x) = \lim_{h \to 0} \frac{f(x + h) - f(x)}{h}$$

- The limit definition can also be expressed as:

$$f'(x) = \lim_{x \to a} \frac{f(x) - f(a)}{x - a}$$

Derivatives and Continuity

- A function is *not differentiable* (the derivative does not exist) at a given value if the function is not continuous at that value.

You need to do this:

- Recognize that the given limit is the definition of the derivative for some function *f*. The big clues are that $h \rightarrow 0$, the numerator contains "+ *h*," and there is an *h* in the denominator.

- Determine what the primary function is—is it a polynomial function, a square root function, a trig function, a logarithmic function, or some other recognizable function?

- Determine the point (the *x*-value) at which the derivative is being taken. This may require rewriting the second term in the numerator to match the primary function.

- Take the derivative of the primary function and plug in the point from the previous step.

Answer and Explanation:

B

The limit represents the definition of the derivative, so look for the primary function. You should immediately notice the square root symbol in the first term of the numerator. Rewrite the numerator so that both terms are square roots. This is a very quick step, but don't skip it because it confirms the point at which the derivative is being taken.

$$\lim_{h \to 0} \frac{\sqrt{16+h} - 4}{h} = \lim_{h \to 0} \frac{\sqrt{16+h} - \sqrt{16}}{h}$$

Here, the derivative is being taken at $x = 16$ and the primary function is $f(x) = \sqrt{x}$. To take the derivative of the function, rewrite the radical using a fractional exponent, then use the power rule. Finally, plug in 16 for *x* and simplify:

$$f(x) = \sqrt{x} = x^{\frac{1}{2}}$$

$$f'(x) = \frac{1}{2}x^{-\frac{1}{2}}$$

$$= \frac{1}{2\sqrt{x}}$$

$$f'(16) = \frac{1}{2\sqrt{16}} = \frac{1}{2(4)} = \frac{1}{8}$$

(B) is correct.

PRACTICE SET

1. $\lim\limits_{h \to 0} \dfrac{\ln(3+h) - \ln(3)}{h} =$

   (A) 0

   (B) $\dfrac{1}{3}$

   (C) 1

   (D) e^3

2. $f(x) = \begin{cases} 2x - 8 & \text{for } x < 5 \\ x^2 - 4x & \text{for } x \geq 5 \end{cases}$

   Let f be the piecewise function defined above. What is the value of $f'(5)$?

   (A) 2

   (B) 5

   (C) 6

   (D) The derivative does not exist at $x = 5$.

3. Which of the following statements are true about $\lim\limits_{x \to 2} \dfrac{\sqrt{x+2} - 2}{x - 2}$?

   I. $\lim\limits_{x \to 2} \dfrac{\sqrt{x+2} - 2}{x - 2}$ represents the slope of the tangent to $f(x) = \sqrt{x+2}$ at any given value of x

   II. $\lim\limits_{x \to 2} \dfrac{\sqrt{x+2} - 2}{x - 2} = \dfrac{1}{4}$

   III. $\lim\limits_{x \to 2} \dfrac{\sqrt{x+2} - 2}{x - 2}$ represents $f'(2)$ if $f(x) = \sqrt{x+2}$

   (A) I only

   (B) I and II only

   (C) I and III only

   (D) II and III only

4. If $\lim\limits_{h \to 0} \dfrac{\arctan(a+h) - \arctan(a)}{h} = \dfrac{1}{10}$, which of the following could be the value of a ?

   (A) $\dfrac{\sqrt{2}}{2}$

   (B) $\dfrac{\sqrt{3}}{2}$

   (C) 2

   (D) 3

Derivatives

ANSWERS AND EXPLANATIONS

1. B

The limit represents the definition of the derivative, so look for the primary function. You should immediately notice the natural log in both terms of the numerator. Here, the derivative is being taken at $x = 3$ and the primary function is $f(x) = \ln x$. The derivative of $\ln x$ is $\frac{1}{x}$, so plug in 3 for x to arrive at the correct answer, $\frac{1}{3}$, which is **(B)**.

2. D

The key to answering this question is recalling that a derivative does not exist at a point where a function is not continuous. Continuity should always come to mind when dealing with a piecewise function. Here, f is not continuous at $x = 5$ because substituting 5 into each piece of the function yields $2(5) - 8 = 2$ and $5^2 - 4(5) = 5$. The values are not the same, so **(D)** is correct. Also, because the derivative is actually a limit, the derivative from the right of 5 must equal the derivative from the left of 5 (they are not the same).

3. D

You need to recognize $\lim\limits_{x \to 2} \dfrac{\sqrt{x+2}-2}{x-2}$ as a form of one of the expressions for the definition of a derivative at a point:

$$f'(a) = \lim_{x \to a} \frac{f(x) - f(a)}{x - a}$$

The left side of the equal sign equals the right side if $f(x) = \sqrt{x+2}$ and $a = 2$, because

$$f'(2) = \lim_{x \to 2} \frac{\sqrt{x+2}-\sqrt{2+2}}{x-2} = \lim_{x \to 2} \frac{\sqrt{x+2}-2}{x-2},$$

making statement III true.

$f'(2)$ represents the slope of the tangent to $f(x) = \sqrt{x+2}$ at $x = 2$. It does not, however, give the slope for any other location, so statement I is false. This means **(D)** must be correct. On Test Day, if you're short on time, mark **(D)** and move on.

To check statement II, evaluate the given limit:

$$\lim_{x \to 2} \frac{\sqrt{x+2}-2}{x-2} \text{(multiply by the conjugate)}$$

$$= \lim_{x \to 2} \frac{\sqrt{x+2}-2}{x-2} \cdot \frac{\sqrt{x+2}+2}{\sqrt{x+2}+2}$$

$$= \lim_{x \to 2} \frac{x+2-4}{(x-2)(\sqrt{x+2}+2)}$$

$$= \lim_{x \to 2} \frac{(x-2)}{(x-2)(\sqrt{x+2}+2)}$$

$$= \lim_{x \to 2} \frac{1}{(\sqrt{x+2}+2)} = \frac{1}{\sqrt{4}+2} = \frac{1}{4}$$

Thus, statement II is true and **(D)** is indeed correct.

Note that you could also use L'Hôpital's rule to evaluate the limit.

4. D

This question looks a bit more difficult because of the inverse trig function and because you don't know the value of a. However, it still just amounts to recognizing that the limit is a derivative and then taking the derivative.

The primary function here is $y = \arctan a$, the derivative of which is $y' = \dfrac{1}{1+a^2}$. You're given that this equals $\dfrac{1}{10}$ (via the limit), so set the two quantities equal and solve for a:

$$\frac{1}{1+a^2} = \frac{1}{10}$$
$$10 = 1 + a^2$$
$$9 = a^2$$
$$a = \pm 3$$

Only positive 3 is one of the choices, so **(D)** is the correct answer.

7.2 ESTIMATING DERIVATIVES FROM GRAPHS AND TABLES

To answer a question like this:

Consider the differentiable function $f(x)$. Some known values of f are given in the table below:

x	2.0	2.2	2.4	2.6
$f(x)$	3	4	6	9

Use the given values to estimate $f'(2.1)$.

(A) −2.5

(B) 0.5

(C) 3.5

(D) 5

You need to know this:

Interpreting the derivative as slope:

The definition of the derivative can be interpreted as an instantaneous rate of change geometrically, on a graph. The rate of change of y between two points x_0 and x_1 is given by the difference quotient $\dfrac{f(x_1) - f(x_0)}{x_1 - x_0}$, which you may recognize as the slope formula. This difference quotient represents the slope of the secant line between the points $(x_0, f(x_0))$ and $(x_1, f(x_1))$.

As the points x_0 and x_1 get closer and closer together (until they are almost the same point) the secant line between the points becomes a better and better approximation of the tangent. Thus, you can use points that are close to a given point and the slope formula to estimate the slope of the tangent line at that given point. This means you can estimate the derivative of the function at that point.

What the derivative tells you about a function:

- The derivative represents the slope of the tangent to a curve at some value of x. So, if $f'(x) > 0$ over a given interval, the function is increasing on that interval. That is, as the x-values increase, the y-values also increase.

- If $f'(x) < 0$ over a given interval, the function is decreasing on that interval. That is, as the x-values increase, the y-values decrease.

- If $f'(x) = 0$ at a given point, then the function may have a local maximum or a local minimum at this x-value.

You need to do this:

- Make sure the function is differentiable at the point of interest.

- Use the closest points to the point of interest.

- Plug the points into the difference quotient (the slope formula) and simplify.

Answer and Explanation:

D

The question stem states that the function is differentiable. However, you don't have the actual function, so it is not possible to calculate an exact value for the derivative at 2.1. What you do know is that $f'(2.1)$ is the slope of the tangent to f at $x = 2.1$. Because you cannot find the exact value using the given information, estimate the slope of the tangent with the slope of the closest secant line: in this case, the secant on the closed interval [2, 2.2]. The slope of this secant line is:

$$\frac{f(2.2) - f(2)}{2.2 - 2} = \frac{4 - 3}{0.2} = \frac{1}{0.2} = 5$$

So the correct answer is **(D)**.

PRACTICE SET

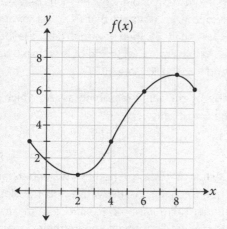

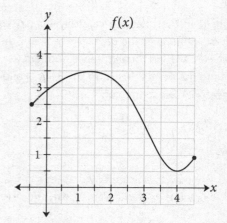

1. Consider the differentiable function f shown in the graph above. Which of the following is the best estimate for $f'(3)$?

   (A) 0

   (B) 1

   (C) 1.5

   (D) 2.5

x	0	5	10	15	20
$f(x)$	−1	4	−3	5	12

2. Some known values of a differentiable function f are given in the table above. Based on the data, which of the following statements must be true?

   (A) $f'(x) > 0$ for all values of x over the open interval $(0, 20)$.

   (B) $f'(x) > 0$ for all values of x over the open interval $(15, 20)$.

   (C) $f'(x) < 0$ for all values of x over the open interval $(0, 10)$.

   (D) None of the above.

3. The graph of a differentiable function f is shown above. For which of the following values of x is $f'(x)$ the least?

   (A) 1

   (B) 2

   (C) 3

   (D) 4

x	$f(x)$
−1	2
0	5
1	8

4. Using only the table of values shown above, which of the following best describes $f'(x)$ and $f''(x)$ over the open interval $(-1, 1)$?

   (A) $f'(x) < 0; f''(x) < 0$

   (B) $f'(x) < 0; f''(x) = 0$

   (C) $f'(x) > 0; f''(x) < 0$

   (D) $f'(x) > 0; f''(x) = 0$

ANSWERS AND EXPLANATIONS

1. B

Be sure to read the question carefully—you're looking for the best estimate of the *derivative* at 3, not the functional value at 3. Using the difference quotient (the slope formula) and the points closest to $x = 3$, $(2, 1)$ and $(4, 3)$, gives a reasonable estimate of:

$$m = \frac{3-1}{4-2} = 1$$

That's **(B)**.

Note that you could also carefully draw in a tangent line, estimate two points on the line, and use the slope formula.

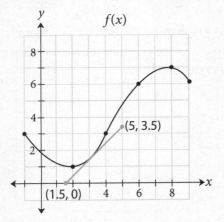

Using this method, the slope is approximately:

$$m = \frac{3.5-0}{5-1.5} = 1$$

Note that the two methods won't always yield exactly the same answer (as they did here), but they should be close enough that either would lead you to the correct choice.

2. D

When answer choices include the phrase "for all values of x," be especially careful. The question asks which statement *must* be true, so examine each statement with caution.

(A): This statement is false: $f'(x) > 0$ means the function is increasing. Between $x = 5$ and $x = 10$, the function decreases from 4 to −3, which means $f'(x) < 0$ for at least some part of the interval (0, 20). Eliminate (A).

(B): This statement may or may not be true. The function is increasing overall between $x = 15$ and $x = 20$, but you don't

have any information about what happens between these values. Suppose, for example, that $f(17) = 1$ (or any other number less than 5); then f would decrease before it increases. In that case, $f'(x)$ would be less than 0 between $x = 15$ and $x = 17$. Eliminate (B).

(C) is false by the same reasoning as (A), which means **(D)** is the correct answer.

3. C

This question is testing whether you understand the connection between the derivative and the slope of a tangent line. There are no calculations to make; simply sketch in tangent lines and compare their slopes.

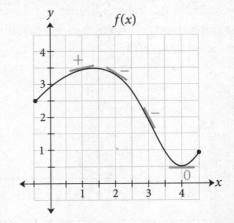

You can eliminate (A) and (D) right away because a positive slope is greater than a negative slope, and a 0 slope is also greater than a negative slope. To choose between (B) and (C), compare the steepness of the two negative slopes. The slope at $x = 3$ is steeper (in the negative direction) than the slope at $x = 2$, so $f'(3) < f'(2)$, making **(C)** the correct answer.

4. D

Examine the values in the table: $f(x)$ increases as x gets larger, which indicates that $f'(x)$, the slope of the function, is positive. This means $f'(x) > 0$, so eliminate (A) and (B).

To choose between (C) and (D), take a closer look at the slopes. The slope between the first pair of points is 3, and the slope between the second pair of points is also 3, so $f'(x)$ is constant. This means $f''(x)$, which is the derivative of $f'(x)$, must be 0. **(D)** is correct.

7.3 THE POWER RULE

To answer a question like this:

If $f(x) = x^5 - 2x^2 + \dfrac{3}{x}$, then $f'(x) =$

(A) $5x^4 - 4x + \ln x$

(B) $5x^4 - 4x + \dfrac{3}{x^2}$

(C) $5x^4 - 4x - \dfrac{3}{x^2}$

(D) $5x^4 + 3x^2 - 4x$

You need to know this:

Theorem: The Derivative of a Constant Function

For any constant function $f(x) = c$, $\dfrac{d}{dx}\big(f(x)\big) = \dfrac{d}{dx}(c) = 0$.

Hence if $y = 5$, then $y' = 0$.

Theorem: The Power Rule

If n is a constant, then $\dfrac{d}{dx}\big(x^n\big) = nx^{n-1}$. Hence, if $y = x^4$, then $y' = 4x^3$.

In other words, to take the derivative of a power function, multiply the entire function by the original exponent and subtract 1 from the exponent.

Applying the power rule to non-polynomial functions:

- You can rewrite rational functions using negative exponents and then apply the power rule. For example:

$$\frac{d}{dx}\left(\frac{1}{x^2}\right) = \frac{d}{dx}\big(x^{-2}\big) = -2x^{-3} = -\frac{2}{x^3}$$

- You can rewrite radical functions using fractional exponents and then apply the power rule. For example:

$$\frac{d}{dx}\big(\sqrt{x}\big) = \frac{d}{dx}\left(x^{\frac{1}{2}}\right) = \frac{1}{2}x^{\frac{1}{2}-1} = \frac{1}{2}x^{-\frac{1}{2}} = \frac{1}{2\sqrt{x}}$$

Other useful rules:

- **The Constant Multiple Rule:** If c is a constant, then $\dfrac{d}{dx}\big(cf(x)\big) = c \cdot \dfrac{d}{dx}\big(f(x)\big)$ If $y = 5x^7$, then $y' = 5 \cdot 7x^6 = 35x^6$.

- **The Sum and Difference Rule:** If f and g are differentiable functions with derivatives f' and g' respectively, then $[f(x) \pm g(x)]' = f'(x) \pm g'(x)$.

You need to do this:

- Rewrite all rational and/or radical terms in exponent form.
- Apply the power rule and any other rule (e.g., the constant multiple rule or the sum and difference rule).
- Simplify if possible.
- Depending on the answer choices, you may have to put some of the terms back in rational or radical form to find a match.

Answer and Explanation:

C

First, write the rational term, $\dfrac{3}{x}$, using a negative exponent. Then, use the power rule, the sum and difference rule, and the constant multiple rule to differentiate f. Finally, put the term with the negative exponent back in rational form to find a match.

$$f(x) = x^5 - 2x^2 + \frac{3}{x} = x^5 - 2x^2 + 3x^{-1}$$

$$f'(x) = 5x^4 - 2 \cdot 2x + 3(-1)x^{-2} = 5x^4 - 4x - \frac{3}{x^2}$$

A perfect match for **(C)**!

PRACTICE SET

1. If $f(x) = \dfrac{2}{3}x^3 - 2x^2 + 4$, then $f'(x) =$

   (A) $\dfrac{2}{9}x^2 - 4x$

   (B) $2x^2 - 4x$

   (C) $\dfrac{2}{9}x^2 - 4x + 4$

   (D) $2x^2 - 4x + 4$

2. If $f(x) = \dfrac{1}{3(x-2)^3}$, then $f'(x) =$

   (A) $-\dfrac{1}{(x-2)^4}$

   (B) $-\dfrac{2}{3(x-2)^2}$

   (C) $-\dfrac{9}{(x-2)^4}$

   (D) $\dfrac{9}{(x-2)^2}$

3. If $f(x) = \sqrt{x}(x+1)^2$, then $f'(x) =$

   (A) $\dfrac{-2x+2}{\sqrt{x}}$

   (B) $-\dfrac{x+1}{2\sqrt{x}}$

   (C) $\dfrac{x+1}{\sqrt{x}}$

   (D) $\dfrac{5x^2 + 6x + 1}{2\sqrt{x}}$

4. If $f(x) = \dfrac{k}{\sqrt{x}}$ and $f'(x) = -\dfrac{1}{x\sqrt{x}}$, then what is the value of k?

   (A) -2

   (B) -1

   (C) 1

   (D) 2

Derivatives

ANSWERS AND EXPLANATIONS

1. B

Use the power rule and other applicable rules to differentiate *f*. Remember, when you use the power rule, you multiply by the *original* exponent, not the new exponent.

$$f(x) = \frac{2}{3}x^3 - 2x^2 + 4$$

$$f'(x) = 3 \cdot \frac{2}{3}x^{3-1} - 2 \cdot 2x^{2-1} + 0$$

$$= 2x^2 - 4x$$

That's **(B)**.

2. A

Start by moving the $\frac{1}{3}$ (the coefficient) out to the front, just to get it out of the way:

$$f(x) = \frac{1}{3(x-2)^3} = \frac{1}{3} \cdot \frac{1}{(x-2)^3}$$

Next, rewrite the rational expression using a negative exponent:

$$f(x) = \frac{1}{3}(x-2)^{-3}$$

Finally, apply the power rule, being careful of negative signs, and simplify the result if possible:

$$f'(x) = -3 \cdot \frac{1}{3}(x-2)^{-3-1}$$

$$= -1(x-2)^{-4} = -\frac{1}{(x-2)^4}$$

Thus, **(A)** is correct.

Note that you must always think about whether you need to use the chain rule. Here, the derivative of the inside, $(x - 2)$, is 1, so it does not affect the overall derivative.

3. D

Don't jump right into using the product rule (which is covered in the next section). In this problem, it's much quicker to rewrite each quantity in exponent form, multiply using rules of exponents, and then take the derivative:

$$f(x) = \sqrt{x}(x+1)^2$$

$$= x^{\frac{1}{2}}(x^2 + 2x + 1) = x^{\frac{5}{2}} + 2x^{\frac{3}{2}} + x^{\frac{1}{2}}$$

$$f'(x) = \frac{5}{2}x^{\frac{3}{2}} + \frac{3}{2} \cdot 2x^{\frac{1}{2}} + \frac{1}{2} \cdot x^{-\frac{1}{2}}$$

$$= \frac{5}{2}x^{\frac{3}{2}} + 3x^{\frac{1}{2}} + \frac{1}{2x^{\frac{1}{2}}}$$

That's not one of the answer choices, so you'll need to perform some algebraic manipulations. Write each term with a common denominator and then simplify:

$$f'(x) = \frac{5x^{\frac{3}{2}} \cdot x^{\frac{1}{2}}}{2 \cdot x^{\frac{1}{2}}} + \frac{3x^{\frac{1}{2}} \cdot 2x^{\frac{1}{2}}}{2x^{\frac{1}{2}}} + \frac{1}{2x^{\frac{1}{2}}}$$

$$= \frac{5x^2 + 6x + 1}{2\sqrt{x}}$$

That's **(D)**.

4. D

You may be tempted to plug in each value for *k* to determine which is correct, but you should only use this strategy as a last resort because it will certainly use up valuable time. Instead, rewrite the radical (which also happens to be in the denominator) using a negative, fractional exponent. Treat *k* as a constant and take the derivative of *f*:

$$f(x) = \frac{k}{\sqrt{x}} = kx^{-\frac{1}{2}}$$

$$f'(x) = -\frac{1}{2}kx^{-\frac{1}{2}-1} = -\frac{1}{2}kx^{-\frac{3}{2}}$$

$$= -\frac{k}{2x^{\frac{3}{2}}} = -\frac{k}{2x\sqrt{x}}$$

Now, set the result equal to the given expression for *f′*, and solve for *k*:

$$-\frac{k}{2x\sqrt{x}} = -\frac{1}{x\sqrt{x}}$$

$$kx\sqrt{x} = 2x\sqrt{x}$$

$$k = 2$$

Hence, **(D)** is correct.

7.4 THE PRODUCT AND QUOTIENT RULES

High-Yield

To answer a question like this:

If $f(x) = (x^3 + 7)\left(5x^2 - \dfrac{3}{2}\right)$, then $f'(x) =$

(A) $30x^2$

(B) $25x^4 + 35x$

(C) $25x^4 - \dfrac{9}{2}x^2 + 35x$

(D) $25x^4 - \dfrac{9}{2}x^2 + 70x$

You need to know this:

The Product Rule

If f and g are differentiable at a point x, then their product is differentiable at x and the derivative is $[f \cdot g]' = f'g + fg'$.

> ✔ **AP Expert Note**
>
> Note that the order in which you combine the derivative and the non-derivative doesn't matter in the product rule because the terms are separated by a plus sign. However, it does matter in the quotient rule, so to keep things straight, remember to always start by taking the derivative of the first function.

The Quotient Rule

If f and g are differentiable at a point x and $g(x) \neq 0$, then their quotient is differentiable at x and the derivative is $\left[\dfrac{f}{g}\right]' = \dfrac{f'g - fg'}{g^2}$.

> ✔ **AP Expert Note**
>
> As the rules get more complicated, rewriting them in verbal form will help you to remember them. We can write the product rule as follows: *The derivative of the first times the second, plus the first times the derivative of the second.* Similarly, remember the quotient rule as: *The derivative of the top times the bottom, minus the top times the derivative of the bottom, all over the bottom squared.*

You need to do this:

- Decide how the pieces that contain the variable are attached (addition, subtraction, multiplication, division) and which rule(s) applies. Then carry out the rule(s).

- Before simplifying the result, check the answer choices to make sure you're not doing extra work.

Answer and Explanation:

D

$$\left[\left(x^3+7\right)\left(5x^2-\frac{3}{2}\right)\right]' = \left[x^3+7\right]'\left(5x^2-\frac{3}{2}\right)+\left(x^3+7\right)\left[5x^2-\frac{3}{2}\right]'$$

$$= 3x^2\left(5x^2-\frac{3}{2}\right)+\left(x^3+7\right)(10x)$$

$$= 15x^4-\frac{9}{2}x^2+10x^4+70x$$

$$= 25x^4-\frac{9}{2}x^2+70x$$

You can also find the derivative of *f* by multiplying (FOILing) and then computing the derivative of the resulting polynomial:

$$\left[\left(x^3+7\right)\left(5x^2-\frac{3}{2}\right)\right]' = \left[5x^5-\frac{3}{2}x^3+35x^2-\frac{21}{2}\right]'$$

$$= 25x^4-\frac{9}{2}x^2+70x$$

Both methods give the same answer of **(D)**, which is expected. In general, if you can multiply something out or do some other algebraic manipulation to make the problem simpler *before* computing the derivative, it's a good idea to go ahead and do it.

PRACTICE SET

1. If $f(x) = (x^2 + 1)(1 + x)^2$, then $f'(x) =$

   (A) $4x^2 + 4x$

   (B) $3x^2 + 2x + 1$

   (C) $2x^3 + 4x^2 + 2x$

   (D) $4x^3 + 6x^2 + 4x + 2$

2.

x	$f(x)$	$f'(x)$	$g(x)$	$g'(x)$
1	4	−3	3	2

Several values for two differentiable functions, f and g, and their derivatives are given in the table above. Based on these values, what is $\left[\dfrac{f}{g}\right]'(1)$?

   (A) $-\dfrac{17}{9}$

   (B) $-\dfrac{3}{2}$

   (C) $\dfrac{3}{2}$

   (D) $\dfrac{17}{9}$

3. If $y = (x + 2)(\sqrt{x - 1})$, then $\dfrac{dy}{dx} =$

   (A) $-\dfrac{x}{2\sqrt{x - 1}}$

   (B) $-\dfrac{2x}{\sqrt{x - 1}}$

   (C) $\dfrac{2x}{3\sqrt{x - 1}}$

   (D) $\dfrac{3x}{2\sqrt{x - 1}}$

4. If $f(x) = \dfrac{3 - \dfrac{1}{x}}{x + 5}$, then $f'(x) =$

   (A) $\dfrac{-3x^2 + 2x + 5}{x^4 + 25x^2}$

   (B) $\dfrac{3x^2 - 2x - 5}{x^4 + 25x^2}$

   (C) $\dfrac{-3x^2 + 2x + 5}{\left(x^2 + 5x\right)^2}$

   (D) $\dfrac{3x^2 - 2x - 5}{\left(x^2 + 5x\right)^2}$

ANSWERS AND EXPLANATIONS

1. D

Use the power rule and the product rule:

$$\left[\left(x^2+1\right)(1+x)^2\right]'$$

$$=\left[x^2+1\right]'(1+x)^2+\left(x^2+1\right)\left[(1+x)^2\right]'$$

$$=2x\left(1+2x+x^2\right)+\left(x^2+1\right)\cdot 2(1+x)$$

$$=2x+4x^2+2x^3+\left(x^2+1\right)(2+2x)$$

$$=2x+4x^2+2x^3+2x^3+2x^2+2x+2$$

$$=4x^3+6x^2+4x+2$$

A perfect match for **(D)**!

2. A

This question is simply testing whether you know how to apply the quotient rule. Take your time and plug the values from the table into the quotient rule carefully:

$$\left[\frac{f}{g}\right]'(1)=\frac{f'(1)\cdot g(1)-f(1)\cdot g'(1)}{\left(g(1)\right)^2}$$

$$=\frac{(-3)\cdot 3-4\cdot 2}{3^2}$$

$$=-\frac{17}{9}$$

(A) is correct.

3. D

The notation $\dfrac{dy}{dx}$ simply means find the derivative. First, rewrite the radical expression in exponent form. Then, apply the power rule and the product rule:

$$f(x)=(x+2)(x-1)^{\frac{1}{2}}$$

$$f'(x)=[x+2]'(x-1)^{\frac{1}{2}}+(x+2)\left[(x-1)^{\frac{1}{2}}\right]'$$

$$=1(x-1)^{\frac{1}{2}}+(x+2)\cdot\frac{1}{2}(x-1)^{-\frac{1}{2}}$$

$$=\frac{2(x-1)^{\frac{1}{2}}}{2}+\frac{(x+2)(x-1)^{-\frac{1}{2}}}{2}$$

$$=\frac{(x-1)^{-\frac{1}{2}}\left(2(x-1)+x+2\right)}{2}$$

$$=\frac{2x-2+x+2}{2\sqrt{x-1}}=\frac{3x}{2\sqrt{x-1}}$$

That's **(D)**.

4. C

You don't want to have a fraction inside a quotient, so simplify the expression first:

$$f(x)=\frac{3-\dfrac{1}{x}}{x+5}$$

$$=\frac{x\left(3-\dfrac{1}{x}\right)}{x(x+5)}$$

$$=\frac{3x-1}{x^2+5x}$$

Next, apply the quotient rule:

$$f'(x)=\frac{[3x-1]'\left(x^2+5x\right)-(3x-1)\left[x^2+5x\right]'}{\left(x^2+5x\right)^2}$$

$$=\frac{3\left(x^2+5x\right)-(3x-1)(2x+5)}{\left(x^2+5x\right)^2}$$

$$=\frac{3x^2+15x-\left(6x^2+13x-5\right)}{\left(x^2+5x\right)^2}$$

$$=\frac{-3x^2+2x+5}{\left(x^2+5x\right)^2}$$

Don't FOIL the denominator until you've checked the answer choices. You already have a match for **(C)**.

7.5 THE CHAIN RULE

To answer a question like this:

If $y = 3x\sqrt{x^2 + 1}$, then $\dfrac{dy}{dx}$ $x = 2$ is

(A) $\dfrac{9}{2\sqrt{5}}$

(B) $3\sqrt{5} + \dfrac{3}{\sqrt{5}}$

(C) $3\sqrt{5} + \dfrac{12}{\sqrt{5}}$

(D) $9\sqrt{5} + \dfrac{12}{\sqrt{5}}$

You need to know this:

The chain rule, as its name implies, is a chain of steps for computing the derivative of a composition of functions (functions inside other functions). Using this rule, you can break down a function such as $f(x) = (x^2 + 4x + 7)^{14}$ into a chain of $g(x) = x^2 + 4x + 7$ and $h(x) = x^{14}$, which makes $f(x) = h(g(x))$.

Theorem: The Chain Rule

If g is differentiable at x and h is differentiable at $g(x)$, then $h(g(x))$ is differentiable at x and its derivative is:

$$\left[h(g(x))\right]' = \underset{\substack{\text{the derivative of the}\\ \text{outside function evaluated}\\ \text{at the inside function}}}{h'(g(x))} \cdot \underset{\substack{\text{the derivative}\\ \text{of the inside}\\ \text{function}}}{g'(x)}$$

Note that you don't need to break the composition into separate pieces—just think of the function as the outside and the inside, and be sure to take the derivative of each.

Typical functions that involve the chain rule include:

- Power functions and rational functions that have a quantity inside the parentheses; for example, $f(x) = (4x + 1)^6$ or $f(x) = \dfrac{1}{(4x + 1)^8}$

- Radical functions that have a quantity inside the radical; for example $f(x) = \sqrt[3]{4x + 1}$

- Trigonometric and logarithmic functions that have a quantity inside the argument; for example $f(x) = \sin(4x + 1)$ or $f(x) = \ln(4x + 1)$

- Exponential functions for which the exponent is a quantity; for example $f(x) = e^{4x + 1}$

You need to do this:

- Determine which derivative rules apply (e.g., power rule, product rule, etc.).
- Determine whether the chain rule applies.
- Rewrite radicals and/or rational quantities in exponent form.
- Carefully carry out each of the rules that applies.
- Simplify the answer ONLY if you absolutely need to. Be sure to check the answer choices before you do this.

Answer and Explanation:

C

This is one of the more straightforward questions on the AP Calc exam, but you'll need to take your time and not skip steps. The function is a product, and one of the factors is a radical with a quantity inside. Thus you'll need to use the power rule, the product rule, and the chain rule. First, rewrite the radical part in exponent form.

$$y = 3x\sqrt{x^2 + 1} = 3x \cdot \left(x^2 + 1\right)^{\frac{1}{2}}$$

Next, to differentiate the function, write out the product rule with the functions inserted:

$$\frac{dy}{dx} = [3x]'\left(x^2 + 1\right)^{\frac{1}{2}} + (3x)\left[\left(x^2 + 1\right)^{\frac{1}{2}}\right]'$$

Compute each piece of the sum carefully, and don't forget to use the chain rule when you take the derivative of the last expression:

$$\frac{dy}{dx} = 3\left(x^2 + 1\right)^{\frac{1}{2}} + (3x) \cdot \frac{1}{2}\left(x^2 + 1\right)^{-\frac{1}{2}} \cdot 2x$$

$$= 3\sqrt{x^2 + 1} + \frac{3x^2}{\sqrt{x^2 + 1}}$$

Stop here—there is no need to simplify the expression because the question asks for the derivative at a specific value of x. This means you can plug the value in and then simplify the result (but only if you don't find a match in the answer choices). So, $\frac{dy}{dx}$ at $x = 2$ is:

$$3\sqrt{(2)^2 + 1} + \frac{3(2)^2}{\sqrt{(2)^2 + 1}} = 3\sqrt{5} + \frac{12}{\sqrt{5}}$$

Don't simplify! You already have a match for **(C)**.

PRACTICE SET

1. If $f(x) = \dfrac{1}{(5-3x)^5}$, then $f'(x) =$

   (A) $-\dfrac{3}{(5-3x)^4}$

   (B) $-\dfrac{3}{(5-3x)^6}$

   (C) $-\dfrac{5}{(5-3x)^6}$

   (D) $\dfrac{15}{(5-3x)^6}$

2. If f is a differentiable function such that $f(1) = 4$ and $f'(1) = -3$, and $g(x) = \sqrt{f(x)}$, then $g'(1) =$

   (A) $-\dfrac{\sqrt{3}}{2}$

   (B) $-\dfrac{3}{4}$

   (C) $2\sqrt{3}$

   (D) 6

3. If $f(x) = (2x+1)(1-3x)^4$, then $f'(x) =$

   (A) $-10(1-3x)^3(1+3x)$

   (B) $-6(1-3x)^3(2x+1)$

   (C) $2(1-3x)^3(3+x)$

   (D) $6(1-3x)^3(2x+1)$

4. If $y = x^3 - 2x$ and $x = 2t+1$, then $\dfrac{dy}{dt} =$

   (A) $\dfrac{3x^2 - 2}{2}$

   (B) $\dfrac{3x^2 - 4}{2}$

   (C) $3x^2 - 4$

   (D) $6x^2 - 4$

ANSWERS AND EXPLANATIONS

1. D

This is a classic chain rule question. Start by rewriting the rational function in exponent form (using a negative exponent). Then, apply the chain rule by taking the derivative of the power part (using the power rule) and multiplying by the derivative of what's inside the parentheses:

$$f(x) = (5 - 3x)^{-5}$$
$$f'(x) = -5(5 - 3x)^{-6} \cdot (-3)$$
$$= \frac{15}{(5 - 3x)^6}$$

(D) is correct.

2. B

This question is testing your ability to interpret the chain rule without knowing one of the functions. If $g(x) = \sqrt{f(x)}$, then the *outside* function is the square root function, and the *inside* function is f, so:

$$g'(x) = \left[\sqrt{f(x)}\right]' = \underbrace{\frac{1}{2\sqrt{f(x)}}}_{\substack{\text{the derivative of} \\ \sqrt{x} \text{ evaluated at} \\ f(x)}} \cdot f'(x)$$

$$g'(1) = \frac{1}{2\sqrt{4}} \cdot (-3) = -\frac{3}{4}$$

That's **(B)**.

3. A

The test makers love questions like this one because they test multiple rules simultaneously. Here, you'll need to apply the product rule, the power rule, and the chain rule. Take your time and don't skip steps. Also, try to write as neatly as possible so you can keep track of your computations:

$$f(x) = (2x + 1)(1 - 3x)^4$$
$$f'(x) = [2x + 1]'(1 - 3x)^4 + (2x + 1)[(1 - 3x)^4]'$$
$$= 2(1 - 3x)^4 + (2x + 1) \cdot 4(1 - 3x)^3(-3)$$
$$= 2(1 - 3x)^4 - 12(2x + 1)(1 - 3x)^3$$

This isn't one of the answer choices, so you'll need to simplify a bit. To start, factor out $(1 - 3x)^3$:

$$f'(x) = 2(1 - 3x)^4 - 12(2x + 1)(1 - 3x)^3$$
$$= (1 - 3x)^3(2(1 - 3x) - 12(2x + 1))$$
$$= (1 - 3x)^3(2 - 6x - 24x - 12)$$
$$= (1 - 3x)^3(-10 - 30x)$$
$$= -10(1 - 3x)^3(1 + 3x)$$

Phew! That's **(A)**.

Note: If you weren't sure how to simplify the derivative, let the answer choices guide you. Each choice has a factor of $(1 - 3x)^3$, which is a clue to factor that quantity out.

4. D

This is a twist on the chain rule. Because you are asked for the derivative with respect to t, each time you differentiate a term that contains x, you must multiply by the derivative of "what's inside." The derivative of x is simply $\frac{dx}{dt}$.

$$y = x^3 - 2x$$
$$\frac{dy}{dt} = 3x^2\frac{dx}{dt} - 2\frac{dx}{dt}$$

Now, the derivative of x with respect to t (or $\frac{dx}{dt}$) is just 2, so:

$$\frac{dy}{dt} = 3x^2(2) - 2(2) = 6x^2 - 4$$

(D) is correct.

7.6 DERIVATIVES OF LOGARITHMIC AND EXPONENTIAL FUNCTIONS

To answer a question like this:

If $f(x) = e^{\frac{1}{x}}$, then $f'(x) =$

(A) $-e^{\frac{1}{x}}$

(B) $-\dfrac{e^{\frac{1}{x}}}{x^2}$

(C) $\dfrac{e^{\frac{1}{x}}}{x}$

(D) $\dfrac{e^{\frac{1}{x}}}{x^2}$

You need to know this:

You will definitely see exponential and logarithmic functions on Test Day, so be sure to memorize the following formulas.

- The derivative of e^x: $\dfrac{d}{dx}\left(e^x\right) = e^x$

- Applying the chain rule: $\dfrac{d}{dx}\left(e^{f(x)}\right) = e^{f(x)} \cdot f'(x)$

- The derivative of other exponential functions: $\dfrac{d}{dx}\left(b^x\right) = \ln b \cdot b^x$

- The derivative of ln x: $\dfrac{d}{dx}\left(\ln x\right) = \dfrac{1}{x}$

- Applying the chain rule: $\dfrac{d}{dx}\ln\left(f(x)\right) = \dfrac{1}{f(x)} \cdot f'(x)$

Useful logarithm properties:

- $\ln\left(\dfrac{a}{b}\right) = \ln a - \ln b$

- $\ln\left(ab\right) = \ln a + \ln b$

- $\ln a^b = b \ln a$

✔ AP Expert Note

When taking derivatives of logs, it's almost always easier to expand the log (if possible) before applying rules of derivatives, such as the product, quotient, or chain rule.

You need to do this:

- Recall the rule: The derivative of e raised to a power is e raised to that same power times the derivative of the power.

- To take the derivative of the power, rewrite $\frac{1}{x}$ as x^{-1}. Then apply the power rule.

Answer and Explanation:

B

$$f(x) = e^{\frac{1}{x}} = e^{x^{-1}}$$

$$f'(x) = \left(e^{\frac{1}{x}}\right) \cdot \left[x^{-1}\right]'$$

$$= e^{\frac{1}{x}} \cdot -1\left(x^{-2}\right)$$

$$f'(x) = -\frac{e^{\frac{1}{x}}}{x^2}$$

(B) is correct.

PRACTICE SET

1. If $f(x) = e^{x^2 - 3}$, then $f'(x) =$

   (A) $e^{x^2 - 2}$

   (B) $2x\,e^{x^2 - 3}$

   (C) $\dfrac{1}{x^2 - 3}$

   (D) $\dfrac{2x}{x^2 - 3}$

2. If $f(x) = x^2 \ln x$, then $f'(x) =$

   (A) $2x \cdot \ln x$

   (B) x

   (C) $2x \cdot \ln x + x$

   (D) $2x \cdot \ln x + x^2 e^x$

3. If $f(x) = x \cdot 2^x$, then $f'(x) =$

   (A) $2^x(x + \ln 2)$

   (B) $2^x(1 + \ln 2)$

   (C) $x \cdot 2^x \cdot \ln 2$

   (D) $2^x(1 + x \cdot \ln 2)$

4. If $f(x) = \ln\left(\dfrac{x(x^2 + 1)^2}{\sqrt{2x^3 - 1}}\right)$, then $f'(x) =$

   (A) $\dfrac{3\sqrt{2x^3 - 1}}{2x(x^2 + 1)^2}$

   (B) $\dfrac{x}{x^2 + 1} - \dfrac{1}{4x^3 - 2}$

   (C) $\dfrac{1}{x} + \dfrac{4x}{x^2 + 1} - \dfrac{3x^2}{2x^3 - 1}$

   (D) $\dfrac{1}{x} + \dfrac{2}{4x(x^2 + 1)} - \dfrac{\sqrt{2x^3 - 1}}{12x}$

ANSWERS AND EXPLANATIONS

1. B

This is a straightforward application of the rule for the derivative of e and the chain rule: The derivative of e raised to a power is e raised to that same power times the derivative of the power:

$$f(x) = e^{x^2 - 3}$$
$$f'(x) = e^{x^2 - 3} \cdot 2x$$
$$= 2x\, e^{x^2 - 3}$$

Thus, **(B)** is correct.

2. C

Use the product rule $[f \cdot g] = f'g + fg'$:

$$f(x) = x^2 \ln x$$
$$f'(x) = [x^2]' \ln x + x^2 [\ln x]'$$
$$f'(x) = 2x \cdot \ln x + x^2 \cdot \frac{1}{x}$$
$$= 2x \cdot \ln x + x$$

(C) is correct.

3. D

Use the product rule $[f \cdot g] = f'g + fg'$ and the derivative $[b^x]' = \ln b \cdot b^x$ to differentiate f:

$$f(x) = x \cdot 2^x$$
$$f'(x) = [x]' \cdot 2^x + x \cdot [2^x]'$$
$$= 1 \cdot 2^x + x \cdot (\ln 2 \cdot 2^x)$$
$$= 2^x (1 + x \cdot \ln 2)$$

That's **(D)**.

4. C

You definitely don't want to perform the product rule, the power rule, the chain rule, and the quotient rule all at one time (if you can avoid it), so rewrite the logarithm using properties of logs before you take the derivative. Don't forget to rewrite the radical in exponent form as well.

$$f(x) = \ln\left(\frac{x(x^2 + 1)^2}{\sqrt{2x^3 - 1}}\right)$$
$$= \ln x + \ln(x^2 + 1)^2 - \ln(2x^3 - 1)^{\frac{1}{2}}$$
$$= \ln x + 2\ln(x^2 + 1) - \frac{1}{2}\ln(2x^3 - 1)$$

Next, use the rule for taking the derivative of the natural log of a quantity: "one over the quantity times the derivative of the quantity":

$$f'(x) = \frac{1}{x} + \left(2\left(\frac{1}{x^2 + 1}\right) \cdot 2x\right) - \left(\frac{1}{2}\left(\frac{1}{2x^3 - 1}\right) \cdot 6x^2\right)$$
$$= \frac{1}{x} + \frac{4x}{x^2 + 1} - \frac{3x^2}{2x^3 - 1}$$

That matches **(C)**.

7.7 L'HÔPITAL'S RULE

To answer a question like this:

$$\lim_{x\to\infty} x^2 e^{-x} =$$

(A) $-\infty$

(B) 0

(C) 1

(D) ∞

You need to know this:

Some basic algebra:

- A quantity raised to a negative power can be rewritten as the reciprocal of the quantity raised to a positive power. For example, $a^{-2} = \dfrac{1}{a^2}$.

Theorem: L'Hôpital's Rule

- If you have a limit of the type $\dfrac{0}{0}$ or $\dfrac{\infty}{\infty}$, then you can use L'Hôpital's rule. (These limits are in *indeterminate form*.)

- To apply L'Hôpital's rule, take the derivative of the top, take the derivative of the bottom, and then re-evaluate the limit. Note that this is NOT the same as using the quotient rule for derivatives. You take each derivative separately.

- You can apply L'Hôpital's rule multiple times if needed, as long as each limit is in the form $\dfrac{0}{0}$ or $\dfrac{\infty}{\infty}$.

When L'Hôpital's rule can save you time:

- Evaluating limits that involve quotients that can be simplified by factoring, multiplying by a conjugate, or using special trig identities; L'Hôpital's rule is almost always faster.

- Evaluating limits that involve quotients that can't be simplified.

- Evaluating limits of functions that involve quantities raised to negative powers (because you can rewrite such a function as a quotient).

- Looking for horizontal asymptotes (because the process involves plugging in ∞).

You need to do this:

- Rewrite any quantities that are raised to negative exponents.

- As with every limit, first plug in the value that x is approaching. If the result is $\frac{0}{0}$ or $\frac{\infty}{\infty}$, apply L'Hôpital's rule: Take the derivative of the top, take the derivative of the bottom, and then re-evaluate the limit.

- If the new limit still yields $\frac{0}{0}$ or $\frac{\infty}{\infty}$, apply L'Hôpital's rule again. Continue until you arrive at a numerical answer.

Answer and Explanation:

B

Start by rewriting $x^2 e^{-x}$ as $\frac{x^2}{e^x}$ and then plug ∞ in for x.

$$\lim_{x \to \infty} x^2 e^{-x} = \lim_{x \to \infty} \frac{x^2}{e^x} = \frac{\infty}{\infty}$$

The limit is in indeterminate form, so apply L'Hôpital's rule:

$$\lim_{x \to \infty} \frac{2x}{e^x} \rightarrow \lim_{x \to \infty} \frac{\frac{d}{dx}\left(x^2\right)}{\frac{d}{dx}\left(e^x\right)} = \lim_{x \to \infty} \frac{2x}{e^x} = \frac{\infty}{\infty}$$

The limit is still in indeterminate form, so apply L'Hôpital's rule again:

$$\lim_{x \to \infty} \frac{2x}{e^x} \rightarrow \lim_{x \to \infty} \frac{\frac{d}{dx}(2x)}{\frac{d}{dx}\left(e^x\right)} = \lim_{x \to \infty} \frac{2}{e^x} = \frac{2}{\infty}$$

The limit is no longer in indeterminate form, so think about how extremely large numbers work: a fixed value divided by a huge number approaches 0, so the limit is equal to 0, which is **(B)**.

PRACTICE SET

1. $\lim\limits_{x \to 3} \dfrac{x^2 - 8x + 15}{x^2 - 9} =$

   (A) -3

   (B) $-\dfrac{1}{3}$

   (C) 0

   (D) nonexistent

2. $\lim\limits_{x \to \infty} \dfrac{e^x - 1}{3x} =$

   (A) $-\dfrac{2}{3}$

   (B) 0

   (C) $\dfrac{1}{3}$

   (D) 3

3. $\lim\limits_{x \to \infty} e^{-2x} \ln x =$

   (A) 0

   (B) 1

   (C) e

   (D) e^2

4. $\lim\limits_{x \to 0} \dfrac{\sin^2 x}{1 - \cos x} =$

   (A) -2

   (B) -1

   (C) 0

   (D) 2

ANSWERS AND EXPLANATIONS

1. B

As with all limit questions, start by plugging the value that x is approaching into the function. Here, the result is:

$$\lim_{x \to 3} \frac{x^2 - 8x + 15}{x^2 - 9} = \frac{9 - 24 + 15}{9 - 9} = \frac{0}{0}$$

The denominator is a difference of squares, which is a hint that you can probably factor and cancel, but now that you know L'Hôpital's rule and you've determined that the limit is of the form $\frac{0}{0}$, you can answer this question much more quickly.

$$\lim_{x \to 3} \frac{x^2 - 8x + 15}{x^2 - 9} \to \lim_{x \to 3} \frac{2x - 8}{2x}$$
$$= \frac{6 - 8}{6} = \frac{-2}{6} = -\frac{1}{3}$$

(B) is correct.

2. C

As with any limit, start by plugging in the value that x is approaching:

$$\lim_{x \to 0} \frac{e^x - 1}{3x} = \frac{e^0 - 1}{3(0)} = \frac{1 - 1}{0} = \frac{0}{0}$$

The limit is in indeterminate form, so use L'Hôpital's rule, then re-evaluate the limit. Also, recall that the derivative of e^x is simply e^x:

$$\lim_{x \to 0} \frac{e^x - 1}{3x} \to \lim_{x \to 0} \frac{e^x}{3} = \frac{e^0}{3} = \frac{1}{3}$$

(C) is correct.

3. A

To calculate the limit, first rewrite the function as a quotient:

$$\lim_{x \to \infty} e^{-2x} \ln x = \lim_{x \to \infty} \frac{\ln x}{e^{2x}}$$

Next, check to see whether the limit is in indeterminate form:

$$\lim_{x \to \infty} \frac{\ln x}{e^{2x}} = \frac{\infty}{\infty}$$

This means you can use L'Hôpital's rule. Don't forget to use the chain rule when you take the derivative of e^{2x}:

$$\lim_{x \to \infty} \frac{\ln x}{e^{2x}} \to \frac{\frac{d}{dx}(\ln x)}{\frac{d}{dx}(e^{2x})} = \lim_{x \to \infty} \frac{\frac{1}{x}}{2e^{2x}} = \lim_{x \to \infty} \frac{1}{2xe^{2x}} = \frac{1}{\infty} = 0$$

That's **(A)**.

4. D

You might be tempted here to try to make a trig substitution or multiply by the conjugate of the denominator, but if L'Hôpital's rule applies, it will definitely save you some valuable time. So, as always, start by plugging in 0 for x:

$$\lim_{x \to 0} \frac{\sin^2 x}{1 - \cos x} = \frac{\sin^2 0}{1 - \cos 0} = \frac{0}{1 - 1} = \frac{0}{0}$$

The limit is in indeterminate form, so apply L'Hôpital's rule. Be careful when taking the derivative of the numerator—you'll need to use the power rule and the chain rule.

$$\lim_{x \to 0} \frac{\sin^2 x}{1 - \cos x} = \lim_{x \to 0} \frac{(\sin x)^2}{1 - \cos x} \to \lim_{x \to 0} \frac{2 \sin x \cdot \cos x}{0 - (-\sin x)}$$
$$= \lim_{x \to 0} \frac{2 \sin x \cos x}{\sin x} = \lim_{x \to 0} 2 \cos x$$
$$= 2 \cos 0 = 2(1) = 2$$

That's **(D)**.

7.8 THE INTERMEDIATE VALUE, EXTREME VALUE, AND MEAN VALUE THEOREMS

To answer a question like this:

x	-5	0	5	10	15	20
$f(x)$	-10	-8	1	11	15	18

Several values for a function f are given in the table above. Which of the following statements must be true given that f is continuous over the closed interval $[-5, 20]$?

(A) f attains its maximum value at $x = 20$.

(B) $f(c) = 0$ at least twice on the closed interval $[-5, 5]$.

(C) $f(c) = 2$ for some value of c on the closed interval $[5, 10]$.

(D) $f'(c) = 2$ for some value of c on the closed interval $[5, 10]$.

You need to know this:

The ability to draw continuous functions without having to lift a pencil makes them definable by several theorems.

Intermediate Value Theorem (IVT): If f is continuous on the closed interval $[a, b]$, then for any number k between $f(a)$ and $f(b)$, there is some $c \in [a, b]$ with $f(c) = k$. That is, f takes on every value between $f(a)$ and $f(b)$ on the interval $[a, b]$.

- As an example: If f is a continuous function on the closed interval $[1, 8]$ where $f(1) = 2$ and $f(8) = 11$, then between $x = 1$ and $x = 8$, f takes on every y-value between 2 and 11.

- A special case of the IVT involves the zeros of a function. If f is continuous on the closed interval $[a, b]$, and $f(a)$ and $f(b)$ have opposite signs, then f has a zero (that is, $f(c) = 0$) for $x = c$ between a and b (because to get from a negative number to a positive number, you must cross through $y = 0$).

Extreme Value Theorem (EVT): If f is continuous on the closed interval $[a, b]$, then f must attain a maximum and a minimum value, one of each, on $[a, b]$. Translation: The function *must* attain its highest value and its lowest value either at the endpoints of the interval or somewhere in between.

Mean Value Theorem (MVT): If a function f is continuous on the closed interval $[a, b]$ and differentiable on (a, b), then there is at least one point $c \in (a, b)$ such that the slope of the secant line between a and b is equal to the slope of the tangent line at c. That is, $f'(c) = \dfrac{f(b) - f(a)}{b - a}$.

You need to do this:

- Decide which theorem applies to the given situation. Take your cues from what the question is asking, or from the answer choices.

- If the question or answer choices involve f taking on a certain value or having a zero in a given interval, apply the IVT.

- If the question or answer choices involve f taking on a maximum or minimum value in a given interval, apply the EVT.

- If the question or answer choices involve the derivative of f taking on a certain value, apply the MVT. For this particular theorem, you'll also be told in the question stem that f is differentiable on the given interval.

Answer and Explanation:

C

Skim through the answer choices—the question stem doesn't say f is differentiable over the given interval, so you can't draw any conclusions about the derivative at a specified point. This means you can eliminate (D) right way.

Evaluate the remaining choices one at a time.

(A): Although $f(20) = 18$ is the maximum of all the data provided in the table, this does not mean there couldn't be a higher value somewhere else on the interval $[-5, 20]$. Eliminate (A).

(B): A special case of the intermediate value theorem guarantees that, if a function is continuous on the closed interval $[a, b]$, and $f(a)$ and $f(b)$ have opposite signs, then the function must equal 0 on the interval $[a, b]$. There is only one such "sign change" shown in the table between -5 and 5, so only one zero is guaranteed on the interval $[-5, 5]$, not two. Eliminate (B).

This means **(C)** must be correct as it is the only remaining option. In case you want to double-check: You're given that $f(5) = 1$ and $f(10) = 11$. By the intermediate value theorem, this means there must be some value of c between $x = 5$ and $x = 10$ such that the y-value is equal to 2 (because 2 is between 1 and 11). The statement is true.

PRACTICE SET

1. Suppose a function f is defined for all x in the closed interval $[0, 5]$. If f does not attain a maximum value on $[0, 5]$, which of the following must be true?

   (A) f is not continuous on $[0, 5]$.

   (B) f does not take on a minimum value on $[0, 5]$.

   (C) The graph of f must have a vertical asymptote in the interval $[0, 5]$.

   (D) The derivative of f does not equal 0 in the interval $[0, 5]$.

2. If f is continuous on the closed interval $[-2, 6]$ where $f(-2) = 1$ and $f(6) = 20$, then which of the following must be true?

   I. $f(c) = 0$ for some c in $(-2, 6)$

   II. $f'(c) = 0$ for some c in $(-2, 6)$

   III. f has a minimum somewhere in $[-2, 6]$

   (A) I only

   (B) II only

   (C) III only

   (D) I and III only

3. Suppose f is continuous on the closed interval $[0, 4]$ and suppose $f(0) = 1, f(1) = 2, f(2) = 0,$ $f(3) = -3,$ and $f(4) = 3$. Which of the following statements about the zeros of f on $[0, 4]$ is always true?

   (A) f has exactly one zero on $[0, 4]$.

   (B) f has more than one zero on $[0, 4]$.

   (C) f has more than two zeros on $[0, 4]$.

   (D) f has exactly two zeros on $[0, 4]$.

4. Suppose $f(a) = 2$ and $f(b) = -3$. Which of the following statements is always true?

   (A) f takes on a maximum value on the closed interval $[a, b]$.

   (B) If f is continuous, then f has a zero on the closed interval $[a, b]$.

   (C) If f is continuous on the closed interval $[a, b]$, then the maximum value of f on the closed interval $[a, b]$ is less than 2.

   (D) If f is continuous on the closed interval $[a, b]$, then f' is always negative on the closed interval $[a, b]$.

ANSWERS AND EXPLANATIONS

1. A

Decide which theorem applies. The question involves a maximum value, so the extreme value theorem applies. This theorem states that if f is continuous on $[a, b]$, then it *must* attain both a maximum and a minimum value on $[a, b]$. Because this function doesn't attain a maximum value on $[0, 5]$, you can conclude that the function is not continuous on that interval, making (**A**) the correct choice.

Note that one or more of the other options *may* be true, but the EVT only guarantees that (**A**) *must* be true.

2. C

Decide which theorem, or theorems, apply.

The question stem doesn't state that the function is differentiable on $[-2, 6]$, so the mean value theorem does not apply and you don't know anything for sure about f'. This means you can't confirm the validity of statement II, which means you can eliminate (B).

Statement I is a special case of the intermediate value theorem. For statement I to be true, $f(-2)$ and $f(6)$ would have to have opposite signs, which they don't. This means you can eliminate both (A) and (D), leaving (**C**) as the correct choice.

To confirm, apply the extreme value theorem. Because f is continuous on $[-2, 6]$, it *must* attain a maximum and a minimum somewhere in that interval, so statement III must be true.

Thus, of the three statements, the only one that is definitely true is statement III, making (**C**) the correct answer.

3. B

Decide which theorem applies. The question asks about the zeros of f, so apply the special case of the intermediate value theorem. Because f is continuous on the closed interval $[0, 4]$, it follows from the IVT and the data given that f has a zero between $x = 1$ and $x = 3$ (because the functional values have opposite signs). You are given that $f(2) = 0$, but it is not guaranteed that there will be another zero between $x = 1$ and $x = 3$; this may be the only one. From the given data, you also know that f has a zero between $x = 3$ and $x = 4$. Therefore, the intermediate value theorem guarantees that there are at least two (maybe more) zeros of f on the interval $[0, 4]$; that is, f has more than one zero on $[0, 4]$, which is (**B**).

4. B

Evaluate each statement, one at a time. Keep in mind that the question stem doesn't say anything about f being continuous, so that condition will need to be stated in the answer choice. This means you can eliminate (A) right away. The extreme value theorem guarantees that a function takes on a maximum value on $[a, b]$ only if the function is *continuous* over that same interval.

(**B**) is true. By the intermediate value theorem, if a function is continuous on the closed interval $[a, b]$, and $f(a)$ and $f(b)$ have opposite signs, then the function must cross the x-axis on the interval $[a, b]$; that is, f has a zero on $[a, b]$.

You could stop there, but if you're not sure, evaluate each of the other choices.

(C) is not always true; the extreme value theorem guarantees that if a function is continuous on a closed interval, it *does* achieve a maximum value on that interval, but it makes no conclusion about the value of the maximum.

(D) is false; none of the three theorems discussed in this section guarantees that a derivative is always positive or negative.

RAPID REVIEW

If you take away only 9 things from this chapter:

1. The limit definition of the derivative: $f'(x) = \lim\limits_{h \to 0} \dfrac{f(x+h) - f(x)}{h}$

2. The difference quotient, $\dfrac{f(x_1) - f(x_0)}{x_1 - x_0}$, represents the slope of the secant line between the points $(x_0, f(x_0))$ and $(x_1, f(x_1))$.

3. The power rule: If n is a constant, then $\dfrac{d}{dx}\left(x^n\right) = nx^{n-1}$.

4. The product rule: If f and g are differentiable at a point x, then their product is differentiable at x and the derivative is $[f \cdot g]' = f'g + fg'$.

5. The quotient rule: If f and g are differentiable at a point x and $g(x) \neq 0$, then their quotient is differentiable at x and the derivative is $\left[\dfrac{f}{g}\right]' = \dfrac{f'g - fg'}{g^2}$.

6. The chain rule: If g is differentiable at x and h is differentiable at $g(x)$, then $h(g(x))$ is differentiable at x and its derivative is $h'(g(x)) \cdot g'(x)$.

7. Derivatives of exponential and logarithmic functions:

   - $\dfrac{d}{dx}\left(e^{f(x)}\right) = e^{f(x)} \cdot f'(x)$

   - $\dfrac{d}{dx}\left(b^x\right) = \ln b \cdot b^x$

   - $\dfrac{d}{dx}\ln\left(f(x)\right) = \dfrac{1}{f(x)} \cdot f'(x)$

8. If you have a limit of the type $\dfrac{0}{0}$ or $\dfrac{\infty}{\infty}$, then you can use L'Hôpital's rule. To apply L'Hôpital's rule, take the derivative of the top, take the derivative of the bottom, and then re-evaluate the limit.

9. Intermediate value theorem (IVT): If f is continuous on the closed interval $[a, b]$, then for any number k between $f(a)$ and $f(b)$, there is some $c \in [a, b]$ with $f(c) = k$. That is, f takes on every value between $f(c)$ and $f(b)$ on the interval $[a, b]$.

   Extreme value theorem (EVT): If f is continuous on the closed interval $[a, b]$, then f must attain a maximum and a minimum value, one of each, on $[a, b]$.

   Mean value theorem (MVT): If a function f is continuous on a closed interval $[a, b]$ and differentiable on an open interval (a, b), then there is at least one point $c \in (a, b)$ such that $f'(c) = \dfrac{f(b) - f(a)}{b - a}$.

Derivatives

TEST WHAT YOU LEARNED

1. If $f(x) = 16\sqrt{x} + \frac{1}{2}x^2 - 7$, then $f'(4) =$

    (A) 8   (B) 16   (C) 33   (D) 40

x	$f(x)$
2	3
5	6.3
8	8.7

2. The table above gives selected values of a function f. The function is twice differentiable with $f''(x) < 0$. Which of the following could be the value of $f'(5)$?

    (A) 0.8   (B) 0.9   (C) 1.1   (D) 2.3

3. If $y = \left(\dfrac{3}{x^2} - 2\sqrt{x}\right)^5$, then $y' =$

    (A) $5\left(\dfrac{3}{x^2} - 2\sqrt{x}\right)^4 + \left(\dfrac{6}{x^3} - \dfrac{1}{\sqrt{x}}\right)$

    (B) $5\left(\dfrac{3}{x^2} - 2\sqrt{x}\right)^4 \cdot \left(-\dfrac{6}{x^3} - \dfrac{1}{\sqrt{x}}\right)$

    (C) $\left(-\dfrac{6}{x^3} - \dfrac{1}{\sqrt{x}}\right)^5$

    (D) $5\left(\dfrac{3}{2x} - x\right)^4$

Derivatives

4. $\displaystyle\lim_{h \to 0} \frac{4(2+h)^3 - 4(2)^3}{h} =$

    (A) 0   (B) 8   (C) 48   (D) nonexistent

5. If $f(x) = \ln(2x^3) + \dfrac{1}{e^{4x}}$, then $f'(x) =$

    (A) $e^{6x^2} \cdot \ln 4x$

    (B) $\dfrac{1}{2x^3} - \ln 4x$

    (C) $\dfrac{1}{2x^3} - 4$

    (D) $\dfrac{3}{x} - 4e^{-4x}$

6. If $y = \dfrac{2x^2}{x+1}$, then $\dfrac{dy}{dx} =$

    (A) $\dfrac{2x^2 + 4x}{(x+1)^2}$

    (B) $\dfrac{4x}{(x+1)}$

    (C) $\dfrac{2x^2 + 4x}{(x+1)^2}$

    (D) $\dfrac{-2x^2 - 4x}{(x+1)^2}$

Derivatives

7. $\lim\limits_{x \to 0} \dfrac{\sin x - 3x}{2x^2 + \sin(6x)} =$

(A) $-\dfrac{1}{2}$

(B) $-\dfrac{1}{3}$

(C) 0

(D) $\dfrac{1}{6}$

8. A function f is continuous on the closed interval $[-4, 3]$ with $f(-4) = -1$ and $f(3) = -1$. Which of the following conditions, if true, guarantees there is a number c in the open interval $(-4, 3)$ such that $f'(c) = 0$?

(A) $f(k) = -1$.

(B) $\displaystyle\int_{-4}^{3} x \, dx$ exists.

(C) f is differentiable on the open interval $(-4, 3)$.

(D) No additional conditions are necessary.

THIS PAGE LEFT INTENTIONALLY BLANK.
Answers and explanations to the practice set are located on the next page.

Answer Key

Test What You Already Know			Test What You Learned		
1.	B	Learning Objective: 7.1	1.	A	Learning Objective: 7.3
2.	D	Learning Objective: 7.2	2.	B	Learning Objective: 7.2
3.	A	Learning Objective: 7.3	3.	B	Learning Objective: 7.5
4.	D	Learning Objective: 7.4	4.	C	Learning Objective: 7.1
5.	D	Learning Objective: 7.5	5.	D	Learning Objective: 7.6
6.	C	Learning Objective: 7.6	6.	A	Learning Objective: 7.4
7.	B	Learning Objective: 7.7	7.	B	Learning Objective: 7.7
8.	C	Learning Objective: 7.8	8.	C	Learning Objective: 7.8

 **REFLECTION**

Test What You Already Know score: _____

Test What You Learned score: _____

Use this section to evaluate your progress. After working through the pre-quiz, check off the boxes in the "Pre" column to indicate which Learning Objectives you feel confident about. Then, after completing the chapter, including the post-quiz, do the same to the boxes in the "Post" column. Keep working on unchecked Objectives until you're confident about them all!

Pre | Post

☐ ☐ **7.1** Recognize the derivative as the limit of a difference quotient

☐ ☐ **7.2** Use graphs and tables to estimate derivatives or functional values

☐ ☐ **7.3** Use the power rule to find the derivative of a polynomial, rational, or radical function

☐ ☐ **7.4** Use the product and quotient rules to find the derivative of a function

☐ ☐ **7.5** Apply the chain rule to calculate derivatives of basic composite functions

☐ ☐ **7.6** Calculate derivatives of logarithmic and exponential functions

☐ ☐ **7.7** Apply L'Hôpital's rule to evaluate limits of the form $\frac{0}{0}$ and $\frac{\infty}{\infty}$

☐ ☐ **7.8** Apply the intermediate value, extreme value, and/or mean value theorems

 FOR MORE PRACTICE

Complete more practice online at kaptest.com. Haven't registered your book yet? Go to kaptest.com/booksonline to begin.

ANSWERS AND EXPLANATIONS

Test What You Already Know

1. B Learning Objective: 7.1

Memorize the limit definition of the derivative before Test Day: $f'(x) = \lim\limits_{h \to 0} \dfrac{f(x+h) - f(x)}{h}$. Notice that both terms in the numerator are being squared, so $\lim\limits_{h \to 0} \dfrac{(5+h)^2 - 5^2}{h}$ is the derivative of the function $f(x) = x^2$ at the point where $x = 5$. For $f(x) = x^2$, the derivative is $f'(x) = 2x$, so the value of the given limit is $f'(5) = 2(5) = 10$.

(B) is correct.

Note that you could also use L'Hôpital's rule to find the limit.

2. D Learning Objective: 7.2

Being able to read a table and knowing what derivatives tell you about a function are the keys to understanding this question. You're given that $f'(x) > 0$, which indicates that the slope of the function is positive. A positive slope means that $f(x)$ should increase as x increases; eliminate (A) because the values of $f(x)$ are decreasing, not increasing. The second derivative is the derivative of the first derivative, so $f''(x) > 0$ indicates that $f'(x)$, the slope itself, is increasing. Use the slope formula to find the slope between the pairs of the values in the tables. In (B), the slope is constant $(m = 2)$, and in (C), the slope is also constant $(m = 1)$. The slope is not increasing because, in each case, the slope between the points stays the same. In (D), the slope between the first pair of points is 2 and the slope between the second pair of points is 3, so the slope is increasing, and **(D)** is therefore correct.

3. A Learning Objective: 7.3

First, rewrite the function as $f(x) = 9x^{-1} - \dfrac{1}{2}x^4 + 2$.

Then, use the power rule to find the derivative:

$$f'(x) = -9x^{-2} - \dfrac{1}{2} \cdot 4x^3 + 0$$
$$= \dfrac{-9}{x^2} - 2x^3$$

The question asks for the derivative at a specific point, so plug in 3 for x and simplify:

$$f'(3) = \dfrac{-9}{3^2} - 2(3^3)$$
$$= \left(\dfrac{-9}{9}\right) - (2 \cdot 27)$$
$$= -1 - 54 = -55$$

(A) is correct.

4. D Learning Objective: 7.4

The notation $\dfrac{dy}{dx}$ simply means find the derivative. To take the derivative, use the quotient rule, which states that, for functions f and g, the derivative is given by:

$$\dfrac{d}{dx}\left(\dfrac{f}{g}\right) = \dfrac{f'g - fg'}{(g)^2}$$

So:

$$\dfrac{dy}{dx} = \dfrac{4 \cdot (x^2 - 3) - 4x \cdot (2x)}{(x^2 - 3)^2} = \dfrac{-4x^2 - 12}{(x^2 - 3)^2}$$

(D) is correct.

Note that (C) is a very common wrong answer that results from taking the derivative of the denominator first. One way to avoid this pitfall is to always take the derivative of the first function you see, first.

5. D Learning Objective: 7.5

To find the derivative, use the chain rule. Before you do that, rewrite the radical term using a fractional exponent, $\sqrt{x} = x^{\frac{1}{2}}$, so you can apply the power rule.

Next, to apply the chain rule, take the derivative of the outermost function (the 6th power), then multiply by the derivative of what's inside:

$$y' = 6\left(x^{\frac{1}{2}} + 2x^3\right)^5 \cdot \left(\dfrac{1}{2}x^{-\frac{1}{2}} + 6x^2\right)$$

This is not one of the answer choices, so rewrite the terms with the fractional exponents as radicals and simplify:

$$y' = 6\left(\sqrt{x} + 2x^3\right)^5 \cdot \left(\dfrac{1}{2\sqrt{x}} + 6x^2\right)$$

This is a perfect match for **(D)**.

Derivatives

6.   C   Learning Objective: 7.6

Rather than using both the chain rule and the quotient rule, simplify the function using rules of logs first:

$$\ln \frac{a}{b} = \ln a - \ln b$$

Here, the function can be written as:

$$f(x) = \ln(4x - 5) - \ln(x + 7)$$

Now use rules of derivatives (don't forget the chain rule if it applies):

$$(\ln u)' = \frac{1}{u} \cdot u'$$

So:

$$f'(x) = \frac{1}{4x - 5}(4) - \frac{1}{x + 7}(1)$$

$$= \frac{4}{4x - 5} - \frac{1}{x + 7}$$

(C) is correct.

7.   B   Learning Objective: 7.7

Substituting π into the expression yields:

$$\frac{\cos(2\pi) - 1}{\sin(4\pi)} = \frac{1 - 1}{0} = \frac{0}{0}$$

This means you can use L'Hôpital's rule, which applies to limits of the form $\frac{0}{0}$ or $\frac{\infty}{\infty}$. To use L'Hôpital's rule, take the derivative of the numerator and the derivative of the denominator, and reevaluate the limit. Don't forget to use the chain rule when necessary:

$$\lim_{x \to \pi} \frac{\cos(2x) - 1}{\sin(4x)} = \lim_{x \to \pi} \frac{(-\sin(2x) \cdot 2) - 0}{\cos(4x) \cdot 4}$$

$$= \frac{-\sin(2x) \cdot 2}{\cos(4x) \cdot 4}$$

$$= \frac{0 \cdot 2}{1 \cdot 4} = \frac{0}{4} = 0$$

(B) is correct.

8.   C   Learning Objective: 7.8

Think of the two functional values that are given as points on a graph: (0, −2) and (5, 13). The intermediate value theorem guarantees that, because the function is continuous from 0 to 5, there must be at least one point on the graph that has a y-value between −2 and 13. Thus, there must be a value of x in the open interval (0, 5) such that $f(x) = 3$ (because 3 is between −2 and 13). **(C)** is correct.

Tip: Knowing that the intermediate value theorem ALWAYS involves $f(x)$ allows you to eliminate two of the choices right away, (B) and (D). (The mean value theorem ALWAYS involves $f'(x)$—or the slope of a line tangent to the function.)

Test What You Learned

1.   A   Learning Objective: 7.3

First, rewrite the function as $f(x) = 16x^{\frac{1}{2}} + \frac{1}{2}x^2 - 7$.

Then use the power rule to find the derivative:

$$f'(x) = \frac{1}{2} \cdot 16x^{-\frac{1}{2}} + 2 \cdot \frac{1}{2}x - 0$$

$$= 8x^{-\frac{1}{2}} + x$$

$$= \frac{8}{\sqrt{x}} + x$$

Next, because the question asks for the derivative specifically at $x = 4$, plug in 4 for x and simplify:

$$f'(4) = \frac{8}{\sqrt{4}} + 4$$

$$= \frac{8}{2} + 4$$

$$= 4 + 4 = 8$$

(A) is correct.

2.   B   Learning Objective: 7.2

The slope between (2, 3) and (5, 6.3) is 1.1, so by the mean value theorem, there must be some point on the curve between (2, 3) and (5, 6.3) for which the slope is 1.1. The slope between (5, 6.3) and (8, 8.7) is 0.8, so by the MVT, there must be some point on the curve between (5, 6.3) and (8, 8.7) for which the slope is 0.8.

Because the second derivative is less than 0, the slopes are decreasing, so the slope at the point (5, 6.3) must be between 1.1 and 0.8. The value 0.9 is the only one that meets this criterion, so **(B)** is correct.

3.   B   Learning Objective: 7.5

To find the derivative, use the chain rule. Before you do that, rewrite the function so you can apply the power rule:

$$y = \left(3x^{-2} - 2x^{\frac{1}{2}}\right)^5$$

Next, to apply the chain rule, take the derivative of the outermost function (the 5th power), leave the inside, then multiply by the derivative of what's inside:

$$y = 5\left(3x^{-2} - 2x^{\frac{1}{2}}\right)^4 \cdot \left(-6x^{-3} - x^{-\frac{1}{2}}\right)$$

$$= 5\left(\frac{3}{x^2} - 2\sqrt{x}\right)^4 \cdot \left(-\frac{6}{x^3} - \frac{1}{\sqrt{x}}\right)$$

This is a perfect match for **(B)**.

4.   C   Learning Objective: 7.1

The definition of the derivative is $\lim_{h \to 0} \frac{f(x+h) - f(x)}{h}$.

Therefore, $\lim_{h \to 0} \frac{4(2+h)^3 - 4(2)^3}{h}$ is the derivative of the function $f(x) = 4x^3$ at the point where $x = 2$. For $f(x) = 4x^3$, the derivative is $f'(x) = 12x^2$, so the value of the given limit is $f'(2) = 12(2)^2 = 48$.

(C) is correct.

5.   D   Learning Objective: 7.6

Simplify the function before taking the derivative:

$$f(x) = \ln\left(2x^3\right) + e^{-4x}$$

Now use rules of derivatives (don't forget the chain rule if it applies):

$$(\ln u)' = \frac{1}{u} \cdot u'$$

$$\left(e^u\right)' = e^u \cdot u'$$

So:

$$f'(x) = \frac{1}{2x^3} \cdot \left(6x^2\right) + e^{-4x}(-4)$$

$$= \frac{3}{x} - 4e^{-4x}$$

(D) is correct.

6.   A   Learning Objective: 7.4

The notation $\frac{dy}{dx}$ means the derivative. Use the quotient rule, which states that, for functions f and g, the derivative is given by:

$$\frac{d}{dx}\left(\frac{f}{g}\right) = \frac{f'g - fg'}{(g)^2}$$

Here, f is $2x^2$ and g is $(x + 1)$. So:

$$\frac{dy}{dx} = \frac{4x(x+1) - 2x^2(1)}{(x+1)^2}$$

$$= \frac{4x^2 + 4x - 2x^2}{(x+1)^2}$$

$$= \frac{2x^2 + 4x}{(x+1)^2}$$

(A) is correct.

Note that (D) is a very common wrong answer that results from taking the derivative of the denominator first. Make sure to always take the derivative of the first function, first.

7. B Learning Objective: 7.7

Substituting 0 into the expression yields:

$$\frac{\sin(0) - 3 \cdot 0}{2(0)^2 + \sin(0)} = \frac{0}{0}$$

This means you can use L'Hôpital's rule, which applies to limits of the form $\frac{0}{0}$ or $\frac{\infty}{\infty}$. To use L'Hôpital's rule, take the derivative of the numerator and the derivative of the denominator, then reevaluate the limit. Don't forget to use the chain rule when necessary:

$$\lim_{x \to 0} \frac{\sin x - 3x}{2x^2 + \sin(6x)} = \lim_{x \to 0} \frac{\cos x - 3}{4x + \cos(6x) \cdot 6}$$

$$= \frac{\cos(0) - 3}{4(0) + \cos(0) \cdot 6}$$

$$= \frac{1 - 3}{0 + 1 \cdot 6} = \frac{-2}{6} = -\frac{1}{3}$$

The correct answer is **(B)**.

8. C Learning Objective: 7.8

The mean value theorem states that if a function is defined and continuous on a closed interval $[a, b]$ and if that function is differentiable on the open interval (a, b), there exists a value c in that interval such that $f'(c) = \frac{f(b) - f(a)}{b - a}$. Applied to this problem:

$$f'(c) = \frac{f(-4) - f(3)}{-4 - 3}$$

$$= \frac{-1 - (-1)}{-7}$$

$$= \frac{0}{-7} = 0$$

In this question, the missing condition of the mean value theorem is that f is differentiable on the interval $(-4, 3)$, so the correct answer is **(C)**. This special case of the MVT when $f'(c) = 0$ is called Rolle's theorem.

CHAPTER 8

More Advanced Derivatives

LEARNING OBJECTIVES

8.1   Calculate derivatives of trig functions

8.2   Calculate derivatives of inverse trig functions

8.3   Determine higher order derivatives

8.4   Use implicit differentiation to find derivatives

8.5   Calculate derivatives of inverse functions

TEST WHAT YOU ALREADY KNOW

1. Which of the following is the derivative of $y = 3x^2 \sin x$?

    (A) $3x^2 \cos x - 6x \sin x$

    (B) $6x \sin x + 3x^2 \cos x$

    (C) $3x^2 (\sin x + \cos x)$

    (D) $6x \cos x$

2. Let $f(x) = \left(\arcsin\left(x^3\right)\right)^4$. Then $f'(x) =$

    (A) $\dfrac{-12x^2 \left(\arccos\left(x^3\right)\right)^3}{\sqrt{1 - x^6}}$

    (B) $\dfrac{-12x^2 \left(\arcsin\left(x^3\right)\right)^3}{\sqrt{1 - x^2}}$

    (C) $\dfrac{12x^2 \left(\arccos\left(x^3\right)\right)^3}{\sqrt{1 - x^2}}$

    (D) $\dfrac{12x^2 \left(\arcsin\left(x^3\right)\right)^3}{\sqrt{1 - x^6}}$

3. If $y = \sin^2 x$, what is $\dfrac{d^3y}{dx^3}$?

    (A) $2 \sin x \cos x$

    (B) $-2 \sin x$

    (C) $-8 \sin x \cos x$

    (D) $-4\left(\sin^2 x - \cos^2 x\right)$

---

4. If $x^3 - 2xy + y^2 = 5$, then $\dfrac{dy}{dx} =$

    (A) $\dfrac{-3x^2}{2y - 2}$

    (B) $\dfrac{-x^3 + 2x}{2y}$

    (C) $\dfrac{2y - 3x^2}{2y - 2x}$

    (D) $\dfrac{8 - 3y^2 - 2x}{xy}$

---

5. Let a function f be defined as $f(x) = x^3 - 3x + 1$ for $x \geq 1$. Let g be the inverse function of f and note that $f(2) = 3$. The value of $g'(3) =$

    (A) -1 (B) $\dfrac{1}{9}$ (C) $\dfrac{1}{3}$ (D) 3

Answers to this quiz can be found at the end of this chapter.

Derivatives

8.1 DERIVATIVES OF TRIG FUNCTIONS

To answer a question like this:

If $f(x) = \sin \sqrt{2x + 3}$, then $f'(x) =$

(A) $-\dfrac{2 \cos \sqrt{2x + 3}}{\sqrt{2x + 3}}$

(B) $\dfrac{\cos \sqrt{2x + 3}}{\sqrt{2x + 3}}$

(C) $2 \cos \sqrt{2x + 3}$

(D) $2\sqrt{2x + 3} \cos \sqrt{2x + 3}$

You need to know this:

Derivative rules for trig functions:

$f(x)$	$f'(x)$
$\sin x$	$\cos x$
$\cos x$	$-\sin x$
$\tan x$	$\sec^2 x$
$\cot x$	$-\csc^2 x$
$\sec x$	$\sec x \tan x$
$\csc x$	$-\csc x \cot x$

✔ AP Expert Note

You'll need to memorize the derivatives given in the table before Test Day. Notice that all the co's (cosine, cotangent, and cosecant) are negative, while the others are positive. This fact alone can help you eliminate one or more answer choices for most trig-related derivative questions. Also, look for patterns in the derivatives to help you remember (i.e., tan and sec are related).

You need to do this:

- Recall the rule for taking the derivative of sin x: The derivative of sin x is cos x.

- Determine what other derivative rules apply. Here, you'll need to use the power rule and the chain rule.

- As with any function, rewrite the radical portion using a fractional exponent.

- Take the derivative, then simplify ONLY if you don't find a match in the answer choices.

Answer and Explanation:

B

Use the chain rule to find the derivative. First, take the derivative of the outside function, sine:

$$f'(x) = \cos \sqrt{2x+3} \cdot \frac{d}{dx}\left(\sqrt{2x+3}\right)$$

Next, rewrite the square root using a fractional exponent, then use the power rule followed by the chain rule (again) to differentiate the inside function (the radical).

$$\frac{d}{dx}\left[(2x+3)^{\frac{1}{2}}\right] = \frac{1}{2}(2x+3)^{-\frac{1}{2}} \cdot \frac{d}{dx}(2x+3)$$

$$= \frac{1}{2}(2x+3)^{-\frac{1}{2}} \cdot (2)$$

$$= \frac{1}{\sqrt{2x+3}}$$

Putting the two pieces of the derivative together yields

$$f'(x) = \frac{\cos \sqrt{2x+3}}{\sqrt{2x+3}}, \text{ which is } \textbf{(B)}.$$

PRACTICE SET

1. If $f(x) = \tan(4x)$, then $f'(x) =$

   (A) $4\sec^2(4x)$

   (B) $\dfrac{\sec^2(4x)}{4}$

   (C) $\dfrac{4\sin(4x)}{\cos(4x)}$

   (D) $-\dfrac{\sec^2(4x)}{4}$

2. If $f(x) = \cos(\ln x)$ for $x > 0$, then $f'(x) =$

   (A) $-\sin(\ln x)$

   (B) $-\dfrac{\sin(\ln x)}{x}$

   (C) $\dfrac{\sin(\ln x)}{x}$

   (D) $\sin\left(\dfrac{\ln x}{x}\right)$

3. If $f(x) = \dfrac{\sin\sqrt{x}}{\sqrt{x}}$, then $f'(x) =$

   (A) $\dfrac{\cos\sqrt{x}}{2x} - \dfrac{\sin\sqrt{x}}{2x\sqrt{x}}$

   (B) $\dfrac{\cos\sqrt{x} - \sin\sqrt{x}}{2x}$

   (C) $\dfrac{\sqrt{x}\cos\sqrt{x} - \dfrac{\sin\sqrt{x}}{2\sqrt{x}}}{x}$

   (D) $\dfrac{\cos\sqrt{x}}{x}$

4. If $f(x) = \cos^2(\sin(2x))$, then $f'\left(\dfrac{\pi}{8}\right) =$

   (A) $\sqrt{2}\,\sin\left(\dfrac{\sqrt{2}}{2}\right)$

   (B) $\cos\left(\sin\left(\dfrac{\sqrt{2}}{2}\right)\right)$

   (C) $-\sqrt{2}\,\cos\left(\dfrac{\sqrt{2}}{2}\right)$

   (D) $-2\sqrt{2}\,\cos\left(\dfrac{\sqrt{2}}{2}\right)\sin\left(\dfrac{\sqrt{2}}{2}\right)$

ANSWERS AND EXPLANATIONS

1. A

Use the chain rule and the derivative $(\tan x)' = \sec^2 x$ to differentiate f:

$$f(x) = \tan(4x)$$
$$f'(x) = \sec^2(4x) \cdot 4$$
$$= 4\sec^2(4x)$$

(A) is correct.

2. B

Use the chain rule and the derivatives $[\cos x]' = -\sin x$ and $[\ln x]' = \dfrac{1}{x}$ to differentiate f:

$$f'(x) = \underbrace{-\sin(\ln x)}_{\substack{\text{the derivative of} \\ \cos x \text{ evaluated} \\ \text{at } \ln x}} \cdot \underbrace{\frac{1}{x}}_{\substack{\text{the derivative} \\ \text{of } \ln x}} = -\frac{\sin(\ln x)}{x}$$

That's **(B)**.

3. A

Use the quotient rule, the chain rule, and the derivative $[\sin x]' = \cos x$ to differentiate f. Write out the quotient rule so you don't forget the pieces or the order in which they come. It may help to write the rule using t and b for *top* and *bottom*:

$$f' = \frac{t'b - tb'}{b^2}$$

$$t = \sin\sqrt{x} \rightarrow t' = \cos\sqrt{x} \cdot \frac{1}{2\sqrt{x}}$$

$$b = \sqrt{x} \rightarrow b' = \frac{1}{2\sqrt{x}}$$

Now, put all the pieces together:

$$f'(x) = \frac{\cos\sqrt{x} \cdot \dfrac{1}{2\sqrt{x}} \cdot \sqrt{x} - \sin\sqrt{x} \cdot \dfrac{1}{2\sqrt{x}}}{\left(\sqrt{x}\right)^2}$$

$$= \frac{\dfrac{\cos\sqrt{x}}{2} - \dfrac{\sin\sqrt{x}}{2\sqrt{x}}}{x}$$

$$= \frac{\cos\sqrt{x}}{2x} - \frac{\sin\sqrt{x}}{2x\sqrt{x}}$$

The final result is **(A)**.

4. D

In this problem, you have to use the chain rule three times. It may also help to rewrite f as $f(x) = (\cos(\sin(2x)))^2$. In words, to differentiate the function, you need to: (1) apply the power rule, (2) take the derivative of the outer trig function, (3) take the derivative of the inner trig function, and (4) take the derivative of the argument. Following these steps yields:

$$f'(x) = 2\big(\cos(\sin(2x))\big)^1 \cdot \big(-\sin(\sin(2x))\big) \cdot \cos(2x) \cdot 2$$
$$= -4\cos(\sin(2x)) \cdot \sin(\sin(2x)) \cdot \cos(2x)$$

Don't waste time trying to simplify the expression, because you're asked for the derivative at a given value of x. This means you can simply plug in the value $x = \dfrac{\pi}{8}$ and use the fact that $\sin\dfrac{\pi}{4} = \cos\dfrac{\pi}{4} = \dfrac{\sqrt{2}}{2}$ to get:

$$f'\left(\frac{\pi}{8}\right) = -4\cos\left(\sin\left(2 \times \frac{\pi}{8}\right)\right) \cdot \sin\left(\sin\left(2 \times \frac{\pi}{8}\right)\right) \cdot \cos\left(2 \times \frac{\pi}{8}\right)$$

$$= -4\cos\left(\sin\left(\frac{\pi}{4}\right)\right) \cdot \sin\left(\sin\left(\frac{\pi}{4}\right)\right) \cdot \cos\left(\frac{\pi}{4}\right)$$

$$= -4\cos\left(\frac{\sqrt{2}}{2}\right) \cdot \sin\left(\frac{\sqrt{2}}{2}\right) \cdot \frac{\sqrt{2}}{2}$$

$$= -2\sqrt{2}\cos\left(\frac{\sqrt{2}}{2}\right) \cdot \sin\left(\frac{\sqrt{2}}{2}\right)$$

Be sure to glance back at the answer choices. Without a calculator, you can't find the cosine or sine of $\dfrac{\sqrt{2}}{2}$, but you don't need to. You already have a match for **(D)**.

8.2 DERIVATIVES OF INVERSE TRIG FUNCTIONS

To answer a question like this:

Let $y = \arccos(2x)$. Then $\dfrac{dy}{dx} =$

(A) $\dfrac{-2}{1 - 4x^2}$

(B) $\dfrac{-2}{\sqrt{1 - 2x^2}}$

(C) $\dfrac{-2}{\sqrt{1 + 4x^2}}$

(D) $\dfrac{-2}{\sqrt{1 - 4x^2}}$

You need to know this:

Derivative rules for inverse trig functions:

$\dfrac{d}{dx}(\arcsin x) = \dfrac{1}{\sqrt{1 - x^2}}$	$\dfrac{d}{dx}(\arccos x) = \dfrac{-1}{\sqrt{1 - x^2}}$
$\dfrac{d}{dx}(\arctan x) = \dfrac{1}{1 + x^2}$	$\dfrac{d}{dx}(\text{arccot } x) = \dfrac{-1}{1 + x^2}$
$\dfrac{d}{dx}(\text{arcsec } x) = \dfrac{1}{|x|\sqrt{x^2 - 1}}$	$\dfrac{d}{dx}(\text{arccsc } x) = \dfrac{-1}{|x|\sqrt{x^2 - 1}}$

✔ **AP Expert Note**

For the Calculus AB exam, learning the formulas for the derivatives of arcsin x, arccos x, and arctan x is sufficient.

Note that the inverse of a trigonometric function may also be indicated using inverse function notation (e.g., $\sin^{-1} x = \arcsin x$).

Derivatives

You need to do this:

- Identify which inverse trig function is involved and quickly jot down the rule for its derivative.

- Just as with derivatives of plain trig functions, all the co's (cosine, cotangent and cosecant) are negative.

- Each rule includes an x^2, so plug the quantity inside the parentheses in for x and square the entire quantity.

- Note that you may have to apply other derivative rules, such as the power, product, quotient, and/or chain rule, to arrive at the final answer.

- Simplify the result ONLY if you don't find your answer among the choices.

Answer and Explanation:

D

The derivative of arccos x is $\dfrac{-1}{\sqrt{1-x^2}}$. Use this derivative and the chain rule to differentiate the function. Don't forget to square the entire quantity inside the parentheses:

$$\frac{d}{dx}\arccos(2x) = \frac{-1}{\sqrt{1-(2x)^2}} \cdot 2 = \frac{-2}{\sqrt{1-4x^2}}$$

Thus, **(D)** is correct.

PRACTICE SET

1. Let $f(x) = \arcsin(3x)$. Then $f'(x) =$

    (A) $\dfrac{1}{\sqrt{1-3x^2}}$

    (B) $\dfrac{3}{\sqrt{1-3x^2}}$

    (C) $\dfrac{3}{\sqrt{1-9x^2}}$

    (D) $\dfrac{3}{\sqrt{1-x^2}}$

2. If $f(x) = \arctan\left(\dfrac{x}{2}\right)$, then $f'(x) =$

    (A) $\dfrac{2}{x^2+4}$

    (B) $\dfrac{2}{4x^2+1}$

    (C) $\dfrac{4}{4x^2+1}$

    (D) $\dfrac{4}{x^2+1}$

3. If $f(x) = \arcsin\left(e^{2x}\right)$, then $f'(x) =$

    (A) $\dfrac{-e^{2x}}{\sqrt{1-e^{4x}}}$

    (B) $\dfrac{e^{2x}}{\sqrt{1-e^{4x}}}$

    (C) $\dfrac{2e^{2x}}{\sqrt{1-e^{2x}}}$

    (D) $\dfrac{2e^{2x}}{\sqrt{1-e^{4x}}}$

4. If $y = \arcsin x + x\sqrt{1-x^2}$, then $y' =$

    (A) $2\sqrt{1-x^2}$

    (B) $\dfrac{2x^2+1}{2\sqrt{1-x^2}}$

    (C) $\dfrac{1}{\sqrt{1-x^2}} + \dfrac{x}{2\sqrt{1-x^2}}$

    (D) $\dfrac{1}{\sqrt{1-x^2}} + \dfrac{2x^2}{2\sqrt{1-x^2}}$

ANSWERS AND EXPLANATIONS

1. C

The derivative of arcsin x is $\dfrac{1}{\sqrt{1-x^2}}$. Use this derivative and the chain rule to differentiate the function. Don't forget to square the entire quantity inside the parentheses:

$$\frac{d}{dx}\arcsin(3x) = \frac{1}{\sqrt{1-(3x)^2}} \cdot 3 = \frac{3}{\sqrt{1-9x^2}}$$

Hence, **(C)** is correct.

2. A

The derivative of arctan x is $\dfrac{1}{1+x^2}$. Use this derivative and the chain rule to differentiate the function. Don't forget to square the entire quantity inside the parentheses, including the coefficient (which is $\frac{1}{2}$):

$$\frac{d}{dx}\arctan\left(\frac{x}{2}\right) = \frac{1}{1+\left(\frac{x}{2}\right)^2} \cdot \frac{1}{2} = \frac{1}{2\left(1+\frac{x^2}{4}\right)}$$

This isn't one of the answer choices, so simplify further. Distribute the 2, write the denominator as a single fraction, then take the reciprocal:

$$\frac{1}{2\left(1+\frac{x^2}{4}\right)} = \frac{1}{2+\frac{x^2}{2}} = \frac{1}{\frac{4+x^2}{2}} = \frac{2}{4+x^2}$$

Reverse the order of the terms in the denominator and you have a match for **(A)**.

3. D

The derivative of arcsin x is $\dfrac{1}{\sqrt{1-x^2}}$; the derivative of e raised to a power is e raised to that same power times the derivative of the power. Use these two rules and the chain rule to differentiate the function:

$$\frac{d}{dx}\arcsin\left(e^{2x}\right) = \frac{1}{\sqrt{1-\left(e^{2x}\right)^2}} \cdot \frac{d}{dx}\left(e^{2x}\right)$$

$$= \frac{1}{\sqrt{1-\left(e^{2x}\right)^2}} \cdot e^{2x} \cdot 2 = \frac{2e^{2x}}{\sqrt{1-e^{4x}}}$$

(D) is the appropriate match.

4. A

After differentiating arcsin x, use the product rule and then do some algebraic manipulations. You may want to leave this question for last. Differentiate each term (each side of the plus sign) separately, then add the results:

$$[\arcsin x]' = \frac{1}{\sqrt{1-x^2}}$$

$$\left[x\sqrt{1-x^2}\right]' = [x]'\sqrt{1-x^2} + x\left[\sqrt{1-x^2}\right]'$$

$$= (1)\cdot\sqrt{1-x^2} + \left[x\cdot\frac{1}{2}(1-x^2)^{-\frac{1}{2}}\cdot(-2x)\right]$$

$$= \sqrt{1-x^2} - \frac{x^2}{\sqrt{1-x^2}}$$

To find y', put the two derivatives together, find a common denominator, add, and simplify:

$$\frac{1}{\sqrt{1-x^2}} + \sqrt{1-x^2} - \frac{x^2}{\sqrt{1-x^2}}$$

$$= \frac{1}{\sqrt{1-x^2}} + \frac{(1-x^2)^1}{\sqrt{1-x^2}} - \frac{x^2}{\sqrt{1-x^2}}$$

$$= \frac{1+1-x^2-x^2}{\sqrt{1-x^2}}$$

$$= \frac{2-2x^2}{\sqrt{1-x^2}} = \frac{2(1-x^2)}{\sqrt{1-x^2}}$$

$$= \frac{2\sqrt{1-x^2}\,\sqrt{1-x^2}}{\sqrt{1-x^2}} = 2\sqrt{1-x^2}$$

That's **(A)**. You could also just add the first and third term because they already have a common denominator, then simplify.

8.3 HIGHER ORDER DERIVATIVES

To answer a question like this:

If $y = \ln\left(x^2 + 3\right)$, then $\dfrac{d^2 y}{dx^2} =$

(A) $\dfrac{-1}{x^2 + 3}$

(B) $\dfrac{6 - 2x}{x^2 + 3}$

(C) $\dfrac{2x}{\left(x^2 + 3\right)^2}$

(D) $\dfrac{6 - 2x^2}{\left(x^2 + 3\right)^2}$

You need to know this:

Higher order derivatives:

- If you take the derivative of a derivative, it is called the *second derivative* of *f*, and you write $f''(x)$ or y'' or $\dfrac{d^2 y}{dx^2}$.
- All the same rules apply (power, product, quotient, chain rules).
- Always simplify the first derivative as much as possible before taking a second or third derivative. Calculus is difficult enough without trying to do it with overly complex problems.

> **AP Expert Note**
>
> Don't worry too much about the notation; if you see two primes or d^2, you need to take the second derivative. For the Free-Response section, stick with the simpler notation (f'' or y'').

Applications of the second derivative:

- Concavity and inflection points of a graph (covered in Chapter 9)
- Acceleration of a particle moving along a straight path (covered in Chapter 10)

You need to do this:

- Take the first derivative using all applicable rules. Here, apply the rule for finding the derivative of a natural log and the chain rule.

- Simplify the result, if possible.

- Take the second derivative using all applicable rules. Here, apply the quotient rule.

- Simplify the result, if necessary, until you find a match among the answer choices.

Answer and Explanation:

D

To find the first derivative, $\dfrac{dy}{dx}$, use the natural log rule: The derivative of "ln (quantity)" is 1 over that quantity times the derivative of the quantity.

$$y = \ln\left(x^2 + 3\right)$$

$$y' = \frac{1}{x^2 + 3} \cdot \frac{d}{dx}\left(x^2 + 3\right)$$

$$= \frac{1}{x^2 + 3} \cdot (2x) = \frac{2x}{x^2 + 3}$$

To find the second derivative, use the quotient rule, $\left[\dfrac{t}{b}\right]' = \dfrac{t'b - tb'}{b^2}$, where t stands for *top* and b stands for *bottom*:

$$y' = \frac{2x}{x^2 + 3}$$

$$y'' = \frac{[2x]'\left(x^2 + 3\right) - (2x)\left[x^2 + 3\right]'}{\left(x^2 + 3\right)^2}$$

$$= \frac{2\left(x^2 + 3\right) - (2x)(2x)}{\left(x^2 + 3\right)^2}$$

$$= \frac{2x^2 + 6 - 4x^2}{\left(x^2 + 3\right)^2} = \frac{6 - 2x^2}{\left(x^2 + 3\right)^2}$$

Therefore, **(D)** is correct.

PRACTICE SET

1. If $f(x) = 4x^3 - 5x^2 + 2x - \dfrac{1}{x}$, then $f''(x) =$

   (A) $-\dfrac{2}{x^3} + 24x - 10$

   (B) $\dfrac{2}{x^2} + 24x - 10$

   (C) $24x - 10 - (\ln x)^2$

   (D) $24x - 10 - \dfrac{1}{x}$

2. If $y = (x^2 + 1)^3$, then $y'' =$

   (A) $25x^4$

   (B) $6x^2 + 6$

   (C) $27x^4 + 8x^2 + 1$

   (D) $30x^4 + 36x^2 + 6$

3. If $f(x) = \sqrt{x}\left(x^2 + 2x + \sqrt{x}\right)$, then $f''(x) =$

   (A) $\dfrac{5x + 3}{4\sqrt{x}}$

   (B) $\dfrac{15x + 6}{4\sqrt{x}}$

   (C) $\dfrac{15x + 3}{2\sqrt{x}}$

   (D) $\dfrac{15x + 6}{2\sqrt{x}}$

4. If $y = \cos^3(4x)$, what is $\dfrac{d^2y}{dx^2}$?

   (A) $-96\cos(4x)$

   (B) $-48\sin^2(4x)\cos^2(4x)$

   (C) $24\left(\sin^2(4x) - \cos^3(4x)\right)$

   (D) $96\cos(4x)\sin^2(4x) - 48\cos^3(4x)$

ANSWERS AND EXPLANATIONS

1. A

To find f', rewrite the rational term in exponent form, then apply the power rule:

$$f(x) = 4x^3 - 5x^2 + 2x - x^{-1}$$
$$f'(x) = 3 \cdot 4x^2 - 2 \cdot 5x + 2 - (-1)x^{-2}$$
$$= 12x^2 - 10x + 2 + x^{-2}$$

To find the second derivative, f'', apply the power rule again to each term. Don't forget—the derivative of a constant is 0.

$$f'(x) = 12x^2 - 10x + 2 + x^{-2}$$
$$f''(x) = 2 \cdot 12x - 10 + 0 + (-2)x^{-3}$$
$$= 24x - 10 - \frac{2}{x^3}$$

Rearrange the terms and you have a match for **(A)**.

2. D

You could use the power rule and the chain rule a couple of times to find the third derivative, but by looking at the answer choices, you can see that all of the terms are added and there are not any parentheses. The quicker approach would be to expand $(x^2 + 1)^3$ by using FOIL or Pascal's Triangle, then take the derivative. This is what the FOIL method would look like:

$$y = (x^2 + 1)(x^2 + 1)(x^2 + 1)$$
$$= (x^4 + 2x^2 + 1)(x^2 + 1)$$
$$= x^6 + x^4 + 2x^4 + 2x^2 + x^2 + 1$$
$$= x^6 + 3x^4 + 3x^2 + 1$$
$$y' = 6x^5 + 12x^3 + 6x$$
$$y'' = 30x^4 + 36x^2 + 6$$

That's **(D)**.

3. B

This is another example where multiplying the product out first will make finding the second derivative easier than trying to use the product rule. To carry out the multiplication, write $\sqrt{x} = x^{\frac{1}{2}}$ and use rules of exponents:

$$f(x) = x^{\frac{1}{2}}\left(x^2 + 2x + x^{\frac{1}{2}}\right)$$
$$= x^{\frac{4}{2} + \frac{1}{2}} + 2x^{\frac{2}{2} + \frac{1}{2}} + x^{\frac{1}{2} + \frac{1}{2}} = x^{\frac{5}{2}} + 2x^{\frac{3}{2}} + x^1$$

Now, carefully apply the product rule to find f' and then f'':

$$f'(x) = \frac{5}{2} \cdot x^{\frac{5}{2} - 1} + \frac{3}{2} \cdot 2x^{\frac{3}{2} - 1} + 1 = \frac{5}{2}x^{\frac{3}{2}} + 3x^{\frac{1}{2}} + 1$$
$$f''(x) = \frac{3}{2} \cdot \frac{5}{2}x^{\frac{1}{2}} + \frac{1}{2} \cdot 3x^{-\frac{1}{2}} + 0 = \frac{15\sqrt{x}}{4} + \frac{3}{2\sqrt{x}}$$

This isn't one of the answer choices, so find a common denominator and simplify:

$$f''(x) = \frac{15\sqrt{x}}{4}\left(\frac{\sqrt{x}}{\sqrt{x}}\right) + \frac{3}{2\sqrt{x}}\left(\frac{2}{2}\right) = \frac{15x + 6}{4\sqrt{x}}$$

Thus, **(B)** is correct.

4. D

The notation $\frac{d^2y}{dx^2}$ means find the second derivative, or y''. Use the power rule and the chain rule to get the first derivative. It may help to rewrite the equation as $y = (\cos(4x))^3$ to make the chain rule easier.

$$y' = 3(\cos(4x))^2 \cdot [\cos(4x)]'$$
$$= 3(\cos(4x))^2 \cdot (-\sin(4x)) \cdot (4)$$
$$= -12\sin(4x) \cdot (\cos(4x))^2$$

Now use the product rule $(f \cdot g)' = f'g + fg'$ to take the derivative again. Use the chain rule where necessary:

$$y'' = [-12\sin(4x)]'(\cos(4x))^2 + (-12\sin(4x))[(\cos(4x))^2]'$$
$$= -12\cos(4x)(4) \cdot \cos^2(4x) -$$
$$12\sin(4x) \cdot 2(\cos(4x))^1(-\sin(4x))(4)$$
$$= -48\cos^3(4x) + 96\sin^2(4x)\cos(4x)$$
$$= 96\sin^2(4x)\cos(4x) - 48\cos^3(4x)$$

After switching the order of the sum, **(D)** is a perfect match.

8.4 IMPLICIT DIFFERENTIATION

To answer a question like this:

If $x^2 + 4y = xy$, then $\dfrac{dy}{dx} =$

(A) $\dfrac{2x}{3}$

(B) $\dfrac{x - 4}{2y}$

(C) $\dfrac{2x - y}{x - 4}$

(D) $\dfrac{2x + y}{x - 4}$

You need to know this:

Some mathematical relationships are defined explicitly, meaning that one variable is given in terms of other variables. For example, $y = 2x + 3$ and $f(x) = \sin^2 x$ are explicitly defined functions.

A mathematical relationship can also be defined *implicitly*, meaning that you are given an equation relating the variables, but the equation is *not* solved for one of the variables. The equation $x^2 + y^2 = 2$ is an example of an implicitly defined relationship.

Using the chain rule, you can differentiate equations that relate y and x. Think of x as the independent variable and y as a function of x, keeping in mind that the derivative of y is $\dfrac{dy}{dx}$, even though you don't have an explicit expression for y in terms of x. You can then differentiate both sides of the equation that relate y and x and find an implicit relationship between x, y, and $\dfrac{dy}{dx}$.

Solving for $\dfrac{dy}{dx}$ (isolating it on one side of the equation) gives the derivative in terms of x and/or y.

✔ AP Expert Note

Sometimes, it is quicker (and easier) to solve for y and then find the derivative explicitly. For example, $3x^2 + 2y = 1$ can be rewritten as $y = \dfrac{-3}{2}x^2 - \dfrac{1}{2}$, the derivative of which is $y' = -3x$.

You need to do this:

Step 1: Differentiate both sides of the equation with respect to x. Each time you take the derivative of a term that contains y, multiply by $\frac{dy}{dx}$, because the derivative of y is $\frac{dy}{dx}$.

Step 2: Collect all the terms that contain a $\frac{dy}{dx}$ on one side of the equation and move all other terms to the opposite side.

Step 3: Factor out the $\frac{dy}{dx}$.

Step 4: Solve for $\frac{dy}{dx}$ by dividing both sides of the equation by whatever remains after factoring.

Answer and Explanation:

C

Differentiate $x^2 + 4y = xy$ implicitly and solve for $\frac{dy}{dx}$. Use the product rule to differentiate xy on the right-hand side of the equal sign:

$$x^2 + 4y = x \cdot y$$

Step 1 $\qquad 2x + \boxed{4\dfrac{dy}{dx}} = 1 \cdot y + \boxed{x \cdot \dfrac{dy}{dx}}$

Step 2 $\qquad 2x - y = x\dfrac{dy}{dx} - 4\dfrac{dy}{dx}$

Step 3 $\qquad 2x - y = \dfrac{dy}{dx}(x - 4)$

Step 4 $\qquad \dfrac{2x - y}{x - 4} = \dfrac{dy}{dx}$

(C) is thus correct.

PRACTICE SET

1. If $y^3 + y^2 - 5y - x^2 = -4$, then $\dfrac{dy}{dx} =$

   (A) $\dfrac{2x}{3y^2 + 2y - 5}$

   (B) $\dfrac{2x - 4}{3y^2 + 2y - 5}$

   (C) $\dfrac{x^2 - 4}{3y^2 + 2y - 5}$

   (D) $\dfrac{x^2 - 4}{y^3 + y^2 - 5y}$

2. If $(x + 2y)\dfrac{dy}{dx} = 2x - y$, what is the value of $\dfrac{d^2y}{dx^2}$ at the point $(5, 0)$?

   (A) -2

   (B) $-\dfrac{1}{5}$

   (C) 0

   (D) $\dfrac{4}{5}$

3. If $x^2y + xy^3 = 8 - 2y$, then the value of $\dfrac{dy}{dx}$ at the point $(2, 1)$ is

   (A) $-\dfrac{5}{6}$

   (B) $-\dfrac{5}{12}$

   (C) $\dfrac{5}{12}$

   (D) $\dfrac{5}{6}$

4. What is $\dfrac{d^2y}{dx^2}$ if $xy + \pi = 2y^2$?

   (A) $\dfrac{y}{4y - x}$

   (B) $\dfrac{y^2}{(4y - x)^2}$

   (C) $\dfrac{1}{4\left(\dfrac{y}{4y - x}\right) - x}$

   (D) $\dfrac{2y - \dfrac{4y^2}{4y - x}}{(4y - x)^2}$

ANSWERS AND EXPLANATIONS

1. A

Follow the four steps for implicit differentiation:

$$y^3 + y^2 - 5y - x^2 = -4$$

Step 1: $\quad 3y^2 \dfrac{dy}{dx} + 2y \dfrac{dy}{dx} - 5 \dfrac{dy}{dx} - 2x = 0$

Step 2: $\quad 3y^2 \dfrac{dy}{dx} + 2y \dfrac{dy}{dx} - 5 \dfrac{dy}{dx} = 2x$

Step 3: $\quad \dfrac{dy}{dx}\left(3y^2 + 2y - 5\right) = 2x$

Step 4: $\quad \dfrac{dy}{dx} = \dfrac{2x}{3y^2 + 2y - 5}$

That's **(A)**.

2. A

Solve for $\dfrac{dy}{dx}$ in the given relationship to find the first derivative:

$$\left(x + 2y\right)\dfrac{dy}{dx} = 2x - y$$

$$\dfrac{dy}{dx} = \dfrac{2x - y}{x + 2y}$$

You will need the value of $\dfrac{dy}{dx}$ for the second derivative, so use the given point, (5, 0), to find its numerical representation:

$$\dfrac{dy}{dx} = \dfrac{2(5) - 0}{5 + 2(0)} = \dfrac{10}{5} = 2$$

Now use implicit differentiation again, along with the quotient rule, $\left[\dfrac{t}{b}\right]' = \dfrac{t'b - tb'}{b^2}$, where t stands for *top* and b stands for *bottom*, to find the second derivative, $\dfrac{d^2y}{dx^2}$:

$$\dfrac{d^2y}{dx^2} = \dfrac{\left(2 - \dfrac{dy}{dx}\right)\left(x + 2y\right) - \left(2x - y\right)\left(1 + 2\dfrac{dy}{dx}\right)}{\left(x + 2y\right)^2}$$

There is no need to algebraically simplify the expression because you know the values of x, y, and $\dfrac{dy}{dx}$, which are 5, 0, and 2, respectively. Plug these values in and simplify:

$$\dfrac{d^2y}{dx^2} = \dfrac{(2 - 2)\left(5 + 2(0)\right) - \left(2(5) - 0\right)\left(1 + 2(2)\right)}{\left(5 + 2(0)\right)^2}$$

$$= \dfrac{0 - (10)(5)}{25} = \dfrac{-50}{25} = -2$$

Hence, **(A)** is correct.

3. B

The xs and ys are mixed in the equation, so use implicit differentiation. You'll need to use the product rule on each of the first two terms:

$$x^2 \cdot y + x \cdot y^3 = 8 - 2y$$

$$2xy + x^2 \dfrac{dy}{dx} + 1 \cdot y^3 + x \cdot 3y^2 \dfrac{dy}{dx} = 0 - 2\dfrac{dy}{dx}$$

$$x^2 \dfrac{dy}{dx} + 3xy^2 \dfrac{dy}{dx} + 2\dfrac{dy}{dx} = -2xy - y^3$$

$$\dfrac{dy}{dx}\left(x^2 + 3xy^2 + 2\right) = -2xy - y^3$$

$$\dfrac{dy}{dx} = \dfrac{-2xy - y^3}{x^2 + 3xy^2 + 2}$$

The question asks for the value of the derivative at a specific point, (2, 1), so plug in $x = 2$ and $y = 1$, and then simplify the result.

$$\dfrac{dy}{dx}(2,1) = \dfrac{-2(2)(1) - (1)^3}{(2)^2 + 3(2)(1)^2 + 2}$$

$$= \dfrac{-4 - 1}{4 + 6 + 2}$$

$$= -\dfrac{5}{12}$$

(B) is the appropriate match. It is often easiest to plug in the given x- and y-values and the value of $\dfrac{dy}{dx}$ right after you take the derivative (before you solve for $\dfrac{dy}{dx}$). Try plugging these values into the first equation that contains $\dfrac{dy}{dx}$ and see if you get the same answer.

4. D

Use implicit differentiation to find the first derivative:

$$x \cdot y + \pi = 2y^2$$

$$\underbrace{1 \cdot y + x \frac{dy}{dx}}_{\text{product rule}} + 0 = 4y \frac{dy}{dx}$$

$$y = 4y \frac{dy}{dx} - x \frac{dy}{dx}$$

$$y = (4y - x) \frac{dy}{dx}$$

$$\frac{y}{4y - x} = \frac{dy}{dx}$$

To find the second derivative, you take the derivative of $\frac{dy}{dx}$. So, use the quotient rule and implicit differentiation again to find the second derivative:

$$\frac{dy}{dx} = \frac{y}{4y - x}$$

$$\frac{d^2y}{dx^2} = \frac{\frac{dy}{dx}(4y - x) - y\left(4\frac{dy}{dx} - 1\right)}{(4y - x)^2} \quad \left(\text{substitute for } \frac{dy}{dx}\right)$$

$$y'' = \frac{\left(\dfrac{y}{4y - x}\right)(4y - x) - y\left(4\left(\dfrac{y}{4y - x}\right) - 1\right)}{(4y - x)^2}$$

$$= \frac{y - \dfrac{4y^2}{4y - x} + y}{(4y - x)^2}$$

$$= \frac{2y - \dfrac{4y^2}{4y - x}}{(4y - x)^2}$$

That's a perfect match for **(D)**.

Derivatives

8.5 DERIVATIVES OF INVERSE FUNCTIONS

To answer a question like this:

If $f(x) = 2x^3 - 4x + 5$, then $f(1) = 3$. Given that h is the inverse of f, what is the value of $h'(3)$?

(A) $\dfrac{1}{50}$ (B) $\dfrac{1}{2}$ (C) 2 (D) 50

You need to know this:

The Inverse Function Theorem

$$\left[f^{-1}(x)\right]' = \frac{1}{\underbrace{f'\left[f^{-1}(x)\right]}_{\substack{\text{the derivative of} \\ f \text{ evaluated at } f^{-1}(x)}}}$$

The biggest obstacle to understanding the inverse function theorem is understanding the notation used to state it, so it may help to assign a different name to the inverse function.

Suppose h is the inverse of a function f. Then:

$$h'(x) = \frac{1}{f'(h(x))}$$

You'll most likely be given a point on f as part of the question. To find the equivalent point on the inverse function, simply swap the x- and y-values.

Translation: Given a point (x, y) on a function, the derivative of the inverse function at y is 1 over the derivative of the original function evaluated at x.

You need to do this:

- Take the ordered pair you're given for the original function and swap the coordinates to find the corresponding ordered pair for the inverse function. For example, if $f(a) = b$, then $h(b) = a$.

- Take the derivative of the original function.

- Evaluate the derivative at $h(b)$, which happens to be the original a given in the problem.

- Plug the result into the theorem: $h'(x) = \dfrac{1}{f'(h(x))}$.

- In other words, find f' of the original x-value and then take its reciprocal.

Answer and Explanation:

B

Because h is the inverse of f, and because $f(1) = 3$, you know that $h(3) = 1$ (swap the values). The derivative of f is $f'(x) = 6x^2 - 4$. To find $h'(3)$, evaluate f' at $h(3) = 1$, and then take the reciprocal:

$$h'(3) = \frac{1}{f'(h(3))} = \frac{1}{f'(1)} = \frac{1}{6(1)^2 - 4} = \frac{1}{2}$$

Note that if you're not fond of memorizing theorems, you could also use facts about inverse functions and implicit differentiation to answer a question like this. To do so, follow these steps:

1. Write the given function as $y = \dots$ and then swap the xs and the ys (this is the first step in finding an inverse function).

2. Take the derivative with respect to x. Don't forget to multiply by $\dfrac{dy}{dx}$ whenever you take the derivative of a term that contains y.

3. Solve for $\dfrac{dy}{dx}$ (which represents the derivative of the inverse function because you swapped the variables).

4. Plug in the original x-value given in the question stem (which is the y-value for the inverse function) and simplify.

$$y = 2x^3 - 4x + 5$$
$$x = 2y^3 - 4y + 5$$
$$1 = 6y^2 \frac{dy}{dx} - 4\frac{dy}{dx} + 0$$
$$1 = \frac{dy}{dx}\left(6y^2 - 4\right) \rightarrow \frac{dy}{dx} = \frac{1}{6y^2 - 4}$$
$$\frac{dy}{dx} \text{ at } y = 1 \quad \rightarrow \quad \frac{1}{6(1)^2 - 4} = \frac{1}{2}$$

Therefore, **(B)** is correct.

PRACTICE SET

x	$f(x)$	$g(x)$	$f'(x)$
-4	2	1	$\dfrac{1}{3}$
1	-4	-2	4

1. The table above gives values of the differentiable functions f and g, and f', the derivative of f, at selected values of x. If g is the inverse of f, what is the value of $g'(-4)$?

   (A) $-\dfrac{1}{2}$

   (B) $-\dfrac{1}{4}$

   (C) $\dfrac{1}{4}$

   (D) 1

2. Suppose f is a differentiable function given by the equation $f(x) = \dfrac{1}{2}x^3 + x - 1$, and h is the inverse of f. If $f(2) = 5$, then what is the value of $h'(5)$?

   (A) $\dfrac{2}{77}$

   (B) $\dfrac{1}{7}$

   (C) 7

   (D) $\dfrac{77}{2}$

3. Suppose f is given by $f(x) = e^x + 3x$, and $g(x) = f^{-1}(x)$. If the point $(0, 1)$ is on the graph of f, what is the value of $g'(1)$?

   (A) $\dfrac{1}{e + 3}$

   (B) $\dfrac{1}{4}$

   (C) 4

   (D) $e + 3$

4. Let $f(x) = x^3 - x + 2$ for certain values of x. If h is the inverse of f over these same values, then $h'(2)$ could be

   (A) $\dfrac{1}{26}$

   (B) $\dfrac{1}{2}$

   (C) 2

   (D) 26

ANSWERS AND EXPLANATIONS

1. C

Because *f* and *g* are inverses, use the inverse function theorem:

$$g'(x) = \frac{1}{f'(g(x))}$$

Now plug $x = -4$ into this expression and evaluate it using the table of values given:

x	$f(x)$	$g(x)$	$f'(x)$
-4	2	1	$\frac{1}{3}$
1	-4	-2	4

$$g'(-4) = \frac{1}{f'(g(-4))}$$

From the table, $g(-4)$ is 1, so this becomes:

$$g'(-4) = \frac{1}{f'(1)} = \frac{1}{4}$$

(C) is correct.

2. B

Because *h* is the inverse of *f*, and because $f(2) = 5$, you know that $h(5) = 2$ (swap the values). The derivative of *f* is $f'(x) = \frac{3}{2}x^2 + 1$. To find $h'(5)$, evaluate f' at $h(5)$, or 2, and then take the reciprocal:

$$h'(5) = \frac{1}{f'(h(5))} = \frac{1}{f'(2)} = \frac{1}{\frac{3}{2}(2)^2 + 1} = \frac{1}{6+1} = \frac{1}{7}$$

That's **(B)**.

3. B

Because *g* is the inverse of *f*, and because $f(0) = 1$, you know that $g(1) = 0$ (swap the coordinates of the given point, $(0, 1)$). The derivative of *f* is $f'(x) = e^x + 3$. To find $g'(1)$, evaluate f' at $g(1)$, or 0, and then take the reciprocal:

$$g'(1) = \frac{1}{f'(g(1))} = \frac{1}{f'(0)} = \frac{1}{e^0 + 3} = \frac{1}{1+3} = \frac{1}{4}$$

That's **(B)**.

4. B

This is a much more difficult question because you're not given a point on *f*. You approach it the same way, but you'll need to fill in some missing information.

If *h* is the inverse of $f(x) = x^3 - x + 2$, then, according to the formula for the derivative of an inverse, $h'(2) = \frac{1}{f'(h(2))}$. To compute $h'(2)$ you must evaluate $f'(h(2))$, so you need to find $f'(x)$ and $h(2)$. The function $f(x) = x^3 - x + 2$ has a derivative of $f'(x) = 3x^2 - 1$.

Because *h* is the inverse of *f*, to find $h(2)$, solve the equation $2 = x^3 - x + 2 \Rightarrow x^3 - x = 0 \Rightarrow x(x+1)(x-1) = 0$ $\Rightarrow x = 0, -1,$ and 1.

Therefore, $h(2) = 0, -1$ or 1. Plug each of these options into the formula for $h'(2)$ above:

$h(2)$	0	-1	1
$f'(h(2))$	-1	2	2
$h'(2) = \frac{1}{f'(h(2))}$	-1	$\frac{1}{2}$	$\frac{1}{2}$

Because -1 is not one of the answer choices, the correct answer is $h'(2) = \frac{1}{2}$, which is **(B)**.

RAPID REVIEW

> **If you take away only 4 things from this chapter:**
>
> 1. You need to memorize the derivative rules for trig functions and inverse trig functions:
>
> | $f(x)$ | $f'(x)$ |
> |---|---|
> | $\sin x$ | $\cos x$ |
> | $\cos x$ | $-\sin x$ |
> | $\tan x$ | $\sec^2 x$ |
> | $\cot x$ | $-\csc^2 x$ |
> | $\sec x$ | $\sec x \tan x$ |
> | $\csc x$ | $-\csc x \cot x$ |
> | $\arcsin x$ | $\dfrac{1}{\sqrt{1 - x^2}}$ |
> | $\arccos x$ | $\dfrac{-1}{\sqrt{1 - x^2}}$ |
> | $\arctan x$ | $\dfrac{1}{1 + x^2}$ |
>
> 2. Higher order derivatives can be indicated using prime notation, such as $f''(x)$ or y'', or by $\dfrac{d^n y}{dx^n}$, where n represents the order of the derivative.
>
> 3. When a function is defined in such a way that it is difficult or impossible to express y explicitly in terms of x, use implicit differentiation to find the derivative of the function.
>
> 4. If h is the inverse of a function f, then:
>
> $$h'(x) = \frac{1}{f'(h(x))}$$
>
> In other words, given a point (x, y) on a function, the derivative of the inverse function at y is 1 over the derivative of the original function evaluated at x.

Derivatives

TEST WHAT YOU LEARNED

1. If $x^2 + xy - 3y^2 = 10$, then $\dfrac{dy}{dx} =$

   (A) $\dfrac{2x + y}{6y - x}$

   (B) $\dfrac{-2x + 2xy^2 - 10}{\left(x - y^2\right)^2}$

   (C) $\dfrac{10 - 6y + 2x}{x - y^2}$

   (D) $\dfrac{2x}{y - 3y^2}$

---

2. If $f(x) = e^{x^2 + 5}$, what is $f''(x)$?

   (A) $4xe^{x^2 + 5}$

   (B) $4x^2 e^{x^2 + 5}$

   (C) $e^{x^2 + 5}\left(1 + 4x^2\right)$

   (D) $2e^{x^2 + 5}\left(1 + 2x^2\right)$

3. Let $f(x) = \frac{1}{2}\left(\arccos\left(x^4\right)\right)^2$. Then $f'(x) =$

(A) $\dfrac{-4x^3 \arcsin\left(x^4\right)}{\sqrt{1-x^2}}$

(B) $\dfrac{-4x^3 \arccos\left(x^4\right)}{\sqrt{1-x^8}}$

(C) $\dfrac{4x^3 \arcsin\left(x^4\right)}{\sqrt{1-x^8}}$

(D) $\dfrac{4x^3 \arccos\left(x^4\right)}{\sqrt{1-x^2}}$

---

4. Which of the following is the derivative of $y = x^3 \cos(5x)$?

(A) $3x^2 \cos(5x) - 5x^3 \sin(5x)$

(B) $15x \sin x - 3x^2 \cos(5x)$

(C) $6x \sin(5x) + x^3 \cos(5x)$

(D) $-15x^2 \sin(5x)$

---

x	$f(x)$	$g(x)$	$f'(x)$
-2	4	3	1
3	-2	-1	$\frac{1}{2}$

5. Let f and g be inverse functions. The table above lists a few values of f, g, and f'. The value of $g'(-2) =$

(A) -1 (B) $-\dfrac{1}{2}$ (C) 2 (D) 3

Derivatives

Answer Key

Test What You Already Know

1. B **Learning Objective:** 8.1

2. D **Learning Objective:** 8.2

3. C **Learning Objective:** 8.3

4. C **Learning Objective:** 8.4

5. B **Learning Objective:** 8.5

Test What You Learned

1. A **Learning Objective:** 8.4

2. D **Learning Objective:** 8.3

3. B **Learning Objective:** 8.2

4. A **Learning Objective:** 8.1

5. C **Learning Objective:** 8.5

 ## REFLECTION

Test What You Already Know score: _____

Test What You Learned score: _____

Use this section to evaluate your progress. After working through the pre-quiz, check off the boxes in the "Pre" column to indicate which Learning Objectives you feel confident about. Then, after completing the chapter, including the post-quiz, do the same to the boxes in the "Post" column. Keep working on unchecked Objectives until you're confident about them all!

Pre | Post

☐ | ☐ **8.1** Calculate derivatives of trig functions

☐ | ☐ **8.2** Calculate derivatives of inverse trig functions

☐ | ☐ **8.3** Determine higher order derivatives

☐ | ☐ **8.4** Use implicit differentiation to find derivatives

☐ | ☐ **8.5** Calculate derivatives of inverse functions

 ## FOR MORE PRACTICE

Complete more practice online at kaptest.com. Haven't registered your book yet? Go to kaptest.com/booksonline to begin.

ANSWERS AND EXPLANATIONS

Test What You Already Know

1.   B   Learning Objective: 8.1

Use the product rule, $(f \cdot g)' = f'g + fg'$, and the fact that the derivative of $\sin x$ is $\cos x$ to differentiate the function:

$$y' = \left(3x^2\right)' \sin x + 3x^2 \left(\sin x\right)'$$
$$y' = 6x \left(\sin x\right) + 3x^2 \left(\cos x\right)$$
$$= 6x \sin x + 3x^2 \cos x$$

That's a perfect match for **(B)**.

2.   D   Learning Objective: 8.2

First, apply the chain rule to the power part of the function:

$$4\left(\arcsin\left(x^3\right)\right)^3 \cdot \frac{d}{dx} \arcsin\left(x^3\right)$$

Next, note that the derivative of $\arcsin x$ is $\dfrac{1}{\sqrt{1-x^2}}$. Use this derivative and the chain rule (again to differentiate the x^3 in the original problem). The result is:

$$\frac{d}{dx} \arcsin\left(x^3\right) = \frac{1}{\sqrt{1-\left(x^3\right)^2}} \cdot 3x^2$$

Combining and simplifying these two pieces of the derivative yields $\dfrac{12x^2 \left(\arcsin\left(x^3\right)\right)^3}{\sqrt{1-x^6}}$, which is **(D)**.

3.   C   Learning Objective: 8.3

The notation $\dfrac{d^3y}{dx^3}$ means find the third derivative, or y'''. First rewrite $\sin^2 x$ as $(\sin x)^2$. You need to use the chain rule multiple times to get:

$$y = (\sin x)^2$$
$$\frac{dy}{dx} = 2 \sin x \cdot \cos x$$
$$\frac{d^2y}{dx^2} = 2 \cos x (\cos x) + 2 \sin x (-\sin x)$$
$$= 2(\cos x)^2 - 2(\sin x)^2$$
$$\frac{d^3y}{dx^3} = 2 \cdot 2 \cos x (-\sin x) - 2 \cdot 2 \sin x (\cos x)$$
$$= -4 \sin x \cos x - 4 \sin x \cos x$$
$$= -8 \sin x \cos x$$

That makes **(C)** the correct answer.

4.   C   Learning Objective: 8.4

This is an exercise in implicit differentiation (because the equation is not already written in terms of y or easily solvable for y). When differentiating both sides of this equation, remember to treat y as a function of x, whereby both the product rule and the chain rule will apply. Be sure to use the product rule to take the derivative of the term $2xy$. Differentiating both sides of the equation gives:

$$3x^2 - \left(2y + 2x\frac{dy}{dx}\right) + 2y\frac{dy}{dx} = 0$$
$$3x^2 - 2y - 2x\frac{dy}{dx} + 2y\frac{dy}{dx} = 0$$
$$(2y - 2x)\frac{dy}{dx} = 2y - 3x^2$$
$$\frac{dy}{dx} = \frac{2y - 3x^2}{2y - 2x}$$

That's **(C)**. When performing implicit differentiation, if you have a hard time remembering that the derivative of y is $\dfrac{dy}{dx}$, just remember that whenever you take the derivative of something that contains a y, you need to multiply by $\dfrac{dy}{dx}$.

5.   B   Learning Objective: 8.5

Because f and g are inverses of each other, you know that:

$$g'(x) = \frac{1}{f'(g(x))} = \frac{1}{f'(y)}$$

The function $f(x) = x^3 - 3x + 1$ has derivative $f'(x) = 3x^2 - 3$.

Because $(2, 3)$ is a point on f, the corresponding point on g is $(3, 2)$.

Thus:

$$g'(3) = \frac{1}{f'(2)} = \frac{1}{3(2)^2 - 3} = \frac{1}{9}$$

(B) is correct.

Test What You Learned

1. A Learning Objective: 8.4

This is an exercise in implicit differentiation (because the equation is not already written as y in terms of x or easily solvable for y). When differentiating both sides of this equation, remember to treat y as a function of x, whereby both the product rule and the chain rule will apply. Differentiating both sides of the equation gives:

$$2x + \left(y + x\frac{dy}{dx}\right) - 6y\frac{dy}{dx} = 0$$

$$2x + y = 6y\frac{dy}{dx} - x\frac{dy}{dx}$$

$$2x + y = \frac{dy}{dx}(6y - x)$$

$$\frac{2x + y}{6y - x} = \frac{dy}{dx}$$

That makes **(A)** the answer.

2. D Learning Objective: 8.3

Use the rule: The derivative of e raised to a power is e raised to that same power times the derivative of the power.

$$f(x) = e^{x^2 + 5}$$

$$f'(x) = e^{x^2 + 5} \cdot \left(x^2 + 5\right)'$$

$$= 2x\, e^{x^2 + 5}$$

Now use the product rule $[(f \cdot g)' = f'g + fg']$ to take the derivative again. Use the chain rule where necessary:

$$f'' = (2x)'\, e^{x^2 + 5} + 2x\left(e^{x^2 + 5}\right)'$$

$$= 2e^{x^2 + 5} + 2x\left(e^{x^2 + 5}\right) \cdot 2x$$

$$= 2e^{x^2 + 5} + 4x^2\left(e^{x^2 + 5}\right)$$

This isn't one of the answer choices, so factor out the GCF, which is $2e^{x^2 + 5}$, to arrive at **(D)**:

$$f''(x) = 2e^{x^2 + 5}\left(1 + 2x^2\right)$$

3. B Learning Objective: 8.2

First, apply the chain rule to see that the derivative of the power part of the equation equals:

$$2 \cdot \frac{1}{2}\left(\arccos\left(x^4\right)\right)' \cdot \frac{d}{dx}\arccos\left(x^4\right)$$

Next, note that the derivative of arccos u is $\dfrac{-1}{\sqrt{1 - u^2}}$.

Apply the chain rule again to differentiate the x^4. The result is:

$$\frac{d}{dx}\arccos\left(x^4\right) = \frac{-1}{\sqrt{1 - \left(x^4\right)^2}} \cdot 4x^3 = \frac{-4x^3}{\sqrt{1 - x^8}}$$

Finally, put the two pieces of the derivative together and simplify:

$$\cancel{2} \cdot \frac{1}{\cancel{2}}\left(\arccos\left(x^4\right)\right)' \cdot \frac{-4x^3}{\sqrt{1 - x^8}} = \frac{-4x^3 \arccos\left(x^4\right)}{\sqrt{1 - x^8}}$$

That makes **(B)** the correct answer.

4. A Learning Objective: 8.1

Use the product rule to find the derivative.

$$y' = 3x^2 \cdot \cos(5x) + x^3 \cdot \left(-\sin(5x)\right)(5)$$

$$= 3x^2 \cos(5x) - 5x^3 \sin(5x)$$

That matches **(A)**.

5. C Learning Objective: 8.5

Because f and g are inverses of each other, you know that:

$$g'(x) = \frac{1}{f'(g(x))}$$

Plug in $x = -2$ and evaluate this expression:

$$g'(-2) = \frac{1}{f'(g(-2))} = \frac{1}{f'(3)} = \frac{1}{\frac{1}{2}} = 2$$

Hence, **(C)** is correct.

CHAPTER 9

Applications of the Derivative

LEARNING OBJECTIVES

9.1 Determine the slope of a curve at a given point

9.2 Solve problems involving tangent lines and normal lines

9.3 Use an equation of the tangent line to make local linear approximations of a function

9.4 Use derivatives to find rates of change

9.5 Use the first derivative to find relative/absolute extrema and to describe intervals of increase or decrease

9.6 Use the second derivative to find points of inflection and to describe concavity

TEST WHAT YOU ALREADY KNOW

1. What is the slope of the curve $f(x) = \dfrac{8}{x^2} + x^3 + x$ at $x = 2$?

   (A) 0   (B) 11   (C) 12   (D) 16

2. At what points on the graph of the function $f(x) = \dfrac{2x}{x+8}$ does the tangent line have a slope of 1?

   (A) $(-4, -2)$ and $(-12, 6)$

   (B) $(-4, 0)$ only

   (C) $(-12, 1)$ only

   (D) $(-4, 1)$ and $(-12, 1)$

3. Suppose for some differentiable function f that $f(4) = 1$ and $f'(x) = \dfrac{x^2 + 5}{x + 3}$. Using a local linear approximation, the value of $f(4.7)$ is best approximated as

   (A) 0   (B) 3   (C) 3.1   (D) 4.7

4. If $f(x) = x^4 + 2x^2 - 3x + 4$, what is the instantaneous rate of change of $f(x)$ at $x = 2$?

(A) 10   (B) 22   (C) 37   (D) 52

5. For $0 < x < 13$, the derivative of the function $f(x)$ is given by $f'(x) = e^{\cos x} + \sin x - 2$. Over what interval(s) is f decreasing?

(A) $0.376 < x < 6.659$

(B) $0 < x < 1.571$ and $5.82 < x < 7.854$

(C) $1.571 < x < 5.82$ and $7.854 < x < 12.103$

(D) $5.82 < x < 7.854$ and $12.103 < x < 13$

Derivatives

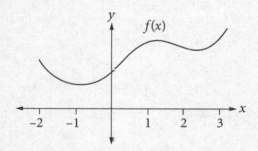

6. The graph of $f(x)$ is given above. Which of the following statements is true about $f''(x)$?

(A) $f''(x) > 0$ for all values of x

(B) $f''(x) < 0$ for all values of x

(C) $f''(x) < 0$ for $-2 < x < -1$ and $f''(x) > 0$ for $-1 < x < 1$ and $x > 2$

(D) $f''(x) > 0$ for $-2 < x < 0$ and $x > 2$, and $f''(x) < 0$ for $0 < x < 2$

Answers to this quiz can be found at the end of this chapter.

9.1 THE SLOPE OF A CURVE AT A POINT

To answer a question like this:

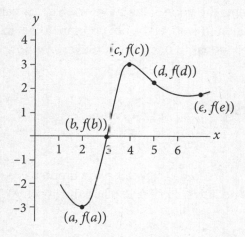

The graph of $f(x)$ is shown above with several points labeled. The values of which of the following are presented in ascending order?

(A) $f(b), f'(a), f'(d), f'(b)$

(B) $f(a), f'(c), f'(e), f(c)$

(C) $f'(a), f'(c), f'(b), f'(d)$

(D) $f(a), f'(b), f'(c), f(d)$

You need to know this:

- The slope of a curve $y = f(x)$ at a point P means the slope of the tangent at the point P.

- The slope of the tangent to a curve at a given point is the first derivative, $f'(x)$, of the function evaluated at that point.

- If a curve is increasing at $x = a$, then $f'(a) > 0$. If a curve is decreasing at $x = a$, then $f'(a) < 0$.

- A horizontal tangent line means the slope at that point is 0.

You need to do this:

- Use the graph to eliminate options one at a time.

- Estimate the functional values and the slopes for each of the indicated points.

- Remember, $f'(a)$ represents the slope of the curve at $x = a$. You can't find exact slopes (because you don't have an equation to work with), but you should be able to determine whether the slopes are approximately equal to 0, *slightly* positive, *very* positive, *slightly* negative, or *very* negative.

Answer and Explanation:

B

(A): $f(b) = 0$ and $f'(a)$ also appears to be 0 (because the graph has a horizontal tangent at $x = a$). The slope of the curve at d is negative, so $f'(d) < 0$ and the list in (A) is therefore not in ascending order. Eliminate (A).

(B): $f(a) < 0$ because the point lies below the x-axis. The graph has a horizontal tangent at $x = c$, so $f'(c) = 0$. The slope of the curve is slightly positive at $x = e$. If you visualize (or sketch) the line tangent to the curve at this point, you can determine that the slope of the line is less than 1, so $0 < f'(e) < 1$. Finally, $f(c)$ appears to equal 3. Therefore, this list is in ascending order, and **(B)** is correct.

There is no need to check the other choices, unless you're not sure about your answer, but just in case you're curious...

(C): The graph has a horizontal tangent at $x = a$; therefore $f'(a) = 0$. The slope of the graph is negative at $x = d$, so $f'(d) < 0$. Therefore, this list is not in ascending order.

(D): The slope of the curve is positive at $x = b$, so $f'(b) > 0$. The graph has a horizontal tangent at $x = c$, so $f'(c) = 0$. Therefore, this list is not in ascending order.

PRACTICE SET

1. The slope of the curve given by $y = \ln(x^2)$ at the point $(e^2, 4)$ is

    (A) $\dfrac{1}{e^4}$

    (B) $\dfrac{2}{e^4}$

    (C) $\dfrac{1}{e^2}$

    (D) $\dfrac{2}{e^2}$

2. Let f be the function given by $f(x) = \dfrac{1}{3}e^{3x^2} - 5x$. For what value of x does the curve have a slope of 0?

    (A) 0.068

    (B) 0.665

    (C) 0.939

    (D) 0.950

3. What is the slope of the curve $y = 4^{\cos x} - 1$ at its first positive x-intercept?

    (A) -1.386

    (B) -1.014

    (C) 0.046

    (D) 1.571

4. A curve is generated by the equation $x^2 + 4y^2 = 16$. How many points on this curve have horizontal tangent lines?

    (A) 0

    (B) 1

    (C) 2

    (D) 4

Derivatives

ANSWERS AND EXPLANATIONS

1. D

To find the slope of the curve at $x = e^2$, take the derivative of y and evaluate it at $x = e^2$. Don't forget to use the chain rule when taking the derivative:

$$y = \ln\left(x^2\right)$$

$$y' = \frac{1}{x^2} \cdot 2x = \frac{2}{x}$$

$$y'\left(e^2\right) = \frac{2}{e^2}$$

(D) is correct. An even faster way to find this derivative would be to use a rule of logs, $\ln (x)^2 = 2 \ln x$, to simplify the function before you take the derivative.

Note that, in this problem, you don't need the y-value of the point to find the slope of the curve.

2. B

Find the derivative of f (which represents slope), set it equal to 0, and solve for x. Don't forget to use the chain rule when you take the derivative of the first term in the equation.

$$f(x) = \frac{1}{3}e^{3x^2} - 5x$$

$$f'(x) = \frac{1}{3}e^{3x^2} \cdot 6x - 5$$

$$0 = 2xe^{3x^2} - 5$$

It is not possible to solve this equation by hand, so graph the equation in your calculator and use the Calculate Zero function. The result is $x = 0.665$, which is **(B)**.

(Using the Calculate Zero function on a graphing calculator is covered in detail in Chapter 19.)

3. A

This slope question has a slight twist but can be solved exactly as the previous two questions were. You just don't know the value of x at which you need to evaluate the derivative. The equation is not easily solvable, so use the Calculate Zero function on your calculator to find the first positive x-intercept of the graph, which is $x = 1.57080$. To find the slope at this value of x, find the value of the derivative at $x = 1.57080$ using the Numerical Derivative feature of your calculator. This gives y' at $x = 1.57080$ is

-1.386, which is **(A)**. It is a good idea to use many decimal places when rounding in the middle steps of a problem because your final answer will need to be correct to 3 decimal places. Use the Store function on your calculator to save all nine or more decimal places.

(Using the Calculate Zero function and the Derivative function on a graphing calculator is covered in Chapter 19.)

4. C

There are two ways to answer this question. The precalculus approach is to realize that the graph of the equation is an ellipse with center (0, 0) and vertices at the points (4, 0), (−4, 0), (0, 2), and (0, −2). Once drawn, you can see that there are exactly two points with horizontal tangent lines (the points (0, 2) and (0, −2)).

The calculus approach is to find critical points via the derivative, because the derivative represents the slope of the tangent line. Use implicit differentiation to find the derivative:

$$x^2 + 4y^2 = 16$$

$$2x + 8y\frac{dy}{dx} = 0$$

$$8y\frac{dy}{dx} = -2x$$

$$\frac{dy}{dx} = \frac{-2x}{8y} = -\frac{x}{4y}$$

A horizontal tangent line has a slope of 0, so set the derivative equal to 0 and solve for x:

$$-\frac{x}{4y} = 0$$

$$x = 0(-4y)$$

$$x = 0$$

Substituting $x = 0$ into the original equation yields $4y^2 = 16$, or $y = \pm 2$. Thus, horizontal tangent lines occur at two points, (0, 2) and (0, −2), making **(C)** the correct answer.

9.2 TANGENT LINES AND NORMAL LINES

To answer a question like this:

What is the y-intercept of the line that is tangent to the curve $f(x) = \sqrt{2x - 3}$ at the point on the curve where $x = 6$?

(A) 0   (B) $\dfrac{1}{3}$ (C) $\dfrac{2}{3}$ (D) 1

You need to know this:

Tangent lines:

- The slope of the tangent to a curve at a given point is the first derivative, $f'(x)$, of the function evaluated at that point.

- If $f'(a) = 0$ for some value of a, then the graph has a horizontal tangent line at $x = a$. For example, $f(x) = x^2 + 3$ has a horizontal tangent at $x = 0$ because $f'(x) = 2x$ and $f'(0) = 2(0) = 0$.

- If $f(b)$ is defined but $f'(b)$ does not exist for some value of b, then the graph either has a *vertical tangent* or a *sharp point* (cusp) at $x = b$. For example, the graph of $f(x) = \sqrt[3]{x}$ has a vertical tangent at $x = 0$ because $f'(x) = \dfrac{1}{3\sqrt[3]{x^2}}$, which is undefined at $x = 0$. (Sharp points will be discussed later.)

Normal lines:

- The normal line is defined as the line that is perpendicular to the tangent line at the point of tangency.

- Because the slopes of perpendicular lines (assuming neither is vertical) are negative reciprocals of each other, the slope of the normal line to the graph of $f(x)$ is $m_{norm} = -\dfrac{1}{f'(x)}$.

Point-slope form of a line:

- $y - y_1 = m(x - x_1)$, where (x_1, y_1) is any point on the line.

You need to do this:

- Find the derivative of the function.
- Plug the given value of x into f' to find the slope of the tangent line.
- Plug the given value of x into the original function, f, to find the corresponding y-value of the point of interest (if it is not already given).
- Plug the point of interest and slope into the point-slope form of a line. Look at the answer choices and simplify only if necessary.

Answer and Explanation:

D

First, find the derivative of $f(x) = \sqrt{2x - 3}$. Start by writing the radical using a fractional exponent, then use the chain rule. The result is:

$$f(x) = (2x - 3)^{\frac{1}{2}}$$
$$f'(x) = \frac{1}{2} \cdot (2x - 3)^{-\frac{1}{2}} \cdot 2$$
$$= \frac{1}{\sqrt{2x - 3}}$$

The slope of the tangent line when $x = 6$ is:

$$f'(6) = \frac{1}{\sqrt{2(6) - 3}} = \frac{1}{\sqrt{9}} = \frac{1}{3}$$

Now use the point-slope form of a line, $y - y_1 = m(x - x_1)$, to write an equation for the tangent line. When $x = 6$, $f(x) = f(6) = \sqrt{9} = 3$. So, $m = \frac{1}{3}$ is the slope and $(6, 3)$ is the point. The equation of the tangent line is $y - 3 = \frac{1}{3}(x - 6)$. To determine the y-intercept, plug in $x = 0$, which yields $y - 3 = \frac{1}{3}(0 - 6) = -2$ or $y = 1$. Therefore, the y-intercept is 1, which is **(D)**.

PRACTICE SET

1. At what point x on the curve $f(x) = \ln(2x - 1)$ does the tangent line have a slope of $\frac{2}{5}$?

   (A) 1

   (B) e

   (C) 3

   (D) 5

2. An equation of the line tangent to the graph of $y = 2\sin\left(2x + \frac{3\pi}{4}\right)$ at $x = \frac{\pi}{8}$ is

   (A) $y = -4x + \frac{\pi}{2}$

   (B) $y = 4x + \frac{\pi}{2}$

   (C) $y = 4x - \frac{\pi}{2}$

   (D) $y = -2x + \frac{\pi}{2}$

3. Let $f(x) = 3x^2 + 1$ and $g(x) = x^3 - 9x$. For how many values of a is the line tangent to f at $x = a$ parallel to the line tangent to g at $x = a$?

   (A) 0

   (B) 1

   (C) 2

   (D) 3

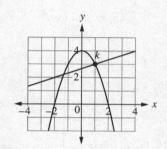

4. Consider the graph of $f(x) = 4 - x^2$ and a normal line to the graph at any point $x = k$, as shown above. In terms of k, what is the y-intercept of the normal line at the point on the curve where $x = k$?

   (A) $3.5 - k^2$

   (B) $1.5 + k^2$

   (C) $2 + 0.5k^2$

   (D) $1.5 + k$

ANSWERS AND EXPLANATIONS

1. C

The slope of the tangent to a curve at a given point is $f'(x)$ evaluated at that point, so start by finding the derivative of $f(x) = \ln(2x - 1)$. By the chain rule:

$$f'(x) = \frac{1}{2x-1} \cdot 2 = \frac{2}{2x-1}$$

Set this derivative equal to the given slope, $\frac{2}{5}$, and solve for x using cross-multiplication:

$$\frac{2}{2x-1} = \frac{2}{5}$$

$$\cancel{2}(5) = \cancel{2}(2x-1)$$

$$5 = 2x - 1$$

$$6 = 2x \rightarrow x = 3$$

Thus, **(C)** is correct.

2. A

To get the equation of the tangent line, you need to know the slope of the tangent line and a point on the tangent line. To calculate the slope of the tangent at $x = \frac{\pi}{8}$, find $y'\left(\frac{\pi}{8}\right)$.

Using the chain rule yields:

$$y'(x) = 2\cos\left(2x + \frac{3\pi}{4}\right) \cdot 2$$

$$= 4\cos\left(2x + \frac{3\pi}{4}\right)$$

$$y'\left(\frac{\pi}{8}\right) = 4\cos\left(\frac{\pi}{4} + \frac{3\pi}{4}\right) = -4$$

Now plug in $\frac{\pi}{8}$ for x in the original function to find the point:

$$y = 2\sin\left(2x + \frac{3\pi}{4}\right)$$

$$y\left(\frac{\pi}{8}\right) = 2\sin\left(2\left(\frac{\pi}{8}\right) + \frac{3\pi}{4}\right)$$

$$= 2\sin\pi = 0$$

$$\text{point} = \left(\frac{\pi}{8}, 0\right)$$

Using the point-slope form of a line yields the equation $y - 0 = -4\left(x - \frac{\pi}{8}\right) = -4x + \frac{\pi}{2}$, which is **(A)**.

3. C

Lines are parallel if their slopes are equal. The slope of the tangent line at a point is the derivative of the function evaluated at that point. Combining these concepts tells you that the solution to this question is the number of solutions to the equation $f'(a) = g'(a)$, so compute these derivatives: $f'(x) = 6x$ and $g'(x) = 3x^2 - 9$. Graph the equations in your calculator to see that the graphs intersect twice, which means $f'(a) = g'(a)$ has two solutions, making **(C)** the correct answer.

Note that you could also set the two derivatives equal to each other and solve the resulting quadratic equation by factoring. The solutions are $x = -1$ and $x = 3$, confirming that there are indeed two solutions.

4. A

Because the question asks for the y-intercept *in terms of* k, don't use the point at k shown on the graph (that is, don't use (1, 3) in your calculations). Instead, use k for x wherever you need to substitute an x-value.

To start, find the derivative of f, which is $f'(x) = -2x$. Next, determine the slope of the normal line to f at the point where $x = k$. The slope of the tangent at $x = k$ is $f'(k) = -2k$. It follows that the slope of the normal line is $m_{\text{norm}} = \frac{1}{2k}$ because the normal line is perpendicular to the tangent line. The point of normalcy (which is the same as the point of tangency) is $(k, f(k))$. Now, use the point-slope form of a line to get the equation of the normal line, $y - f(k) = \frac{1}{2k}(x - k)$. Because $f(k) = 4 - k^2$, the normal line can be written as:

$$y - \left(4 - k^2\right) = \frac{1}{2k}(x - k)$$

Finally, the y-intercept occurs when $x = 0$, which means:

$$y - \left(4 - k^2\right) = \frac{1}{2k}(0 - k)$$

$$y - 4 + k^2 = -\frac{1}{2}$$

$$y = 3.5 - k^2$$

That's **(A)**.

9.3 LOCAL LINEAR APPROXIMATIONS

To answer a question like this:

Suppose a linear approximation for $f(x) = \sqrt[3]{x}$ at $x = 8$ is used to approximate $\sqrt[3]{8.3}$. Which of the following is the resulting approximation?

(A) 2.013   (B) 2.019   (C) 2.025   (D) 2.031

You need to know this:

Using the tangent line to approximate a function:

- The line tangent to a curve at a point, if it exists, is the line that best approximates the curve near that point. This means that if a function is differentiable at a point, then locally it looks like its tangent line. That is to say, if you zoom in far enough with your graphing calculator, the function looks like a straight line—and the line that it looks like is its tangent line. The graph below demonstrates this for the function $y = \sqrt{x}$ at $x = 16$.

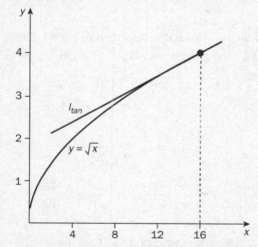

- Thus, at points where a function is differentiable, the tangent line can be used to approximate the function.

- A function does not look like a line close to a point at which it is not differentiable—for instance, if the function has a sharp corner, as the graph of $y = |x|$ does at $x = 0$.

You need to do this:

- Take the derivative of the function, which represents the slope of the tangent line, and evaluate it at the given *x*-value.

- Plug the given *x*-value into the original function, *f*, to find the *y*-value of the point of tangency.

- Use the slope and the point to write the equation of the tangent line.

- Plug in the *x*-value for which you're approximating the function.

AP Expert Note

Note that when finding a linear approximation, it is best NOT to simplify the linear equation, because you will be asked to approximate the value of *f* at an *x*-value that is close to the *x*-value of the point given. By not simplifying, the calculations are easier, which is important because these questions may appear in the no-calculator section of the test.

Answer and Explanation:

C

According to the question stem, you need to approximate $\sqrt[3]{8.3}$ using the function $f(x) = \sqrt[3]{x}$ at $x = 8$ (the closest value to 8.3 that is a perfect cube). Start by computing the derivative, $f'(8)$, or the slope of the tangent line:

$$y = x^{\frac{1}{3}}$$

$$y' = \frac{1}{3}x^{-\frac{2}{3}} = \frac{1}{3\left(\sqrt[3]{x}\right)^2}$$

$$y'(8) = \frac{1}{3\left(\sqrt[3]{8}\right)^2} = \frac{1}{3(2)^2} = \frac{1}{12}$$

Next, find the *y*-value of the point of tangency: $f(8) = 2$, so the line tangent to the curve at $x = 8$ is:

$$y - 2 = \frac{1}{12}(x - 8)$$

$$y = 2 + \frac{1}{12}(x - 8)$$

Using this equation, the linear approximation of $\sqrt[3]{8.3}$ is:

$$\sqrt[3]{8.3} \approx 2 + \frac{1}{12}(8.3 - 8) = 2 + \frac{0.3}{12} = 2.025$$

Hence, **(C)** is correct.

PRACTICE SET

1. Suppose a linear approximation for $f(x) = \sqrt{x}$ at $x = 16$ is used to approximate $\sqrt{17}$. Which of the following is the resulting approximation?

   (A) 4.075

   (B) 4.125

   (C) 4.375

   (D) 4.875

2. Let $f(x)$ be a differentiable function with $f(-1) = 5$ and $f'(-1) = 2$. Use the given information to find the local linear approximation of $f(-0.9)$.

   (A) 4.9

   (B) 5.1

   (C) 5.2

   (D) 5.3

3. A differentiable function f is defined by $f(x) = x^2 \cdot \sin(\pi x) - 4x$. Which of the following is a linear approximation to f at $x = 1$?

   (A) $y = -x - 1$

   (B) $y = -\pi x + \pi$

   (C) $y = (-\pi - 4)x + \pi$

   (D) $y = (2x \sin \pi x + x^2 \pi \cos \pi x - 4)x - 1$

4. The local linear approximation to a function f at $x = -3$ is $y = 2x + 7$. What is the value of $f(-3) + f'(-3)$?

   (A) -1

   (B) 1

   (C) 2

   (D) 3

ANSWERS AND EXPLANATIONS

1. B

According to the question stem, approximate $\sqrt{17}$ using the function $f(x) = \sqrt{x}$, and $x = 16$ (the closest value to 17 that is a perfect square). Start by computing the derivative, $f'(16)$, or the slope of the tangent line:

$$y = x^{\frac{1}{2}}$$

$$y' = \frac{1}{2}x^{-\frac{1}{2}} = \frac{1}{2\sqrt{x}}$$

$$y'(16) = \frac{1}{2\sqrt{16}} = \frac{1}{8}$$

Next, find the y-value of the point of tangency: $f(16) = 4$, so the line tangent to the curve at $x = 16$ is:

$$y - 4 = \frac{1}{8}(x - 16)$$

$$y = 4 + \frac{1}{8}(x - 16)$$

Therefore, a good linear approximation is:

$$\sqrt{17} \approx 4 + \frac{1}{8}(17 - 16) = 4.125$$

Thus, **(B)** is correct.

2. C

To use a local linear approximation, find the equation of the tangent line. You've been given all the information you need in the question stem; just piece it all together. The point on the function is given by $f(-1) = 5$, which translates to the point $(-1, 5)$. The slope of the tangent line at $x = -1$ is given by $f'(-1) = 2$, so the slope is 2. Now the point-slope form of the tangent line is:

$$y - 5 = 2(x - (-1))$$

$$y = 5 + 2(x + 1)$$

Substituting $x = -0.9$ into the equation of the tangent line yields:

$$5 + 2(-0.9 + 1) = 5 + 2(0.1) = 5.2$$

That's **(C)**.

3. C

You're looking for a *linear* approximation, so you can eliminate D right way as it contains an x^2 term. The tangent line to the function at the given point provides a linear approximation to the function at that point, so start by finding the first derivative and evaluating it at $x = 1$; this gives you the slope of the tangent line. Use the product rule, $(f \cdot g)' = f'g + fg'$, to find the derivative of the first term:

$$f(x) = x^2 \cdot \sin \pi x - 4x$$

$$f'(x) = 2x \cdot \sin \pi x + x^2 \cdot \pi \cos \pi x - 4$$

$$f'(1) = 2(1) \sin \pi + (1)^2 \pi \cos \pi - 4$$

$$f'(1) = 2(0) + \pi(-1) - 4 = \boxed{-\pi - 4}$$

Now, don't do more work than is needed to obtain the correct solution—because the answer choices all have different slopes (the coefficient of x), it suffices to know the slope of the tangent line at $x = 1$. In other words, finding $f'(1) = -\pi - 4$ allows you to find the correct answer, which is **(C)**.

4. D

Think creatively to answer this question. The local linear approximation of f at $x = -3$ is found using the equation of the line tangent to f at $x = -3$. This tells you that the tangent line passes through $(-3, f(-3))$ and the slope of the tangent at $x = -3$ is 2 (the coefficient of x in the linear approximation $y = 2x + 7$). The slope of the tangent is also given by $f'(-3)$, so $f'(-3) = 2$. Because *both* the tangent line and the curve pass through the same point on the curve, you can find $f(-3)$ by plugging $x = -3$ into the tangent line: $c = 2(-3) + 7 = 1$. Thus $f(-3) = 1$. Finally, $f(-3) + f'(-3) = 1 + 2 = 3$, which is **(D)**.

9.4 DERIVATIVES AS RATES OF CHANGE

To answer a question like this:

For the function $f(x) = 3x^2 - 7$, if the average rate of change over the closed interval $[0, 2]$ is used to approximate the instantaneous rate of change at $z = 1$, by how much does the average rate of change exceed the instantaneous rate of change?

(A) 0   (B) 0.5   (C) 1   (D) 1.5

You need to know this:

Instantaneous and average rates of change:

- Lots of real-life applications of the derivative involve modeling the rates of change of functions.

- The average rate of change over a closed interval $[a, b]$ is the slope of the secant line between the endpoints of the interval: $m = \dfrac{f(b) - f(a)}{b - a}$.

- The instantaneous rate of change of a function at a point is the slope of the tangent line at that point (or the derivative evaluated at that point).

Increasing/decreasing:

- For a differentiable function f, where f' is the rate of change, the function is increasing where $f' > 0$ and decreasing where $f' < 0$.
- The rate of change, f', is increasing where its derivative $f'' > 0$ and decreasing where $f'' < 0$.

You need to do this:

- To find the average rate of change, plug the endpoints of the interval into the function and then use the slope formula.

- To find the instantaneous rate of change, take the derivative of the function and plug in the given value of x.

- To compare the two rates, subtract the results.

Answer and Explanation:

A

The average rate of change over a closed interval $[a, b]$ is the slope of the secant line between the endpoints of the interval: $m = \dfrac{f(b) - f(a)}{b - a}$. Find $f(0)$ and $f(2)$ by plugging $x = 0$ and $x = 2$ into the original function:

$$f(x) = 3x^2 - 7$$
$$f(0) = 3(0)^2 - 7 = -7$$
$$f(2) = 3(2)^2 - 7 = 5$$

Now find the average rate of change using the points $(0, -7)$ and $(2, 5)$:

$$m = \frac{f(b) - f(a)}{b - a} = \frac{5 - (-7)}{2 - 0} = \frac{12}{2} = 6$$

To find the instantaneous rate of change at $x = 1$, take the derivative of f and evaluate it at $x = 1$:

$$f'(x) = 6x$$
$$f'(1) = 6(1) = 6$$

At $x = 1$, the difference between the average rate of change on the interval $[0, 2]$ and the instantaneous rate of change is $6 - 6 = 0$, so **(A)** is correct.

PRACTICE SET

1. The cost, in dollars, to store old medical x-ray films for a hospital system is modeled by a differentiable function C, where C depends on the weight of the films in pounds. Of the following, which is the best interpretation of $C'(200) = 15$?

   (A) The cost to store 200 pounds of films is $15.

   (B) The average cost to store the films is $\dfrac{15}{200}$ dollars per pound.

   (C) The cost to store the films is increasing at a rate of $15 per pound when the weight of the films is 200 pounds.

   (D) Increasing the weight of the films by 200 pounds will increase the cost to store the films by approximately $15.

2. A water tank holds 6,000 gallons. When the drain is opened, the tank empties in 100 minutes. The volume of the water remaining in the tank t minutes after the drain is opened is modeled by $V(t) = 6,000(1 - 0.01t)^2$. What is the rate of change of the volume of the water in gallons per minute when $t = 60$?

   (A) -48

   (B) -24

   (C) 24

   (D) 48

3. After an injection, the concentration in mg/L (milligrams per liter) of a drug in the bloodstream is modeled by the function $C(t) = K(e^{-at} - e^{-bt})$ where a, b, and K are positive constants with $b > a$ and t is time, measured in hours, after the injection is given. What is the rate of change of the drug's concentration in mg/L per hour exactly one hour after an injection?

   (A) $K(e^{-a} - e^{-b})$

   (B) $K(e^{-b} - e^{-a})$

   (C) $K(be^{-b} - ae^{-a})$

   (D) $K(ae^{-a} - be^{-b})$

4. The rate of change of the temperature in March is modeled by the function $T'(t) = t \cdot \cos\left(\dfrac{t}{7}\right)$. Which of the following best describes the temperature at $t = 7$?

   (A) The temperature is increasing, but the rate of change in the temperature is decreasing.

   (B) The temperature is increasing, and the rate of change in the temperature is also increasing.

   (C) The temperature is decreasing, but the rate of change in the temperature is increasing.

   (D) The temperature is decreasing, and the rate of change in the temperature is also decreasing.

Derivatives

ANSWERS AND EXPLANATIONS

1. C

The derivative of a function at a specific value represents the instantaneous rate of change of the function at that value. Thus, the cost is changing at a rate of $15. Here, cost depends on the weight of the films, so weight is the independent variable. Hence, the 200 in the expression $C'(200) = 15$ represents the specific weight of the films when the cost is changing at a rate of $15. This matches **(C)**.

2. A

The rate of change is the derivative, so
$V'(t) = 12000(1 - 0.01t)^1 \cdot (-0.01) = -120(1 - 0.01t)$.
To find the rate of change at $t = 60$, evaluate
$V'(60) = -120(1 - 0.01 \cdot 60) = -120(1 - 0.6) = -48$.

This matches **(A)**.

3. C

The rate of change is the derivative of $C(t) = K(e^{-at} - e^{-bt})$, so it follows that $C'(t) = K(-ae^{-at} + be^{-bt})$ by the chain rule. Now we need to evaluate at $t = 1$, so
$C'(1) = K(-ae^{-a} + be^{-b}) = K(be^{-b} - ae^{-a})$.
Therefore, **(C)** is correct.

4. A

The temperature, T, is increasing when $T'(t) > 0$. Use your graphing calculator (set to radians mode) to compute $T'(7)$:

$$T'(7) = 7 \cdot \cos\left(\frac{7}{7}\right) = 3.782 > 0.$$

Therefore, the temperature is increasing at $t = 7$.

The rate of change of temperature, $T'(t) > 0$, is increasing when $T''(t) > 0$. Use the Derivative function on your calculator to compute $T''(7)$, which is the derivative of T'; the result is $T''(7) = -0.301 < 0 \Rightarrow T'(t)$ is decreasing, i.e., the rate of change of the temperature is decreasing. **(A)** is correct. You could also graph T' on your calculator and notice that the graph is above the x-axis and has a negative slope.

(Using the Derivative function on your calculator is covered in Chapter 19.)

9.5 FIRST DERIVATIVES AND CURVE SKETCHING

To answer a question like this:

Given the continuous function $y = x^2(x - 3)$, which of the following statements is true?

(A) The function has a local maximum at $x = 0$ and a local minimum at $x = 2$.

(B) The function has a local minimum at $x = 0$ and a local maximum at $x = 2$.

(C) The function has a local maximum at $x = 0$ but no local minimum.

(D) The function has a local minimum at $x = 2$ but no local maximum.

You need to know this:

Where a function is increasing, decreasing, or constant:

- f is increasing on intervals where $f'(x) > 0$.

- f is decreasing on intervals where $f'(x) < 0$.

- f is constant on intervals where $f'(x) = 0$.

Relative extrema and the First Derivative Test:

- Relative extrema occur only where the derivative is zero or where it does not exist (DNE). These x-values are called *critical numbers* and the points are called *critical points*.

- A function f has a relative (local) maximum at x_0 if $f(x_0)$ is defined and there is an open interval on which f changes from increasing on the left of x_0 to decreasing on the right of x_0 (i.e., the derivative changes from positive to negative).

- A function f has a relative (local) minimum at x_0 if $f(x_0)$ is defined and there is an open interval on which f changes from decreasing on the left of x_0 to increasing on the right of x_0 (i.e., the derivative changes from negative to positive).

- The function f does *not* have a relative extremum if the sign of $f'(x)$ does not change on some open interval.

The graphs below show several possibilities for the behavior of a graph at a relative extremum.

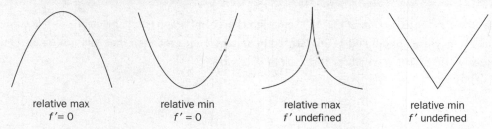

relative max
$f' = 0$

relative min
$f' = 0$

relative max
f' undefined

relative min
f' undefined

Derivatives

You need to do this:

- Multiply the function (because it's easier than applying the product rule).
- Take the derivative of y, set it equal to zero, and solve to find the critical numbers.
- Do a sign analysis around the critical points (using a sign chart or a test line).
- Identify the maximums and minimums based on whether the derivative changes signs.

Answer and Explanation:

A

Multiply the function: $y = x^2(x - 3) = x^3 - 3x^2$

Take the derivative of y and set it equal to zero: $y' = 3x^2 - 6x = 0$

Find the solutions by factoring: $y' = 3x(x - 2) = 0$

Critical Points: $x = 0$ or $x = 2$

Interval	$(-\infty, 0)$	$(0, 2)$	$(2, \infty)$
Test Point	$x = -1$	$x = 1$	$x = 4$
Sign of f'	$f'(-1) = (-3)(-3) = 9$ positive	$f'(1) = (3)(-1) = -3$ negative	$f'(4) = (12)(2) = 24$ positive
Conclusion	f increasing	f decreasing	f increasing

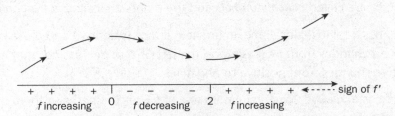

A relative maximum occurs at $x = 0$, because the first derivative changes signs from positive to negative. A relative minimum occurs at $x = 2$, because the first derivative changes signs from negative to positive. This means **(A)** is correct.

✔ AP Expert Note

In the Free-Response section of the AP exam, sign charts and graphs must be clearly and accurately labeled. If they are not labeled or are labeled in an ambiguous manner, they will not be evaluated and you will not receive credit for that part of your work.

PRACTICE SET

1. A function is defined by $f(x) = 2x^3 - 150x$. Over which of the following intervals is the function decreasing?

   (A) $[-5, 5]$ only

   (B) $[5, \infty)$ only

   (C) $(-\infty, 0]$ and $[0, \infty)$

   (D) $(-\infty, -5]$ and $[5, \infty)$

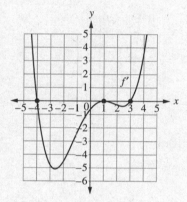

2. The graph of $f'(x)$ is shown above. Which of the following statements is true?

   (A) $f(-4)$ is a local minimum value.

   (B) $f(-4)$ is a local maximum value.

   (C) $f(1)$ is a local minimum value.

   (D) $f(1)$ is a local maximum value.

3. The derivative of a function f is given by $f'(x) = 2^{-x} - 3\cos(x^2)$. On the interval $-3 < x < 3$, at which values of x does f have a local maximum?

   (A) $x \approx -1.737$ and $x \approx 1.794$

   (B) $x \approx -2.454$ and $x \approx 2.508$

   (C) $x \approx 0.441$ and $x \approx 2.508$

   (D) $x \approx -0.937$ and $x \approx 2.188$

4. Suppose that $f'(x) > 0$ on $(-\infty, -1)$, $f'(x) < 0$ on $(-1, 1)$, and $f'(x) > 0$ on $(1, \infty)$. Which of the following functions could be f?

   (A) $f(x) = x^2 - x$

   (B) $f(x) = x^3 - 3x$

   (C) $f(x) = x^3 + 3x$

   (D) $f(x) = x^4 + 4x$

Derivatives

ANSWERS AND EXPLANATIONS

1. A

A function is decreasing where its first derivative is negative, so start by finding $f'(x)$, which is $6x^2 - 150$. To determine where the derivative is negative (or < 0), find the critical numbers by setting $f'(x)$ equal to 0, solving for x, and then testing easy values around the solutions by plugging them into f'. (You don't need to find exact values—you just need to determine the sign of the result.)

$$6x^2 - 150 = 0$$
$$6x^2 = 150$$
$$x^2 = 25$$
$$x = \pm 5$$

Now test values around $+5$ and -5. Use either a sign chart or a test line (graph), whichever you prefer:

$$
\begin{array}{ccc}
600 - 150 & 0 - 150 & 600 - 150 \\
+ & - & +
\end{array}
$$

f' -10 -5 0 5 10

The derivative is negative over the closed interval $[-5, 5]$, so **(A)** is correct.

2. B

This question has a twist because the graph shown is of f', not f, so pay careful attention to what the graph is telling you. Also keep the First Derivative Test in mind: f has a local maximum where the derivative changes from positive to negative (here, from above the x-axis to below the x-axis) and a local minimum where the derivative changes from negative to positive (here, from below the x-axis to above the x-axis). To the left of $x = -4$, the derivative is positive, and to the right of $x = -4$, the derivative is negative. This means $f(-4)$ is a local maximum value, so the answer is **(B)**. All of the other statements are false. There is no maximum or minimum at $x = 1$ because the derivative is negative on both sides of $x = 1$.

3. D

Graph the given equation (which is f', not f) in your calculator to determine where the derivative changes from positive to negative (because you're looking for a local maximum). Looking at the graph of f' below, the only places where f' crosses the x-axis and changes from positive (above the axis) to negative (below the axis) are at $x \approx -0.937$ and $x \approx 2.188$.

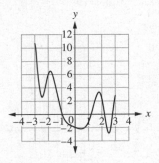

Thus, **(D)** is correct.

4. B

For this question, it will help to see the graph of f. The information given implies that f is increasing (going up) on the interval $(-\infty, -1)$, decreasing (going down) on the interval $(-1, 1)$, and increasing again on the interval $(1, \infty)$. This means the graph has a local maximum at $x = -1$ and a local minimum at $x = 1$. You are given various answer choices that represent choices for f. Use your graphing calculator to graph these functions and check to see which one satisfies the given conditions. **(B)** fits the information perfectly.

9.6 SECOND DERIVATIVES AND CURVE SKETCHING

To answer a question like this:

For $0 < x < 2\pi$, the graph of $f(x) = x + 2 \sin x$ has inflection points at

(A) $x = \dfrac{2\pi}{3}$ only

(B) $x = \dfrac{2\pi}{3}$ and $x = \dfrac{4\pi}{3}$

(C) $x = \pi$ only

(D) $x = \dfrac{\pi}{2}$ and $x = \dfrac{3\pi}{2}$

You need to know this:

Concavity and inflection points:

- Inflection points occur only when the second derivative equals zero or does not exist (DNE).

- The graph of f is concave up where the second derivative is positive ($f'' > 0$). In other words, the graph of f is concave up where the graph of f' is increasing or where the graph of f'' is above the x-axis.

- The graph of f is concave down where the second derivative is negative ($f'' < 0$). In other words, the graph of f is concave down where the graph of f' is decreasing or where the graph of f'' is below the x-axis.

- The graph of f has an inflection point where the concavity changes from positive to negative or vice versa.

- An increasing (or decreasing) function can be either concave up or concave down. Four possibilities are shown below.

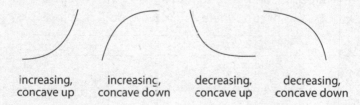

increasing, concave up   increasing, concave down   decreasing, concave up   decreasing, concave down

The Second Derivative Test for Extrema:

- If $f'(c) = 0$ (slope $= 0$) and $f''(c) > 0$ (concave up), then f has a relative minimum at $x = c$.

- If $f'(c) = 0$ (slope $= 0$) and $f''(c) < 0$ (concave down), then f has a relative maximum at $x = c$.

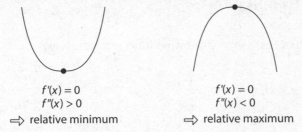

$$f'(x) = 0$$
$$f''(x) > 0$$
$$\Rightarrow \text{relative minimum}$$

$$f'(x) = 0$$
$$f''(x) < 0$$
$$\Rightarrow \text{relative maximum}$$

You need to do this:

- Find f', then f''.

- Determine where f'' is zero or does not exist.

- Do a sign analysis for values of x around the results of the previous step.

- Determine where (if anywhere) the graph of f changes concavity (f'' changes from positive to negative or vice versa).

Answer and Explanation:

C

To find the inflection points of f for $0 < x < 2\pi$, first identify the points in this interval where f'' is zero or does not exist.

$$f(x) = x + 2\sin x$$
$$f'(x) = 1 + 2\cos x$$
$$f''(x) = -2\sin x$$

This function is defined everywhere, so inflection points can only occur at points where the second derivative is zero. Solve the equation $f'' = 0$ to find these points: $f''(x) = -2\sin x = 0 \Rightarrow \sin x = 0$. On the interval $0 < x < 2\pi$, this occurs where $x = \pi$. Now, do a sign analysis:

Interval	$0 < x < \pi$	$\pi < x < 2\pi$
Test Point	$x = \dfrac{\pi}{2}$	$x = \dfrac{3\pi}{2}$
Sign of f''	$f''\left(\dfrac{\pi}{2}\right) = -2\sin\dfrac{\pi}{2}$ $= -2(1) = -2$ negative	$f''\left(\dfrac{3\pi}{2}\right) = -2\sin\dfrac{3\pi}{2}$ $= -2(-1) = 2$ positive
Concavity of a	down	up

Because the sign of f'' changes at $x = \pi$, f has an inflection point at $x = \pi$, making **(C)** the correct answer.

Note that you can also use a test line for f'' if you prefer.

PRACTICE SET

1. The graph of $y = x^3 + 21x^2 - x + 1$ is concave down for

   (A) $x < -7$

   (B) $-7 < x < 7$

   (C) $x > 7$

   (D) $x < -7$ and $x > 7$

 2. Where on the interval $0 \leq x \leq 2\pi$ does the graph of

   $$f(x) = \cos x - \frac{1}{4}x^2 + 3x$$

   change from concave up to concave down?

   (A) $x = \frac{1}{3}\pi$ only

   (B) $x = \frac{4}{3}\pi$ only

   (C) $x = \frac{2}{3}\pi$ and $\frac{4}{3}\pi$

   (D) $x = \frac{2}{3}\pi$ only

3. The graph of $g(x) = x^2 - \frac{8}{x}, x > 0$, has a point of inflection at $x =$

   (A) 1

   (B) 2

   (C) 8

   (D) There are no inflection points.

4. Let $f(x) = x^3 - 3x^2 + 3x + 7$. Which of the following statements is true?

   (A) f has a relative extremum at $x = 1$ and no inflection points.

   (B) f is increasing everywhere and does not change concavity.

   (C) f has no relative extrema but has an inflection point at $x = 1$.

   (D) f has a relative maximum and an inflection point at $x = 1$.

Derivatives

ANSWERS AND EXPLANATIONS

1. A

The function f is concave down for all x such that $f''(x) < 0$ (the second derivative is negative). Use the power rule to find the first derivative, $y' = 3x^2 + 42x - 1$, and the second derivative, $y'' = 6x + 42$. So f is concave down when $6x + 42 < 0$. Solving this inequality yields $x < -7$, which is **(A)**.

2. B

Concavity is determined by the second derivative, so start by finding f', then f''.

$$f(x) = \cos x - \frac{1}{4}x^2 + 3x$$

$$f'(x) = -\sin x - \frac{1}{2}x + 3$$

$$f''(x) = -\cos x - \frac{1}{2}$$

Now, graph f'' in your calculator. Set the window to the given range, $0 \leq x \leq 2\pi$. Because the answer choices are all multiples of $\frac{\pi}{3}$, set the x scale (Xscl) to $\frac{\pi}{3}$ to find that $f''(x) = 0$ at $x = \frac{2}{3}\pi$ and $x = \frac{4}{3}\pi$. Be careful here—you are looking for where the graph changes from concave up to concave down, and f'' changes from positive to negative only at $x = \frac{4}{3}\pi$, making **(B)** the correct answer. You could also find the values without a calculator and find the sign of f'' on each side of these values.

3. B

A point of inflection occurs when g'' changes sign, so start by finding the first and second derivatives. To do this, rewrite the rational term using a negative exponent and then use the power rule:

$$g(x) = x^2 - 8x^{-1}$$

$$g'(x) = 2x + 8x^{-2}$$

$$g''(x) = 2 - 16x^{-3} = 2 - \frac{16}{x^3}$$

$$= \frac{2x^3 - 16}{x^3}$$

Set $g''(x) = 0$ and solve the resulting equation:

$$\frac{2x^3 - 16}{x^3} = 0 \Rightarrow x^3 = 8 \Rightarrow x = 2$$

Use the point $x = 2$ to set up a sign analysis for the second derivative.

Interval	$0 < x < 2$	$x > 2$
Test Point	$x = 1$	$x = 3$
Sign of g''	$g''(1) = 2 - \frac{16}{1} = -14$ negative	$g''(3) = 2 - \frac{16}{27} \approx 1\frac{1}{2}$ positive
Concavity of g	down	up

Because g'' changes from negative to positive at $x = 2$, i.e., because g changes concavity, g has an inflection point at $x = 2$. That's **(B)**. If you were asked to find concavity and you weren't told that $x > 0$, then you would have needed to check the sign of f'' on the left of $x = 0$ (because that is where f'' does not exist).

4. C

Each of the answer choices involves relative extrema and/or concavity/inflection points, so you'll need to examine both the first and second derivatives.

The first derivative is $f'(x) = 3x^2 - 6x + 3$. Solving the equation $0 = 3x^2 - 6x + 3 = 3(x-1)^2$ yields $x = 1$ This means the point $x = 1$ is a critical point. You could use either the First or Second Derivative Test to determine whether the function has a relative extremum at $x = 1$, or if you look carefully at the derivative, $f'(x) = 3(x-1)^2$, you should notice that its graph is a parabola with vertex $(1, 0)$ that opens upward, so $f'(x) \geq 0$ everywhere. Therefore the function is always increasing, so it has no relative extrema and you can eliminate (A) and (D). Next, you need to determine if and where the function changes concavity. The second derivative is $f''(x) = 6x - 6$ and solving the equation $0 = 6x - 6$ yields $x = 1$. Thus, $x = 1$ is a candidate for an inflection point, i.e., the function might change concavity at $x = 1$. A sign analysis determines that $f''(x) > 0$ for $x > 1$, so the function is concave up on this interval; $f''(x) < 0$ for $x < 1$, so the function is concave down on this interval. The function changes concavity, so $x = 1$ is an inflection point, making **(C)** the correct answer.

RAPID REVIEW

If you take away only 10 things from this chapter:

1. The slope of a curve $y = f(x)$ at a point P means the slope of the tangent at the point P. The slope of the tangent to a curve at a given point is the first derivative, $f'(x)$, of the function evaluated at that point.

2. If $f'(a) = 0$ for some value of a, then the graph has a horizontal tangent line at $x = a$. If $f'(b)$ is undefined for some value of b, then the graph has a vertical tangent at $x = b$.

3. The slope of the normal line to the graph of $f(x)$ is $m_{norm} = -\dfrac{1}{f'(x)}$. The normal line at a point is perpendicular to the tangent line at that point.

4. The line tangent to a curve at a point, if it exists, is the line that best approximates the curve near that point. This is known as a local linear approximation.

5. The average rate of change over a closed interval $[a, b]$ is the slope of the secant line between the endpoints of the interval: $m = \dfrac{f(b) - f(a)}{b - a}$.

6. The instantaneous rate of change of a function at a point is the slope of the tangent line at that point (or the derivative evaluated at that point).

7. A function is increasing where $f'(x) > 0$ and decreasing where $f'(x) < 0$.

8. A function has a relative maximum where $f'(x)$ changes from increasing to decreasing and a relative minimum where $f'(x)$ changes from decreasing to increasing.

9. The graph of f is concave up where the second derivative is positive ($f'' > 0$) and concave down where the second derivative is negative ($f'' < 0$).

10. The graph of f has an inflection point where the concavity changes from positive to negative or vice versa.

Derivatives

TEST WHAT YOU LEARNED

1. Find the instantaneous rate of change for the function given by $f(x) = x^4 - 3x^3 + 7x - 4$ at $x = 2$.

   (A) 2   (B) 3   (C) 12   (D) 16

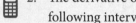 2. The derivative of a function is given by $f'(x) = \ln x - e^{\sin x}$ for $0 < x < 10$. Over which of the following interval(s) is f increasing?

   (A) $0 < x < 3.036$ only

   (B) $3.036 < x < 7.013$ only

   (C) $0 < x < 3.036$ and $7.013 < x < 8.555$

   (D) $3.036 < x < 7.013$ and $8.555 < x < 10$

---

3. If f is given by $f(x) = 4\sqrt{x} + 3x^2 - 2x$, what is the slope of the line tangent to the curve at $x = 4$?

   (A) 5.875   (B) 23   (C) 24.125   (D) 48

Derivatives

Derivatives

4. Suppose for some differentiable function f that $f(2) = 4$ and $f'(x) = 3x^2 - 2$. Using a local linear approximation, the value of $f(2.1)$ is best approximated as

    (A) 5 (B) 10 (C) 11.23 (D) 16

5. The function $f(x)$ is given by $f(x) = \dfrac{2x+4}{x^2+2}$. What is the equation of the line tangent to the graph of $f(x)$ at the point $\left(2, \dfrac{4}{3}\right)$?

    (A) $y = -\dfrac{5}{9}(x-2) - \dfrac{4}{3}$

    (B) $y = -\dfrac{5}{9}(x-2) + \dfrac{4}{3}$

    (C) $y = \dfrac{5}{9}(x-2) - \dfrac{4}{3}$

    (D) $y = \dfrac{5}{9}(x-2) + \dfrac{4}{3}$

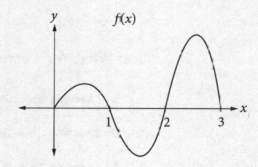

6. Consider the graph of $f(x)$ shown above. Which of the following statements is true about f?

(A) $f''(x) > 0$ for $1 < x < 2$

(B) $f''(x) < 0$ for $1 < x < 2$

(C) $f'(x) > 0$ for $0 < x < 1$ and $2 < x < 3$

(D) $f'(x) < 0$ for $0 < x < 1$ and $2 < x < 3$

Answer Key

Test What You Already Know

1. **B** **Learning Objective:** 9.1

2. **A** **Learning Objective:** 9.2

3. **C** **Learning Objective:** 9.3

4. **C** **Learning Objective:** 9.4

5. **C** **Learning Objective:** 9.5

6. **D** **Learning Objective:** 9.6

Test What You Learned

1. **B** **Learning Objective:** 9.4

2. **D** **Learning Objective:** 9.5

3. **B** **Learning Objective:** 9.1

4. **A** **Learning Objective:** 9.3

5. **B** **Learning Objective:** 9.2

6. **A** **Learning Objective:** 9.6

REFLECTION

Test What You Already Know score: _____

Test What You Learned score: _____

Use this section to evaluate your progress. After working through the pre-quiz, check off the boxes in the "Pre" column to indicate which Learning Objectives you feel confident about. Then, after completing the chapter, including the post-quiz, do the same to the boxes in the "Post" column. Keep working on unchecked Objectives until you're confident about them all!

Pre	Post		
☐	☐	**9.1**	Determine the slope of a curve at a given point
☐	☐	**9.2**	Solve problems involving tangent lines and normal lines
☐	☐	**9.3**	Use an equation of the tangent line to make local linear approximations of a function
☐	☐	**9.4**	Use derivatives to find rates of change
☐	☐	**9.5**	Use the first derivative to find relative/absolute extrema and to describe intervals of increase or decrease
☐	☐	**9.6**	Use the second derivative to find points of inflection and to describe concavity

FOR MORE PRACTICE

Complete more practice online at kaptest.com. Haven't registered your book yet? Go to kaptest.com/booksonline to begin.

ANSWERS AND EXPLANATIONS

Test What You Already Know

1. B   Learning Objective: 9.1

The slope of a curve $y = f(x)$ at a point P refers to the slope of the tangent line at the point P. Find this by taking the first derivative of the function and evaluating it at that point. To find $f'(x)$, rewrite the rational term using a negative exponent, then apply the power rule:

$$f(x) = \frac{8}{x^2} + x^3 + x$$
$$= 8x^{-2} + x^3 + x$$
$$f'(x) = -2 \cdot 8x^{-3} + 3x^2 + 1$$
$$= \frac{-16}{x^3} + 3x^2 + 1$$

Now plug in 2 for x and simplify the result:

$$f'(2) = \frac{-16}{2^3} + 3(2)^2 + 1 = -2 + 12 + 1 = 11$$

Hence, **(B)** is correct.

2. A   Learning Objective: 9.2

Take the first derivative to get the equation for finding the *slope* of the tangent line. Use the quotient rule, $\left(\frac{t}{b}\right)' = \frac{t'b - tb'}{b^2}$, where t stands for *top* and b stands for *bottom*. Apply this rule to find $f'(x)$:

$$f'(x) = \frac{2(x+8) - 2x(1)}{(x+8)^2}$$
$$= \frac{2x + 16 - 2x}{(x+8)^2}$$
$$= \frac{16}{(x+8)^2}$$

The question states that the slope of the tangent is 1, so set $f'(x)$ equal to 1 and solve for x:

$$1 = \frac{16}{(x+8)^2}$$
$$(x+8)^2 = 16$$
$$x + 8 = \pm 4$$
$$x = -4 \text{ or } -12$$

Plug these values of x into $f(x)$ to find the y-coordinates of the points:

$$f(x) = \frac{2x}{(x+8)}$$
$$f(-4) = \frac{2(-4)}{(-4+8)} = \frac{-8}{4} = -2$$
$$f(-12) = \frac{2(-12)}{(-12+8)} = \frac{-24}{-4} = 6$$

The coordinates of the points on f that have a tangent line with a slope of 1 are $(-4, -2)$ and $(-12, 6)$, which is **(A)**.

3. C   Learning Objective: 9.3

The tangent line to the function at the given point provides a linear approximation to the function at that point, so start by finding the equation of the tangent line. Then, you can simply plug in the value of x. Because $f(4) = 1$, you know the coordinates of the point of tangency, $(4, 1)$.

To write the equation of the tangent line, you also need the slope at $x = 4$. The first derivative (which is given) represents this slope:

$$f'(4) = \frac{4^2 + 5}{4 + 3} = \frac{16 + 5}{7} = \frac{21}{7} = 3$$

The equation of the tangent line is therefore $y - 1 = 3(x - 4)$ or $y = 3(x - 4) + 1$. When $x = 4.7$, $y = 3(4.7 - 4) + 1 = 3(0.7) + 1 = 3.1$. This means $f(4.7) \approx 3.1$, which is **(C)**.

Note that when finding a linear approximation, it is best NOT to simplify the linear equation because you will be asked to approximate the value of f at an x-value that is close to the x-value of the point given. By not simplifying, the calculations are easier, which is important because these questions are likely to appear in the no-calculator section of the test.

4. C Learning Objective: 9.4

The instantaneous rate of change of a function at a point is the slope of the tangent line at that point. To find the slope, take the first derivative and evaluate it at $x = 2$:

$$f(x) = x^4 + 2x^2 - 3x + 4$$
$$f'(x) = 4x^3 + 4x - 3$$
$$f'(2) = 4(2^3) + 4(2) - 3$$
$$= 4(8) + 8 - 3$$
$$= 32 + 8 - 3$$
$$= 37$$

That's **(C)**.

5. C Learning Objective: 9.5

The graph of a function is decreasing where its first derivative is negative (or < 0). Here, the first derivative, which is given, cannot be solved by hand, so graph the equation in your calculator and estimate which parts of the graph are negative (below the x-axis). Set the calculator window to the given x-values ($0 < x < 13$) to make it easier to see where the graph crosses the x-axis.

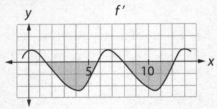

Now compare the graph to the answer choices. The first part of the graph that is below the x-axis begins at about $x = 1.5$ and ends just to the left of $x = 6$, and the second part begins just to the left of $x = 8$ and ends just to the right of $x = 12$. These estimates are a good match for the intervals given in **(C)**.

Note that if the answer choices had been close together, you would need to find the zeros of the graph using the Calculate Zero function on your calculator.

6. D Learning Objective: 9.6

This question is all about knowing what the second derivative tells you about a graph: $f''(x) > 0$ means the second derivative is positive, which tells you the graph of f is concave up (a cup that opens up). Similarly, $f''(x) < 0$ means the second derivative is negative, which tells you the graph of f is concave down (a cup that opens down). The concavity of this graph changes, so eliminate (A) and (B). The graph is

concave up in the range $-2 < x < 0$ and $x > 2$, and concave down in the range $0 < x < 2$, so **(D)** is correct.

Test What You Learned

1. B Learning Objective: 9.4

The instantaneous rate of change of a function at a point is the slope of the tangent line at that point. To find the slope, take the first derivative and evaluate it at $x = 2$:

$$f'(x) = 4x^3 - 9x^2 + 7$$
$$f'(2) = 4(2)^3 - 9(2)^2 + 7$$
$$= 32 - 36 + 7 = 3$$

That's **(B)**.

2. D Learning Objective: 9.5

The graph of a function is increasing where its first derivative is positive ($f' > 0$). Here, the first derivative, which is given, is too complicated to solve by hand, so graph the equation in your calculator and estimate which parts of the graph are positive (above the x-axis). Set the calculator window to the given x-values ($0 < x < 10$) to make it easier to see where the graph crosses the x-axis.

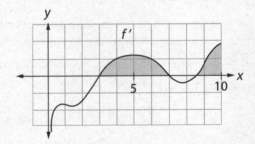

Now compare the graph to the answer choices. The first part of the graph that is above the x-axis begins at about $x = 3$ and ends at about $x = 7$, and the second part begins at about $x = 8.5$ and ends where the graph ends, at $x = 10$. These estimates are a good match for the intervals given in **(D)**.

Note that if the answer choices had been close together, you would need to find the zeros of f' using the Calculate Zero function on your calculator.

3. B Learning Objective: 9.1

The slope of a curve at a point P means the slope of the tangent at P; this is the first derivative of the function evaluated at that point. To find $f'(x)$, rewrite the radical term using a fractional exponent, then apply the power rule:

$$f(x) = 4x^{\frac{1}{2}} + 3x^2 - 2x$$

$$f'(x) = \frac{1}{2} \cdot 4x^{-\frac{1}{2}} + 6x - 2$$

$$= 2x^{-\frac{1}{2}} + 6x - 2$$

$$= \frac{2}{\sqrt{x}} + 6x - 2$$

Now plug in 4 for x and simplify the result:

$$f'(4) = \frac{2}{\sqrt{4}} + 6(4) - 2 = \frac{2}{2} + 24 - 2 = 23$$

Thus, **(B)** is correct.

4. A Learning Objective: 9.3

The tangent line to the function at the given point provides a linear approximation to the function at that point. So start by finding the equation of the tangent line. Then, you can simply plug in the value of x. Because $f(2) = 4$, you know the coordinates of the point of tangency, $(2, 4)$.

To write the equation of the tangent line, you also need the slope at $x = 2$. The first derivative (which is given) represents this slope:

$$f'(2) = 3(2)^2 - 2$$

$$= 3(4) - 2 = 10$$

The equation of the tangent line is therefore:

$$y - 4 = 10(x - 2)$$

$$y = 10(x - 2) + 4$$

Plug 2.1 into this equation for x to find the approximate value of $f(2.1)$:

$$f(2.1) \approx 10(2.1 - 2) + 4 = 10(0.1) + 4 = 1 + 4 = 5$$

(A) is the appropriate match.

5. B Learning Objective: 9.2

To find the equation of the tangent line, start by finding the slope of the tangent line, which is given by the derivative of the function. To take the derivative of $f(x)$, use

the quotient rule, $\left(\dfrac{t}{b}\right)' = \dfrac{t'b - tb'}{b^2}$, where t stands for *top* and b stands for *bottom*:

$$f'(x) = \frac{2(x^2 + 2) - (2x + 4)(2x)}{(x^2 + 2)^2}$$

There is no need to simplify the expression because you're finding the slope of the tangent at a specific point. Plug in 2 for x to find the slope at $x = 2$:

$$f'(2) = \frac{2(4 + 2) - (8)(4)}{(4 + 2)^2}$$

$$= \frac{12 - 32}{36} = -\frac{20}{36} = -\frac{5}{9}$$

Next, use the given point, $\left(2, \dfrac{4}{3}\right)$, and the point-slope form of a line to write the equation of the tangent line:

$$y - \frac{4}{3} = -\frac{5}{9}(x - 2)$$

$$y = -\frac{5}{9}(x - 2) + \frac{4}{3}$$

Before you start to simplify answers, take a quick look back at the answer choices. There is no need to simplify further as you already have a match for **(B)**.

6. A Learning Objective: 9.6

This question is all about knowing what the first and second derivatives tell you about a graph. Examine each statement, one at a time. Cross out false statements as you go.

(A): The given graph is of the function itself. If $f''(x) > 0$, the second derivative is positive, and the graph is concave up (a cup that opens up). This graph is indeed concave up for $1 < x < 2$, so **(A)** is correct.

There is no need to check the other statements unless you're not sure of your answer, but just in case you're curious . . .

(B): $f''(x) < 0$ means the graph is concave down (a cup that opens down). This graph is concave up, not down, between $1 < x < 2$, so this statement is false.

(C): $f'(x) > 0$ means the graph is increasing. This graph increases for approximately the first half of the interval $0 < x < 1$, but then decreases for the second half of the interval, so this statement is false.

(D): $f'(x) < 0$ means the graph is decreasing. This graph increases for approximately the first half of the interval $0 < x < 1$, so this statement is false.

CHAPTER 10

More Advanced Applications

LEARNING OBJECTIVES

10.1 Use the relationship between differentiability and continuity to analyze function behavior

10.2 Use key features of the graphs of f, f', and f'' to analyze functions

10.3 Use derivatives to solve optimization problems

10.4 Use derivatives to solve rectilinear motion problems involving position, speed, velocity, and acceleration

10.5 Use derivatives to solve related rates problems

TEST WHAT YOU ALREADY KNOW

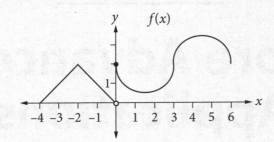

1. The graph of a piecewise-defined function f is shown above. The graph has a vertical tangent at $x = 3$ and horizontal tangent lines at $x = 1.5$ and $x = 4.5$. What are all values of x, for $-4 < x < 6$, at which f is continuous but not differentiable?

   (A) $x = 0$

   (B) $x = -2$ and $x = 3$

   (C) $x = 1.5$ and $x = 4.5$

   (D) $x = -2$, $x = 0$, and $x = 3$

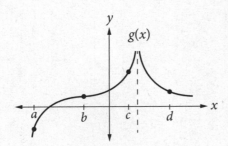

2. The graph of a function g is shown above. At which of the following points is $g'(x)$ positive and decreasing?

   (A) a (B) b (C) c (D) d

3. An employee is cordoning off an area of a local bookstore for a poetry reading. The area is bounded on one side by the wall of the bookstore. The employee has 60 feet of rope to use. What are the dimensions of the largest area the employee can cordon off?

(A) 15 ft × 30 ft

(B) 15 ft × 45 ft

(C) 20 ft × 20 ft

(D) 25 ft × 10 ft

4. The velocity of a particle moving along the x-axis is given by $v(t) = (t-4)(t-7)^2$ for $t \geq 0$. At what value(s) of t is the acceleration of the particle equal to 0?

(A) 4   (B) 7   (C) 4 and 7   (D) 5 and 7

5. The area of a circle is increasing at a rate of 24π cm^2/sec. How fast is the radius of the circle increasing when the radius is 4 cm?

(A) 3 cm/sec   (B) 6 cm/sec   (C) 12 cm/sec   (D) 20 cm/sec

Answers to this quiz can be found at the end of this chapter.

10.1 THE RELATIONSHIP BETWEEN DIFFERENTIABILITY AND CONTINUITY

To answer a question like this:

Let f be a function defined by:

$$f(x) = \begin{cases} ax - x^2 & \text{for} \quad x \leq 2 \\ x^3 - 3x^2 + b & \text{for} \quad x > 2 \end{cases}$$

What are all values of a and b that will make f both continuous and differentiable at $x = 2$?

(A) $a = 0$ and $b = 0$

(B) $a = 1$ and $b = 2$

(C) $a = 4$ and $b = 8$

(D) Any real numbers where $2a = b$

You need to know this:

When a function is differentiable at x_0:

- If the limit in the definition of the derivative exists, we say that f is differentiable at x_0.

- If the limit in the definition of the derivative does not exist, then f is not differentiable at x_0, and f does not have a tangent line at this point.

- Reasons f may not be differentiable at x_c include: f is not defined at x_0, f is defined but not continuous at x_0, or f has a sharp turn at x_0.

How differentiability and continuity are related:

- A continuous function may not be differentiable. Therefore, *continuity does not imply differentiability*.

- The reverse, however, is true. If a function is differentiable at a point x_0, it is continuous at this point, i.e., *differentiability implies continuity*.

You need to do this:

- For a piecewise function, check that the sub-functions are continuous at their shared endpoint ($f(2)$ in this question).

- If the piecewise function contains additional algebraic expressions (such as a and b), set the sub-functions equal to each other and solve.

- Determine the values at which the shared endpoint of the sub-functions will be differentiable by solving $f'(2)$ for each sub-function, and setting them equal to each other.

- Compare the results of the functions and derivatives. Solve for any remaining algebraic expressions.

Answer and Explanation:

C

Start by solving each function for $f(2)$, then set them equal to each other:

$$f(2) = \begin{cases} a \cdot 2 - 2^2 & \text{for } x \le 2 \\ 2^3 - 3 \cdot 2^2 + b & \text{for } x > 2 \end{cases} = \begin{cases} 2a - 4 & \text{for } x \le 2 \\ b - 4 & \text{for } x > 2 \end{cases}$$

$$2a - 4 = b - 4 \Rightarrow 2a = b$$

This gives us the values of a and b for which f will be continuous at $x = 2$. Next, to determine for what values f will be differentiable at $x = 2$, solve $f'(2)$ for each function and set them equal to each other:

$$f'(x) = \begin{cases} a - 2x & x \le 2 \\ 3x^2 - 6x & x > 2 \end{cases}$$

$$f'(2) = \begin{cases} a - 2 \cdot 2 & x \le 2 \\ 3 \cdot 2^2 - 6 \cdot 2 & x > 2 \end{cases} = \begin{cases} a - 4 & x \le 2 \\ 0 & x > 2 \end{cases}$$

$$a - 4 = 0 \Rightarrow a = 4$$

Finally, solve the system and get:

$$\begin{cases} 2a = b \\ a = 4 \end{cases} \Rightarrow \begin{cases} a = 4 \\ b = 8 \end{cases}$$

Thus, **(C)** is correct.

PRACTICE SET

1. Suppose $f(x)$ is not differentiable at $x = a$. Which of the following must be true?

   (A) $f(x)$ is not continuous at $x = a$ but could be defined at $x = a$.

   (B) $f(x)$ is defined at $x = a$ but is not continuous at $x = a$.

   (C) $f(x)$ is continuous and defined at $x = a$.

   (D) $f(x)$ could be defined and continuous at $x = a$.

2. If $f'(6) = -2$, which of the following must be true?

   I. f is decreasing at $x = 6$

   II. f is continuous at $x = 6$

   III. f is concave up at $x = 6$

   (A) I and II only

   (B) I and III only

   (C) II and III only

   (D) I, II and III

3. Which of the following statements must be true for the differentiable function $f(x)$?

   I. The average rate of change of $f(x)$ is given by $f'(x)$.

   II. $\lim\limits_{x \to a} f(x) = f(a)$

   III. $\lim\limits_{x \to a} \dfrac{f(x) - f(a)}{x - a}$ exists for all real values of a.

   (A) I and II only

   (B) I and III only

   (C) II and III only

   (D) I, II and III

4. Let f be a function defined by:
$$f(x) = \begin{cases} 2x - x^2 & \text{for} \quad x \leq 1 \\ x^2 + kx + p & \text{for} \quad x > 1 \end{cases}$$

   What values of k and p will make f both continuous and differentiable at $x = 1$?

   (A) $k = 0$ and $p = 2$

   (B) $k = 1$ and $p = 3$

   (C) $k = -2$ and $p = 3$

   (D) $k = -2$ and $p = 2$

Derivatives

ANSWERS AND EXPLANATIONS

1. D

Consider the validity of each statement, one at a time, crossing off statements that *could* be false as you go. (A) could be false. A function can be continuous and defined at a point where it is not differentiable. For example, the function $y = |x|$ is not differentiable at $x = 0$, but it is continuous and defined there. (B) could be false. A function can be defined and continuous at a point where it is not differentiable. Once again, the function $y = |x|$ is not differentiable at $x = 0$, but it is continuous and defined there. (C) could be false. A function is not differentiable at a point where it is not continuous or not defined. For example, the function $y = \dfrac{1}{x}$ at $x = 0$ is not differentiable at $x = 0$ because it is not defined at that point. **(D)** must be true. There are functions that fail to be differentiable at a point but are nonetheless defined and continuous there. The function $y = |x|$ at $x = 0$ is such an example.

2. A

$f'(6) = -2$ means that f has a slope of -2 at $x = 6$. A negative slope means that the function is decreasing, so statement I is true.

Also, you know that f is differentiable at $x = 6$ because you are told that the derivative at $x = 6$ is -2. Because differentiability implies continuity, statement II must also be true.

Concavity is determined by the second derivative of f. Because you do not know anything about f'', you do not know if $f''(6) > 0$. Thus, statement III *might* be true, but it does not have to be. Therefore, **(A)** is correct.

3. C

Consider each of the three statements individually:

I. $f'(x)$ is the instantaneous rate of change of $f(x)$, which may equal the average rate of change but does not have to. Statement I is false.

II. $\lim\limits_{x \to a} f(x) = f(a)$ implies that f is continuous at $x = a$. Because f is a differentiable function, it must be continuous for all values of x, and therefore statement II is true.

III. $\lim\limits_{x \to a} \dfrac{f(x) - f(a)}{x - a} = f'(a)$, and because f is a differentiable function, $f'(a)$ must exist for all values of x. Thus, statement III is true.

So **(C)** is correct.

4. D

To start, for f to be continuous at $x = 1$, then:

$$f(1) = \begin{cases} 2 \cdot 1 - 1^2 & \text{for } x \leq 1 \\ 1^2 + k \cdot 1 + p & \text{for } x > 1 \end{cases} = \begin{cases} 1 & \text{for } x \leq 1 \\ 1 + k + p & \text{for } x > 1 \end{cases}$$

$$1 + k + p = 1 \Rightarrow k + p = 0$$

Secondly, for f to be differentiable at $x = 1$, then:

$$f'(x) = \begin{cases} 2 - 2x & x \leq 1 \\ 2x + k & x > 1 \end{cases}$$

$$f'(1) = \begin{cases} 2 - 2 \cdot 1 & x \leq 1 \\ 2 \cdot 1 + k & x > 1 \end{cases} = \begin{cases} 0 & x \leq 1 \\ 2 + k & x > 1 \end{cases}$$

$$2 + k = 0 \Rightarrow k = -2$$

Finally, solve the system and get:

$$\begin{cases} k + p = 0 \\ k = -2 \end{cases} \Rightarrow \begin{cases} p = 2 \\ k = -2 \end{cases}$$

(D) is correct.

10.2 USING THE GRAPHS OF *f*, *f'*, AND *f"*

To answer a question like this:

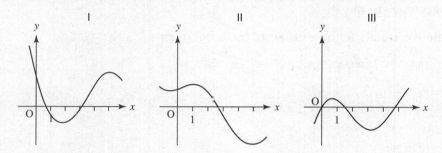

Three graphs labeled I, II, and III are shown above. One is the graph of *f*, one is the graph of *f'*, and one is the graph of *f"*. Which of the following correctly identifies each of the three graphs?

(A)

f	*f'*	*f"*
I	II	III

(B)

f	*f'*	*f"*
II	I	III

(C)

f	*f'*	*f"*
II	III	I

(D)

f	*f'*	*f"*
III	I	II

You need to know this:

Turning Points (Relative Maximums and Minimums)

- A turning point on the graph of *f* corresponds to an *x*-intercept on the graph of *f'*. Reason: The slope of the tangent line at a turning point on curves like these is 0, and the derivative represents the slope of the tangent line.

- A turning point on the graph of *f'* corresponds to an *x*-intercept on the graph of *f"*. Reason: Same as above because *f"* is the derivative of *f'*.

Increasing/Decreasing

- If the *y*-values of *f* are increasing over a given interval, then the corresponding *y*-values of *f'* should be above the *x*-axis. Reason: The slope of an increasing function is positive, and *y*-values above the *x*-axis are positive.

You need to do this:

- Draw a number line for each graph showing the approximate x-values for all turning points.

- Mark the intervals on each side of these points with $+$ and $-$ signs to indicate whether the slope is positive or negative.

- Compare the results to the y-values of the other graphs.

- Eliminate impossible pairings.

Answer and Explanation:

C

There are several approaches that could arrive at the correct answer on this problem as long as appropriate logic is applied when making comparisons between the graphs. One such approach would be to start with a general analysis of the three graphs: Graph I has 2 turning points and 2 x-intercepts; Graph II has 3 turning points and 1 x-intercept; Graph III has 2 turning points and 3 x-intercepts.

Because a turning point on the graph of f corresponds to an x-intercept on the graph of f', Graph II could not be the derivative of either Graph I or III due to having only 1 x-intercept. Hence, Graph II must be f. Graph III in turn must be f' because it has 3 x-intercepts to match the 3 turning points of Graph II. That leaves Graph I as f'', which matches 2 x-intercepts to the 2 turning points of Graph III. Thus, **(C)** is correct.

PRACTICE SET

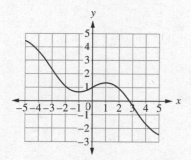

1. The graph of $f(x)$ is shown above. Which of the following is the graph of $f'(x)$?

(A)

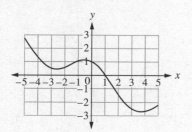

(B)

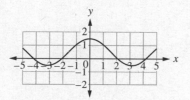

(C)

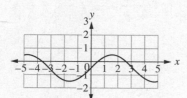

(D)

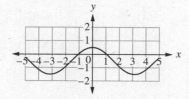

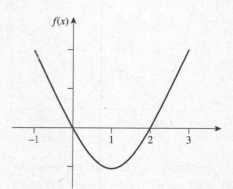

2. The graph of the differentiable function $y = f(x)$ is shown above. Which of the following is true?

(A) $f'(0) > f(0)$

(B) $f'(1) < f(1)$

(C) $f'(2) < f(2)$

(D) $f'(1) = f(0)$

Derivatives

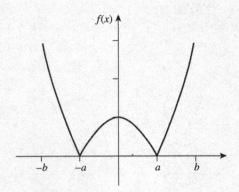

3. The graph of $y = f(x)$ is shown above. Which of the following graphs could be the first derivative of f?

(A)

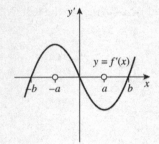

(C)

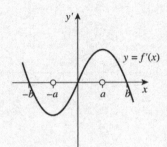

(B)

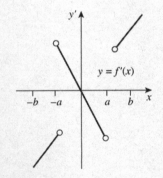

(D)

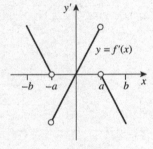

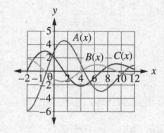

4. The graphs above include a function and its first and second derivatives. Which of the following describes the relationship?

    (A) $A''(x) = B'(x) = C(x)$

    (B) $A''(x) = C'(x) = B(x)$

    (C) $B''(x) = A'(x) = C(x)$

    (D) $B''(x) = C'(x) = A(x)$

ANSWERS AND EXPLANATIONS

1. D

A function has horizontal tangents where its first derivative is equal to 0. The graph of $f(x)$ has horizontal tangents at the points where $x = \pm 1$. Of the choices for the graph $f'(x)$, only **(D)** has zeros at $x = \pm 1$.

2. D

Each of the answer choices compares the values of f and f'. In most questions like this, it is enough to simply determine whether the function is positive (above the x-axis) or negative (below the x-axis), and whether the derivative (slope) is positive (function is increasing) or negative (function is decreasing).

    (A)   False: $f'(0) < 0$ and $f(0) = 0$

    (B)   False: $f'(1) = 0$ and $f(1) < 0$

    (C)   False: $f'(2) > 0$ and $f(2) = 0$

    (D)   True: $f'(1) = 0$ and $f(0) = 0$

Therefore, **(D)** is correct.

3. B

Describe the given graph in terms of its derivative, because this will allow you to eliminate answer choices.

For $x < -a$, the slope of the curve is negative, so the graph of f' lies below the x-axis. Thus, (A) and (D) can be eliminated. For $-a < x < 0$, the slope of the curve is positive, so the graph of f' lies above the x-axis. Thus, (C) and (D) are not possible. You've eliminated all of the answer choices except one, so the correct answer is **(B)**.

4. B

The function C is the derivative of A. The function B is the second derivative of A. On those intervals where the function A is increasing, C is positive, and on those intervals where the function A is decreasing, C is negative. Similarly, on those intervals where the function A is concave up, B is positive, and on those intervals where the function A is concave down, B is negative. It follows that $A''(x) = C'(x) = B(x)$, making **(B)** the correct answer.

10.3 OPTIMIZATION

To answer a question like this:

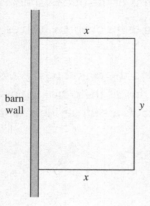

Farmer Fred wants to build a 150-square-meter rectangular pigpen for his pigs. One wall of the pen will be the existing barn wall. The other three walls will be built from mesh fencing that costs \$20/meter. Find the dimensions of the pigpen that will minimize Farmer Fred's fencing cost.

(A) $x = 5$ and $y = 30$

(B) $x = 5\sqrt{3}$ and $y = 10\sqrt{3}$

(C) $x = 10$ and $y = 15$

(D) $x = 5\sqrt{6}$ and $y = 5\sqrt{6}$

You need to know this:

In math, optimization refers to finding the maximum or minimum value of a quantity, usually subject to specific conditions. For example, companies try to maximize profits and minimize costs; this optimization may be subject to constraints such as factory capacity, office space, or other limitations.

Optimization problems are sometimes called applied max/min problems because you are trying to find the maximum or minimum value of a function. This is where calculus comes in. If the quantity you are trying to optimize is described by a continuous function, you can use the information that calculus provides to find the maximum or minimum value of the function. Previously, you saw that the maximum or minimum of a continuous function can only occur at a critical point, or, if the function is continuous and you are looking at a closed interval, at one of the endpoints of the interval.

Derivatives

You need to do this:

The typical procedure for solving an optimization problem:

- Define variables; one variable should be the quantity to optimize. Write a function for the quantity to be optimized in terms of the other variables in the problem.

- Write the constraint equation in terms of your variables. Use the constraint equation to eliminate all but one variable.

- Optimize the function by finding the critical points, determining the nature of each critical point, and evaluating the function at the critical points and endpoints (if applicable). Make sure to answer the question that is asked, paying attention to units and rounding.

Answer and Explanation:

B

Define the value being optimized, the cost of the fencing materials for the pigpen, as the quantity C. Looking at the figure, the total amount of fencing is the sum of the length of the three freestanding sides of the fence. Write an equation that expresses the cost in terms of these lengths. The cost is $20 times the total length of fencing that needs to be purchased, $C = 20(x + y + x) = 20(2x + y)$. This is the *equation to be optimized*.

Express C in terms of just one variable, i.e., eliminate x or y, by defining the variables with an additional equation. Farmer Fred needs the pigpen to have an area of 150 square meters, which can be expressed as $xy = 150$. This is called a *constraint equation*. Rewrite the equation to be optimized by replacing y with $y = \dfrac{150}{x}$: $C = 20(2x + y) = 20\left(2x + \dfrac{150}{x}\right)$.

To find the critical points of C, find its derivative: $C' = 20\left(2 - \dfrac{150}{x^2}\right)$. The critical points are the points where $C' = 0$ or C' is undefined. C' is undefined at $x = 0$, but this is not in the domain of C because $x > 0$. Hence, the only critical points of C occur where $C' = 0$. That is: $20\left(2 - \dfrac{150}{x^2}\right) = 0$. This gives $x = \pm 5\sqrt{3}$.

The point $x = -5\sqrt{3}$ is not in the domain of C, hence the only critical point of C is $x = 5\sqrt{3}$. Verify that this critical point is a minimum by taking the second derivative and evaluating it at the critical point. Because $C''\left(5\sqrt{3}\right) > 0$, it follows from the second derivative test that C has a relative minimum at $x = 5\sqrt{3}$. Therefore $y = \dfrac{150}{5\sqrt{3}} = 10\sqrt{3}$. The dimensions of the pigpen that minimize the cost of the fencing materials are $5\sqrt{3}$ m $\times$ $10\sqrt{3}$ m. Hence, **(B)** is correct.

PRACTICE SET

1. An international shipping container manufacturer has established that the total cost of producing one specific type and size of container can be determined using the equation $C(x) = 6x^2 - 240x + 2{,}800$, where x is the number of units that the company makes. How many of this particular container should the company manufacture in order to minimize the cost?

   (A) 16

   (B) 20

   (C) 32

   (D) 40

2. An advertisement is being designed in the form of a rectangular flyer. The flyer needs to contain 24 square inches of print. The margins at the top and bottom of the page are to be 1.5 inches, and the margin on each side is to be 1 inch. What should the length of the longer side of the flyer be so that the least amount of paper is used?

   (A) 6 inches

   (B) 8 inches

   (C) 9 inches

   (D) 12 inches

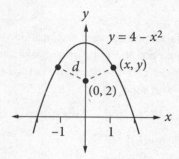

3. The graph of $y = 4 - x^2$ is shown above. What are the x-coordinates of the points on this graph that are closest to the point $(0, 2)$?

   (A) $\pm\sqrt{\dfrac{3}{2}}$

   (B) $\pm\dfrac{3}{2}$

   (C) $\pm\sqrt{\dfrac{5}{2}}$

   (D) $\pm\dfrac{5}{2}$

4. A resort is building a wading pool that will be shaped like an open rectangle topped by a semicircle. The planned perimeter of the pool is 114 feet. What is the approximate radius of the semicircle that will maximize the area of the pool?

   (A) 12 feet

   (B) 16 feet

   (C) 18 feet

   (D) 24 feet

ANSWERS AND EXPLANATIONS

1. B

This is a very straightforward optimization problem—you're given an equation and you're asked to minimize that equation. So, take the derivative of the equation, set it equal to 0, and solve for x:

$$C(x) = 6x^2 - 240x + 2800$$
$$C'(x) = 12x - 240$$
$$0 = 12x - 240$$
$$240 = 12x \rightarrow x = 20$$

This tells us that $x = 20$ is the only critical value. Because this is a multiple choice question (and there must be a correct answer), there is no need to check whether a maximum or minimum occurs at $x = 20$; **(B)** must be correct.

2. C

You want to minimize the area of the paper, so start by writing an equation that represents the area. First, draw a quick sketch.

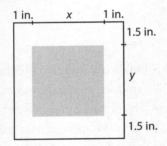

If you call the length x and the width y, the area of the flyer (including margin space) will be $A = (x + 2)(y + 3)$. This is the equation you need to minimize.

Now you need a second equation: The printed area inside the margins is $24 = xy$. Solve this equation in terms of either variable: $x = \dfrac{24}{y}$.

You can now substitute this for x in the equation for the area of the whole flyer:

$$A = \left(\frac{24}{y} + 2\right)(y + 3)$$
$$= 30 + 2y + \frac{72}{y}$$

To minimize A, take the derivative of A, set it equal to 0, and solve:

$$\frac{dA}{dy} = 2 - \frac{72}{y^2} \rightarrow 0 = 2 - \frac{72}{y^2}$$
$$\frac{72}{y^2} = 2 \rightarrow y^2 = 36 \rightarrow y = \pm 6$$

Length can't be negative, so the area of the flyer is at a minimum when $y = 6$. Now, be careful—(A) is *not* the answer. You're asked for the length of the longer side, and you still need to add the margins. If $y = 6$, then based on the sketch, the vertical height of the flyer is $6 + 3 = 9$. Also, if $y = 6$, then $x = \dfrac{24}{6} = 4$ and $x + 2 = 6$. The vertical height (9) is the longer side, making **(C)** the correct choice.

3. A

You need to minimize distance, so write an equation that represents the distance between the point $(0, 2)$ and (x, y): $D = \sqrt{(x - 0)^2 + (y - 2)^2}$.

You're given an equation for y, so substitute this into the distance equation and simplify:

$$D = \sqrt{x^2 + \left(4 - x^2 - 2\right)^2}$$
$$= \sqrt{x^2 + \left(2 - x^2\right)^2}$$
$$= \sqrt{x^2 + 4 - 4x + x^4}$$
$$= \sqrt{x^4 - 3x^2 + 4}$$

Because D is at a minimum when the quantity inside the radical (call it d) is at a minimum, you need only take the derivative of d (without the radical), set it equal to 0, and solve:

$$d' = 4x^3 - 6x$$
$$0 = 2x\left(2x^2 - 3\right)$$
$$x = 0 \text{ and } \pm\sqrt{\frac{3}{2}}$$

Testing these critical values using the First Derivative Test confirms that $x = 0$ produces a maximum, while both $x = \sqrt{\dfrac{3}{2}}$ and $x = -\sqrt{\dfrac{3}{2}}$ produce minimums, so **(A)** is the correct answer.

4. B

You're asked to maximize the area of the pool, so write an equation that represents the area of the rectangle plus the area of the semicircle. First, draw a quick sketch:

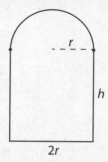

Call the height of the rectangle h. If the radius of the semicircle is r, then the width of the rectangle is $2r$. Thus the area of the wading pool can be expressed as $A = 2rh + \frac{\pi r^2}{2}$.

Now you need a second equation. You're given information about the perimeter, so use that. Refer back to your sketch: The perimeter of the pool (which is given as 114 feet) can be represented by the equation $2r + 2h + \pi r = 114$. Solve this equation for h (you're asked about the radius, so you want to keep that variable), substitute the result into the area formula, and simplify:

$$h = 57 - r - \frac{\pi r}{2}$$

$$
\begin{aligned}
A &= 2rh + \frac{\pi r^2}{2} \\
&= 2r\left(57 - r - \frac{\pi r}{2}\right) + \frac{\pi r^2}{2} \\
&= 114r - 2r^2 - \frac{\pi r^2}{2}
\end{aligned}
$$

To maximize the area, take the derivative, set it equal to 0, and solve:

$$
\begin{aligned}
\frac{dA}{dr} &= 114 - 4r - \pi r \\
0 &= 114 - 4r - \pi r \\
4r + \pi r &= 114 \\
r(4 - \pi) = 114 \rightarrow r &= \frac{114}{4 + \pi} \approx 15.963
\end{aligned}
$$

Therefore, **(B)** is correct. (Because this is a multiple choice question, there is no need to check whether a maximum or minimum occurs at $r = 16$.)

10.4 RECTILINEAR MOTION

To answer a question like this:

A particle moves along the x-axis with its velocity at time t given by $v(t) = bt^2 - ct + 8$, where b and c are constants and b does not equal c. For which of the following values of t is the particle at constant velocity?

(A) $t = \dfrac{c}{2b}$

(B) $t = 2bc$

(C) $t = \dfrac{2b}{c}$

(D) $t = 2b - c$

You need to know this:

Derivatives as rates of change

- Lots of real-life applications of the derivative involve modeling the rates of change of functions.

- Rates of change (derivatives) commonly appear on the AP exam in problems involving motion, where the ideas of position, velocity, and acceleration come into play.

- Recall that velocity is the rate of change of position; acceleration is the rate of change of velocity.

Sign convention

- Position, velocity, and acceleration are signed quantities. For instance, if north is considered the positive direction: a distance of 200 km due north of a starting point is +200 km; traveling 60 km/hr due north to that point is +60 km/hr; decelerating 10 km/hr^2 during that drive is −10 km/hr^2 because deceleration is opposite the direction of motion.

- Speed is the absolute value of velocity, so it has no sign. Speed is decreasing when velocity as a function is going toward zero.

You need to do this:

- Differentiate the equation for velocity with respect to time, $v(t)$, to obtain the equation for acceleration with respect to time, $a(t)$.

- Recall that constant velocity is when acceleration is equal to zero.

- Solve for remaining variables with respect to t.

Answer and Explanation:

A

Take the derivative of the velocity function to obtain the equation for acceleration.

$$v(t) = bt^2 - ct + 8$$
$$v'(t) = a(t) = 2bt - c$$

The particle will be at constant velocity when its acceleration is equal to zero. Set $a(t) = 0$ and solve for the remaining variables with respect to t.

$$0 = 2bt - c$$
$$c = 2bt$$
$$t = \frac{c}{2b}$$

Therefore, **(A)** is the correct answer.

PRACTICE SET

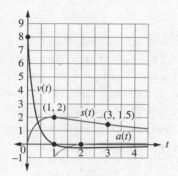

1. A particle moves along a number line (not shown) for $t \geq 0$. Its position function is $s(t)$, its velocity function is $v(t)$, and its acceleration function is $a(t)$. All are graphed with respect to time t in seconds. For the time interval when $1 < t < 2$, the particle is

   (A) not moving.

   (B) moving to the left and slowing down.

   (C) moving to the left and speeding up.

   (D) moving to the right and slowing down

2. If the position of a particle is given by the function $s(t) = t^3 - 12t^2 + 36t + 18$, for $t > 0$, at what value(s) of t does the particle change direction?

   (A) $t = 2$

   (B) $t = 6$

   (C) $t = 2$ and $t = 6$

   (D) The particle never changes direction.

3. Let $x(t) = t^{\frac{2}{3}}$ give the distance of a moving particle from its starting point as a function of time t. For what value of t is the instantaneous velocity of the particle equal to its average velocity over the closed interval $[0, 8]$?

   (A) $\dfrac{8}{27}$

   (B) $\dfrac{27}{64}$

   (C) $\dfrac{64}{27}$

   (D) $\dfrac{64}{9}$

4. A particle is moving along the x-axis with a velocity, $v(t) = \sin t + \cos t$, for $t \geq 0$. What is its maximum acceleration over the closed interval $[0, 2\pi]$?

   (A) $\cos \dfrac{3\pi}{4} + \sin \dfrac{3\pi}{4}$

   (B) $\cos 2\pi + \sin 2\pi$

   (C) $\cos \dfrac{7\pi}{4} + \sin \dfrac{7\pi}{4}$

   (D) $\cos \dfrac{7\pi}{4} - \sin \dfrac{7\pi}{4}$

Derivatives

ANSWERS AND EXPLANATIONS

1. C

For $1 < t < 2$, both the velocity and the acceleration functions are negative, which means the particle is moving to the left (i.e., $v(t) < 0$) and the velocity is becoming more negative (i.e., $a(t) < 0$), so **(C)** is correct.

2. C

The particle changes direction after slowing to a stop, then speeding back up. In other words, the particle changes direction when its velocity is 0 and its acceleration is *not* 0. Velocity is the derivative of position, so take the derivative of the given equation, set it equal to 0, and solve for t:

$$s'(t) = v(t) = 3t^2 - 24t + 36$$
$$0 = 3(t^2 - 8t + 12)$$
$$0 = (t-2)(t-6)$$
$$t = 2 \text{ and } 6$$

Finally, check that the acceleration is not 0:

$$s''(t) = 6t - 24 \rightarrow t = 4$$

Thus, the particle changes direction at $t = 2$ and $t = 6$, which is **(C)**.

3. C

The equation $x(t) = t^{\frac{2}{3}}$ represents the distance of the particle from its starting point as a function of time t. Its average velocity over the closed interval [0, 8] is given by:

$$v_{aver} = \frac{x(8) - x(0)}{8 - 0} = \frac{4 - 0}{8} = \frac{1}{2}$$

Its instantaneous velocity is given by the first derivative of the equation:

$$x'(t) = \frac{2}{3} t^{-\frac{1}{3}} = \frac{2}{3\sqrt[3]{t}}$$

Now set $x'(t) = \frac{2}{3\sqrt[3]{t}} = \frac{1}{2}$ and solve to get:

$$3\sqrt[3]{t} = 4 \Rightarrow \sqrt[3]{t} = \frac{4}{3} \Rightarrow t = \frac{64}{27}$$

That's **(C)**.

4. D

Acceleration is the derivative of velocity, so to find an equation for the acceleration, differentiate the velocity function $v(t) = \sin t + \cos t \Rightarrow v'(t) = a(t) = \cos t - \sin t$. Because the acceleration function is continuous, the maximum acceleration can only occur at the critical points (where $a'(t)$ is zero or undefined) or at one of the endpoints of the interval $[0, 2\pi]$.

$$a(t) = \cos t - \sin t$$
$$a'(t) = -\sin t - \cos t$$
$$0 = -\sin t - \cos t$$
$$\sin t = -\cos t$$
$$\tan t = -1$$

This occurs in the second and fourth quadrants at $t = \frac{3\pi}{4}$ and $\frac{7\pi}{4}$.

Notice that the derivative $a'(t)$ is defined everywhere, so the only critical points are the points at which $a'(t) = 0$. Make a table of the critical points and endpoints and the value of the acceleration $a(t)$ at each of these points.

t	$\frac{3\pi}{4}$	$\frac{7\pi}{4}$	0	2π
$a(t)$	$\cos\frac{3\pi}{4} - \sin\frac{3\pi}{4}$ $= -\frac{\sqrt{2}}{2} - \frac{\sqrt{2}}{2}$ $= -\sqrt{2} \approx -1.414$	$\cos\frac{7\pi}{4} - \sin\frac{7\pi}{4}$ $= \frac{\sqrt{2}}{2} - \left(-\frac{\sqrt{2}}{2}\right)$ ≈ 1.414	$\cos 0 +$ $\sin 0 = 1 +$ $0 = 1$	$\cos 2\pi +$ $\sin 2\pi = 1 +$ $0 = 1$

Of these four choices, the maximum acceleration is $\sqrt{2}$; the maximum acceleration must be one of these four values, so the correct answer is **(D)**.

In this problem, you really have to pay attention to what you are maximizing. Usually, position is $x(t)$ and for this, you need to set $x'(t) = v(t) = 0$ and look for $x''(t) = a(t) < 0$. This problem is changed around, and you are asked to maximize acceleration, so you need to look at $a'(t)$ and $a''(t)$. This concept can be difficult at first, but it becomes easier with practice.

10.5 RELATED RATES

To answer a question like this:

Claire is rollerblading due east toward the mall at a speed of 10km/hr. John is biking due south away from the mall at a speed of 15 km/hr. Let x be the distance between Claire and the mall at time t and let y be the distance between John and the mall at time t. Find the rate of change, in km/hr, of the distance between Claire and John when $x = 5$ km and $y = 12$ km.

(A) 5 km/hr   (B) 10 km/hr   (C) 13 km/hr   (D) 15 km/hr

You need to know this:

Often information is given about the rate at which one quantity changes, and the question asks that the information be used to provide data about changes in another, related quantity. Problems of this type are referred to as related-rates problems.

These steps typify the method for solving related-rates problems:

1. Draw a diagram.

2. Identify the quantities that are changing and assign variables to represent them. Add labels to the diagram.

3. Write down an equation that relates the quantities.

4. Considering each variable as a function of t, use implicit differentiation to differentiate both sides of the equation with respect to t to find the rate of change (derivative) of each quantity.

5. Restate the given problem(s) in terms of your variables and their rates of change (derivatives).

6. Use Step 4 to solve the problem(s) in Step 5.

You need to do this:

Start by making a diagram to help understand the information in the problem:

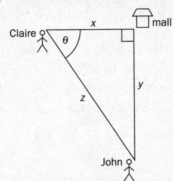

The diagram is a right triangle; therefore, relate x, y, and z with the Pythagorean theorem: $x^2 + y^2 = z^2$.

Answer and Explanation:

B

Given that $x = 5$ and $y = 12$, the Pythagorean theorem can be used to find that $5^2 + 12^2 = z^2 \Rightarrow z = 13$. In other words, the distance between John and Claire is 13 km.

To find an equation relating the velocities, use implicit differentiation to differentiate the equation from the Pythagorean theorem: $2x\dfrac{dx}{dt} + 2y\dfrac{dy}{dt} = 2z\dfrac{dz}{dt}$. Claire is traveling toward the mall at 10 km per hour; her distance from the mall is decreasing, so $\dfrac{dx}{dt} = -10$. John is traveling away from the mall at 15 km/hr. His distance from the mall is increasing, so $\dfrac{dy}{dt} = 15$. To find the rate of change of the distance between John and Claire when $x = 5$ and $y = 12$, cancel the 2s in the equation above and fill in all of the required information:

$$x\frac{dx}{dt} + y\frac{dy}{dt} = z\frac{dz}{dt}$$

$$(5)(-10) + (12)(15) = 13\frac{dz}{dt}$$

$$130 = 13\frac{dz}{dt} \Rightarrow \frac{dz}{dt} = 10\,\text{km/hr}$$

Thus, **(B)** is correct.

PRACTICE SET

1. The volume of a cube is decreasing at the rate of 750 cm^3/min. When the length of one edge of the cube is 5 cm, how fast is the area of one face of the cube changing?

   (A) -100 cm^2/min

   (B) -30 cm^2/min

   (C) 40 cm^2/min

   (D) 50 cm^2/min

2. The base of a triangle is decreasing at a constant rate of 0.2 cm/sec and the height is increasing at 0.1 cm/sec. If the area is increasing, which answer best describes the constraints on the height h at the instant when the base is 3 centimeters?

   (A) $h > 3$

   (B) $h < 1$

   (C) $h > 1.5$

   (D) $h < 1.5$

3. Dashing Dan is driving south on Sunset Boulevard, where the speed limit is 40 miles per hour. He is traveling at a constant speed. He crosses an intersection. A police officer is sitting 33 feet due east of the intersection with her radar tuned to Dan's car. One second after Dan crosses the intersection, Dan is 55 feet away from the police officer and she records Dan going away from her at a speed of 37.5 miles per hour. She issues Dan a ticket because Dan's speed on Sunset Boulevard is

   (A) between 35 and 40 miles per hour.

   (B) between 40 and 45 miles per hour.

   (C) between 45 and 50 miles per hour.

   (D) between 50 and 55 miles per hour.

4. The lengths of the sides of a square are decreasing at a constant rate of 4 ft/min. In terms of the perimeter, P, what is the rate of change of the area of the square in square feet per minute?

   (A) $-2P$

   (B) $2P$

   (C) $-4P$

   (D) $4P$

ANSWERS AND EXPLANATIONS

1. A

This is a related-rates question because you're being asked to relate the rate of change in the volume to the rate of change in the area. Remember, rates of change are derivatives.

Let $V = s^3$. Then $\frac{dV}{dt} = 3s^2 \frac{ds}{dt}$. You're given that the volume of a cube is decreasing at the rate of 750 cm³/min, so when $s = 5$, $-750 = 3 \cdot 5^2 \cdot \frac{ds}{dt}$, so $\frac{ds}{dt} = -10$ cm/min.

Now, turn to area: The area of one face of the cube is $A = s^2$, so:

$$\frac{dA}{dt} = 2s \frac{ds}{dt} = 2 \cdot 5 \cdot -10 = -100 \text{ cm}^2/\text{min}.$$

(A) is correct.

2. D

First, draw a diagram and label the quantities in the problem:

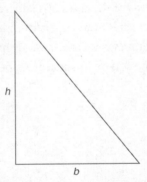

Next, find an equation relating the quantities in the problem; here, the formula for the area of a triangle: $A = \frac{1}{2}bh$. Differentiate both sides of the equation with respect to t using the product rule:

$$A'(t) = \frac{dA}{dt} = \frac{1}{2}\left(b\frac{dh}{dt} + h\frac{db}{dt}\right)$$

Now consider what information is given: $\frac{db}{dt} = -0.2$ cm/sec, $\frac{dh}{dt} = 0.1$ cm/sec, and $b = 3$ cm. Substitute the information into the expression above, which gives:

$$A' = \frac{1}{2}\big(3(+0.1) + h(-0.2)\big)$$

You are also given that the area is increasing, i.e., $A'(t) > 0$; therefore, set the expression for A' greater than zero and solve for h:

$$\frac{1}{2}\big(3(+0.1) + h(-0.2)\big) > 0 \Rightarrow h(-0.2) > -0.3 \Rightarrow h < 1.5$$

Thus, **(D)** is correct.

3. C

First, draw a diagram and identify the quantities in the problem.

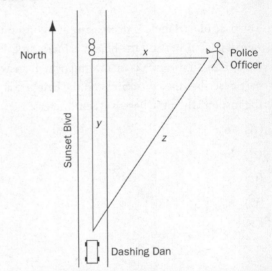

Use the Pythagorean theorem to find an equation relating the variables: $x^2 + y^2 = z^2$. Differentiate both sides of the equation with respect to time: $2x\frac{dx}{dt} + 2y\frac{dy}{dt} = 2z\frac{dz}{dt}$. The police officer's distance from the intersection is given in feet; Dan's speed is given in miles per hour, and you are told this speed at a time t given in seconds. You need to express everything in terms of the same units. Use feet, seconds, and feet per second. Dan is traveling away from the police officer at 37.5 mph, which is:

$$\frac{37.5 \text{ mi}}{\text{hr}} \cdot \frac{5280 \text{ ft}}{\text{mi}} \cdot \frac{1 \text{ hr}}{3600 \text{ sec}} = 55 \text{ feet/second}$$

Thus, $\frac{dz}{dt} = 55$ ft/sec.

You are told that one second after crossing the intersection, Dan is 55 feet away from the police officer.

Use the Pythagorean theorem to find Dan's distance due south from the intersection: $33^2 + y^2 = 55^2 \Rightarrow y = 44$ ft. Fill this information into the equation relating the derivatives of the quantities and solve for $\frac{dy}{dt}$, which is the speed at which Dan is traveling on Sunset Boulevard:

$$2 \cdot 33 \cdot 0 + 2 \cdot 44 \cdot \frac{dy}{dt} = 2 \cdot 55 \cdot 55 \Rightarrow \frac{dy}{dt} = 68.75 \text{ ft/sec}$$

The units here are feet per second, but the answer choices are given in miles per hour, so convert:

$$68.75 \, \frac{\text{ft}}{\text{sec}} \cdot \frac{1 \text{ mi}}{5280 \text{ ft}} \cdot \frac{3600 \text{ sec}}{1 \text{ hr}} = 46.875 \text{ mph}$$

This means **(C)** is correct.

4.   A

This is a related-rates question with a twist. You are given information about the rate of change of the side of the square and you need to answer a question about the rate of change of the area. After you do this, you are done with the related-rates part of the problem and you have to express your answer in terms of the perimeter.

Find the area of the square in terms of the variables: $A = s^2$. Differentiate both sides of this equation with respect to t to find an equation relating $\frac{dA}{dt}$ and $\frac{ds}{dt}$: $\frac{dA}{dt} = 2s \underbrace{\frac{ds}{dt}}_{\text{chain rule}}$. You are given that the lengths of the sides of the square are decreasing at a rate of 4 ft/min, i.e., $\frac{ds}{dt} = -4$. Substitute this information into the equation for $\frac{dA}{dt}$ above and find $\frac{dA}{dt} = 2s(-4) = -8s$.

To finish the problem, you need to express $\frac{dA}{dt}$ in terms of the perimeter. The perimeter of a square is $P = 4s$. Therefore, $s = \frac{P}{4}$. Substituting this expression for s into the expression above yields:

$$\frac{dA}{dt} = -8 \cdot \frac{P}{4} = -2P$$

That's **(A)**.

RAPID REVIEW

If you take away only five things from this chapter:

1. Continuity does not imply differentiability, but differentiability does imply continuity.

2. The x-intercepts of a derivative f' will be equal in number and position on the x-axis to the turning points of the function f. The same holds true for f'' and f'.

3. Optimization is an application of finding the max/min of a function by evaluating critical points over a defined interval. Define variables for the quantity to be optimized and other unknowns. Set up optimization and constraint equations.

4. Velocity is the rate of change (derivative) of position; acceleration is the rate of change (derivative) of velocity.

5. Related-rates problems apply the rate of change of a known quantity to determine the rate of change of an unknown, related quantity. Assign variables to the changing quantities and write an equation that relates them.

Derivatives

TEST WHAT YOU LEARNED

1. If $y = 18 - 3x$, what is the maximum value of xy ?

    (A) 9   (B) 12   (C) 24   (D) 27

2. Sand is being poured into a conical pile at a construction site at a rate of 10 ft³/min. The diameter of the base of the cone is approximately three times the height of the cone. At what rate is the height of the pile changing when the pile is 15 feet high? The volume of a cone is given by the formula $V_{cone} = \frac{1}{3}\pi r^2 h$.

    (A) 0.006 ft/min   (B) 0.012 ft/min   (C) 0.026 ft/min   (D) 0.030 ft/min

Derivatives

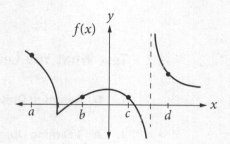

3. The graph of $y = f(x)$ is shown above. For which of the following values of x is $f'(x)$ negative and increasing?

   (A) a (B) b (C) c (B) d

---

4. A function f is defined by $f(x) = |x + 4|$. For what values of x is the graph of f not differentiable?

   (A) $x = -4$

   (B) $x = 0$

   (C) $x = 4$

   (D) The function is differentiable over its entire domain.

---

5. The position of a particle moving along a number line is given by $s(t) = \frac{2}{3}t^3 - 6t^2 + 10t$ where t is time in seconds. For what value(s) of t is the particle at rest?

   (A) 1 second only

   (B) 3 seconds only

   (C) 1 and 2 seconds

   (D) 1 and 5 seconds

Derivatives

Answer Key

Test What You Already Know	Test What You Learned
1. **B** **Learning Objective:** 10.1	1. **D** **Learning Objective:** 10.3
2. **A** **Learning Objective:** 10.2	2. **A** **Learning Objective:** 10.5
3. **A** **Learning Objective:** 10.3	3. **D** **Learning Objective:** 10.2
4. **D** **Learning Objective:** 10.4	4. **A** **Learning Objective:** 10.1
5. **A** **Learning Objective:** 10.5	5. **D** **Learning Objective:** 10.4

 ## REFLECTION

Test What You Already Know score: _____

Test What You Learned score: _____

Use this section to evaluate your progress. After working through the pre-quiz, check off the boxes in the "Pre" column to indicate which Learning Objectives you feel confident about. Then, after completing the chapter, including the post-quiz, do the same to the boxes in the "Post" column. Keep working on unchecked Objectives until you're confident about them all!

Pre	Post		
☐	☐	**10.1**	Use the relationship between differentiability and continuity to analyze function behavior
☐	☐	**10.2**	Use key features of the graphs of f, f', and f'' to analyze functions
☐	☐	**10.3**	Use derivatives to solve optimization problems
☐	☐	**10.4**	Use derivatives to solve rectilinear motion problems involving position, speed, velocity, and acceleration
☐	☐	**10.5**	Use derivatives to solve related rates problems

 ## FOR MORE PRACTICE

Complete more practice online at kaptest.com. Haven't registered your book yet? Go to kaptest.com/booksonline to begin.

ANSWERS AND EXPLANATIONS

Test What You Already Know

1. B Learning Objective: 10.1

The question is asking for points at which f is continuous but not differentiable. That means you need to identify the points where you would not lift up your pencil if you were drawing the graph (so f is continuous), but where the derivative is undefined. This occurs at sharp turns or corners, like those at the vertex of an absolute value function, and at points where the tangent line is vertical. Here, you have two such points: a corner point at $x = -2$ and a vertical tangent at $x = 3$. This means **(B)** is the correct answer.

Be careful of (D). While the function is not differentiable at $x = 0$, it is also not continuous there.

2. A Learning Objective: 10.2

This question is asking where the first derivative is positive and decreasing. A positive first derivative, or $g'(x) > 0$, is represented graphically by a positive slope. Only the points at a and c have positive slopes, so eliminate (B) and (D). For the first derivative to be decreasing, $g''(x)$ must be negative. In other words, the graph must be concave down. This is true at a, but not at c, so **(A)** is the correct answer.

3. A Learning Objective: 10.3

You want to maximize the area, so start by writing an equation that represents the area of the reading space. First, draw a quick sketch.

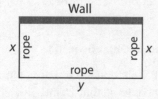

If you call the length y and the width x, the area of the reading space will be $A = xy$.

The employee has a limited amount of rope which represents the perimeter (not including the wall), so write

an equation for the perimeter and solve the equation for the easier variable (here, y because the y term has a coefficient of 1):

$$2x + y = 60$$
$$y = 60 - 2x$$

You can now substitute this for y in the equation for the area of the reading space:

$$A = x(60 - 2x) = 60x - 2x^2$$

To maximize A, take the derivative of A, set it equal to 0, and solve:

$$\frac{dA}{dx} = 60 - 4x = 0 \Rightarrow x = 15$$

Use the equation of y (from before) to find the value of y. You need to do this because there is more than one answer choice that includes 15:

$$y = 60 - 2x$$
$$= 60 - 2(15) = 30$$

The dimensions that maximize the area are 15 ft × 30 ft, which is **(A)**.

4. D Learning Objective: 10.4

Position, velocity, and acceleration are all connected by derivatives and integrals. Acceleration is the derivative of velocity, so you can find $v'(t)$ to determine which values of t result in an acceleration of 0. To find $v'(t)$, you can use the product rule or expand the expression first, whichever you are more comfortable with.

$$v(t) = (t - 4)(t - 7)^2 = (t - 4)(t^2 - 14t + 49)$$
$$v(t) = t^3 - 18t^2 + 105t - 196$$

$$a(t) = v'(t) = 3t^2 - 36t + 105$$
$$0 = 3(t^2 - 12t + 35)$$
$$0 = 3(t - 5)(t - 7) \rightarrow t = 5 \text{ and } 7$$

Thus, when $t = 5$ and 7, the acceleration of the particle equals 0. The answer is **(D)**.

Derivatives

5. **A** **Learning Objective:** 10.5

First, write an expression that relates the area of a circle to its radius: $A = \pi r^2$. Next, jot down what you're given and what you're looking for. Remember, rates are derivatives. It is a good idea to include units so that, if you are answering a free-response question, it will be easier to keep track of the units for the final answer.

Given: $\dfrac{dA}{dt} = 24\pi \text{ cm}^2/s$, $r = 4 \text{ cm}$

Looking for: $\dfrac{dr}{dt} = ?$

Now, take the derivative of the area formula with respect to t, plug in the given information, and solve for $\dfrac{dr}{dt}$:

$$A = \pi r^2$$
$$\frac{dA}{dt} = 2\pi r \frac{dr}{dt}$$
$$24\pi = 2\pi(4)\frac{dr}{dt}$$
$$\frac{24\pi}{8\pi} = \frac{dr}{dt} \rightarrow \frac{dr}{dt} = 3$$

The answer is **(A)**.

Test What You Learned

1. **D** **Learning Objective:** 10.3

Call the product A and substitute the expression given for y into A: $A = xy = x(18 - 3x) = 18x - 3x^2$. If this function has a maximum value, it will occur where $\dfrac{dA}{dx}$ is equal to zero or is undefined, so compute $\dfrac{dA}{dx} = 18 - 6x$. Because the derivative of the function is defined everywhere, if the function has a maximum value it must occur at a point where $\dfrac{dA}{dx} = 0$. Solve the equation:

$$\frac{dA}{dx} = 0 \Rightarrow 18 - 6x = 0 \Rightarrow x = 3$$

Apply the second derivative test to determine whether the function has a *relative* maximum at $x = 3$: $A''(x) = -6 < 0$ for all x; therefore, the function has a relative maximum at $x = 3$. Because A is a quadratic function, its graph is a parabola. The vertex of a parabola is the only relative extremum on the graph, and it is a global extremum. Therefore, A has a global maximum at $x = 3$. To compute the maximum value, substitute $x = 3$ back into the original equation for A: $18 \cdot 3 - 3 \cdot 3^2 = 27$. **(D)** is correct.

2. **A** **Learning Objective:** 10.5

You're given the expression that relates the volume of a cone to its height: $V = \dfrac{1}{3}\pi r^2 h$. Now, jot down what else you're given and what you're looking for. Remember, rates are derivatives. Sometimes you need to look at the units to decide what you are given; volume is measured in units³, so 10 ft³/min must be the rate of change of volume.

Given: $\dfrac{dV}{dt} = 10 \text{ ft}^3/\text{min}$, $d = 3h$, $h = 15 \text{ ft}$

Looking for: $\dfrac{dh}{dt} = ?$

Because radius is half of the diameter, then $r = \dfrac{d}{2} = \dfrac{3}{2}h$. Plug this information in for r and simplify the given volume equation. Then, take the derivative with respect to t, plug in the given information, and solve for $\dfrac{dh}{dt}$:

$$V = \frac{1}{3}\pi r^2 h = \frac{1}{3}\pi\left(\frac{3}{2}h\right)^2 \cdot h$$
$$V = \frac{3\pi}{4}h^3$$
$$\frac{dV}{dt} = \frac{9\pi}{4}h^2\frac{dh}{dt}$$
$$10 = \frac{9\pi}{4}(15)^2\frac{dh}{dt}$$
$$\frac{40}{2025\pi} = \frac{dh}{dt} \rightarrow \frac{dh}{dt} = 0.006$$

The answer is **(A)**.

3. **D** **Learning Objective:** 10.2

This problem is asking where the first derivative of f is negative and increasing. A negative first derivative, or $f'(x) < 0$, is represented graphically by a negative slope. The points at a, c, and d have negative slopes, so eliminate (B). For the first derivative to be increasing, $f''(x)$ must be positive. In other words, the graph must be concave up. This is true at d, but not at a or c, so **(D)** is the correct answer.

4. **A** **Learning Objective:** 10.1

If the graph of a function has a sharp point, the function is not differentiable at that point. This is because the slopes directly to the left and right of the point do not approach the same value.

An absolute value function has a sharp point at its vertex. The graph of $f(x) = |x + 4|$ is a horizontal translation (to the *left* 4 units) of the standard absolute value function, $y = |x|$, which has vertex $(0, 0)$. Thus, the vertex of f is $(-4, 0)$, which means the function is not differentiable at $x = -4$. **(A)** is correct.

5. D Learning Objective: 10.4

A particle is at rest when its velocity is zero. To find a velocity equation, take the first derivative of the position function. Correct application of the power rule yields the following:

$$v(t) = s'(t) = 2t^2 - 12t + 10$$
$$0 = 2(t^2 - 6t + 5)$$
$$0 = 2(t - 1)(t - 5)$$
$$t = 1, 5$$

So, the particle is at rest when $t = 1$ and 5 seconds, which is **(D)**.

Derivatives

PART 4

Integrals and the Fundamental Theorem of Calculus

CHAPTER 11

Integration—The Basics

LEARNING OBJECTIVES

11.1 Use integration rules and algebraic manipulation to find antiderivatives of polynomial, radical, and rational functions

11.2 Find the antiderivative of a function subject to an initial condition

11.3 Use *u*-substitution to find antiderivatives of polynomial, radical, and rational functions

11.4 Integrate trigonometric functions and inverse trigonometric forms

11.5 Integrate exponential and logarithmic functions

TEST WHAT YOU ALREADY KNOW

1. If $\dfrac{dy}{dx} = x^2 + 4\sqrt{x} - \dfrac{1}{x^2}$, then $y =$

   (A) $\dfrac{x^3}{3} + \dfrac{8}{\sqrt{x}} - \dfrac{1}{x} + C$

   (B) $\dfrac{x^3}{3} + \dfrac{8}{3}x^{\frac{3}{2}} + \dfrac{1}{x} + C$

   (C) $3x^3 + \dfrac{8\sqrt{x}}{3} - \dfrac{1}{3x^3} + C$

   (D) $3x^3 + 4x^{\frac{3}{2}} + \dfrac{1}{x^3} + C$

2. If $f(x) = F'(x)$, where $f(x) = 9x^2 + x - 8$ subject to the initial condition $F(0) = 2$, which of the following represents $F(x)$?

   (A) $\dfrac{x^3}{3} + x^2 + 8x + 2$

   (B) $3x^3 + 2x^2 + 8x - 2$

   (C) $3x^3 + \dfrac{x^2}{2} - 8x + 2$

   (D) $18x^3 + x^2 - 8x + 2$

3. $\int \dfrac{\left(x^{\frac{1}{2}} - 8\right)^3}{x^{\frac{1}{2}}}\, dx =$

(A) $\dfrac{1}{4}\left(x - 8x^{\frac{1}{2}}\right)^4 + C$

(B) $\dfrac{1}{2}\left(x^{\frac{1}{2}} - 8\right)^4 + C$

(C) $\dfrac{1}{2}\left(x^{\frac{3}{2}} - 8x\right)^4 + C$

(D) $\dfrac{1}{4}\left(x^{-\frac{1}{2}} - 8\right)^4 + C$

---

4. $\int \sec^2\left(\dfrac{x}{2}\right)\, dx =$

(A) $-\dfrac{1}{2}\sec^3\left(\dfrac{x}{2}\right) + C$

(B) $\dfrac{1}{2}\tan\left(\dfrac{x}{2}\right) + C$

(C) $2\sec^2\left(\dfrac{x}{2}\right) + C$

(D) $2\tan\left(\dfrac{x}{2}\right) + C$

---

5. $\int \dfrac{-x}{e^{3x^2}}\, dx =$

(A) $-\dfrac{e^{3x^2}}{6x} + C$

(B) $\dfrac{1}{6e^{3x^2}} + C$

(C) $\dfrac{1}{2}\ln e^{3x^2} + C$

(D) $\dfrac{3x^2}{e^{3x^2}} + C$

Answers to this quiz can be found at the end of this chapter.

Integrals

11.1 ANTIDERIVATIVES AND THE INDEFINITE INTEGRAL

To answer a question like this:

Which of the following is an antiderivative of $f(x) = \sqrt[3]{x}(x + 2)$?

(A) $\frac{3}{4}x^{\frac{7}{4}} + \frac{2}{3}x^{\frac{3}{4}} + C$

(B) $\frac{7}{3}x^{\frac{7}{3}} + \frac{2}{3}x^{\frac{4}{3}} + C$

(C) $\frac{3}{7}x^{\frac{7}{3}} + \frac{3}{2}x^{\frac{4}{3}} + C$

(D) $\frac{4}{3}x^{\frac{4}{3}} + \frac{3}{2}x^{\frac{1}{3}} + C$

You need to know this:

An antiderivative

- If $F(x)$ is a function whose derivative is $f(x)$, that is, if $\frac{d}{dx}\left(F(x)\right) = f(x)$, $F(x)$ is an antiderivative of $f(x)$.

- It is *an* antiderivative and not *the* antiderivative because an antiderivative is not uniquely associated with one function. It follows from the definition of the antiderivative that if $F(x)$ is an antiderivative of $f(x)$, then for all real numbers C, $F(x) + C$ is also an antiderivative of $f(x)$ because the derivative of a constant is zero.

- If $F(x)$ and $G(x)$ are both antiderivatives of $f(x)$, then $F(x) - G(x) = C$, for some real number C.

The indefinite integral

- The notation for the antiderivative is the symbol $\int dx$, the indefinite integral.

- If $F(x)$ is an antiderivative of $f(x)$, then $\int f(x)\, dx = F(x) + C$.

> **✔ AP Expert Note**
>
> Never forget the $+ C$ when calculating an indefinite integral. Doing so is a great way to lose easy points on the FRQ section and fall for trap answers in the multiple choice questions. The best way to avoid forgetting is to make it second nature. Don't cut corners when you practice. Always write the $+ C$.

You need to do this:

- There are no product or quotient rules when taking antiderivatives, so simplify the expression first and integrate each term.

- When finding derivatives where $r \neq 0$, $\dfrac{d}{dx}\left(x^r\right) = rx^{r-1}$. The reverse process applies for integrals: for $r \neq -1$, $\displaystyle\int x^r \, dx = \dfrac{x^{r+1}}{r+1} + C$.

- From the properties of derivatives or integrals, it follows that $\displaystyle\int m \, dx = mx + C$.

Answer and Explanation:

C

There is no "product rule" for integrating, so simplify the expression first by writing the radical using a fractional exponent and distributing. Then integrate each term to find the antiderivative, F:

$$f(x) = \sqrt[3]{x}(x+2) = x^{\frac{1}{3}}(x+2) = x^{\frac{4}{3}} + 2x^{\frac{1}{3}}$$

$$F(x) = \int x^{\frac{4}{3}} \, dx + 2\int x^{\frac{1}{3}} \, dx$$

$$= \frac{x^{\frac{4}{3}+\frac{3}{3}}}{\frac{7}{3}} + 2\left(\frac{x^{\frac{1}{3}+\frac{3}{3}}}{\frac{4}{3}}\right)$$

$$= \frac{3}{7}x^{\frac{7}{3}} + \frac{3}{2}x^{\frac{4}{3}} + C$$

(C) is correct.

> ✔ **AP Expert Note**
>
> Note that you could also take the derivative of each of the answer choices, but that would take considerably more time.

Integrals

PRACTICE SET

1. Suppose $G(x)$ is an antiderivative of $g(x)$ and $F(x)$ is an antiderivative of $f(x)$. If $f(x) = g(x)$, which of the following statements must be true?

    (A) $F(x) = G(x)$

    (B) $F(x) = -G(x)$

    (C) $F(x) = G(x) + 1$

    (D) $F(x) = G(x) + C$

2. If $F' = f$ and $f(x) = 18x^5 - 12x^3 + 2x$, which of the following could be $F(x)$?

    (A) $3x^6 - 3x^4 + x^2 + 1$

    (B) $3x^6 - 4x^4 + x^2 + 5$

    (C) $18x^6 - 12x^4 + 2x^2$

    (D) $90x^4 - 36x^2 + 2$

3. Which of the following is an antiderivative of
$$f(x) = \frac{x^6 - 2x^4 + 4}{x^3} \ ?$$

    (A) $\dfrac{x^4}{4} - 2x^2 - \dfrac{4}{x^2} + C$

    (B) $\dfrac{x^4}{4} - x^2 - \dfrac{2}{x^2} + C$

    (C) $\dfrac{x^4}{4} - x^2 + \dfrac{2}{x^2} + C$

    (D) $\dfrac{x^4}{4} - 2x^2 + \dfrac{4}{x^2} + C$

4. If $f(t) = 13t^{\frac{8}{5}}$ and $f(t)$ is the derivative of a function $F(t)$, then $\dfrac{F(t)}{t}$ could be

    (A) $5t^{\frac{8}{5}} + 1$

    (B) $5t^{\frac{8}{5}} + \dfrac{1}{t}$

    (C) $5t^{\frac{13}{5}} + 2$

    (D) $5t^{\frac{13}{5}} + \dfrac{5}{t}$

Integrals

ANSWERS AND EXPLANATIONS

1. D

Two antiderivatives of the same function must differ by a constant. Because $f(x) = g(x)$, $F(x)$ and $G(x)$ must differ by a constant—but you don't know what the constant is. Therefore, the answer is **(D)**.

2. A

Because $F' = f$ and $f(x) = 18x^5 - 12x^3 + 2x$, then

$$F(x) = \int \left(18x^5 - 12x^3 + 2x\right) dx$$

$$F(x) = \int \left(18x^5 - 12x^3 + 2x\right) dx$$

$$= \frac{18}{6}x^6 - \frac{12}{4}x^4 + \frac{2}{2}x^2 + C$$

$$= 3x^6 - 3x^4 + x^2 + C$$

The only choice that matches is **(A)**, $3x^6 - 3x^4 + x^2 + 1$.

3. B

There is no "quotient rule" for integrating, so simplify the expression first. Then integrate each term to find the antiderivative, F:

$$f(x) = \frac{x^6 - 2x^4 + 4}{x^3} = x^3 - 2x + 4x^{-3}$$

$$F(x) = \int x^3\, dx - 2\int x\, dx + 4\int x^{-3}\, dx$$

$$= \frac{x^4}{4} - \frac{2x^2}{2} + \frac{4x^{-2}}{-2}$$

$$= \frac{x^4}{4} - x^2 - \frac{2}{x^2} + C$$

This is a perfect match for **(B)**.

4. B

Because $f(t) = 13t^{\frac{8}{5}}$ and $f(t)$ is the derivative of $F(t)$, find the antiderivative (the integral) of $f(t)$:

$$F(t) = \int f(t)\, dt = \int 13t^{\frac{8}{5}}\, dt$$

$$F(t) = 13\int t^{\frac{8}{5}}\, dt = 13 \cdot \frac{5}{13}t^{\frac{13}{5}} + C$$

$$\therefore F(t) = 5t^{\frac{13}{5}} + C$$

This means:

$$\frac{F(t)}{t} = \frac{5t^{\frac{13}{5}}}{t} + \frac{C}{t} = 5t^{\frac{8}{5}} + \frac{C}{t}$$

Only **(B)**, $5t^{\frac{8}{5}} + \frac{1}{t}$, agrees.

11.2 ANTIDERIVATIVES SUBJECT TO AN INITIAL CONDITION

To answer a question like this:

Suppose Carol is a new first-year college student. She begins saving money at a rate of $f(t) = 300t^2$ dollars/year, where t is in years, to buy a car when she graduates. If Carol began college with $5,000, how much money will she have saved by the time she graduates in four years?

(A) $4,800

(B) $6,400

(C) $9,800

(D) $11,400

You need to know this:

Initial Conditions—What Is C ?

- Information that allows for the determination of C in the indefinite integral $\int f(x)\, dx = F(x) + C$ is called an initial condition.

- This can be given as a starting value when $x = 0$, or as some other data point (i.e., money saved after two years).

✔ AP Expert Note

Pay attention to whether the question is asking for an application of the rate of change equation or its antiderivative. Antiderivatives with initial conditions need to be adjusted by the constant C.

Integrals

You need to do this:

- The equation given is a rate of change of money; take the antiderivative to get the equation for the amount of money saved.

- Solve for C based on the initial condition given.

- Apply the antiderivative equation to the condition(s) in the question.

Answer and Explanation:

D

If $f(t)$ is the rate of change of Carol's money, then the amount of money Carol has, $g(t)$, is an antiderivative of f. Which antiderivative? That is,

$$g(t) = \int 300t^2 \, dt = 100t^3 + C.$$

To answer the question, how much money Carol starts with needs to be known, which the question states is $5,000. This is the initial condition; it be can phrased in terms of g: $g(0) = 5,000$, which can be used to solve for C: $g(0) = 5,000 = 100 \cdot 0^3 + C \Rightarrow C = 5,000$.

This completes the equation for the amount of money Carol has t years after starting college: $g(t) = 100t^3 + 5,000$. After four years of school, Carol has $g(4) = 100 \cdot 4^3 + 5,000 = \$11,400$—not enough to buy an SUV, but enough for a nice budget sedan. Therefore, **(D)** is correct.

PRACTICE SET

1. A city's population began increasing at a rate of $p(t) = 900t^2 - 400t$ in 2000. If the population is 100,000 at the start of 2005, what will the population be at the start of 2010 ?

   (A) 186,000

   (B) 280,000

   (C) 347,500

   (D) 380,000

2. Steve is a recent college graduate who has decided to start saving for a down payment on his first home. Steve receives a sign-on bonus for his new job of $20,000, to which he will begin adding money at the rate of $f(t) = 50t + 400$ dollars/month, where t is in months. How many months will it take Steve to reach his goal of having $50,000 for a down payment?

   (A) 24

   (B) 28

   (C) 32

   (D) 36

3. From the months of October through May, cases of pneumonia seen at a town's hospital increase at the rate of $n(t) = 8t^3$, where t is in months. If the hospital has had 300 cases from October 1st through January 1st, how many cases can the hospital expect to see by May 1st?

   (A) 2,744

   (B) 2,882

   (C) 4,802

   (D) 4,940

4. Charles is retiring from his career as a professor and is considering moving his retirement fund. Charles's current account accrues money monthly at the rate of $r(t) = 4{,}000t$, where t is time in months, and he is projected to have $350,000 in 6 months. He is considering moving his money into a different account that will accrue money only once every 3 months at the rate of $s(u) = 9{,}000u^2$, where u is time in 3-month blocks. Which account would give Charles more money at the end of a 12-month period from now?

   (A) His current account would have more money.

   (B) The new account would have more money.

   (C) The accounts would have the same amount of money.

   (D) Impossible to answer without being provided with his current money.

Integrals

ANSWERS AND EXPLANATIONS

1. C

If $p(t)$ is the rate of change of the population, then the total population, $q(t)$, is an antiderivative of p.

$$q(t) = \int 900t^2 - 400t \; dt = 300t^3 - 200t^2 + C$$

The initial condition given in this question is that the population is 100,000 at the start of 2005; in terms of q: $q(5) = 100{,}000$.

$$q(5) = 300 \cdot 5^3 - 200 \cdot 5^2 + C$$
$$100000 = 32500 + C \Rightarrow C = 67500$$

This completes the equation for the total population of the city: $q(t) = 300t^3 - 200t^2 + 67{,}500$. Evaluating for the population in 2010 yields:

$q(10) = 300(10)^3 - 200(10)^2 + 67{,}500 = 347{,}500$. **(C)** is correct.

2. B

If $f(t)$ is the rate of change of Steve's money, then the amount of money saved, $g(t)$, is an antiderivative of f.

$$g(t) = \int 50t + 400 \; dt = 25t^2 + 400t + C$$

The initial condition given in this question is that Steve starts with a sign-on bonus of \$20,000; in terms of g: $g(0) = 20{,}000$.

$$g(0) = 25 \cdot 0^2 + 400 \cdot 0 + C$$
$$20000 = 0 + C \Rightarrow C = 20000$$

This completes the equation for the total money saved: $q(t) = 25t^2 + 400t + 20{,}000$. To determine how many months it will take Steve to reach \$50,000, evaluate the equation $g(t) = 50{,}000$:

$$g(t) = 25t^2 + 400t + C$$
$$50000 = 25t^2 + 400t + 20000$$
$$0 = 25t^2 + 400t - 30000$$

Solving this quadratic equation gives values of $t = -43.6$ and $t = 27.6$. However, only 27.6 is in the domain of $t > 0$. This matches **(B)**, which states that it would take Steve 28 months to reach his \$50,000 goal.

3. D

If $n(t)$ is the rate of new cases of pneumonia seen at the hospital each month, then $p(t)$, the total cases of pneumonia seen at the hospital, is an antiderivative of $n(t)$.

$$p(t) = \int 8t^3 \; dt = 2t^4 + C$$

The initial condition given in this question is that the hospital has had 300 cases by January 1st, which is 3 months into the season. In terms of p this is: $p(3) = 300$.

$$p(3) = 2 \cdot 3^4 + C$$
$$300 = 162 + C \Rightarrow C = 138$$

This completes the equation for the total cases of pneumonia: $p(t) = 2t^4 + 138$. To determine how many cases the hospital will have seen, evaluate the equation $p(7) = 2(7)^4 + 138 = 4{,}940$ cases. **(D)** is correct.

4. A

If $r(t)$ is the rate of change of Charles's money, then the amount of money saved, $R(t)$, is an antiderivative of r.

$$R(t) = \int 4000t \; dt = 2000t^2 + C$$

The initial condition given in this question is that Charles's current account will have \$350,000 in 6 months. In terms of R this is: $R(6) = 350{,}000$.

$$R(6) = 2000 \cdot 6^2 + C$$
$$350000 = 72000 + C \Rightarrow C = 278000$$

This completes the equation for the money in Charles's current account: $R(t) = 2{,}000t^2 + 278{,}000$. To determine how much Charles would have in 12 months, evaluate the equation $R(12) = 2{,}000(12)^2 + 278{,}000 = 566{,}000$ dollars.

To evaluate how much Charles would have after a year in the new account, obtain the antiderivative, $S(u)$, of the new account s.

$$S(u) = \int 9000u^2 \; du = 3000u^3 + C$$

The constant C will have the same value of 278,000 because that's the amount Charles has now (when $t = 0$). Note that one year will only be $u = 4$ because u is in 3-month blocks. Evaluate the equation $S(4) = 3{,}000(4)^3 + 278{,}000 = 470{,}000$ dollars. Therefore, **(A)** is correct.

11.3 U-SUBSTITUTION AND ALGEBRAIC FUNCTIONS

High-Yield

To answer a question like this:

$$\int 6x\sqrt{x^2 - 1} \ dx =$$

(A) $\frac{2}{3}(x^2 - 1)^{\frac{1}{2}} + C$

(B) $2(x^2 - 1)^{\frac{3}{2}} + C$

(C) $2x(x^2 - 1)^{\frac{3}{2}} + C$

(D) $\frac{2}{5}(x^2 - 1)^{\frac{5}{2}} + C$

You need to know this:

Substitution

- There is a method, called *u-substitution*, which works like the chain rule in reverse.

- It allows simplification of complicated antiderivatives and transforms them into something that can be computed directly.

- $\int F'(u(x)) \cdot u'(x) \ dx = F(u(x)) + C$

Manipulating constants

- Not all substitution problems are a straight application of the chain rule in reverse.

- If the antiderivative is off by a constant factor, then that constant may be factored out of the antiderivative.

✔ **AP Expert Note**

You can only pull constants out of the integrand, not functions.

Integrals

You need to do this:

- Think of solving indefinite integrals by this method as the "Substitution Game."

- The only rules of the game are that all of the x's must disappear and there must be a du in the new integral.

- The goal of the game is to wind up with an integral that you can compute directly.

Answer and Explanation:

B

The presence of a function within the root indicates the chain rule would be used when taking the derivative of the antiderivative function. Begin by evaluating if the function within the root is able to be substituted out. Set $u = x^2 - 1$, then obtain the derivative $du = 2x\,dx$. Factor $6x$ into $3(2x)$ and then move the constant 3 out of the integral, as seen below.

$$\int 6x\sqrt{x^2 - 1}\,dx = \int 3 \cdot 2x\sqrt{x^2 - 1}\,dx = 3\int \sqrt{x^2 - 1}\cdot 2x\,dx$$

$$= 3\int \sqrt{u}\,du = 3\int (u)^{\frac{1}{2}}\,du = 3 \cdot \frac{(u)^{\frac{3}{2}}}{\frac{3}{2}} + C$$

$$= 2u^{\frac{3}{2}} + C = 2(x^2 - 1)^{\frac{3}{2}} + C$$

The function with respect to u can be integrated directly. Finish the problem by substituting $x^2 - 1$ back in for u to obtain **(B)**.

PRACTICE SET

1. $\int 3x^2 \left(x^3 + 1\right)^5 dx =$

   (A) $\dfrac{\left(x^3 + 1\right)^6}{6} + C$

   (B) $\dfrac{\left(x^3 + 1\right)^6}{3} + C$

   (C) $\dfrac{\left(x^3 + 1\right)^6}{6x^3} + C$

   (D) $6x^3 \left(x^3 + 1\right)^6 + C$

2. $\int x\sqrt{4 - x^2}\, dx =$

   (A) $\left(x^2 - 4\right)^{\frac{3}{2}} + C$

   (B) $\left(4 - x^2\right)^{\frac{3}{2}} + C$

   (C) $-\dfrac{1}{3}\left(4 - x^2\right)^{\frac{3}{2}} + C$

   (D) $-\dfrac{1}{3}x\left(4 - x^2\right)^{\frac{3}{2}} + C$

3. $\int \dfrac{x}{\left(3x^2 - 5\right)^4}\, dx =$

   (A) $-\dfrac{1}{18\left(3x^2 - 5\right)^3} + C$

   (B) $-\dfrac{1}{2\left(3x^2 - 5\right)^3} + C$

   (C) $-\dfrac{2}{\left(3x^2 - 5\right)^3} + C$

   (D) $-\dfrac{1}{2\left(3x^2 - 5\right)^5} + C$

4. $\int x^2 \sqrt{x - 1}\, dx =$

   (A) $= \dfrac{4}{5}(x - 1)^{\frac{5}{2}} + \dfrac{2}{3}(x - 1)^{\frac{3}{2}} + C$

   (B) $= \dfrac{2}{7}(x - 1)^{\frac{7}{2}} + \dfrac{2}{3}(x - 1)^{\frac{3}{2}} + C$

   (C) $= \dfrac{2}{7}(x - 1)^{\frac{7}{2}} + \dfrac{4}{5}(x - 1)^{\frac{5}{2}} + C$

   (D) $= \dfrac{2}{7}(x - 1)^{\frac{7}{2}} + \dfrac{4}{5}(x - 1)^{\frac{5}{2}} + \dfrac{2}{3}(x - 1)^{\frac{3}{2}} + C$

Integrals

ANSWERS AND EXPLANATIONS

1. A

There is a quantity inside the parentheses that is being multiplied by another term, so this integral is a good candidate for *u*-substitution. Start by moving the extra term next to the *dx*:

$$\int 3x^2 \left(x^3 + 1\right)^5 dx = \left(x^3 + 1\right)^5 \cdot 3x^2 \, dx$$

Now, let $u = x^3 + 1$ (the quantity inside the parentheses), which in turn means $du = 3x^2 \, dx$.

You already have exactly what you need to rewrite the integral in terms of *u*:

$$\int \left(x^3 + 1\right)^5 3x^2 \, dx = \int u^5 \, du$$

Now integrate by raising the exponent by 1 and dividing the expression by the new exponent. The result is $\int u^5 \, du = \dfrac{u^6}{6} + C.$

Finally, write the expression back in terms of *x* to arrive at $\dfrac{\left(x^3 + 1\right)^6}{6} + C,$ making **(A)** the correct choice.

2. C

There is a quantity inside the radical and an "extra" *x*, so this integral is a good candidate for *u*-substitution. Start by rewriting the function without the radical (using a fractional exponent) and moving the "extra" *x* next to the *dx*:

$$\int x\sqrt{4 - x^2} \, dx = \int \left(4 - x^2\right)^{\frac{1}{2}} x \, dx$$

Now, let $u = 4 - x^2$ (the quantity under the radical symbol), which in turn means $du = -2x \, dx$.

You already have the *x*, but not the −2, so multiply the integrand by −2. To offset this, multiply the entire integral by $-\dfrac{1}{2}$ (because $-\dfrac{1}{2} \cdot -2 = 1$ and you haven't actually changed the integral). You now have:

$$-\frac{1}{2} \int \left(4 - x^2\right)^{\frac{1}{2}} (-2x) \, dx = -\frac{1}{2} \int u^{\frac{1}{2}} \, du$$

Now integrate to get:

$$-\frac{1}{2} \int u^{\frac{1}{2}} \, du = -\frac{1}{2} \cdot \frac{2}{3} u^{\frac{3}{2}} + C = -\frac{1}{3} u^{\frac{3}{2}} + C$$

Finally, write the expression back in terms of *x* to arrive at $-\dfrac{1}{3}\left(4 - x^2\right)^{\frac{3}{2}} + C.$ **(C)** is correct.

3. A

There is a quantity raised to a power in the denominator and an extra *x* in the numerator, so this integral is a good candidate for *u*-substitution. Start by rewriting the function without the denominator (using a negative exponent) and moving the "extra" *x* next to the *dx*:

$$\int \frac{x}{\left(3x^2 - 5\right)^4} = \int \left(3x^2 - 5\right)^{-4} x \, dx$$

Now, let $u = 3x^2 - 5$ (the quantity inside the parentheses), which in turn means $du = 6x \, dx$.

You already have the *x*, but not the 6, so multiply the integrand by 6. To offset this, multiply the entire integral by $\dfrac{1}{6}$ (because $\dfrac{1}{6} \cdot 6 = 1$ and you haven't actually changed the integral). You now have:

$$\frac{1}{6} \int \left(3x^2 - 5\right)^{-4} (6x) \, dx = \frac{1}{6} \int u^{-4} \, du$$

Now integrate to get:

$$\frac{1}{6} \int u^{-4} \, du = \frac{1}{6} \cdot \frac{1}{-3} \cdot u^{-3} + C = -\frac{1}{18u^3} + C$$

Finally, write the expression back in terms of *x* to arrive at $-\dfrac{1}{18\left(3x^2 - 5\right)^3} + C.$ **(A)** is correct.

Integrals

4. **D**

The substitution $u = x^2$ would blow the rules of the substitution game, because there would be no $2x\,dx$ to sub out. What about $u = x - 1$?

$$\int x^2 \sqrt{\underset{u}{x-1}} \, \underset{du}{dx}$$

Well, that takes care of the $\sqrt{x-1}$ and the dx, but what about the x^2? This is where the rules (or lack of rules) can help. If $u = x - 1$, then $u + 1 = x$. So we can replace x^2 with $(u+1)^2$ in what is called a backsubstitution. That is:

$$\int \underset{(u+1)^2}{x^2} \sqrt{\underset{u}{x-1}} \, \underset{du}{dx} = \int (u+1)^2 \cdot \sqrt{u} \, du$$

This expression can be integrated, but first rewrite

$$\int (u+1)^2 \cdot \sqrt{u} \, du = \int (u^2 + 2u + 1) \cdot u^{\frac{1}{2}} \, du$$
$$= \int u^{\frac{5}{2}} + 2u^{\frac{3}{2}} + u^{\frac{1}{2}} \, du$$

Compute this integral directly.

$$\int u^{\frac{5}{2}} + 2u^{\frac{3}{2}} + u^{\frac{1}{2}} \, du$$
$$= \frac{2}{7} u^{\frac{7}{2}} + 2 \cdot \frac{2}{5} u^{\frac{5}{2}} + \frac{2}{3} u^{\frac{3}{2}} + C$$
$$= \frac{2}{7} u^{\frac{7}{2}} + \frac{4}{5} u^{\frac{5}{2}} + \frac{2}{3} u^{\frac{3}{2}} + C$$

Now we substitute back and arrive at **(D)**.

$$= \frac{2}{7}(x-1)^{\frac{7}{2}} + \frac{4}{5}(x-1)^{\frac{5}{2}} + \frac{2}{3}(x-1)^{\frac{3}{2}} + C$$

11.4 INTEGRATION OF TRIG FUNCTIONS

To answer a question like this:

$$\int 2 \sec^2 x \, dx =$$

(A) $2 \sin x$

(B) $2 \cos x$

(C) $2 \tan x$

(D) $2 \sec^2 x \tan^2 x$

You need to know this:

Because the derivative is...	The indefinite integral is...
$\dfrac{d}{dx}(\sin x) = \cos x$ $\dfrac{d}{dx}(\tan x) = \sec^2 x$ $\dfrac{d}{dx}(\sec x) = \sec x \tan x$	$\displaystyle\int \cos x \, dx = \sin x + C$ $\displaystyle\int \sec^2 x \, dx = \tan x + C$ $\displaystyle\int \sec x \tan x \, dx = \sec x + C$
$\dfrac{d}{dx}(-\cos x) = \sin x$ $\dfrac{d}{dx}(-\cot x) = \csc^2 x$ $\dfrac{d}{dx}(-\csc x) = \csc x \cot x$	$\displaystyle\int \sin x \, dx = -\cos x + C$ $\displaystyle\int \csc^2 x \, dx = -\cot x + C$ $\displaystyle\int \csc x \cot x \, dx = -\csc x + C$
$\dfrac{d}{dx}(\arctan x) = \dfrac{1}{1+x^2}$ $\dfrac{d}{dx}(\arcsin x) = \dfrac{1}{\sqrt{1-x^2}}$ $\dfrac{d}{dx}(\text{arcsec } x) = \dfrac{1}{\lvert x\rvert\sqrt{x^2-1}}$	$\displaystyle\int \dfrac{1}{1+x^2}\, dx = \arctan x + C$ $\displaystyle\int \dfrac{1}{\sqrt{1-x^2}}\, dx = \arcsin x + C$ $\displaystyle\int \dfrac{1}{x\sqrt{x^2-1}}\, dx = \text{arcsec }\lvert x\rvert + C$

Integrals

You need to do this:

- Build up a library of basic functions and their direct antiderivatives.

- For other, more complicated functions, develop more advanced techniques, such as substitution.

Answer and Explanation:

C

Integration of trigonometric functions requires the same memorization as taking their derivatives. Recognizing the derivatives of basic trig functions is key to being able to take their antiderivatives. As seen in the table, $\sec^2 x$ is the derivative of $\tan x$, so factor out the 2 and take the antiderivative of $\sec^2 x$.

$$\int 2 \sec^2 x \, dx = 2 \int \sec^2 x \, dx = 2 \tan x$$

Thus, **(C)** is correct.

PRACTICE SET

1. $\int \left(\beta + \sec^2 \beta\right) d\beta =$

   (A) $\frac{1}{2}\beta^2 + \frac{1}{3}\sec^3 \beta + C$

   (B) $\beta^2 + \frac{1}{3}\sec^3 \beta + C$

   (C) $\frac{1}{2}\beta^2 + \frac{1}{3}\tan \beta + C$

   (D) $\frac{1}{2}\beta^2 + \tan \beta + C$

2. If $\tan^2 x$ is an antiderivative of $f(x)$, which of the following *cannot* be an antiderivative of $f(x)$?

   (A) $\sec^2 x$

   (B) $\tan^2 x + 3$

   (C) $\dfrac{\sin^2 x}{\cos^2 x}$

   (D) $\dfrac{\sin^2 x}{1 - \cos^2 x}$

3. Let $G'(x) = g(x)$ and $g(x) = \sin x + \cos x$. Which of the following could be $G(x) + 1$?

   (A) $\sin x + \cos x + 1$

   (B) $\sin x - \cos x + 2$

   (C) $\cos x - \sin x - 1$

   (D) $-\sin x - \cos x - 2$

4. $\int \sin(2x + 1)\, dx =$

   (A) $\cos(2x + 1) + C$

   (B) $-\cos(2x + 1) + C$

   (C) $-2\cos(2x + 1) + C$

   (D) $-\frac{1}{2}\cos(2x + 1) + C$

ANSWERS AND EXPLANATIONS

1. D

Integrate each term individually. Then look carefully for a match among the answer choices.

$$\int \left(\beta + \sec^2 \beta\right) d\beta = \int \beta \, d\beta + \int \sec^2 \beta \, d\beta$$
$$= \frac{1}{2}\beta^2 + \tan \beta + C$$

(D) is correct.

2. D

Two antiderivatives of the same function must differ by a constant. You can use trig identities to assess each answer option.

A: $\sec^2 x = \tan^2 x + 1$ is an antiderivative of $f(x)$.

B: $\tan^2 x + 3$ is an antiderivative of $f(x)$.

C: $\dfrac{\sin^2 x}{\cos^2 x} = \tan^2 x$ is an antiderivative of $f(x)$.

By process of elimination, **(D)** is correct; you can also see this directly using trig identities: $\dfrac{\sin^2 x}{1 - \cos^2 x} = \dfrac{\sin^2 x}{\sin^2 x} = 1.$

3. B

Because $G'(x) = g(x)$ and $g(x) = \sin x + \cos x$:

$$G(x) = \int g(x) \, dx = \int (\sin x + \cos x) \, dx$$
$$G(x) = \int (\sin x + \cos x) \, dx$$
$$G(x) = \int \sin x \, dx + \int \cos x \, dx$$
$$G(x) = -\cos x + \sin x + C$$
$$G(x) = \sin x - \cos x + C$$

It follows that $G(x) + 1 = \sin x - \cos x + C + 1$. If $C = 1$, then **(B)** would agree with $G(x) + 1 = \sin x - \cos x + 2.$

4. D

You know that $\int \sin x \, dx = -\cos x + C$, so eliminate (A). Now, think about the derivatives of the other answer choices, and which one could possibly yield $\sin(2x + 1)$. Because of the chain rule, you can differentiate (D) to get:

$$\frac{d}{dx}\left(-\frac{1}{2}\cos(2x + 1) + C\right)$$
$$= -\frac{1}{2}\left(-\sin(2x + 1)\right) \cdot 2$$
$$= \sin(2x + 1)$$

So, **(D)** must be correct.

11.5 INTEGRATION OF EXPONENTIAL AND LOGARITHMIC FUNCTIONS

To answer a question like this:

$$\int 6^x \ln 6 \, dx =$$

(A) $6^x + C$

(B) $\dfrac{6^x}{x} + C$

(C) $\dfrac{6^x}{6x} + C$

(D) $\dfrac{6^x}{\ln 6} + C$

You need to know this:

Because the derivative is...	The indefinite integral is...
$\dfrac{d}{dx}(e^x) = e^x$	$\int e^x \, dx = e^x + C$
$\dfrac{d}{dx}(\ln x) = \dfrac{1}{x}$	$\int \dfrac{1}{x} \, dx = \ln|x| + C$
$\dfrac{d}{dx}\left(\dfrac{b^x}{\ln b}\right) = b^x$ for $b > 0$	$\int b^x \, dx = \dfrac{b^x}{\ln b} + C$

Building the library

- Just as was seen with trigonometric functions, exponential and logarithmic functions require a degree of memorization.

- Look for these patterns in indefinite integral problems, but remember to use substitution and constant manipulation where needed to get the function in the appropriate form.

You need to do this:

- Get the equation in the appropriate form, b^x, to integrate.

- Recall ln is simply taking the log base e of a number. Therefore, ln 6 is a real number of constant value and as such can be moved out of the integral.

Answer and Explanation:

A

$$\int 6^x \ln 6\, dx = \ln 6 \int 6^x\, dx$$
$$= \ln 6\, \frac{6^x}{\ln 6} + C$$
$$= 6^x + C$$

When the appropriate form of the integral and the solution for that integral are known, the problem can be quickly solved to obtain **(A)**.

PRACTICE SET

1. Let $H(x)$ be an antiderivative of $h(x)$ and $h(x) = e^x + x$. It follows that $H(x) + x =$

   (A) $xe^{x-1} + x^2 + C$

   (B) $e^x + x^2 + C$

   (C) $e^x + \dfrac{1}{2}x^2 + C$

   (D) $e^x + \dfrac{1}{2}x^2 + x + C$

2. $\displaystyle\int \frac{(\ln x)^3}{x} =$

   (A) $\dfrac{(\ln x)^4}{4} + C$

   (B) $\dfrac{(\ln x)^4}{4x} + C$

   (C) $\dfrac{(\ln x)^4}{4} + \dfrac{1}{x^2} + C$

   (D) $\dfrac{(\ln x)^4}{4} - \dfrac{1}{x^2} + C$

3. Which of the following is an antiderivative of $f(x) = \ln x$?

   (A) $F(x) = \dfrac{1}{x}$

   (B) $F(x) = x \ln x$

   (C) $F(x) = x \ln x - x$

   (D) $F(x) = \ln x - \dfrac{1}{x}$

4. $\displaystyle\int \left(2e^x + \frac{5}{x^2}\right) dx =$

   (A) $e^{2x} - \dfrac{5}{x} + C$

   (B) $e^{2x} - \dfrac{10}{x^3} + C$

   (C) $2e^x - \dfrac{5}{x} + C$

   (D) $e^{2x} - \dfrac{10}{x^3} + C$

ANSWERS AND EXPLANATIONS

1. D

Because $H(x)$ is an antiderivative of $h(x)$, it follows that:

$$H(x) = \int h(x)\,dx = \int \left(e^x + x\right) dx$$
$$H(x) = \int e^x\,dx + \int x\,dx$$
$$H(x) = e^x + \frac{1}{2}x^2 + C$$

Thus, $H(x) + x = e^x + \frac{1}{2}x^2 + x + C$. That's **(D)**.

2. A

Because the quantity inside the parentheses is natural log, and because $\frac{d}{dx}\ln x = \frac{1}{x}$, this integral is a good candidate for u-substitution. Let $u = \ln x$, which means $du = \frac{1}{x}\,dx$. This is already part of the integrand, so you're ready to write the integral in terms of u:

$$\int \frac{(\ln x)^3}{x}\,dx = \int (\ln x)^3 \cdot \frac{1}{x}\,dx = \int u^3\,du$$

Now integrate:

$$\int u^3\,du = \frac{u^4}{4} + C$$

Replacing u with $\ln x$ gives $\frac{(\ln x)^4}{4} + C$, which is **(A)**.

Don't forget—if you're not sure how to integrate, you can always take the derivative of each of the answer choices.

3. C

You do not need to compute the antiderivative of $f(x)$ to answer this question. By definition, $F(x)$ is an antiderivative of $f(x)$ if $\frac{d}{dx}(F(x)) = f(x)$. This means you can differentiate the answer choices to determine which is the desired antiderivative.

A: $\dfrac{d}{dx}\left(\dfrac{1}{x}\right) = \dfrac{d}{dx}\left(x^{-1}\right) = -\dfrac{1}{x^2}$ Eliminate A.

B: $\dfrac{d}{dx}(x \ln x) = \ln x + 1$ Eliminate B.

(C): $\dfrac{d}{dx}(x \ln x - x) = \ln x + 1 - 1 = \boxed{\ln x}$

D: $\dfrac{d}{dx}\left(\ln x - \dfrac{1}{x}\right) = \dfrac{1}{x} + \dfrac{1}{x^2}$ Eliminate D.

The correct answer is therefore **(C)**.

4. C

Use properties of indefinite integrals to rewrite the given integral as:

$$\int \left(2e^x + \frac{5}{x^2}\right) dx = 2\int e^x\,dx + 5\int x^{-2}\,dx$$

The antiderivative of e^x is e^x, because $\frac{d}{dx}(e^x) = e^x$, so:

$$2\int e^x\,dx = 2e^x + C$$

The antiderivative of x^{-2} is $-x^{-1}$ because $\frac{d}{dx}(-x^{-1}) = x^{-2}$, so:

$$5\int x^{-2}\,dx = 5(-x^{-1}) + D = -\frac{5}{x} + D$$

Combining these two indefinite integrals and writing the sum of the two constants as a single constant yields:

$$\int \left(2e^x + \frac{5}{x^2}\right) dx = 2\int e^x\,dx + 5\int x^{-2}\,dx$$
$$= 2e^x - \frac{5}{x} + C$$

This matches **(C)**.

RAPID REVIEW

If you take away only three things from this chapter:

1.

Because the derivative is...	The indefinite integral is...
$\dfrac{d}{dx}(x) = 1$	$\displaystyle\int 1\, dx = x + C$
$\dfrac{d}{dx}\left(\dfrac{x^{r+1}}{r+1}\right) = x^r, r \neq -1$	$\displaystyle\int x^r\, dx = \dfrac{x^{r+1}}{r+1} + C$
$\dfrac{d}{dx}(\sin x) = \cos x$	$\displaystyle\int \cos x\, dx = \sin x + C$
$\dfrac{d}{dx}(\tan x) = \sec^2 x$	$\displaystyle\int \sec^2 x\, dx = \tan x + C$
$\dfrac{d}{dx}(\sec x) = \sec x\,\tan x$	$\displaystyle\int \sec x\,\tan x\, dx = \sec x + C$
$\dfrac{d}{dx}(-\cos x) = \sin x$	$\displaystyle\int \sin x\, dx = -\cos x + C$
$\dfrac{d}{dx}(-\cot x) = \csc^2 x$	$\displaystyle\int \csc^2 x\, dx = -\cot x + C$
$\dfrac{d}{dx}(-\csc x) = \csc x\,\cot x$	$\displaystyle\int \csc x\,\cot x\, dx = -\csc x + C$
$\dfrac{d}{dx}\left(e^x\right) = e^x$	$\displaystyle\int e^x\, dx = e^x + C$
$\dfrac{d}{dx}(\ln x) = \dfrac{1}{x}$	$\displaystyle\int \dfrac{1}{x}\, dx = \ln|x| + C$
$\dfrac{d}{dx}\left(\dfrac{b^x}{\ln b}\right) = b^x$ for $b > 0$	$\displaystyle\int b^x\, dx = \dfrac{b^x}{\ln b} + C$
$\dfrac{d}{dx}(\arctan x) = \dfrac{1}{1+x^2}$	$\displaystyle\int \dfrac{1}{1-x^2}\, dx = \arctan x + C$
$\dfrac{d}{dx}(\arcsin x) = \dfrac{1}{\sqrt{1-x^2}}$	$\displaystyle\int \dfrac{1}{\sqrt{1-x^2}}\, dx = \arcsin x + C$
$\dfrac{d}{dx}(\text{arcsec } x) = \dfrac{1}{|x|\sqrt{x^2-1}}$	$\displaystyle\int \dfrac{1}{x\sqrt{x^2-1}}\, dx = \text{arcsec }|x| + C$

2. An initial condition is information that allows for the determination of C in the indefinite integral $\int f(x)\, dx = F(x) + C$.

3. The rules of the substitution game say that any x disappears, and that there must be a du. Where applicable, replace any additional x values with u values through back-substitution (i.e., $u = x - 1$, then $u + 1 = x$).

TEST WHAT YOU LEARNED

x	$F(x)$
1	$-\dfrac{5}{2}$
4	10

1. Let $f(x) = F'(x)$. The table above lists some values of x and F. If $f(x) = 5x\sqrt{x} - 7x + 1$, which of the following represents $F(x)$?

   (A) $2x^{\frac{5}{2}} - \dfrac{7}{2}x^2 + x$

   (B) $10x^{\frac{1}{2}} - \dfrac{7}{2}x^2 + 4$

   (C) $10x^{\frac{1}{2}} - \dfrac{7}{2}x^2$

   (D) $2x^{\frac{5}{2}} - \dfrac{7}{2}x^2 + x - 2$

2. Which of the following is an antiderivative of $\displaystyle\int \dfrac{x^2 + 3x + 7}{\sqrt{x}}\, dx$?

   (A) $\dfrac{2}{5}x^{\frac{5}{2}} + \dfrac{20}{3}x^{\frac{3}{2}} + C$

   (B) $\dfrac{2}{5}x^{\frac{5}{2}} + 2x^{\frac{3}{2}} + 14x^{\frac{1}{2}} + C$

   (C) $\dfrac{5}{2}x^{\frac{5}{2}} + 2x^{\frac{3}{2}} + \dfrac{7}{2}x^{\frac{1}{2}} + C$

   (D) $2\left(\dfrac{x^{\frac{7}{2}}}{3} + x^{\frac{5}{2}} + 7x^{\frac{3}{2}}\right) + C$

3.

$$\int \frac{3}{\sqrt{1-9x^2}}\, dx =$$

(A) $\sin^{-1}(3x) + C$

(B) $3 \tan^{-1}(3x) + C$

(C) $\tan^{-1}(3x) + C$

(D) $\frac{1}{3} \sin^{-1}(3x) + C$

4. Which of the following is an antiderivative of $\int (x^2 + 2x)\sqrt{x^3 + 3x^2 + 5}\, dx$?

(A) $\left(\frac{x^3}{3} + x^2\right)\left(\frac{2}{5}x^{\frac{5}{2}} + \frac{3}{2}x^2 + 5x\right) + C$

(B) $\frac{2}{9}\left(x^3 + 3x^2 + 5\right)^{\frac{3}{2}} + C$

(C) $2\left(x^3 + 3x^2 + 5\right)^{\frac{5}{2}} + C$

(D) $\frac{9\sqrt{x^3 + 3x^2 + 5}}{x^2 + 2} + C$

5.

$$\int \frac{6x^2 - 1}{2x^3 - x}\, dx =$$

(A) $\frac{4x^3 - 2x}{x^4 - x^2} + C$

(B) $\ln\left|6x^2 - 1\right| + C$

(C) $\ln\left|2x^3 - x\right| + C$

(D) $2 \ln |x| + C$

Integrals

Integrals

Answer Key

Test What You Already Know

1. **B** **Learning Objective:** 11.1

2. **C** **Learning Objective:** 11.2

3. **B** **Learning Objective:** 11.3

4. **D** **Learning Objective:** 11.4

5. **B** **Learning Objective:** 11.5

Test What You Learned

1. **D** **Learning Objective:** 11.2

2. **B** **Learning Objective:** 11.1

3. **A** **Learning Objective:** 11.4

4. **B** **Learning Objective:** 11.3

5. **C** **Learning Objective:** 11.5

 # REFLECTION

Test What You Already Know score: _____

Test What You Learned score: _____

Use this section to evaluate your progress. After working through the pre-quiz, check off the boxes in the "Pre" column to indicate which Learning Objectives you feel confident about. Then, after completing the chapter, including the post-quiz, do the same to the boxes in the "Post" column. Keep working on unchecked Objectives until you're confident about them all!

Pre	Post		
☐	☐	**11.1**	Use integration rules and algebraic manipulation to find antiderivatives of polynomial, radical, and rational functions
☐	☐	**11.2**	Find the antiderivative of a function subject to an initial condition
☐	☐	**11.3**	Use *u*-substitution to find antiderivatives of polynomial, radical, and rational functions
☐	☐	**11.4**	Integrate trigonometric functions and inverse trigonometric forms
☐	☐	**11.5**	Integrate exponential and logarithmic functions

 # FOR MORE PRACTICE

Complete more practice online at kaptest.com. Haven't registered your book yet? Go to kaptest.com/booksonline to begin.

ANSWERS AND EXPLANATIONS

Test What You Already Know

1. B Learning Objective: 11.1

To go from $\frac{dy}{dx}$ to y, integrate the expression:

$$y = \int \left(x^2 + 4\sqrt{x} - \frac{1}{x^2} \right) dx$$

$$= \int \left(x^2 + 4x^{\frac{1}{2}} - x^{-2} \right) dx$$

$$= \frac{x^3}{3} + 4 \cdot \frac{2}{3} x^{\frac{3}{2}} - \frac{x^{-1}}{-1} + C$$

$$= \frac{x^3}{3} + \frac{8}{3} x^{\frac{3}{2}} + \frac{1}{x} + C$$

That's a perfect match for **(B)**.

2. C Learning Objective: 11.2

To find the function F, integrate f:

$$\int \left(9x^2 + x - 8 \right) dx = 9 \cdot \frac{x^3}{3} + \frac{x^2}{2} - 8x + C$$

Simplify, then use the initial condition to solve for C:

$$F(x) = 3x^3 + \frac{x^2}{2} - 8x + C$$

$$F(0) = 3 \cdot 0^3 + \frac{0^2}{2} - 8 \cdot 0 + C$$

$$2 = C$$

Substitute $C = 2$ into the expression for $F(x)$ and you have **(C)**.

Note that you could narrow down the choices here by checking the initial condition, $F(0) = 2$. You could also find the answer by taking the derivative of each answer choice, but that would take longer than integrating the function f.

3. B Learning Objective: 11.3

When you see a complicated expression raised to a power, think u-substitution. Here, let $u = x^{\frac{1}{2}} - 8$, which means $du = \frac{1}{2} x^{-\frac{1}{2}} dx$. Compensate for the extra $\frac{1}{2}$ that is multiplied to the dx portion of the integrand by multiplying the entire integral by 2, then rearrange to get:

$$2 \int \left(x^{\frac{1}{2}} - 8 \right)^3 \times \frac{1}{2} x^{-\frac{1}{2}} dx$$

Written in terms of u, this is:

$$2 \int u^3 \, du = 2 \cdot \frac{u^4}{4} + C = \frac{1}{2} u^4 + C$$

Once you write your answer back in terms of x, you get: $\frac{1}{2} \left(x^{\frac{1}{2}} - 8 \right)^4 + C$, which is **(B)**.

4. D Learning Objective: 11.4

Be sure to memorize the derivatives of the six basic trig functions and their inverses before Test Day. Then write each function and its derivative on the test before you even start (this way you won't forget during the test). The derivative of tan u is $\sec^2 u$, so:

$$\int \sec^2 u \, du = \tan u + C$$

Here, if $u = \frac{x}{2}$, then $du = \frac{1}{2} dx$. You already have the dx, but not the $\frac{1}{2}$, so multiply the integrand by $\frac{1}{2}$, and to offset this, multiply the entire integral by 2. This means **(D)** is correct.

5. B Learning Objective: 11.5

Write the integrand so that e is no longer in the denominator:

$$\int -xe^{-3x^2} dx$$

Because the power of the x in front of the e is one less than the power of the x in the exponent, this is a good candidate for u-substitution. Let $u = -3x^2$, which means $du = -6x \, dx$. You already have the $-x \, dx$, but not the 6, so multiply the integrand by 6 and offset it by multiplying the entire integral by $\frac{1}{6}$. Then integrate:

$$\frac{1}{6} \int e^{-3x^2} (-6x) \, dx = \frac{1}{6} \int e^u \, du = \frac{1}{6} e^u + C$$

Replace u with $-3x^2$, flip the e term back to the bottom, and that's **(B)**.

Integrals

Test What You Learned

1. D Learning Objective: 11.2

To find F, integrate f. Rewrite $x\sqrt{x} = x^1 \cdot x^{\frac{1}{2}} = x^{\frac{3}{2}}$ to make the function easier to integrate:

$$\int \left(5x^{\frac{3}{2}} - 7x + 1\right) dx = 5\left(\frac{2}{5}x^{\frac{5}{2}}\right) - 7\left(\frac{1}{2}x^2\right) + x + C$$

Simplify, then use one of the values of F given in the table to solve for C.

$$F(x) = 2x^{\frac{5}{2}} - \frac{7}{2}x^2 + x + C$$

$$F(1) = 2 \cdot 1^{\frac{5}{2}} - \frac{7}{2} \cdot 1^2 + 1 + C$$

$$-\frac{5}{2} = 2 - \frac{7}{2} + 1 + C$$

$$\frac{7}{2} - \frac{5}{2} = 3 + C$$

$$1 - 3 = C$$

$$-2 = C$$

Substitute $C = -2$ into the expression, and that gives you **(D)**.

2. B Learning Objective: 11.1

To find the antiderivative, first rewrite $\sqrt{x}$ with a negative (fractional) exponent. Then distribute and evaluate each term separately:

$$\int \frac{x^2 + 3x + 7}{\sqrt{x}} dx = \int \left(x^2 + 3x^1 + 7\right) x^{-\frac{1}{2}} dx$$

$$= \int \left(x^{\frac{3}{2}} + 3x^{\frac{1}{2}} + 7x^{-\frac{1}{2}}\right) dx$$

$$= \frac{x^{\frac{3}{2}+1}}{\frac{3}{2}+1} + 3\left(\frac{x^{\frac{1}{2}+1}}{\frac{1}{2}+1}\right) + 7\left(\frac{x^{-\frac{1}{2}+1}}{-\frac{1}{2}+1}\right) + C$$

$$= \frac{2}{5}x^{\frac{5}{2}} + 3\left(\frac{2}{3}x^{\frac{3}{2}}\right) + 7\left(2x^{\frac{1}{2}}\right) + C$$

$$= \frac{2}{5}x^{\frac{5}{2}} + 2x^{\frac{3}{2}} + 14x^{\frac{1}{2}} + C$$

That's a perfect match for **(B)**.

3. A Learning Objective: 11.4

In most integrals that look like this, think u-substitution first. Unfortunately, it won't work this time. Next, if you see perfect squares in the denominator of an integral, think "inverse trig function." Here, use the fact that $\frac{d}{du}\sin^{-1} u = \frac{1}{\sqrt{1-u^2}}$. Let $u = 3x$, which means $du = 3\,dx$, which already appears in the integrand.

So:

$$\int \frac{3}{\sqrt{1-9x^2}} dx = \int \frac{1}{\sqrt{1-(3x)^2}} 3\,dx$$

$$= \int \frac{1}{\sqrt{1-u^2}} du$$

$$= \sin^{-1} u + C$$

$$= \sin^{-1}(3x) + C$$

(A) is correct.

4. B Learning Objective: 11.3

First, note that $3x^2 + 6x = 3(x^2 + 2x)$ is the derivative of $x^3 + 3x^2 + 5$. That makes this a good candidate for u-substitution, with $u = x^3 + 3x^2 + 5$.

$$du = \left(3x^2 + 6x\right) dx = 3\left(x^2 + 2x\right) dx$$

Multiply the integrand by 3 and offset that by multiplying the entire integral by $\frac{1}{3}$. Now integrate:

$$\int \left(x^2 + 2x\right)\left(\sqrt{x^3 + 3x^2 + 5}\right) dx$$

$$= \frac{1}{3} \int \left(\sqrt{x^3 + 3x^2 + 5}\right) \cdot 3\left(x^2 + 2x\right) dx$$

$$= \frac{1}{3} \int \sqrt{u}\, du = \frac{1}{3} \int u^{\frac{1}{2}}\, du$$

$$= \frac{1}{3}\left(\frac{2}{3}u^{\frac{3}{2}}\right) + C$$

$$= \frac{2}{9}u^{\frac{3}{2}} + C$$

Finally, replace u with $x^3 + 3x^2 + 5$ for the final answer. You've got a match for **(B)**.

Note: If you didn't think of using substitution here, you could take the derivative of each of the answer choices until you find one that matches the integrand. Of course, this will take considerably longer!

5. C Learning Objective: 11.5

First, recognize that the derivative of the denominator is the numerator:

$$\frac{d}{dx}\left[2x^3 - x\right] = 6x^2 - 1$$

This means you can use u-substitution with $u = 2x^3 - x$ and $du = (6x^2 - 1)\,dx$. After the substitution, you're ready to integrate:

$$\int \frac{6x^2 - 1}{2x^3 - x}\,dx = \int \frac{1}{u}\,du$$
$$= \ln|u| + C$$
$$= \ln\left|2x^3 - x\right| + C$$

That makes **(C)** the correct answer.

CHAPTER 12

The Definite Integral

LEARNING OBJECTIVES

12.1 Apply the first and second fundamental theorems of calculus to definite integrals

12.2 Use Riemann sums to approximate definite integrals

12.3 Use properties to evaluate definite integrals

12.4 Evaluate a definite integral using u-substitution

12.5 Rewrite a definite integral in an equivalent form

TEST WHAT YOU ALREADY KNOW

1. $\dfrac{d}{dx}\displaystyle\int_{3}^{x^{2}}\ln\left(1+t^{2}\right)dt =$

    (A) $2x\ln\left(1+x^{4}\right)$

    (B) $2x\ln\left(1+x^{4}\right)-\ln 10$

    (C) $4x^{3}\ln\left(1+x^{4}\right)-\ln 10$

    (D) $4x^{3}\ln\left(1+x^{4}\right)+\ln 10$

x	0	3	6	9	12	15
$f(x)$	7	20	30	18	24	35

2. The table above provides data points for the continuous function $f(x)$. Use a right Riemann sum with 5 subdivisions to approximate the area under the curve of $f(x)$ on the closed interval [0, 15].

    (A) 127   (B) 134   (C) 381   (D) 402

$$f(x) = \begin{cases} 3 & \text{for } x \le 4 \\ x + 2 & \text{for } x > 4 \end{cases}$$

3. For the function f defined above, what is the value of $\int_1^{10} f(x)\, dx$?

(A) 9   (B) 54   (C) 63   (D) 70

---

4. $\int_1^2 \dfrac{dx}{3x + 5} =$

(A) $3 \ln\left(\dfrac{11}{8}\right)$

(B) $\dfrac{1}{3}(\ln 3)$

(C) $3(\ln 3)$

(D) $\dfrac{1}{3}\ln\left(\dfrac{11}{8}\right)$

5. $\int_{2}^{5} f(4x + 2)\, dx$ can also be written as:

(A) $4\int_{10}^{22} f(u)\, du$

(B) $\frac{1}{4}\int_{10}^{22} f(u)\, du$

(C) $\int_{2}^{5} f(u)\, du$

(D) $-4\int_{2}^{5} f(u)\, du$

Answers to this quiz can be found at the end of this chapter.

12.1 THE FUNDAMENTAL THEOREMS OF CALCULUS

To answer a question like this:

For $x > 0$, if $f(x) = \int_{x}^{x^2} \sqrt{t}\ dt$, which of the following gives $f'(x)$?

(A) $x - \sqrt{x}$

(B) $2x^2 - \sqrt{x}$

(C) $-\dfrac{1}{2\sqrt{x}}$

(D) $-\dfrac{1}{2x} + \dfrac{1}{2\sqrt{x}}$

You need to know this:

First Fundamental Theorem of Calculus

- If f is continuous on the closed interval $[a\ b]$ and F is any antiderivative of f, then

$$\int_{a}^{b} f(x)\,dx = F(b) - F(a).$$

- The FTC provides a connection between derivatives and integrals—it works like a bridge between the two legs of calculus.

- Graphically, this would be the area under the curve from a to b, an accumulation function for that interval.

Second Fundamental Theorem of Calculus

- The function $G(x) = \int_{a}^{x} f(t)\,dt = F(x) - F(a)$ is differentiable and $\dfrac{d}{dx}(G(x)) = f(x)$, that is,

$$\frac{d}{dx}\int_{a}^{x} f(t)\,dt = f(x).$$

- The difference between the function defined here and the accumulation function is that now there is a concrete way of computing $G(x)$: $G(x) = F(x) - F(a)$ where $F(x)$ is an antiderivative of $f(x)$.

You need to do this:

- This question is an application of the second fundamental theorem taken one step further, combining the SFT and the chain rule.

- This isn't computing the definite integral from a constant to a variable, but from a variable to a *function*.

- Set $H(u) = \int_a^u f(t)\, dt$ and rewrite $G(x) = H(u(x))$. By the chain rule, $G'(x) = H'(u(x)) \cdot u'(x)$. According to the second fundamental theorem, $H'(u) = f(u)$, so $G'(x) = f(u) \cdot u'(x)$.

✔ AP Expert Note

Think of derivatives and integrals in terms of a staircase: think of computing derivatives as going down a stair and computing integrals as going up a stair.

Answer and Explanation:

B

To find $f'(x)$, you need to take the derivative of $f(x)$. It follows from the second fundamental theorem that:

$$\frac{d}{dx}\left(\int_x^{x^2} \sqrt{t}\, dt \right) = \frac{d}{dx}\left(\int_c^{x^2} \sqrt{t}\, dt \right) - \frac{d}{dx}\left(\int_c^{x} \sqrt{t}\, dt \right)$$

$$= \sqrt{(x^2)} \cdot \frac{d}{dx}\left(x^2 \right) - \sqrt{x} \cdot \frac{d}{dx}\left(x \right)$$

$$= x \cdot 2x - \sqrt{x} \cdot 1$$

$$= 2x^2 - \sqrt{x}$$

That's a perfect match for **(B)**.

PRACTICE SET

x	0	1	2	3
$f(x)$	-1	0	1	-2
$F(x)$	4	3	A	8

1. Suppose $f(x)$ is a continuous function and f is the derivative of the function $F(x)$. Given the table of values above, what is $\int_1^3 f(x)\, dx$?

   (A) -2

   (B) 5

   (C) 8

   (D) 19

2. Find the derivative of $f(x) = \int_0^{x^2} \cos(t^2)\, dt$.

   (A) $\cos(x^4)$

   (B) $\cos(x^2)$

   (C) $2x \cos(x^2)$

   (D) $2x \cos(x^4)$

3. Let $f(x) = \int_{-2}^{x} \left(3t^2 - 5t + 1\right) dt$. Then $f'(2) =$

   (A) -2

   (B) 0

   (C) 3

   (D) 6

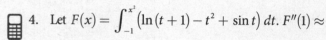 4. Let $F(x) = \int_{-1}^{x^2} \left(\ln(t+1) - t^2 + \sin t\right) dt$. $F''(1) \approx$

   (A) -3.256

   (B) -2.991

   (C) -2.770

   (D) -1.919

Integrals

ANSWERS AND EXPLANATIONS

1. B

Because you are given that F is the antiderivative of f, the first fundamental theorem of calculus tells you:

$$\int_1^3 f(x)\,dx = F(3) - F(1)$$

Filling in the values for $F(3)$ and $F(1)$ from the table yields:

$$\int_1^3 f(x)\,dx = F(3) - F(1) = 8 - 3 = 5$$

That's **(B)**.

2. D

Set $f(x) = \int_0^x \cos\left(t^2\right)dt$. Therefore, $f\left(x^2\right) = \int_0^{x^2} \cos\left(t^2\right)dt$.

By the second fundamental theorem of calculus, $f'(x) = \cos(x^2)$. Applying the chain rule yields:

$$\frac{d}{dx}\left(\int_0^{x^2} \cos\left(t^2\right)dt\right) = \frac{d}{dx}\,f\left(x^2\right)$$

$$= f'\left(x^2\right)\cdot 2x$$

$$= \cos\left(\left(x^2\right)^2\right)\cdot 2x$$

$$= 2x\cos\left(x^4\right)$$

(D) is correct.

3. C

Using the second fundamental theorem of calculus yields:

$$f'(x) = 3x^2 - 5x + 1$$

$$f'(2) = 3(2)^2 - 5(2) + 1 = 3$$

Thus, **(C)** is correct.

4. C

Use the second fundamental theorem of calculus along with the chain rule to find $F'(x)$. Then use the Derivative function on your graphing calculator to find $F''(1)$.

$$F(x) = \int_{-1}^{x^2}\left(\ln(t+1) - t^2 + \sin t\right)dt$$

$$F'(x) = \left(\ln\left(x^2 + 1\right) - \left(x^2\right)^2 + \sin\left(x^2\right)\right)\cdot 2x$$

$$F''(1) = \frac{d}{dx}\left(F'(x)\right)\big|_{x=1} \approx -2.770$$

That's, **(C)**.

12.2 RIEMANN SUMS AND DEFINITE INTEGRALS

To answer a question like this:

x	2	3	8	10	14
$f(x)$	3	7	9	5	1

The table above gives values for a continuous function $f(x)$. Which of the following gives a left Riemann approximation using 4 subintervals of $\int_{2}^{14} f(x)\, dx$?

(A) 68   (B) 71   (C) 72   (D) 76

You need to know this:

Numerical approximation rules:

- Riemann sums are approximations of the area under the curve by the rectangles determined by points along the closed interval $[a, b]$ in n equal pieces each of length $\frac{b-a}{n}$.

- The height of each rectangle can be measured using the height of its left edge (left Riemann sum), its right edge (right Riemann sum), or the average of the two (midpoint Riemann sum).

- Instead of approximating the area under the curve using a sum of rectangles, n trapezoidal regions, denoted by T_n, can be used in what is known as *trapezoidal approximation*.

Nature of the approximations:

- A left Riemann sum will underestimate portions of the curve that are increasing and overestimate portions that are decreasing. A right Riemann sum will overestimate portions of the curve that are increasing and underestimate portions that are decreasing.

- A trapezoidal approximation will always lie under the curve if the curve is concave down (and therefore be an underestimate) and above the curve if it is concave up (and therefore be an overestimate).

You need to do this:

- Left Riemann sum:

$$\underset{\text{left endpoint}}{R_n} \approx \sum_{i=1}^{n} f(x_{i-1})\Delta x$$

$$= \left(f(x_0) + f(x_1) + \ldots + f(x_{i-1})\right)\frac{b-a}{n}$$

- Right Riemann sum:

$$\underset{\text{right endpoint}}{R_n} \approx \sum_{i=1}^{n} f(x_i)\Delta x$$

$$= \left(f(x_1) + f(x_2) + \ldots + f(x_n)\right)\frac{b-a}{n}$$

- Midpoint Riemann sum:

$$\underset{\text{midpoint}}{R_n} \approx \sum_{i=1}^{n} f\left(x_{i-1} + \frac{b-a}{2n}\right)\Delta x$$

$$= \left(f\left(x_0 + \frac{b-a}{2n}\right) + f\left(x_1 + \frac{b-a}{2n}\right) + \ldots + f\left(x_{i-1} + \frac{b-a}{2n}\right)\right)\frac{b-a}{n}$$

- Trapezoidal approximation:

$$T_n = \left(\frac{1}{2}(f(x_0) + f(x_1)) \cdot \Delta x\right) + \left(\frac{1}{2}(f(x_1) + f(x_2)) \cdot \Delta x\right) + \cdots + \left(\frac{1}{2}(f(x_{n-1}) + f(x_n)) \cdot \Delta x\right)$$

$$= \left(\frac{1}{2}f(x_0) + f(x_1) + f(x_2) + \cdots + f(x_{n-1}) + \frac{1}{2}f(x_n)\right)\Delta x$$

Answer and Explanation:

D

To calculate a left Riemann sum, choose the left endpoint for each interval and multiply by the width of the interval. Based on the table of values, this gives:

$$\int_2^{14} f(x)\, dx \approx 3(3-2) + 7(8-3) + 9(10-8) + 5(14-10)$$

$$= 3(1) + 7(5) + 9(2) + 5(4)$$

$$= 3 + 35 + 18 + 20$$

$$= 76$$

(D) is correct.

PRACTICE SET

x	0	2	4	6	8
f(x)	7	4	11	5	5

1. Let f be a continuous function on the closed interval $[0, 8]$. If the values of f at five points are given in the table above, the trapezoidal approximation of $\int_0^8 f(x)\, dx$ using four subintervals of equal length is

   (A) 36

   (B) 42

   (C) 48

   (D) 52

x	2	5	8	11	14
f(x)	3	7	9	5	1

2. Which of the following gives a midpoint Riemann approximation using 2 subintervals of $\int_2^{14} f(x)\, dx$ where f has values as given in the table above?

   (A) 68

   (B) 71

   (C) 72

   (D) 76

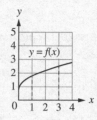

3. Consider the function $y = f(x)$ whose graph is shown above, for which $f'(x) > 0$ and $f'(x) > 0$ for all $x > 0$. Which of the following will overestimate the value of $\int_1^3 f(x)\, dx$?

   I. A left Riemann sum

   II. A right Riemann sum

   III. A trapezoidal approximation

   (A) I only

   (B) II only

   (C) III only

   (D) II and III only

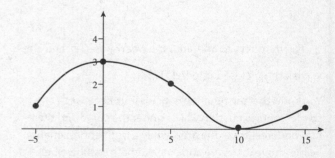

4. The graph of a differentiable function f over the closed interval $[-5, 15]$ is shown above. Find the trapezoidal approximation of the area under the curve using four equal subintervals.

   (A) 24

   (B) 26

   (C) 28

   (D) 30

Integrals

ANSWERS AND EXPLANATIONS

1. D

The formula for a trapezoidal approximation of $\int_a^b f(x)\,dx$ with n subintervals is:

$$T_n = \Delta x\left(\frac{1}{2}f(x_0) + f(x_1) + f(x_2) + \cdots + f(x_{n-1}) + \frac{1}{2}f(x_n)\right)$$

where $\Delta x = \dfrac{b-a}{n}$. Therefore:

$$\int_0^8 f(x)\,dx \approx \frac{8-0}{4}\left(\left(\frac{1}{2}\cdot 7\right) + 4 + 11 + 5 + \left(\frac{1}{2}\cdot 5\right)\right) = 52$$

That's **(D)**.

2. C

To calculate a midpoint Riemann sum, choose the midpoint for each interval and multiply by the width of the interval. This yields:

$$\int_2^{14} f(x)\,dx \approx 7(8-2) + 5(14-8)$$
$$= 7(6) + 5(6)$$
$$= 42 + 30$$
$$= 72$$

(C) is correct.

3. B

$\int_1^3 f(x)\,dx$ refers to the area between $y = f(x)$ and the

x-axis on the closed interval $[1, 3]$.

You know that the function is increasing, because $f'(x) > 0$, so a left approximation will lie under the curve and therefore underestimate the area, while a right approximation will lie above the curve and overestimate as shown below:

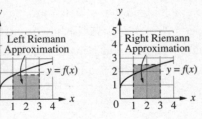

Therefore, statement I is false and statement II is true. This means you have only to consider the trapezoidal approximation. A trapezoidal approximation will always lie under the curve if the curve is concave down, and above the curve if it is concave up. Here, you know that $f''(x) < 0$; therefore, the curve is concave down and the trapezoidal approximation will underestimate the area as shown below:

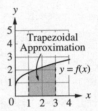

Therefore, statement III is false, making **(B)** the correct answer.

4. D

Use the formula for the fourth trapezoidal approximation:

$$T_4 = \Delta x\left(\frac{1}{2}f(x_0) + f(x_1) + f(x_2) + f(x_3) + \frac{1}{2}f(x_4)\right)$$

where $\Delta x = \dfrac{b-a}{4}$. Here, $\Delta x = \dfrac{15-(-5)}{4} = 5$ and:

Sampling point	x_0	x_1	x_2	x_3	x_4
x_i	−5	0	5	10	15
$f(x_i)$	1	3	2	0	1

where you read the values $f(x_i)$ from the graph provided. Thus:

$$T_4 = 5\left(\frac{1}{2}\cdot 1 + 3 + 2 + 0 + \frac{1}{2}\cdot 1\right) = 5 \cdot 6 = 30$$

(D) is correct.

12.3 PROPERTIES OF DEFINITE INTEGRALS

To answer a question like this:

Suppose $\int_1^2 f(x)\ dx = 2$, $\int_0^3 f(x)\ dx = 1$, and $\int_1^3 3f(x)\ dx = -6$. What is $\int_0^1 f(x)\ dx$?

(A) -2 (B) 2 (C) 3 (B) 5

You need to know this:

Because the definite integral is defined in terms of a limit of Riemann sums, the rules for manipulation of limits come into play when evaluating definite integrals. For these theorems, assume that the functions f and g are basically continuous, meaning that they are continuous except for a few point or jump discontinuities.

- $\int_a^b f(x) \pm g(x)\ dx = \int_a^b f(x)\ dx \pm \int_a^b g(x)\ dx$

- $\int_a^b c \cdot f(x)\ dx = c \int_a^b f(x)\ dx$

- $\int_a^b f(x)\ dx = -\int_b^a f(x)\ dx$

- $\int_a^b f(x)\ dx = \int_a^c f(x)\ dx + \int_c^b f(x)\ dx$

Area under the x-axis is negative

- Because the definite integral is defined as the area of a sum of rectangles whose heights are the value of the function at various points, the definite integral considers any area that lies *under* the x-axis to be negative.

- If part of the curve is above the x-axis and part of the curve is below the x-axis, then the positive and negative areas will partially or totally cancel each other out.

Integrals

You need to do this:

- Use properties of the definite integral to solve this problem.

- Like with limits and indefinite integrals, constants in the definite integral can be manipulated.

- Because the definite integral is defined in terms of the area under the curve, it makes sense that areas can be added. If f is integrable on the closed interval $[a, b]$ and c is some point between a and b, then the function evaluated from a to b will be the same as a to $c + c$ to b.

- In fact, because of the way definite integrals are defined, this statement is true even if c is not between a and b.

Answer and Explanation:

C

Using properties of the definite integral:

$$\int_1^3 3\, f(x)\, dx = -6 \Rightarrow 3\int_1^3 f(x)\, dx = -6 \Rightarrow \int_1^3 f(x)\, dx = -2$$

Using the additivity of the definite integral:

$$\int_0^3 f(x)\, dx = \int_0^1 f(x)\, dx + \int_1^3 f(x)\, dx$$

You are given that $\int_0^3 f(x)\, dx = 1$, and you found above that $\int_1^3 f(x)\, dx = -2$. Substitute these values into the equation above and solve:

$$1 = \int_0^1 f(x)\, dx + (-2) \Rightarrow \int_0^1 f(x)\, dx = 3$$

That's **(C)**.

PRACTICE SET

1. If $\int_2^7 f(x)\,dx = 11$ and $\int_7^4 f(x)\,dx = 5$, then

   $$\int_2^4 f(x)\,dx =$$

   (A) -16

   (B) -6

   (C) 6

   (D) 16

2. Let f be a continuous function with the following properties:

   $$A = \int_a^b -f(x)\,dx, \quad B = \int_b^a f(x)\,dx$$

   $$C = \int_b^a -f(x)\,dx \quad D = \int_a^b f(x)\,dx$$

   Which of following statements is true?

   (A) $A = -B$

   (B) $C = D$

   (C) $A = C$

   (D) All three statements are true.

3. Let $a < b < c < d$ and let a continuous function f have the following properties:

   $$\int_a^b f(x)\,dx = 12, \int_a^c f(x)\,dx = -5 \text{ and}$$

   $$\int_b^d f(x)\,dx = -9$$

   Then, $\int_c^d f(x)\,dx =$

   (A) -2

   (B) 2

   (C) 7

   (D) 8

4. The function f, continuous for all real numbers x, has the following properties:

   I. $\int_1^3 f(x)\,dx = 7$

   II. $\int_1^5 f(x)\,dx = 10$

   What is the value of k if $\int_3^5 k\,f(x)\,dx = 33$?

   (A) -11

   (B) -3

   (C) 3

   (D) 11

Integrals

ANSWERS AND EXPLANATIONS

1. D

Because $\int_2^7 f(x)\ dx = 11$, it follows that:

$$\int_2^4 f(x)\ dx + \int_4^7 f(x)\ dx = 11$$

Given $\int_7^4 f(x)\ dx = 5$, it follows that $\int_4^7 f(x)\ dx = -5$.

This means that $\int_2^4 f(x)\ dx - 5 = 11$. So $\int_2^4 f(x)\ dx = 16$, which is **(D)**.

2. B

Using the properties of the definite integral, you can express each of A, B, and C in terms of D.

$$A = \int_a^b -f(x)\ dx = -\int_a^b f(x)\ dx = -D$$

$$B = \int_b^a f(x)\ dx = -\int_b^a f(x)\ dx = -D$$

$$C = \int_b^a -f(x)\ dx = -\int_b^a f(x)\ dx$$
$$= -\left(-\int_a^b f(x)\ dx\right) = \int_a^b f(x)\ dx = D$$

It follows that **(B)** is correct.

3. D

Using the properties of the definite integral, you can piece together a path from c to d using the other given integrals.

$$\int_c^d f(x)\ dx = \int_c^a f(x)\ dx + \int_a^b f(x)\ dx + \int_b^d f(x)\ dx$$
$$= -(-5) + 12 + (-9)$$
$$= 8$$

(D) is correct.

4. D

Because f is continuous and $\int_1^3 f(x)\ dx = 7$ and $\int_1^5 f(x)\ dx = 10$, it follows that:

$$\int_1^5 f(x)\ dx = \int_1^3 f(x)\ dx + \int_3^5 f(x)\ dx = 10$$

By substitution, $7 + \int_3^5 f(x)\ dx = 10$, so $\int_3^5 f(x)\ dx = 3$.

Now, consider $\int_3^5 k\ f(x)\ dx = 33$, which means that $k\int_3^5 f(x)\ dx = 33$ and by substitution $3k = 33$. Finally, $k = 11$, which is **(D)**.

12.4 U-SUBSTITUTION AND DEFINITE INTEGRALS

To answer a question like this:

$$\int_0^{\frac{\pi^2}{4}} \frac{\cos \sqrt{x}}{\sqrt{x}} \, dx =$$

(A) 0   (B) $\frac{\pi}{2}$ (C) 2   (D) π

You need to know this:

When substituting with definite integrals, there are two options available for solving the problem.

Substitute back to original variable

- With this method, use substitution to find the antiderivative of the original function, substitute back, and plug in the original limits of integration.

Change of variables

- With definite integrals, there is also the option of putting *everything*, including the limits of integration, in terms of *u* and getting rid of the *x*s completely.

✔ **AP Expert Note**

Both methods are equally valid. Choose whichever is more comfortable or quicker for the problem (i.e., if the limits of *x* readily convert to *u*).

You need to do this:

- After making the appropriate *u*-substitution, select either to substitute back to the original variable after finding the antiderivative or convert the limits of integration to *u* terms as well.

Answer and Explanation:

C

Rewrite and rearrange the original integral. Make the *u*-substitution $u = \sqrt{x}$ and $du = \dfrac{1}{2\sqrt{x}}\,dx$.

$$\int_0^{\frac{\pi^2}{4}} 2 \cdot \underbrace{\frac{1}{2}}_{\substack{\text{multiply}\\\text{by 1}}} \frac{\cos\sqrt{x}}{\sqrt{x}}\,dx = 2\int_0^{\frac{\pi^2}{4}} \cos\underbrace{\sqrt{x}}_{u} \cdot \underbrace{\frac{1}{2\sqrt{x}}\,dx}_{du}$$

$$= 2\int_{x=0}^{x=\frac{\pi^2}{4}} \cos u \,du$$

Substitute back to the original variable: Write the limits of integration as $x = 0$ and $x = \dfrac{\pi^2}{4}$ as a reminder that the values refer to *x* and not to *u*. Compute the antiderivative of cos *u* (it is sin *u*) and then substitute back.

$$2\int_{x=0}^{x=\frac{\pi^2}{4}} \cos u \,du = 2\left(\sin u \,\Big|_{x=0}^{x=\frac{\pi^2}{4}}\right) = 2\left(\sin\sqrt{x}\,\Big|_{x=0}^{x=\frac{\pi^2}{4}}\right)$$

$$= 2\left(\sin\sqrt{\frac{\pi^2}{4}} - \sin\sqrt{0}\right) = 2\sin\frac{\pi}{2} - 2\sin 0 = 2$$

Change variables: Use the equation relating *x* and *u* to change the limits of integration to be in terms of *u*. Because $u = \sqrt{x}$, the lower limit of integration $x = 0$ becomes $u = \sqrt{0} = 0$. The upper limit of integration $x = \dfrac{\pi^2}{4}$ becomes $u = \sqrt{\dfrac{\pi^2}{4}} = \dfrac{\pi}{2}$. When substituting *u* and *du* into the original integral, also replace the limits of integration with their corresponding *u*-values. The integral becomes:

$$2\int_{u=0}^{u=\frac{\pi}{2}} \cos u \,du = 2\left(\sin u \,\Big|_{u=0}^{u=\frac{\pi}{2}}\right) = 2\left(\sin\frac{\pi}{2} - \sin 0\right) = 2$$

This matches **(C)**.

PRACTICE SET

1. $\displaystyle\int_0^{\frac{\pi}{2}} \sqrt{\sin x}\cos x \ dx =$

   (A) 1

   (B) $\sqrt{\pi}$

   (C) $\dfrac{2}{3}$

   (D) $\dfrac{2\pi}{3}$

2. $\displaystyle\int_0^{\ln\frac{\pi}{2}} e^x \sin\left(e^x + \pi\right) dx =$

   (A) -1

   (B) 0

   (C) 1

   (D) π

3. $\displaystyle\int_{\frac{\pi}{3}}^{\frac{\pi}{2}} \frac{2\sin x}{\sqrt{1 - \cos x}} \ dx =$

   (A) $-4 + 2\sqrt{2}$

   (B) 1

   (C) $4 - 2\sqrt{2}$

   (D) $4 + 2\sqrt{2}$

4. $\displaystyle\int_0^1 \frac{e^x}{e^x + 1} \ dx =$

   (A) $\ln \dfrac{e+1}{2}$

   (B) $\ln (e + 1)$

   (C) $\ln (2e + 2)$

   (D) $\ln 2$

ANSWERS AND EXPLANATIONS

1. C

This integral involves two different trig functions, so it is a good candidate for u-substitution. If $u = \sin x$, then $du = \cos x \, dx$. Rewrite the integral as:

$$\int_0^{\frac{\pi}{2}} \sqrt{\sin x} \cos x \, dx = \int u^{\frac{1}{2}} \, du = \frac{2}{3} u^{\frac{3}{2}}$$

$$\frac{2}{3}(\sin x)^{\frac{3}{2}} \Big|_0^{\frac{\pi}{2}} = \frac{2}{3}(1)^{\frac{3}{2}} - \frac{2}{3}(0)^{\frac{3}{2}} = \frac{2}{3}$$

(C) is correct.

2. A

To perform a substitution with change of variables on this problem, substitute $u = e^x + \pi$, $du = e^x \, dx$, the lower limit $x = \ln \frac{\pi}{2}$ with $u = e^{\ln \frac{\pi}{2}} + \pi = \frac{\pi}{2} + \pi = \frac{3\pi}{2}$, and the upper limit $x = \ln \pi$ with $u = e^{\ln \pi} + \pi = + \pi = 2\pi$.

$$\int_{\ln \frac{\pi}{2}}^{\ln \pi} e^x \sin\left(e^x + \pi\right) dx$$

$$= \int_{u = \frac{3\pi}{2}}^{u = 2\pi} \sin u \, du = -\cos u \Big|_{\frac{3\pi}{2}}^{2\pi}$$

$$= -\left(\cos(2\pi) - \cos\left(\frac{3\pi}{2}\right)\right) = -(1 - 0) = -1$$

This matches **(A)**.

3. C

This integral involves two different trig functions, so it is a good candidate for u-substitution. Start by pulling the 2 out of the integrand:

$$\int_{\frac{\pi}{3}}^{\frac{\pi}{2}} \frac{2 \sin x}{\sqrt{1 - \cos x}} \, dx = 2 \int_{\frac{\pi}{3}}^{\frac{\pi}{2}} \frac{\sin x}{\sqrt{1 - \cos x}} \, dx$$

Now, let $u = 1 - \cos x$, which means $du = \sin x \, dx$. Rearrange the integral so that:

$$2 \int_{\frac{\pi}{3}}^{\frac{\pi}{2}} \frac{1}{\sqrt{1 - \cos x}} \sin x \, dx = 2 \int_{x = \frac{\pi}{3}}^{x = \frac{\pi}{2}} \frac{1}{\sqrt{u}} \, du$$

Next, change the bounds of integration. When $x = \frac{\pi}{3}$, $u = 1 - \cos \frac{\pi}{3} = \frac{1}{2}$. So $u = \frac{1}{2}$. When $x = \frac{\pi}{2}$, $u = 1 - \cos \frac{\pi}{2} = 1$. So $u = 1$.

$$2 \int_{u = \frac{1}{2}}^{u = 1} \frac{1}{\sqrt{u}} \, du = 2 \int_{\frac{1}{2}}^{1} u^{-\frac{1}{2}} \, du$$

$$= 2 \cdot \frac{u^{\frac{1}{2}}}{\frac{1}{2}} \Big|_{\frac{1}{2}}^{1} = 4u^{\frac{1}{2}} \Big|_{\frac{1}{2}}^{1}$$

$$= 4\left(\sqrt{1} - \sqrt{\frac{1}{2}}\right) = 4 - 2\sqrt{2}$$

That's **(C)**.

4. A

Let $u = e^x + 1 \Rightarrow du = e^x \, dx$.

Change the bounds of integration so:

$$x = 0 \Rightarrow u = e^0 + 1 = 2$$
$$x = 1 \Rightarrow u = e^1 + 1 = e + 1$$

Rewrite the integral:

$$\int_0^1 \frac{e^x}{e^x + 1} \, dx = \int_0^1 \frac{1}{e^x + 1} e^x \, dx$$

$$= \int_2^{e+1} \frac{1}{u} \, du$$

$$= \ln |u| \Big|_2^{e+1}$$

$$= \ln(e + 1) - \ln 2$$

$$= \ln \frac{e + 1}{2}$$

(A) is correct.

12.5 EQUIVALENT FORMS OF DEFINITE INTEGRALS

To answer a question like this:

Which of the following definite integrals has the same value as $\int_1^2 x^2 \cdot 3^{x^3} \, dx$?

(A) $\dfrac{1}{3} \int_1^2 3^u \, du$

(B) $\dfrac{1}{3} \int_1^8 3^u \, du$

(C) $3 \int_1^2 3^u \, du$

(D) $3 \int_1^8 3^u \, du$

You need to know this:

Finding equivalent forms

- This is nothing new! Finding equivalent forms is an application of the various rules of definite integrals.

- Often, but not always, a changing of variable is involved.

- Sometimes the problem requires solving the integral in the question and comparing it to the solutions of the integrals in the answer choices.

✔ **AP Expert Note**

Equivalent forms meshes the ability to solve definite integrals of all forms (exponentials, logarithmics, trigonometrics, etc.) with the application of the rules of definite integrals (changing variables substitution, back substitution, manipulating constants, additive properties).

You need to do this:

- Compare the definite integral in the question to the integrals in the answer choices. Note the use of x in the question and u in the answer choices.

- Determine what rule of definite integrals needs to be applied. To get from everything being in terms of x to everything being in terms of u, a change of variables needs to take place.

- Solve in terms of u for x, dx, and the upper and lower limits of integration.

✔ **AP Expert Note**

Sometimes comparing answer choices can provide insight into the potential pitfalls encountered when solving. Will the limits remain 1 to 2, or become 1 to 8? Will the constant $\frac{1}{3}$ or 3 be factored out?

Answer and Explanation:

B

$$\int_1^2 x^2 \cdot 3^{x^3} \, dx = \int_1^2 \frac{1}{3} \cdot 3x^2 \cdot 3^{x^3} \, dx$$

$$= \frac{1}{3} \int_{x=1}^{x=2} 3^u \, du$$

$$= \frac{1}{3} \int_{u=1}^{u=8} 3^u \, du$$

The solution above shows the application of changing variables substitution to the integral from the question. Note how u was put in place of x^3. Recall, $du = u' \, dx = 3x^2 \, dx$. The integral is multiplied by $\frac{1}{3} \cdot 3$, effectively 1, to provide the necessary constant for substitution. The $\frac{1}{3}$ is taken out of the integral. The limits must also undergo the variable change $u = x^3$, so $x = 1$ becomes $u = 1$ and $x = 2$ becomes $u = 8$. This matches **(B)**.

PRACTICE SET

1. Which of the following definite integrals is equivalent to $\int_0^2 f(x)\,dx$?

    I. $\int_2^4 f(x-2)\,dx$

    II. $\int_{-2}^0 f(x-2)\,dx$

    III. $\int_{-3}^{-1} f(x+3)\,dx$

    (A) I only

    (B) II only

    (C) I and II only

    (D) I and III only

2. If f is a function such that $\int_2^4 f(6x)\,dx = 8$, then which of the following must also be true?

    (A) $\int_2^4 f(u)\,du = 8$

    (B) $\int_{12}^{24} f(u)\,du = 8$

    (C) $\int_2^4 f(u)\,du = 48$

    (D) $\int_{12}^{24} f(u)\,du = 48$

3. If the substitution $u = \sqrt{x+1}$ is used, then $\int_0^3 \dfrac{dx}{x\sqrt{x+1}}$ is equivalent to

    (A) $\int_1^2 \dfrac{du}{u^2 - 1}$

    (B) $\int_1^2 \dfrac{2\,du}{u^2 - 1}$

    (C) $2\int_0^3 \dfrac{du}{u^2 - 1}$

    (D) $2\int_0^3 \dfrac{du}{u(u-1)}$

4. The u-substitution $u = \sqrt{x}$ transforms the indefinite integral $\int \sin \sqrt{x}\,dx$ into which of the following indefinite integrals?

    (A) $\int \sin u\,du$

    (B) $2\int \sin u\,du$

    (C) $\int u \sin u\,du$

    (D) $2\int u \sin u\,du$

Integrals

ANSWERS AND EXPLANATIONS

1. D

To answer this question, you need to compare the horizontal shifts in the function to the shifts in the limits of integration. In statement I, the function is shifted right 2 units and the limits of integration are also shifted right 2 units, so statement I is equivalent to the given integral. In II, the function is shifted right 2 units, but the limits of integration are shifted left 2 units, so II is not equivalent. Eliminate (B) and (C). In III, the function is shifted left 3 units and the limits of integration are also shifted left 3 units, so III is true. This means **(D)** is the correct answer.

2. D

Because $u = 6x$, $du = 6\,dx \Rightarrow \frac{1}{6}\,du = dx$. Convert $x = 2$ to $u = 12$ and $x = 4$ to $u = 24$.

$$8 = \int_2^4 f(6x)\,dx$$

$$= \int_{12}^{24} f(u)\frac{1}{6}\,du$$

$$= \frac{1}{6}\int_{12}^{24} f(u)\,du$$

$$48 = \int_{12}^{24} f(u)\,du$$

Thus, **(D)** is correct.

3. B

You're given that $u = \sqrt{x+1}$. The initial integral is written in terms of x, so rewrite x in terms of u (by squaring both sides of the equation and solving for x). This yields $u^2 = x + 1 \Rightarrow x = u^2 - 1$. Now, find dx so you can substitute that as well:

$$\frac{dx}{du} = 2u \Rightarrow dx = 2u\,du$$

Converting the bounds of integration yields:

$$x = 0 \Rightarrow u = \sqrt{0+1} = 1$$
$$x = 3 \Rightarrow u = \sqrt{3+1} = 2$$

Finally, put all the pieces together:

$$\int_0^3 \frac{dx}{x\sqrt{x+1}} = \int_1^2 \frac{2u\,du}{(u^2-1)\cdot u} = \int_1^2 \frac{2\,du}{u^2-1}$$

A perfect match for **(B)**.

4. D

This is an example of a sneaky substitution. If $u = \sqrt{x}$, then $du = \frac{1}{2\sqrt{x}}\,dx$.

Remembering the rules of substitution, you can rewrite the original integral to squeeze in a du:

$$\int \sin\sqrt{x}\,dx = \int \underbrace{2\sqrt{x}\cdot\frac{1}{2\sqrt{x}}}_{\text{multiply by 1}} \sin\sqrt{x}\,dx$$

$$= \int 2\underbrace{\sqrt{x}}_{u}\cdot\sin\underbrace{\sqrt{x}}_{u}\underbrace{\frac{1}{2\sqrt{x}}}_{du}\,dx$$

An additional term of $\sqrt{x}$ was introduced, but that's okay; you can swallow it up with a u.

Transform the integral to:

$$\int 2u\sin u\,du = 2\int u\sin u\,du$$

The correct answer is **(D)**.

Integrals

RAPID REVIEW

If you take away only six things from this chapter:

1. First fundamental theorem of calculus: If f is continuous on the closed interval $[a, b]$ and F is any antiderivative of f, then $\int_a^b f(x)\, dx = F(b) - F(a)$.

2. Second fundamental theorem of calculus: $G(x) = \int_a^x f(t)\, dt = F(x) - F(a)$

   is differentiable and $\dfrac{d}{dx}\big(G(x)\big) = f(x)$, that is, $\dfrac{d}{dx}\displaystyle\int_a^x f(t)\, dt = f(x)$.

3. Approximation of definite integrals by taking a left Riemann sum, a right Riemann sum, a midpoint Riemann sum, and a trapezoidal approximation.

4. Properties of definite integrals:

   - $\displaystyle\int_a^b f(x) \pm g(x)\, dx = \int_a^b f(x)\, dx \pm \int_a^b g(x)\, dx$

   - $\displaystyle\int_a^b c \cdot f(x)\, dx = c\ dx \int_a^b f(x)$

   - $\displaystyle\int_a^b f(x)\, dx = -\int_b^a f(x)\, dx$

   - $\displaystyle\int_a^b f(x)\, dx = \int_a^c f(x)\, dx + \int_c^b f(x)\, dx$

5. There are two methods for u-substitution with definite integrals: substituting back to the original variable or changing variables.

6. Equivalent forms meshes the ability to solve definite integrals of all forms (exponentials, logarithmics, trigonometrics, etc.) with application of the rules of definite integrals (changing variables substitution, back substitution, manipulating constants, additive properties).

Integrals

TEST WHAT YOU LEARNED

x	0	2	8	10
$g(x)$	2	7	1	4

1. Given the above data points for the continuous function $g(x)$, approximate the value of $\int_0^{10} g(x)\,dx$ using trapezoids with 3-subintervals.

   (A) 30   (B) 38   (C) 40   (D) 68

2. If $u = \sin(3x)$ is substituted into the definite integral, $\int_{\frac{\pi}{6}}^{\frac{\pi}{3}} \sin^4(3x)\cos(3x)\,dx$ can be rewritten as:

   (A) $-3\int_0^{\frac{1}{2}} u^4\,du$

   (B) $-3\int_{\frac{1}{2}}^{1} u^4\,du$

   (C) $-\dfrac{1}{3}\int_0^1 u^4\,du$

   (D) $\dfrac{1}{3}\int_{\frac{1}{2}}^{\frac{\sqrt{3}}{2}} u^4\,du$

3. Given $\int_0^8 f(x)\, dx = 7$, $\int_2^8 f(x)\, dx = 5$, and $\int_2^0 g(x)\, dx = 9$, what is the value of

$\int_0^2 \big(3f(x) + g(x)\big)\, dx$?

(A) -7 (B) -3 (C) 2   (D) 14

---

4. $\int_{-2}^3 \big(3x^2 + 6x - 4\big)\, dx =$

(A) 12   (B) 22   (C) 30   (D) 42

---

5. The function f is continuous and $\int_{-4}^{20} f(u)\, du = 5$. What is the value of $\int_1^5 x \cdot f\big(x^2 - 5\big)\, dx$?

(A) 2.5   (B) 5   (C) 10   (D) 25

Answer Key

Test What You Already Know	Test What You Learned
1. A **Learning Objective:** 12.1	1. B **Learning Objective:** 12.2
2. C **Learning Objective:** 12.2	2. C **Learning Objective:** 12.5
3. C **Learning Objective:** 12.3	3. B **Learning Objective:** 12.3
4. D **Learning Objective:** 12.4	4. C **Learning Objective:** 12.1
5. B **Learning Objective:** 12.5	5. A **Learning Objective:** 12.4

 # REFLECTION

Test What You Already Know score: _____

Test What You Learned score: _____

Use this section to evaluate your progress. After working through the pre-quiz, check off the boxes in the "Pre" column to indicate which Learning Objectives you feel confident about. Then, after completing the chapter, including the post-quiz, do the same to the boxes in the "Post" column. Keep working on unchecked Objectives until you're confident about them all!

Pre	Post		
☐	☐	**12.1**	Apply the first and second fundamental theorems of calculus to definite integrals
☐	☐	**12.2**	Use Riemann sums to approximate definite integrals
☐	☐	**12.3**	Use properties to evaluate definite integrals
☐	☐	**12.4**	Evaluate a definite integral using u-substitution
☐	☐	**12.5**	Rewrite a definite integral in an equivalent form

 # FOR MORE PRACTICE

Complete more practice online at kaptest.com. Haven't registered your book yet? Go to kaptest.com/booksonline to begin.

ANSWERS AND EXPLANATIONS

Test What You Already Know

1.   A   Learning Objective: 12.1

When you see the derivative of an integral and one of the limits of integration is a variable, remember to apply the second fundamental theorem of calculus. Don't forget to use the chain rule if it applies.

$$\frac{d}{dx} \int_3^{x^2} \ln\left(1+t^2\right) dt$$
$$= \ln\left(1+\left(x^2\right)^2\right) \cdot (2x) - \ln\left(1+3^2\right)(0)$$
$$= 2x \ln\left(1+x^4\right) - 0$$

(A) is correct.

2.   C   Learning Objective: 12.2

The first step to approaching this problem is to look at the width of the rectangles—all of them are 3 because this is the distance between the x-values. The height of each rectangle is the y-value on the right side of each subinterval (because the question asks for a *right* Riemann sum). For a rectangle, $A = lw$. Add the five areas to find the total area under the curve.

The most straightforward way to approach this is to add the required heights (right sides of the rectangles), then multiply by the width (3).

$$f(3) + f(6) + f(9) + f(12) + f(15)$$
$$= 20 + 30 + 18 + 24 + 35$$
$$= 127$$

The approximate area under the curve is therefore $3 \times 127 = 381$, which is **(C)**.

An alternative way of solving this problem would be to multiply the height and width of each rectangle and then add the products, but this approach may take more time.

3.   C   Learning Objective: 12.3

To evaluate the definite integral of a piecewise function, split the integral into parts using the "for x . . ." pieces of the function's definition. For this function, split the integral at 4 to get:

$$\int_1^{10} f(x) \, dx = \int_1^4 f(x) \, dx + \int_4^{10} f(x) \, dx$$
$$= \int_1^4 3 \, dx + \int_4^{10} (x+2) \, dx$$

Next, evaluate each of the definite integrals and add the results:

$$\int_1^4 3 \, dx = 3x\Big|_1^4 = 12 - 3 = 9$$
$$\int_4^{10} (x+2) \, dx = \frac{x^2}{2} + 2x\Big|_4^{10} = \left(\frac{100}{2}+20\right) - \left(\frac{16}{2}+8\right)$$
$$= 70 - 16 = 54$$
$$\int_1^{10} f(x) \, dx = 9 + 54 = 63$$

(C) is correct.

4.   D   Learning Objective: 12.4

The best way to approach this problem would be to use u-substitution by changing the variables from x to u. Change all the x's, including dx and the limits of integration.

$$u = 3x + 5 \Rightarrow du = 3 \, dx \Rightarrow dx = \frac{1}{3} \, du$$

The limits of integration are therefore:

$$u = 3(1) + 5 = 8$$
$$u = 3(2) + 5 = 11$$

So:

$$\frac{1}{3} \int_8^{11} \frac{1}{u} \, du = \frac{1}{3} \ln|u|\Big|_8^{11} = \frac{1}{3}\left(\ln 11 - \ln 8\right) = \frac{1}{3} \ln\left(\frac{11}{8}\right)$$

(D) is correct.

5.   B   Learning Objective: 12.5

This problem requires using u-substitution to rewrite the given integral in an equivalent form. Change all the x's to u's, including the dx and the limits of integration:

$$u = 4x + 2 \Rightarrow du = 4 \, dx \Rightarrow dx = \frac{1}{4} \, du$$

The limits of integration therefore become:

$$u = 4(2) + 2 = 10$$

$$u = 4(5) + 2 = 22$$

Therefore the equivalent integral is:

$$\int_2^5 f(4x + 2)\, dx = \frac{1}{4}\int_{10}^{22} f(u)\, du$$

That's **(B)**.

Test What You Learned

1. B Learning Objective: 12.2

For a trapezoid, area $= \frac{1}{2}(\text{base}_1 + \text{base}_2)\cdot(\text{height})$. Within each subinterval, the Δx is the height and the y-values on the right and left are the bases. The given x-values are not equally spaced, so the area of each trapezoid must be found separately.

$$\int_0^{10} g(x)\, dx$$

$$\approx \frac{1}{2}(2 + 7)(2) + \frac{1}{2}(7 + 1)(6) + \frac{1}{2}(1 + 4)(2)$$

$$= \frac{1}{2}(9)(2) + \frac{1}{2}(8)(6) + \frac{1}{2}(5)(2)$$

$$= 9 + 24 + 5$$

$$= 38$$

This makes **(B)** the correct answer.

2. C Learning Objective: 12.5

Start by finding du and the new limits of integration.

$$u = \sin(3x) \Rightarrow du = 3\cos(3x)\, dx \Rightarrow \frac{1}{3}du = \cos(3x)\, dx$$

$$u = \sin\left(3 \times \frac{\pi}{6}\right) = \sin\frac{\pi}{2} = 1$$

$$u = \sin\left(3 \times \frac{\pi}{3}\right) = \sin\pi = 0$$

Replace all x's with u's:

$$\int_{\frac{\pi}{6}}^{\frac{\pi}{3}} \sin^4(3x)\cos(3x)\, dx = \frac{1}{3}\int_1^0 u^4\, du$$

Flip the limits of integration and change the sign in front to get **(C)**.

3. B Learning Objective: 12.3

First evaluate the integral of $f(x)$ from $x = 0$ to $x = 2$.

$$\int_0^8 f(x)\, dx = \int_0^2 f(x)\, dx + \int_2^8 f(x)\, dx$$

$$7 = \int_0^2 f(x)\, dx + 5$$

$$2 = \int_0^2 f(x)\, dx$$

Next, evaluate the integral of $g(x)$ from $x = 0$ to $x = 2$. When you flip the limits of integration, you are changing the order that you plug in the x-values. The result is the negative of the original value of the integral.

$$\int_0^2 g(x)\, dx = -\int_2^0 g(x)\, dx = -9$$

Plug these values into the definite integral and simplify:

$$\int_0^2 (3f(x) + g(x))\, dx = 3\int_0^2 f(x)\, dx + \int_0^2 g(x)\, dx$$

$$= 3(2) + (-9) = -3$$

(B) is correct.

4. C Learning Objective: 12.1

Use the first fundamental theorem of calculus: $\int_a^b f'(x) = f(b) - f(a)$. Take the antiderivative of the integrand, then plug in the limits of integration:

$$\int_{-2}^3 (3x^2 + 6x - 4)\, dx = x^3 + 3x^2 - 4x\Big|_{-2}^3$$

$$= (27 + 3(9) - 4(3)) - ((-8) + 3(4) - 4(-2))$$

$$= (27 + 27 - 12) - (-8 + 12 + 8)$$

$$= 42 - 12 = 30$$

That's **(C)**.

5. A Learning Objective: 12.4

Start by identifying u and du. Then replace the x, dx, and the limits of integration with u's:

$$u = x^2 - 5 \Rightarrow du = 2x\, dx \Rightarrow \frac{1}{2}du = x\, dx$$

$$\int_1^5 x \cdot f(x^2 - 5)\, dx = \frac{1}{2}\int_{-4}^{20} f(u)\, du$$

$$= \frac{1}{2}(5) = 2.5$$

(A) is correct.

CHAPTER 13

Geometric Applications of Integration

LEARNING OBJECTIVES

13.1 Find (or use) the area under a given curve

13.2 Find the area between or bounded by curves

13.3 Find volumes of solids with known cross sections

13.4 Find volumes of solids of revolution using discs and/or washers

TEST WHAT YOU ALREADY KNOW

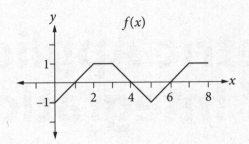

1. The graph of f is shown above. What is the value of $\int_0^8 f(x)\, dx$?

   (A) -1 (B) 0 (C) 1 (D) 2

2. What is the area of the region bounded by the graph of $f(x) = \frac{1}{2}x^3$, the line $x = 2$, and the x-axis?

   (A) 1 (B) 2 (C) 4 (D) 8

Integrals

3. The base of a solid region is bordered in the first quadrant by the x-axis, the y-axis, and the line $y = 4 - \frac{4}{3}x$. Cross-sections perpendicular to the x-axis are squares. What is the volume of the solid?

(A) 8   (B) 10   (C) 12   (D) 16

4. An urn has a shape determined by revolving the curve $y = 3 - 2 \cos x$ from $x = 0$ to $x = 6$ about the x-axis, where x and y are measured in inches. What is the volume, in cubic inches, of the urn?

(A) 58.304   (B) 68.816   (C) 216.193   (D) 679.191

Answers to this quiz can be found at the end of this chapter.

13.1 AREA UNDER A CURVE

To answer a question like this:

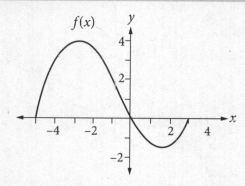

Part of the graph of a function f is shown above. The area of the region between f and the x-axis on the closed interval $[-5, 0]$ is 16. The area of the region between f and the x-axis on the closed interval $[0, 3]$ is 3. What is the value of $\int_{-5}^{3} f(x)\, dx$?

(A) -13 (B) 10   (C) 13   (D) 19

You need to know this:

Because the definite integral is defined as the area of a sum of rectangles whose heights are the value of the function at various points, the definite integral considers any area that lies *under* the x-axis to be negative.

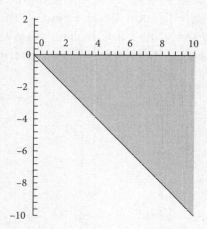

Consider the figure above. Its height, given by $f(10)$, is negative. You can compute its area:

$\frac{1}{2}bh = \frac{1}{2}(10 - 0)(-10) = -50$, that is, $\int_{0}^{10} -x\, dx = -50$.

You need to do this:

If part of the curve is above the x-axis and part of the curve is below the x-axis, then the positive and negative areas will partially or totally cancel each other out.

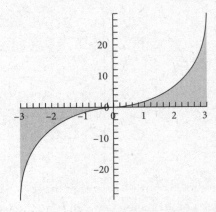

The curve $y = x^3$ is symmetric around the y-axis. Because all odd functions have the same rotational symmetry, this finding can be generalized into a time-saving tip! The integral of any odd function over an interval that is symmetric about the y-axis (−3 to 3, −1 to 1, etc.) will be 0. Thus, the same holds true for $y = x^5$, $y = x^7$, and even $y = x$.

Answer and Explanation:

C

A definite integral is used to find the sum of an infinite number of infinitely small terms. When you are using an integral to find the area under a curve, you are adding the areas of an infinite number of extremely thin rectangles; area = (length)(width) = $(y)(dx)$. The integral from $x = -5$ to $x = 0$ is positive because the graph is above the x-axis, making the y-values positive. The integral from $x = 0$ to $x = 3$ is negative because the graph is below the x-axis, making the y-values negative.

$$\int_{-5}^{3} f(x)\, dx = \int_{-5}^{0} f(x)\, dx - \int_{0}^{3} f(x)\, dx$$

$$= 16 - 3 = 13$$

(C) is correct.

PRACTICE SET

1. What is the value of the definite integral
$\int_0^6 2x - 2 \ dx$?

   (A) -24

   (B) -18

   (C) 24

   (D) 36

3. $\int_{-1}^4 |2x - 4| \ dx =$

   (A) -5

   (B) 5

   (C) $\frac{19}{2}$

   (D) 13

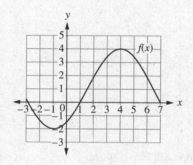

2. Given the graph of f shown above, which of the following statements are true?

   I. $\int_{-1}^5 f(x) \ dx < \int_0^5 f(x) \ dx$

   II. $\int_{-2}^2 f(x) \ dx < \int_2^{-2} f(x) \ dx$

   III. $2\int_1^2 f(x) \ dx < \int_2^3 f(x) \ dx$

   (A) I only

   (B) I and II only

   (C) II and III only

   (D) I, II, and III

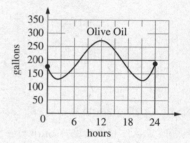

4. The rate of production of olive oil, in gallons per hour, at a processing facility on a given day is given by the graph shown above. Which value best approximates the total number of gallons of olive oil produced in the 24-hour period?

   (A) 1,800

   (B) 2,400

   (C) 3,600

   (D) 4,800

Integrals

ANSWERS AND EXPLANATIONS

1. C

The definite integral is the area under the curve, where the area above the x-axis is positive and the area below the x-axis is negative. The graph of this function, shown below, has two regions: a small triangle below the x-axis and a large triangle above the x-axis.

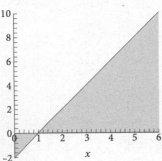

The area of the small triangle is $\frac{1}{2} \cdot 1 \cdot 2 = 1$; the area of the big triangle is $\frac{1}{2} \cdot 5 \cdot 10 = 25$. Count the area of the small triangle as negative, so:

$$\int_0^6 2x - 2 \ dx = -(\text{Area of the small triangle})$$
$$+ (\text{Area of the big triangle}) = -1 + 25 = 24$$

The answer is **(C)**.

2. D

This problem requires that you compare integral values from the graph.

I. True. $\int_{-1}^5 f(x) \ dx$ is approximately 1.8 units less than $\int_0^5 f(x) \ dx$ because the region $\int_{-1}^0 f(x) \ dx \approx -1.8$.

II. True. To start, recognize that $\int_{-2}^2 f(x) \ dx$ and $\int_2^{-2} f(x) \ dx$ are opposites. From the graph, it is clear that $\int_{-2}^2 f(x) \ dx < 0$, so $\int_2^{-2} f(x) \ dx > 0$. A negative number is always less than a positive number.

III. True. $2\int_1^2 f(x) \ dx \approx 2$ as compared to $\int_2^3 f(x) \ dx \approx 2.8$.

All three statements are true. The answer is **(D)**.

3. D

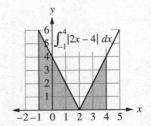

This integral can be interpreted as the area shown in the figure above. Find the sum of the areas of the two triangles.

$$\int_{-1}^4 |2x - 4| \ dx = \int_{-1}^2 |2x - 4| \ dx + \int_2^4 |2x - 4| \ dx$$
$$= \frac{1}{2}(3)(6) + \frac{1}{2}(2)(4) = 13$$

The answer is **(D)**.

4. D

The solution is the area under the rate curve between 0 and 24 hours. To approximate this area, draw a rectangle that is close in value to the area under the curve as shown below.

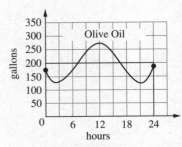

The area of the rectangle is $200 \cdot 24 = 4,800$. The answer is **(D)**.

13.2 AREA BETWEEN OR BOUNDED BY CURVES

To answer a question like this:

 The area bounded by the curves $y = x^2 + 6$ and $y = -3x + 1$ between $x = -1$ and $x = 4$ equals

(A) 69.033   (B) 69.167   (C) 69.333   (D) 70.000

You need to know this:

Area between two curves

- The definite integral can be used to find the area *between* two curves, either between specified limits of integration or between points where the curves intersect.

- The area between two curves is the integral of the length of the typical cross section.

- Sometimes the curves switch places on the interval of integration; when this happens, break the integral into pieces. Sum the absolute values of the pieces.

Area bounded by two curves

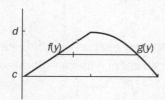

- A typical question on the AP exam asks for the area of the region bounded by given curves; often a graph of the curves is provided as part of the problem setup. This is a twist on finding the area between curves.

- The area of this region can also be found with a single integral expression by using *horizontal* cross-sectional segments that sweep through the region from bottom to top. Rewrite the original curves from $y = f(x)$ and $y = g(x)$ to $x = f(y)$ and $x = g(y)$, and travel through the region from the horizontal boundaries $y = c$ to $y = d$. The basic setup is as follows:

$$\text{Area} = \int_c^d (\text{function on right} - \text{function on left})\, dy.$$

You need to do this:

- To compute the area between two curves between $x = a$ and $x = b$, evaluate the integral $\int_a^b$ (function on top – function on bottom) dx.

- It doesn't matter if one or both of the curves lies under the x-axis; the computation remains the same.

- The area between two curves is always positive, while the definite integral is a signed area.

- The area between a curve and the x-axis is given by the integral $\int_a^b |f(x)|\ dx$.

✔ **AP Expert Note**

It should be noted here that sometimes there isn't a choice regarding the horizontal or vertical setup, so it's important to know them both.

Answer and Explanation:

B

You should first determine the graphical relationship between the two curves, $y = x^2 + 6$ and $y = -3x + 1$. Using your graphing calculator, you can quickly see that $y = x^2 + 6$ is always above $y = -3x + 1$ on the closed interval $[-1, 4]$.

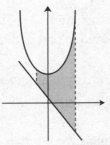

Therefore, the area in question is found by computing the value of:

$$\int_{-1}^4 \left((x^2 + 6) - (-3x + 1)\right) dx = \int_{-1}^4 (x^2 + 3x + 5)\ dx$$

This can be done by hand or by calculator, but the calculator may save you time. To evaluate the integral on the calculator, enter the simplified integrand into $\boxed{y=}$ and use the Integrate function [Calc, #7]. You'll also need to enter the limits of integration (here, the lower limit is −1 and the upper limit is 4). The result is approximately 69.167, or **(B)**.

PRACTICE SET

1. What is the area enclosed by the curve $y = e^{2x}$ and the lines $x = 1$ and $y = 1$?

   (A) $\dfrac{2 - e^2}{2}$

   (B) $\dfrac{e^2 - 3}{2}$

   (C) $\dfrac{3 - e^2}{2}$

   (D) $\dfrac{e^2 - 2}{2}$

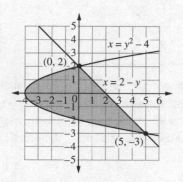

2. Which of the following gives the area of the shaded region above?

   (A) $\displaystyle\int_{-4}^{5} (y^2 + y - 6)\, dy$

   (B) $\displaystyle\int_{-3}^{2} (y^2 + y - 6)\, dy$

   (C) $\displaystyle\int_{-4}^{5} (6 - y - y^2)\, dy$

   (D) $\displaystyle\int_{-3}^{2} (6 - y - y^2)\, dy$

 3. There are two regions enclosed by the curves $y = \ln x$ and $y = -(x - 1)(x - 2)(x - 4)$. What is the sum of the areas of the two regions?

   (A) 1.47

   (B) 1.53

   (C) 1.77

   (D) 2.17

4. Find the area between the curves $f(x) = x^2 - 3x + 2 = (x - 1)(x - 2)$ and $g(x) = -(x^2 - 3x + 2) = -(x - 1)(x - 2)$ on the interval from $x = 0$ to $x = 2$.

   (A) 0

   (B) 1

   (C) $\dfrac{4}{3}$

   (D) 2

Integrals

ANSWERS AND EXPLANATIONS

1. B

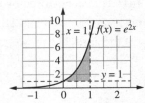

The graph of the region is shown above. The shaded region is bounded above by $y = e^{2x}$ and below by the line $y = 1$. The area of the region is given by

$$A = \int_0^1 (e^{2x} - 1)\ dx = \frac{1}{2}e^{2x} - x\Big|_0^1$$

$$= \frac{1}{2}e^2 - 1 - \frac{1}{2} = \frac{1}{2}e^2 - \frac{3}{2} = \frac{e^2 - 3}{2}$$

The answer is **(B)**.

2. D

The area of the shaded region is bounded to the right by $x = 2 - y$ and to the left by $x = y^2 - 4$. For $-3 \le y \le 2$, you know that $2 - y \ge y^2 - 4$, so it follows that the area of the shaded region is given by

$$\int_{-3}^2 (2 - y - (y^2 - 4))\ dy = \int_{-3}^2 (6 - y - y^2)\ dy.$$

The answer is **(D)**.

3. C

This is a calculator problem, so let $Y_1 = \ln x$ and $Y_2 = -(x-1)(x-2)(x-4)$. The graphs intersect at points where $x = 1, 2.389$, and 3.719. The area of the region to the left is $\int_1^{2.389} (Y_1 - Y_2)\ dx$ and to the right $\int_{2.389}^{3.719} (Y_2 - Y_1)\ dx$. The sum of the areas is

$$\int_1^{2.389} (Y_1 - Y_2)\ dx + \int_{2.389}^{3.719} (Y_2 - Y_1)\ dx = 1.77$$

or with the absolute value function you get

$$\int_1^{3.719} |Y_2 - Y_1|\ dx = 1.77$$

The answer is **(C)**.

Note: Chapter 19.3 goes through using a graphing calculator to solve definite integrals, or you can visit the online PDF *More Advanced Calculator Techniques*, available in your Online Resources.

4. D

Start by sketching the graphs of these curves on the same set of axes. They are each a parabola that intersects the x-axis at $x = 1$ and $x = 2$. The parabola $f(x)$ opens up and the parabola $g(x)$ opens down.

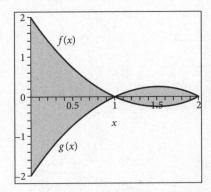

The area between the curves is the integral of the typical cross section, i.e., the top curve minus the bottom curve. From $x = 0$ to $x = 1$, the curve $f(x)$ is on top; from $x = 1$ to $x = 2$, $g(x)$ is on top. The area is therefore given by two integrals:

$$\text{Area} = \int_1^2 g(x) - f(x)\ dx + \int_0^1 f(x) - g(x)\ dx$$

Notice that $g(x) = -f(x)$. You can therefore simplify the calculation:

$$\text{Area} = \int_0^1 f(x) - (-f(x))\ dx + \int_1^2 (-f(x)) - f(x)\ dx$$

$$= \int_0^1 2f(x)\ dx + \int_1^2 -2f(x)\ dx$$

$$= 2\int_0^1 f(x)\ dx - 2\int_1^2 f(x)\ dx$$

$$= 2\int_0^1 (x^2 - 3x + 2)\ dx - 2\int_1^2 (x^2 - 3x + 2)\ dx$$

$$= 2\left(\frac{x^3}{3} - \frac{3x^2}{2} + 2x\right)\Big|_0^1 - 2\left(\frac{x^3}{3} - \frac{3x^2}{2} + 2x\right)\Big|_1^2$$

$$= 4\left(\frac{1}{3} - \frac{3}{2} + 2\right) - 2\left(\frac{8}{3} - \frac{12}{2} + 4\right)$$

$$= 4\left(\frac{5}{6}\right) - 2\left(\frac{2}{3}\right) = 2$$

The answer is **(D)**.

13.3 VOLUMES OF SOLIDS WITH KNOWN CROSS SECTIONS

To answer a question like this:

 Suppose R is the region in the following plane, bounded by the curves $f(x) = e^x - 5$, $g(x) = (x-1)^3$, and the y-axis, and R is the base of a solid.

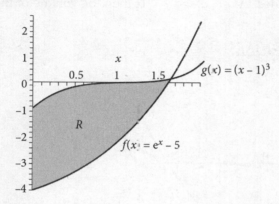

If each cross section of the solid perpendicular to the x-axis is an isosceles right triangle with one leg in the base, what is the volume of the solid?

(A) 5.271   (B) 5.461   (C) 5.681   (D) 5.841

You need to know this:

Expand upon the idea of the area between two curves as being swept out by the typical cross section to thinking of volumes the same way. Think of a volume of a solid as being swept out by the typical cross section, i.e., an area. Just as the area between two curves is the integral of (the length of) the typical cross section, the volume of a solid is the integral of (the area of) the typical cross section.

The solids above show cross sections that move along the x-axis and cross sections that move along the y-axis. If the cross sections are moving along the x-axis, that is, if the cross sections are perpendicular to the x-axis, then integrate in the x-direction for the area of the typical cross section expressed as a function of x. If the cross sections are moving along the y-axis, integrate in the y-direction for the area of the typical cross section expressed as a function of y.

Integrals

You need to do this:

- Let S be a solid, bounded region in space.

- If the area can be expressed as a cross section of S in a plane perpendicular to the x-axis as $A(x)$ and if S is bounded by $x = a$ and $x = b$, then the volume of the solid S is $\int_a^b A(x)\, dx$.

- If the area can be expressed as a cross section of S in a plane perpendicular to the y-axis as $A(y)$ and if S is bounded by $y = a$ and $y = b$, then the volume of the solid S is $\int_a^b A(y)\, dy$.

Answer and Explanation:

A

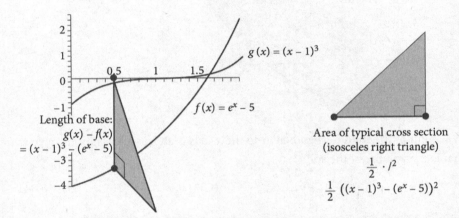

Looking at the graph of the region R, cross sections of R perpendicular to the x-axis are bounded by the curve $g(x) = (x - 1)^3$ above and $f(x) = e^x - 5$ below. These are the legs of the isosceles right triangles that form the cross sections of the solid. Therefore, the length of the leg in terms of x is $l = (x - 1)^3 - (e^x - 5)$, so the area of the typical cross section is given by:

$$A(x) = \frac{1}{2}\left((x - 1)^3 - \left(e^x - 5\right)\right)^2$$

The left boundary a of the region is $x = 0$, and the right boundary of the region is the point where the curves $g(x) = (x - 1)^3$ and $f(x) = e^x - 5$ intersect. Use a graphing calculator to determine this point of intersection: ≈ 1.667. Set up the definite integral expressing the volume of the solid and use a graphing calculator to compute the definite integral:

$$\int_0^{1.667} \frac{1}{2}\left((x - 1)^3 - \left(e^x - 5\right)\right)^2 dx = 5.271$$

This matches **(A)**.

Integrals

PRACTICE SET

1. The region R is the base of a solid. R is bounded by the parabola $y = \frac{1}{3}x^2$ and the line $y = 3$. Each cross section of the solid perpendicular to the y-axis is a square. Which integral would give the volume of the solid?

    (A) $\int_0^3 2\sqrt{3y}\ dy$

    (B) $2\int_0^3 3y\ dy$

    (C) $2\int_0^3 6y\ dy$

    (D) $\int_{-3}^3 12y\ dy$

2. The function $f(x) = -x^2 + x + 12$ and the x- and y-axes define the base of a solid with height 3. Find the volume of the solid.

    (A) 104

    (B) 144

    (C) 168

    (D) 232

3. The base of a solid is the region bounded by $y = \sqrt{x}$ and the line $y = \frac{1}{2}x$, and each cross section perpendicular to the x-axis is an equilateral triangle. The volume of the solid is

    (A) 0.051

    (B) 0.231

    (C) 0.408

    (D) 5.33

4. The region R is defined by $f(x) = 3\sin(x)$ and $g(x) = \dfrac{\sin(3x)}{2}$. R is the base of a solid S. Cross sections perpendicular to the x-axis are circles with diameters defined by R. What is the volume of the solid S?

    (A) 3.632

    (B) 7.265

    (C) 11.412

    (D) 22.823

ANSWERS AND EXPLANATIONS

1.   C

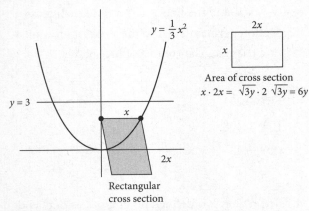

Area of cross section
$x \cdot 2x = \sqrt{3y} \cdot 2\sqrt{3y} = 6y$

Rectangular
cross section

Start by drawing a graph of the two functions.

You see that the functions intersect at $x = -3$ and $x = 3$ on the range from $y = 0$ to $y = 3$. The cross sections are perpendicular to the y-axis, so the square cross sections are defined by the difference between the x-values of the function $y = \frac{1}{3}x^2$ at the same y-value. This means you will be integrating with respect to y. Because of symmetry, you can take the volume of the solid in the first quadrant and multiply it by 2 to get the total volume. The area of each rectangle (half of the square cross sections) will be:

$$y = \frac{1}{3}x^2$$
$$x = \sqrt{3y}$$
$$A = x \times 2x$$
$$= \sqrt{3y} \times 2\sqrt{3y}$$
$$= 6y$$

Set up the integral of this area from $y = 0$ to $y = 3$ and multiply it by 2 to find the total volume of the solid.

$$2\int_0^3 6y \; dy$$

The answer is **(C)**.

2.   A

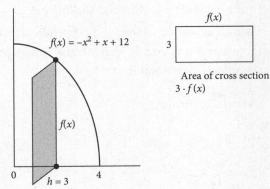

Area of cross section
$3 \cdot f(x)$

First, draw the figure.

The base of this solid is bound by $f(x)$ in the first quadrant. The function gives x as the independent variable, so you should take vertical cross sections. This means you will be integrating from 0, the y-axis, to 4, the intercept of $f(x)$ and the x-axis. Volume is base times height, and the height of this solid is a constant 3, so you can just multiply the integral, which gives you the base, by 3 to find the volume of the solid.

$$V = 3\int_0^4 \left(-x^2 + x + 12\right) dx$$
$$= 3\left(\frac{-x^3}{3} + \frac{x^2}{2} + 12x\right)\Big|_0^4$$
$$= 3\left(\frac{-4^3}{3} + \frac{4^2}{2} + 12(4)\right) - 3(0)$$
$$= 104$$

The answer is **(A)**.

Integrals

3. B

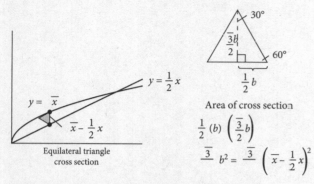

Equilateral triangle cross section

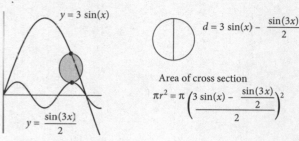

Area of cross section

$$\frac{1}{2}(b)\left(\frac{\sqrt{3}}{2}b\right)$$

$$\frac{\sqrt{3}}{4}b^2 = \frac{\sqrt{3}}{4}\left(\sqrt{x}-\frac{1}{2}x\right)^2$$

Start by drawing a graph of the two functions.

You see that the functions intersect at $x = 0$ and $x = 4$ and that $y = \sqrt{x}$ is above $y = \frac{1}{2}x$. The cross sections are perpendicular to the x-axis, so the base of those triangles is defined by the difference between the y-values of the functions at the same x-value. The area of an equilateral triangle is half the base times the height, so each cross section has an area:

$$A = \left(\frac{1}{2}\right)\left(\sqrt{x}-\frac{1}{2}x\right)\left(\frac{\sqrt{3}}{2}\right)\left(\sqrt{x}-\frac{1}{2}x\right)$$

$$= \left(\frac{\sqrt{3}}{4}\right)\left(\sqrt{x}-\frac{1}{2}x\right)^2$$

Use your calculator to integrate this area from $x = 0$ to $x = 4$ to find the volume of the solid.

$$\int_0^4 \left(\frac{\sqrt{3}}{4}\right)\left(\sqrt{x}-\frac{1}{2}x\right)^2 dx = \frac{2}{5\sqrt{3}} \approx 0.231$$

The answer is **(B)**.

4. C

First, draw the region. You can use your calculator.

The functions intersect at $x = 0$ and $x = \pi$, so you will be integrating the area of the cross sections from 0 to π. The vertical difference between the functions is the diameter of each cross section, so half of each difference is the radius. The area of each of the circular cross sections is then:

$$\pi r^2 = \pi \left(\frac{3\sin(x) - \frac{\sin(3x)}{2}}{2}\right)^2$$

Use your calculator to evaluate the integral:

$$\pi \int_0^\pi \left(\frac{3\sin x - \frac{\sin 3x}{2}}{2}\right)^2 dx = 11.412$$

The answer is **(C)**.

Integrals

13.4 VOLUMES OF SOLIDS OF REVOLUTION   High-Yield

To answer a question like this:

The region bounded by $y = x^2$, $x = 1$, $x = 2$, and the x-axis is revolved around the x-axis. What is the volume of the resulting solid?

(A) 15.354   (B) 17.916   (C) 19.478   (D) 21.241

You need to know this:

Solids of revolution

- Created when a region is rotated around a line, usually the x-axis, the y-axis, or a line parallel to one of these axes.

- The cross sections are always circles or washers (i.e., a circle with a hole in the middle).

- Find the radius of the circle to find the area of the typical cross section. Substitute the radius length into $V = \int_a^b \pi r^2 \, dx$.

Revolving around a line parallel to the x-axis

- A solid of revolution is obtained when a curve is revolved around *any* line, not just the x-axis.

- The typical cross section here is not a circle; it is a washer, or, if you prefer, a donut. The area of a washer is the area of the outer circle minus the area of the inner circle. Call the radius of the outer circle R and the radius of the inner circle r. The area of the washer is:

$$A_{\text{washer}} = A_{\text{outside circle}} - A_{\text{inside circle}} = \pi R^2 - \pi r^2.$$

- It is also important to be able to find the volume of a solid that is generated when a plane region is revolved about a vertical line (the y-axis, or any line parallel to the y-axis).

Integrals

You need to do this:

- Draw a picture of the region R that is being rotated.

- Consider drawing a picture of the solid of revolution if it helps with visualization.

- Draw the typical cross section.

- Determine its shape (circle or washer).

- Determine the radius of the circle (or, for a washer, the radii of the outer and inner circle).

- Determine a formula for $A(x)$, the area of the typical cross section in terms of x.

- Determine the boundaries of the region R, $x = a$, and $x = b$.

- Compute the definite integral of $A(x)$ with limits a and b.

Answer and Explanation:

C

Graph the solid to get a sense of what the solid of revolution looks like. It isn't necessary to draw the three-dimensional solid. Because this is a solid of revolution and there is no hole in the middle, the typical cross section is a circle. The radius of the typical cross section is x^2:

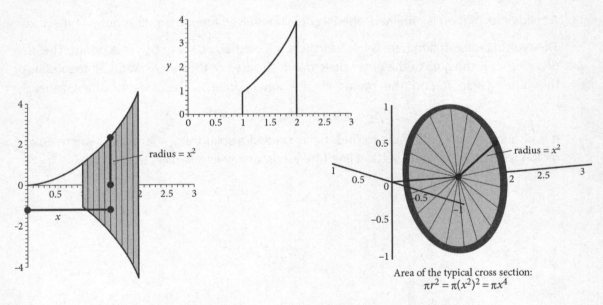

Area of the typical cross section:
$\pi r^2 = \pi (x^2)^2 = \pi x^4$

The boundaries of the solid are $x = 1$ and $x = 2$, so the volume of the solid is:

$$\int_1^2 \pi x^4 \; dx = \frac{31\pi}{5} \approx 19.478$$

This matches **(C)**.

PRACTICE SET

1. Which integral represents the volume of the solid of revolution determined by rotating the region bounded by the graphs of $f(x) = x^3$ and $g(x) = 4x$ about the x-axis?

   (A) $\int_0^2 \pi\left(\left(x^3\right)^2 + \left(4x\right)^2\right) dx$

   (B) $\int_0^2 \pi\left(\left(x^3\right)^2 - \left(4x\right)^2\right) dx$

   (C) $\int_0^2 \left(\left(\pi 4x\right)^2 + \left(\pi x^3\right)^2\right) dx$

   (D) $\int_0^2 \pi\left(\left(4x\right)^2 - \left(x^3\right)^2\right) dx$

2. The region R is defined by the x-axis, y-axis, and the line $y = 3 - \frac{1}{2}x$. Find the volume of the solid obtained by rotating R about the x-axis.

   (A) 6π

   (B) 9π

   (C) 18π

   (D) 21π

3. The region R is defined by the curves $y = \frac{7}{2}x$ and $y = x^2$. Which integral represents the volume of the solid obtained by rotating R about the line $y = -1$?

   (A) $\int_0^{\frac{7}{2}} \left(\pi(x-1)^2 - \pi\left(x^2 + 1\right)^2\right) dx$

   (B) $\int_0^{\frac{7}{2}} \left(\pi\left(\frac{7}{2}x + 1\right)^2 - \pi\left(x^2 + 1\right)^2\right) dx$

   (C) $\int_{-1}^{\frac{7}{2}} \left(\pi\left(\frac{7}{2}x + 1\right)^2 - \pi(x-1)^2\right) dx$

   (D) $\int_0^{\frac{7}{2}} \left(\pi\left(\frac{7}{2}x + 1\right)^2 + \pi\left(x^2 + 1\right)^2\right) dx$

4. What is the volume of the solid formed when the region bounded by $x = \sqrt{\dfrac{y-1}{2}}$, $x = 0$, and $y = 4$ is revolved about the y-axis?

   (A) 2.449

   (B) 4.544

   (C) 6.706

   (D) 7.069

ANSWERS AND EXPLANATIONS

1. D

First, draw the region. The graph of $f(x)$ is a cubic in standard position. It is below the graph of $g(x)$, which is a line through the origin with a slope of 4. The graphs intersect at $x = 0$ and $x = 2$ (set the equations equal to each other to find where they intersect). The graph looks like the following:

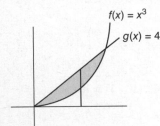

Area of typical cross section
$$\pi r_2^2 - \pi r_1^2 = \pi(r_2^2 - r_1^2)$$
$$\pi\left((4x)^2 - (x^3)^2\right)$$

Use the washer method to find the volume. Each washer will have an area of $\pi((4x)^2 - (x^3)^2)$, and the regions go from $x = 0$ to $x = 2$. The problem asks only for the integral, so set up the integral to find the volume, but do not evaluate it.

$$\int_0^2 \pi\left((4x)^2 - (x^3)^2\right)\, dx$$

This integral matches **(D)**.

2. C

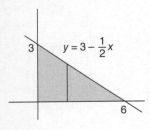

Area of typical cross section
$$\pi r^2 = \pi\left(3 - \frac{1}{2}x\right)^2$$

Each cross-section is a circle with radius $y = 3 - \frac{1}{2}x$, and the area of a circle is given by $A = \pi r^2$. The line intersects the x-axis at $x = 6$, so evaluate the integral below:

$$\int_0^6 \pi r^2\, dx = \pi \int_0^6 \left(3 - \frac{1}{2}x\right)^2 dx$$
$$= \pi \int_0^6 9 - 3x + \frac{1}{4}x^2 \, dx$$
$$= \pi\left(9x - \frac{3}{2}x^2 + \frac{1}{12}x^3\right)\Big|_0^6$$
$$= \pi(54 - 54 + 18) - \pi(0) = 18\pi$$

That's **(C)**.

3. B

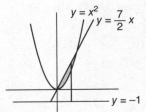

Area of typical cross section
$$\pi(r_2^2 - r_1^2) = \pi\left(\left(\frac{7}{2}x + 1\right)^2 - \left(x^2 + 1\right)^2\right)$$

First, graph both curves and the line in your calculator, so you can refer to the graph as you set up your integral.

This is a "washer" problem, where the volume is the sum of infinitely many rings or "washers" bounded by the curves. Because the region is being rotated around the line $y = -1$, the inner diameter of each washer is $x^2 + 1$ and the outer diameter of each washer is $\frac{7}{2}x + 1$. This means the area of each washer is $\pi\left(\frac{7}{2}x + 1\right)^2 - \pi\left(x^2 + 1\right)^2$, or $\pi\left(\frac{57}{4}x^2 + 7x - x^4\right)$. To find the volume, integrate this area from $x = 0$, where the curves first intersect, to $x = \frac{7}{2}$, where the curves intersect and close the region.

The question is asking you just for the integral:

$$\int_0^{\frac{7}{2}}\left(\pi\left(\frac{7}{2}x + 1\right)^2 - \pi\left(x^2 + 1\right)^2\right) dx.$$ The answer is **(B)**.

4. D

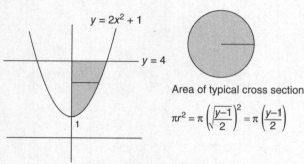

$y = 2x^2 + 1$

$y = 4$

Area of typical cross section

$$\pi r^2 = \pi\left(\sqrt{\frac{y-1}{2}}\right)^2 = \pi\left(\frac{y-1}{2}\right)$$

First, draw the graphs of the curve and lines in the problem. For a quick view on a graphing calculator, plot the lines $y = 4$ and $y = 2x^2 + 1$.

Because the question asks to rotate this region around the y-axis, the discs forming this region are horizontal. Therefore, the discs formed when the region is rotated are bound by $x = 0$ and $x = \sqrt{\frac{y-1}{2}}$ and the discs stack from $y = 1$ to $y = 4$. The area of each disc is

$\pi\left(\sqrt{\frac{y-1}{2}}\right)^2 - \pi(0)^2 = \pi\left(\frac{y-1}{2}\right)$, so to find the volume, integrate that area from 1 to 4 with respect to y:

$$\int_1^4 \pi\left(\frac{y-1}{2}\right) dy = \frac{9\pi}{4} = 7.069$$

The answer is **(D)**. Note: Chapter 19.3 reviews using a graphing calculator to evaluate a definite integral.

 ## RAPID REVIEW

If you take away only five things from this chapter:

1. The definite integral considers any area that lies *under* the x-axis to be negative. If part of the curve is above the x-axis and part of the curve is below the x-axis, then the positive and negative areas will partially or totally cancel each other out.

2. While the definite integral is signed, the area between two curves is always positive. The area between two curves is given by $\int_a^b$ (function on top – function on bottom) dx.

3. The area bounded by two curves is determined using horizontal cross sections. Rewrite the original curves from $y = f(x)$ and $y = g(x)$ to $x = f(y)$ and $x = g(y)$, from the horizontal boundaries $y = c$ to $y = d$. The area bounded by the curves is given by $\int_c^d$ (function on right – function on left) dy.

4. The volume of solids can be expressed as a definite integral of their area with respect to the x- or y-axis as either $\int_a^b A(x)\, dx$ or $\int_a^b A(y)\, dy$.

5. Solids of revolution are created when a region is rotated around a line, usually the x-axis, the y-axis, or a line parallel to one of these axes. This creates a circular cross section, which can be used to solve for the volume of the solid with $V = \int_a^b \pi r^2\, dx$.

   If the cross section is a washer, then integrate using the area given by:

   $$A_{washer} = A_{outside\ circle} - A_{inside\ circle} = \pi R^2 - \pi r^2.$$

TEST WHAT YOU LEARNED

1. $\int_{-3}^{4} \left(|x - 3| - 1 \right) dx =$

(A) 11.5   (B) 12   (C) 12.5   (D) 13

2. The area of the region bounded by the graph of $y = 2x^3 + x + 3$ and the x-axis from $x = 2$ to $x = 3$ is

(A) 38   (B) 45   (C) 54   (D) 70

3. The base of a solid is the region bounded by the parabola $x^2 = 8y$ and the line $y = 4$, and each cross section perpendicular to the y-axis is an equilateral triangle. The volume of the solid is

(A) 32   (B) $\dfrac{64\sqrt{3}}{3}$ (C) $32\sqrt{3}$ (D) $64\sqrt{3}$

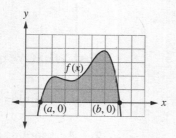

4. The shaded region shown above is rotated about the line $y = -3$. Which of the integrals is guaranteed to yield the volume of the resulting solid?

(A) $\pi \int_a^b \left((f(x))^2 + 9 \right) dx$

(B) $\pi \int_a^b (f(x) - 3)^2 \, dx$

(C) $\pi \int_a^b \left((f(x) + 3)^2 + 9 \right) dx$

(D) $\pi \int_a^b \left((f(x) + 3)^2 - 9 \right) dx$

Answer Key

Test What You Already Know	Test What You Learned

Test What You Already Know

1. **D** **Learning Objective:** 13.1

2. **B** **Learning Objective:** 13.2

3. **D** **Learning Objective:** 13.3

4. **C** **Learning Objective:** 13.4

Test What You Learned

1. **A** **Learning Objective:** 13.1

2. **A** **Learning Objective:** 13.2

3. **D** **Learning Objective:** 13.3

4. **D** **Learning Objective:** 13.4

 REFLECTION

Test What You Already Know score: _____

Test What You Learned score: _____

Use this section to evaluate your progress. After working through the pre-quiz, check off the boxes in the "Pre" column to indicate which Learning Objectives you feel confident about. Then, after completing the chapter, including the post-quiz, do the same to the boxes in the "Post" column. Keep working on unchecked Objectives until you're confident about them all!

Pre	Post	
☐	☐	**13.1** Find (or use) the area under a given curve
☐	☐	**13.2** Find the area between or bounded by curves
☐	☐	**13.3** Find volumes of solids with known cross sections
☐	☐	**13.4** Find volumes of solids of revolution using discs and/or washers

 FOR MORE PRACTICE

Complete more practice online at kaptest.com. Haven't registered your book yet? Go to kaptest.com/booksonline to begin.

Integrals

ANSWERS AND EXPLANATIONS

Test What You Already Know

1.   D   Learning Objective: 13.1

Graphically, a definite integral is the sum of the areas bounded by the curve and the x-axis. Areas above the x-axis are given a positive sign, and areas below the x-axis are given a negative sign. Here, you could divide the figure into triangles and trapezoids, and use area formulas. However, because the boundaries of the figure divide the regions into whole squares and half squares, it is easier to simply count:

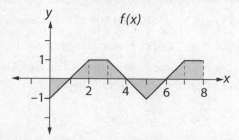

There are a total of 3.5 squares in the upper region and 1.5 squares in the lower region (area in the lower region gets a negative sign), so the definite integral has a value of $3.5 - 1.5 = 2$. **(D)** is correct.

2.   B   Learning Objective: 13.2

You are asked to find the area between curves, which means you will be integrating. Drawing a quick sketch will help you to construct your integral.

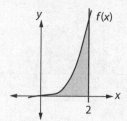

The graph shows you that you need to integrate $y = \frac{1}{2}x^3$ from $x = 0$ to $x = 2$.

$$\int_0^2 \frac{1}{2}x^3 \, dx = \frac{1}{2}\int_0^2 x^3 \, dx$$
$$= \frac{1}{2}\left(\frac{x^4}{4}\right)\Big|_0^2$$
$$= \frac{1}{2}(4 - 0) = 2$$

The area is 2, and the answer is **(B)**.

3.   D   Learning Objective: 13.3

In calculus, finding a volume almost always involves integration. Draw a quick sketch to get an idea of what you need to integrate.

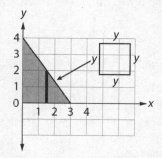

Each cross-section is a square with side length y and you are given that $y = 4 - \frac{4}{3}x$. The area of a square is given by $A = (\text{side})^2$, so the volume of the solid is:

$$\int_0^3 \left(4 - \frac{4}{3}x\right)^2 dx$$

Save yourself some time by using your calculator to evaluate the definite integral. The result is 16, which is **(D)**.

4.   C   Learning Objective: 13.4

As always, your first step is to draw a sketch to help you visualize the shape that is being generated.

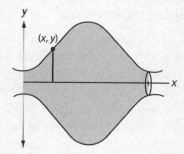

You are rotating this curve around the x-axis, forming discs. The radius of each disc is the y-value of the curve and the area of a disc is $A = \pi r^2$, or in this problem, $A = \pi(3 - 2\cos(x))^2$ This means the volume can be found by computing $\int_0^6 \pi(3 - 2\cos(x))^2 \, dx$.

Use the Integrate function on your calculator to save time. The integral is roughly 216.193, so **(C)** is the answer.

Integrals

Test What You Learned

1. **A** **Learning Objective:** 13.1

You could split the equation into a piecewise function to integrate the absolute value, but it's quicker to draw the graph of the integrand function, $y = |x - 3| - 1$, then calculate areas. Transform the graph of $y = |x|$ by shifting right 3 units and down 1 unit. The graph is shown below.

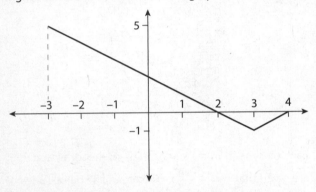

A definite integral is the sum of the areas bounded by the curve and the x-axis. The integral of a curve that is above the x-axis is given a positive sign and a curve below the x-axis is given a negative sign. Note that the triangle between $x = -3$ and $x = 2$ has an area of $\frac{1}{2}(5)(5) = 12.5$, so its contribution to the integral is 12.5 because the graph is above the x-axis. The area of the triangle below the x-axis is $\frac{1}{2}(2)(1) = 1$, so the value of the definite integral is $12.5 - 1 = 11.5$. **(A)** is correct.

2. **A** **Learning Objective:** 13.2

Finding the area between curves means evaluating a definite integral. First, graph the function in your calculator and see that the curve is entirely above the x-axis on the range $x = 2$ to $x = 3$. This means that if you calculate the definite integral from 2 to 3, you will find the area (because area is always positive). Now set up the integral and evaluate it:

$$\int_2^3 (2x^3 + x + 3)\, dx = \frac{x^4}{2} + \frac{x^2}{2} + 3x \Big|_2^3$$

$$\frac{3^4}{2} + \frac{3^2}{2} + 3(3) - \left(\frac{2^4}{2} + \frac{2^2}{2} + 3(2)\right)$$

$$\frac{81}{2} + \frac{9}{2} + 9 - (8 + 2 + 6) = 54 - 16 = 38$$

The area is 38, making **(A)** the correct answer. (Note that because this question is marked as a calculator question, you could also use the Integrate function on your calculator to evaluate the integral.)

3. **D** **Learning Objective:** 13.3

Start by drawing a sketch:

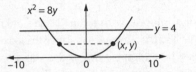

The dotted line in the top left diagram represents a side of one of the equilateral triangles, so its length must be $2x$. The top right diagram shows one of the equilateral triangles with its dimensions labeled. Because the area of a triangle is $A = \frac{1}{2}bh$, then one of these triangles has an area of $A = \frac{1}{2}(2x)(x\sqrt{3}) = x^2\sqrt{3}$. Because the cross-sections are perpendicular to the y-axis, you need to integrate in terms of y. Because $x^2 = 8y$, then

$$A = x^2\sqrt{3} = (8y)\sqrt{3} = 8\sqrt{3}y$$

Integrate $(\text{area}) \cdot dx$ to find the volume of the solid.

$$V = \int_0^4 8\sqrt{3}y\, dy = \frac{8\sqrt{3}}{2}y^2\Big|_0^4 = \frac{8\sqrt{3}}{2}(16 - 0) = 64\sqrt{3}$$

That is **(D)**.

4. **D** **Learning Objective:** 13.4

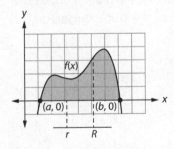

To find the volume, use the washer method:

$$V = \pi\int_a^b (R^2 - r^2)\, dx,\text{ where } R \text{ is the outer radius and } r$$

is the inner radius. Because $y = -3$ is the axis of rotation, $R = f(x) + 3$ and $r = 3$. This gives:

$$\pi\int_a^b \left((f(x) + 3)^2 - 3^2\right) dx$$

This matches **(D)**.

CHAPTER 14

Further Applications of Integration

LEARNING OBJECTIVES

14.1 Find the average value of a function

14.2 Determine the net change of a function over an interval

14.3 Use integration to solve problems involving motion along a line

14.4 Solve differential equations with given initial conditions

14.5 Solve problems involving exponential growth and decay

14.6 Match a slope field with its corresponding differential equation

TEST WHAT YOU ALREADY KNOW

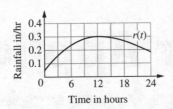

1. The graph of the rate of rainfall $r(t)$, in inches per hour, for a 24-hour period is shown. Which statement is false?

   (A) The rain fell hardest in the middle hours of the 24-hour period.

   (B) During the 24-hour period, the total rainfall in inches is $\int_0^{24} r(t)\, dt$.

   (C) During the 24-hour period, the total rainfall is between 1 and 7 inches.

   (D) During the 24-hour period, the average rate of rainfall is $\dfrac{r(24) - r(0)}{24}$ in/hr.

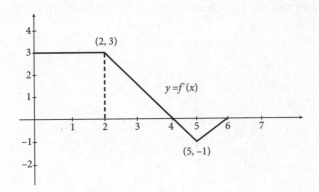

2. The graph of the derivative of f, f', is shown above. If $f(0) = 7$, what is $f(6)$?

   (A) 9   (B) 12   (C) 14   (D) 15

3. A particle moves along the x-axis so that its acceleration at any time t is given by $a(t) = 9t^2 - 12t$. At time $t = 0$, the velocity v of the particle is 15 units/sec. What is the velocity when $t = 3$?

   (A) 27 units/sec   (B) 36 units/sec   (C) 42 units/sec   (D) 45 units/sec

4. At every point (x, y) on a curve, the slope of the curve is $4x^3 y$. If the curve contains the point $(0, 4)$, then its equation is

   (A) $y = 4e^{x^4}$

   (B) $y = e^{x^4} + 3$

   (C) $y = \ln(x + 1) + 4$

   (D) $y = x^4 + 4$

5. Timmy is investing his first \$100. Account 1 offers him a bonus \$100 for signing up and has a growth constant of 0.005/day. Account 2 offers a \$50 sign-up bonus and has a growth constant of 0.008/day. How many days will it take for the value of account 2 to surpass that of account 1 ?

   (A) 87 days   (B) 96 days   (C) 114 days   (D) 152 days

Integrals

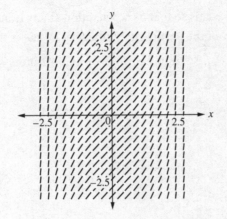

6. Which of the following could be a path through the slope field created by the differential equation $\frac{dy}{dx} = \sec^2(0.5x)$?

   (A) $y = 2\tan x + 1$

   (B) $y = \tan(0.5x)$

   (C) $y = 2\tan(0.5x) + 3$

   (D) $y = \tan(0.5x) + 2$

Answers to this quiz can be found at the end of this chapter.

14.1 AVERAGE VALUE OF A FUNCTION

High-Yield

To answer a question like this:

For $t \geq 0$, the height of a particle is given by $h(t) = 3\sin(4t) - 2t + 1$. What is the average height of the particle on the interval $1 \leq t \leq 4$?

(A) -5.632 (B) -3.924 (C) -1.531 (D) -0.792

You need to know this:

To find the average of a list of numbers, add the numbers to get their total and divide that total by the number of numbers in the list:

$$Average = \frac{a_1 + a_2 + \cdots a_n}{n}$$

To find the average value of a function on a closed interval $[a, b]$, find the area under the graph of the function (that's the "total" in the average) and divide by the length of the interval $b - a$. That is, if f is continuous on the closed interval $[a, b]$, then the average value of f is given by:

$$C = \frac{\int_a^b f(x)\, dx}{b - a}$$

> ✔ **AP Expert Tip**
>
> Questions relating to the average value of a function appear regularly on the AP exam.

Interpret this definition as saying there is some point C such that the area of the rectangle with base $b - a$ and height C is the same as the area under $f(x)$ on $[a, b]$.

Integrals

You need to do this:

- Assign the appropriate function and limit values to the average value integral: $\frac{1}{b-a}\int_a^b f(x)\, dx$.

- Use your graphing calculator to integrate the function $h(t)$ on the closed interval $[1, 4]$.

Answer and Explanation:

B

The average value of a function f on the closed interval $[a, b]$ is defined by: $\frac{1}{b-a}\int_a^b f(x)\, dx$.

Therefore,

$$\frac{1}{3}\int_1^4 h(t)\, dt \approx -3.924.$$

Thus, **(B)** is correct.

PRACTICE SET

1. What is the average value of $f(x) = e^{kx}$ on the closed interval $[0, k]$?

   (A) $\dfrac{1}{k^2} e^{k^2}$

   (B) $\dfrac{1}{k} e^{k^2}$

   (C) $\dfrac{1}{k^2} e^{k^2} - 1$

   (D) $\dfrac{1}{k^2} e^{k^2} - \dfrac{1}{k^2}$

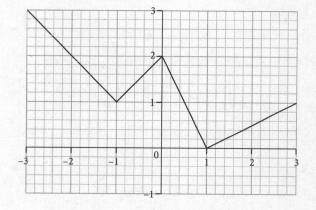

2. What is the average value of $y = \tan^2 x$ over the interval from $x = 0$ to $x = \dfrac{\pi}{4}$?

   (A) $\sqrt{2}$

   (B) $1 + \dfrac{\pi}{4}$

   (C) $\dfrac{4}{\pi} - 1$

   (D) $1 - \dfrac{4}{\pi}$

3. The function f is continuous on the closed interval $[-3, 3]$. The graph of f shown above consists of four line segments. What is the average value of f on the closed interval $[-3, 3]$?

   (A) 1

   (B) $\dfrac{5}{4}$

   (C) $\dfrac{4}{3}$

   (D) $\dfrac{5}{2}$

4. If the average value of $f(x) = \sqrt{x - 3}$ on the closed interval $[3, k]$ is 8, what is k ?

   (A) 8

   (B) 24

   (C) 147

   (D) 151

Integrals

355 | K

ANSWERS AND EXPLANATIONS

1. D

Use the average value formula: $\frac{1}{b-a}\int_a^b f(x)\,dx$.

Then evaluate the integral and simplify the result:

$$\frac{1}{k-0}\int_0^k e^{kx}\,dx = \frac{1}{k^2}e^{kx}\Big|_0^k = \frac{1}{k^2}\left(e^{k^2}-1\right).$$

Distribute the $\frac{1}{k^2}$ and that's a perfect match for **(D)**.

2. C

To find the average value, evaluate

$$\frac{1}{\frac{\pi}{4}-0}\int_0^{\frac{\pi}{4}}\tan^2 x\,dx = \frac{4}{\pi}\int_0^{\frac{\pi}{4}}\tan^2 x\,dx.$$

Recall that

$$\tan^2 x + 1 = \sec^2 \Rightarrow \tan^2 x = \sec^2 x - 1.$$

Now use substitution and the first fundamental theorem of calculus to evaluate the integral:

$$\frac{4}{\pi}\int_0^{\frac{\pi}{4}}\tan^2 x\,dx = \frac{4}{\pi}\int_0^{\frac{\pi}{4}}\left(\sec^2 x - 1\right)dx$$

$$= \frac{4}{\pi}\left(\tan x - x\right)\Big|_0^{\frac{\pi}{4}}$$

$$= \frac{4}{\pi}\left(1 - \frac{\pi}{4}\right)$$

$$= \frac{4}{\pi} - 1$$

The answer is **(C)**.

3. B

Use geometry to find the value of $\int_{-3}^{3} f(x)\,dx$, then use the formula for average value of a function.

To find the area between f and the x-axis, break the shapes into geometric regions that are easy to find the areas of, such as rectangles and triangles. Alternatively, you can try counting squares and half-squares in the graph to arrive at an estimate of the area between f and the x-axis. Remember to count areas below the x-axis as negative.

In this case, counting or using triangles and rectangles both yield an area of 7.5. So the average value is

$$\frac{1}{b-a}\int_a^b f(x)\,dx = \frac{7.5}{6}$$

That simplifies to $\frac{5}{4}$, making **(B)** the correct answer.

4. C

The average value of $f(x) = \sqrt{x-3}$ over $[3, k]$ is given by $\frac{1}{k-3}\int_3^k \sqrt{x-3}\,dx$. Evaluate the integral and use the FTC:

$$\frac{1}{k-3}\int_3^k \sqrt{x-3}\,dx = 8$$

$$\frac{2}{3(k-3)}\cdot(x-3)^{\frac{3}{2}}\Big|_3^k = 8$$

$$\frac{2}{3(k-3)}\cdot(k-3)^{\frac{3}{2}} = 8$$

Next, simplify the left-hand side of the equation and solve for k:

$$\frac{2}{3}(k-3)^{\frac{1}{2}} = 8 \Rightarrow (k-3)^{\frac{1}{2}} = 12$$

$$\Rightarrow k-3 = 144 \Rightarrow k = 147$$

So **(C)** is the correct answer.

14.2 NET CHANGE OVER AN INTERVAL

To answer a question like this:

Let $G(x) = \int_a^x f(t)\, dt$ where the graph of $f(t)$ is shown below.

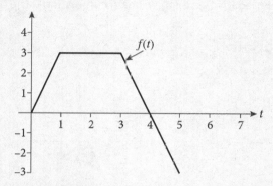

Suppose $f(t)$ gives the rate of change of the water level in Lake Esmerelda where the height of the water is described in meters above sea level and the time t is given in months. If the level of the lake is 110 m above sea level at the starting time, what is the level of the lake at time $t = 5$ months?

(A) 115 m
(B) 117.5 m
(C) 119 m
(D) 120.5 m

You need to know this:

The total change in a quantity over a time period is the definite integral of its rate of change over that time period. Use the definite integral to think about change thus far or to evaluate how much of a total quantity has accumulated at a certain point in time.

Suppose f is a function that gives the rate of change of some quantity. The total change in the quantity over the time period $t = a$ to $t = b$ is the definite integral $\int_a^b f(t)\, dt$. Use the definite integral to answer questions about how much stuff has accumulated *so far*. If f is a function describing the rate of change of some quantity, then we can define a function $G(x)$: $G(x) = \int_0^x f(t)\, dt$.

You need to do this:

- The function $G(x)$ is the area under the curve $f(t)$ from some specified starting time a to time x.

- It tells how much stuff accumulates from the starting time to time x. Functions like these are called *accumulation functions*.

- Over intervals where f is negative, nothing will be accumulating—the quantity will be decreasing.

Answer and Explanation:

B

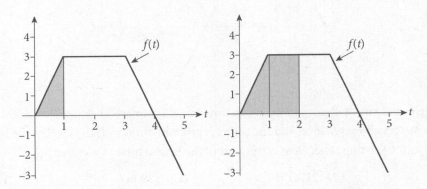

$G(1)$ is the area shown on the left, which is a triangle with area $\frac{1}{2}bh = \frac{1}{2} \cdot 1 \cdot 3 = \frac{3}{2}$. To compute $G(2)$, add the area of the adjacent rectangle. The area of this rectangle is $1 \cdot 3 = 3$, so the total shaded area is the area of the triangle plus the area of the rectangle: $G(2) = \frac{3}{2} + 3 = \frac{9}{2}$. Continuing in this way, compute $G(3) = \frac{9}{2} + 3 = \frac{15}{2}$; $G(4) = \frac{15}{2} + \frac{3}{2} = 9$. When computing $G(5)$, subtract the area that lies below the x-axis.

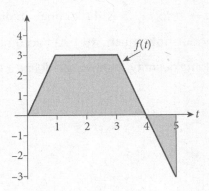

$G(5) = 9 - \frac{3}{2} = \frac{15}{2}$. This matches **(B)**, 117.5 m.

PRACTICE SET

 1. Suppose gasoline is being added to a tank at a rate of $r(t) = 100\left(1 - e^{-0.02t}\right)$ liters per minute, where t is the number of minutes past 3:00 pm. If the tank contained 250 liters of gasoline at 3:00 pm, how much water is in the tank at 3:15 pm? Round your answer to the nearest liter.

   (A) 454 liters

   (B) 1,776 liters

   (C) 2,592 liters

   (D) 3,046 liters

2. If the rate of change of a quantity over the closed interval $[0, 2]$ is given by $f'(x) = x^2\sqrt{x^3 + 1}$, then the net change of the quantity over the interval $[0, 2]$ is

   (A) $\dfrac{4\sqrt{2}}{9}$

   (B) $\dfrac{12}{5}$

   (C) $\dfrac{52}{9}$

   (D) $\dfrac{82}{3}$

3. Water flows into a pond at a rate of $300\sqrt{t}$ gallons/hour and flows out at a rate of 400 gallons/hour. After 1 hour there are 10,000 gallons of water in the pond. How much water is in the pond after 9 hours?

   (A) 10,000

   (B) 11,000

   (C) 12,000

   (D) 14,000

4. A cup of coffee is heated and then placed outside to cool off at time $t = 0$ minutes. The temperature of the coffee is changing at a rate of $\left(-t^2 e^{-0.25t}\right)\ °\text{F}$ per minute, for $t > 0$. At $t = 10$ minutes, the temperature of the coffee is $140\ °\text{F}$. What is the temperature of the coffee at time $t = 20$ minutes?

   (A) $27.955\ °\text{F}$

   (B) $81.608\ °\text{F}$

   (C) $86.347\ °\text{F}$

   (D) $100.869\ °\text{F}$

Integrals

ANSWERS AND EXPLANATIONS

1. A

Total change from one time to another is equal to the accumulation of small changes during that time. Here, the function $r(t)$ is a rate of change of gasoline in the tank, and the question asks for the ending amount after 15 minutes. Add the amount that has accumulated during the 15 minutes to the amount at 3:00:

$$250 + \int_0^{15} r(t)\, dt = 250 + \int_0^{15} 100\left(1 - e^{-0.02t}\right) dt$$

Use your calculator to evaluate the integral and arrive at the total amount in the tank at 3:15.

$$250 + \int_0^{15} 100\left(1 - e^{-0.02t}\right) dt \approx 454.091$$

That makes **(A)** the correct answer.

2. C

The definite integral of the rate of change of a quantity over an interval gives the net change of that quantity over that interval, so find the antiderivative of $f'(x)$ first. Using u-substitution with $u = x^3 + 1$, and $du = 3x^2\, dx$, the result is:

$$\int_0^2 x^2 \sqrt{x^3 + 1}\, dx = \frac{\frac{1}{3}\left(x^3 + 1\right)^{\frac{3}{2}}\Big|_0^2}{\frac{3}{2}}$$

$$= \frac{2}{9}\left(9^{\frac{3}{2}} - 1^{\frac{3}{2}}\right)$$

$$= \frac{2(27 - 1)}{9}$$

$$= \frac{52}{9}$$

That matches **(C)**.

3. C

The total rate of change of the volume of water in the pond is $r(t) = $ *Rate in* $-$ *Rate out*.

In this problem, you are told that the rate in is $300\sqrt{t}$ gallons/hr and the rate out is 400 gallons per hour. Therefore, $r(t) = 300\sqrt{t} - 400$.

Let $A(t)$ be the amount of water in the pond at time t. Then A is an antiderivative of r. Approach this problem as an accumulation function. Because you are given the volume of water at $t = 1$ and you are asked to find the volume of water at $t = 9$, you can set the problem up as $A(9) = A(1) + $ *change in volume on* $1 \leq t \leq 9$. The change in the volume of water is given by the definite integral

$$\int_1^9 r(t)\, dt = \int_1^9 300\sqrt{t} - 400\, dt$$

so:

$$A(9) = 10{,}000 + \int_1^9 \left(300\sqrt{t} - 400\right) dt$$

$$= 10{,}000 + \int_1^9 \left(300t^{\frac{1}{2}} - 400\right) dt$$

$$= 10{,}000 + \left(\frac{2}{3} \cdot 300t^{\frac{3}{2}} - 400t\right)\Big|_1^9$$

$$= 10{,}000 + \left(200t^{\frac{3}{2}} - 400t\right)\Big|_1^9$$

$$= 10{,}000 + \left(200 \cdot 9^{\frac{3}{2}} - 400 \cdot 9\right)$$

$$- \left(200 \cdot 1^{\frac{3}{2}} - 400 \cdot 1\right) = 12{,}000$$

That makes **(C)** the correct answer.

4. C

Let $f(t)$ be the temperature function at time t. The net change in temperature from $t = 10$ to $t = 20$ is equal to $\int_{10}^{20} f'(t)\, dt$. Solve for $f(20)$:

$$f(20) = f(10) + \int_{10}^{20} -t^2 e^{-0.25t}\, dt$$

Use your calculator for the definite integral, and use the fact that $f(10) = 140$ to arrive at **(C)**.

Integrals

14.3 MOTION ALONG A LINE

To answer a question like this:

A particle moves along the x-axis with initial position $x(0) = 1$. The velocity of the particle at time $t \geq 0$ is given by $v(t) = \sin \sqrt{t}$. What is the position of the particle at time $t = 15$?

(A) 4.429   (B) 4.929   (C) 5.429   (D) 5.929

You need to know this:

Particle traveling along a straight line

- Typically, information is given about the particle's velocity and questions ask about the particle's position and its acceleration.

- Use derivatives to go from a particle's position function, $x(t)$, to its velocity and acceleration functions, $v(t)$ and $a(t)$, respectively. Use $v(t) = x'(t)$ and $a(t) = v'(t) = x''(t)$.

- The definite integral of the velocity function can give either the total distance a particle travels or its displacement.

✔ **AP Expert Note**

These types of problems are another application of the definite integral. They can be presented numerically with data tables, graphically, or algebraically with functions.

You need to do this:

- Use integration to work in reverse!

- If given a velocity or acceleration function, take the integrals to work backwards to a position function.

- $v(t) = \int_a^t a(t)\, dt$ and $x(t) = \int_a^t v(t)\, dt$

Answer and Explanation:

C

Velocity is the rate of change of position, so by thinking of the change in position as the accumulated change in the velocity, find the change in the particle's position using a calculator to compute the definite integral:

$$\int_0^{15} \sin \sqrt{t}\, dt \approx 4.429$$

The position of the particle at time $t = 15$ is the particle's position at the start plus the change in position over the time period. It was given that the particle's position at time $t = 0$ is $x(0) = 1$. Therefore, the position of the particle at time $t = 15$ is its initial position plus its change in position over the time period $0 \leq t \leq 15$: $x(15) = 1 + 4.429 = 5.429$. This matches **(C)**.

PRACTICE SET

 1. The velocity of a particle is given by $v(t) = 2t - 3\cos(t^2) - 3$, for $t \geq 0$. If the position of the particle at time $t = 2$ is -4, then what is the position of the particle at time $t = 3$?

   (A) -3.057

   (B) -2.724

   (C) 1.276

   (D) 9.418

2. A particle moves along the x-axis with a velocity given by $v(t) = 2 + \sin t$. When $t = 0$, the particle is at $x = -2$. Where is the particle when $t = \pi$?

   (A) $\pi - 2$

   (B) $\pi - 1$

   (C) π

   (D) 2π

3. A particle moves along the x-axis so that its acceleration at any time t is given by $a(t) = 6t - 18$. At time $t = 0$, the velocity v of the particle is 24 units/sec and at the time $t = 1$, the position x of the particle is 20. Which statement is false?

   (A) $v(t) = 3t^2 - 18t + 24$ for all $t \geq 0$.

   (B) The particle is moving to the left when $2 < t < 4$.

   (C) The starting position of the particle is $x(0) = 4$.

   (D) The total distance traveled by the particle for $0 \leq t \leq 4$ is $\int_0^4 (3t^2 - 18t + 24)\, dt$.

4. A jogger's acceleration is given by $a(t) = -kt$ where k is a positive constant. At time $t = 0$, the jogger is running at a velocity of 192 meters per minute. If the jogger comes to a stop in 8 minutes, what is her total distance covered in meters?

   (A) 960

   (B) 1,024

   (C) 1,440

   (D) 1,820

Integrals

ANSWERS AND EXPLANATIONS

1. B

The position at $t = 3$ is the initial position (-4) plus the change from the initial position. Use your calculator to find the value:

$$s(3) = -4 + \int_2^3 \left(2t - 3\cos(t^2) - 3\right) dt$$
$$= -2.724$$

So **(B)** is the correct answer.

2. D

The position of the particle $x(t)$ is an antiderivative of the velocity function, so start by computing the indefinite integral of the velocity:

$$v(t) = 2 + \sin t$$
$$x(t) = \int v(t)\, dt = \int 2 + \sin t\, dt$$
$$x(t) = 2t - \cos t + C$$

Next, use the initial condition to determine C. Because $x(0) = -2$,

$$-2 = 2 \cdot 0 - \cos 0 + C$$
$$-2 = -1 + C$$
$$C = -1$$

So $x(t) = 2t - \cos t - 1$. Finally, evaluate the position function at time $t = \pi$:

$$x(\pi) = 2\pi - \cos \pi - 1$$
$$= 2\pi - (-1) - 1$$
$$= 2\pi$$

That's a match for **(D)**.

3. D

All of the statements are true except **(D)**. The total distance traveled by the particle for $0 \le t \le 4$ is $\int_0^4 \left| 3t^2 - 18t + 24 \right| dt$ and not $\int_0^4 \left(3t^2 - 18t + 24\right) dt$, which represents the displacement (or "net change") for $0 \le t \le 4$.

Note: the particle changes direction at $t = 2$.

4. B

This problem requires two antidifferentiations with resolution of the constants along the way. First, because $v(T) = \int_0^T -kt\, dt$, it follows that $v(T) = -k\dfrac{T^2}{2} + C$. Next, solve for C with the initial condition $v(0) = 192$ to get

$$v(T) = -k\frac{T^2}{2} + 192.$$

Given $v(8) = 0$, substitute $-k\dfrac{8^2}{2} + 192 = 0$, $-32k + 192 = 0$, so $k = 6$.

This means $v(T) = -6 \cdot \dfrac{T^2}{2} + 192 = -3T^2 + 192$.

Finally, evaluate the definite integral of the velocity function to obtain the jogger's total distance.

$$s(8) = \int_0^8 \left(-3T^2 + 192\right) dT$$
$$= \left(-T^3 + 192T\right)\Big|_0^8$$
$$= 1{,}024$$

That makes **(B)** the right answer.

14.4 DIFFERENTIAL EQUATIONS

To answer a question like this:

If $\dfrac{dy}{dx} = xy^2$ with the initial condition $y(0) = 1$, then $y =$

(A) $\dfrac{2}{2 - x^2}$

(B) $\dfrac{1}{1 - x^2}$

(C) $\dfrac{2}{2 + x^2}$

(D) $e^{\frac{x^2}{2}}$

You need to know this:

Differential equations

- Equations containing derivatives are differential equations.

- In a differential equation, information is given about the rate of change of a function and you are asked to find the function.

- Because solutions to differential equations involve taking antiderivatives, their solutions contain constants.

- When given initial conditions that allow for determination of the constants, the solution is called a *particular solution* to the differential equation.

- If not given the initial conditions, then the solution is expressed with unknown constants, termed a *general solution*.

Integrals

You need to do this:

- The most basic type of differential equation tells that the rate of change of a function is $f(x)$.
- In the language of differential equations, this is $\frac{dy}{dx} = f(x)$.
- Take the integral to obtain the amount the rate applies to.
- If initial conditions are given, apply them to the antiderivative to obtain the constant C.

Answer and Explanation:

A

The question provides the differential equation $\frac{dy}{dx} = xy^2$, which rearranges to the form
$$\frac{1}{y^2}\,dy = x\,dx \Rightarrow \int \frac{1}{y^2}\,dy = \int x\,dx.$$

From here, antidifferentiate the equation on each side and rearrange to solve for y:
$$\frac{-1}{y} = \frac{1}{2}x^2 + C_1 \Rightarrow \frac{1}{y} = C_2 - \frac{1}{2}x^2 \Rightarrow y = \frac{1}{C_2 - \frac{1}{2}x^2}.$$

Next, apply the initial condition from the question $y(0) = 1$ to solve for the constant: $1 = \frac{1}{C_2} \Rightarrow C_2 = 1$.
This means $y = \dfrac{1}{1 - \frac{1}{2}x^2} = \dfrac{2}{2 - x^2}$, matching **(A)**.

Integrals

PRACTICE SET

1. Consider the differential equation $\dfrac{dy}{dx} = \dfrac{e^x - 1}{2y}$.
   If $y = 4$ when $x = 0$, what is the value of y when $x = 1$?

   (A) $\sqrt{e + 14}$

   (B) $\sqrt{e + 15}$

   (C) $\sqrt{e^2 + 11}$

   (D) $\sqrt{e^2 + 15}$

2. If $\dfrac{dy}{dx} = y \sec^2 x$ and $y = 8$ when $x = 0$, then $y =$

   (A) $\tan x + 8$

   (B) $8e^{\tan x}$

   (C) $e^{\tan x} + 7$

   (D) $e^x \tan x + 8$

3. Consider the differential equation
   $\dfrac{dy}{dx} = (1 - 2x) \cdot y$. If $y = 10$ when $x = 1$, find an equation for y.

   (A) $y = e^{x - x^2}$

   (B) $y = 10 + e^{x - x^2}$

   (C) $y = 10 \cdot e^{x - x^2}$

   (D) $y = x - x^2 + 10$

4. If $\dfrac{dy}{dx} = \dfrac{1}{4}y$ and $y(0) = 5$, then $y(4) =$

   (A) $5 + e$

   (B) $10 + e$

   (C) $5e$

   (D) $10e$

Integrals

ANSWERS AND EXPLANATIONS

1.   A

To solve this differential equation, use the technique of separation of variables:

$$\frac{dy}{dx} = \frac{e^x - 1}{2y}$$

$$2y\,dy = \left(e^x - 1\right)dx$$

$$\int (2y)\,dy = \int \left(e^x - 1\right)dx$$

$$y^2 = e^x - x + C$$

Use the initial condition $y(0) = 4$ to find C by plugging the information into the solution above:

$$4^2 = e^0 - 0 + C$$

$$16 = 1 + C$$

$$15 = C$$

With this value of C, you have $y^2 = e^x - x + 15$.

Now substitute $x = 1$ into the equation, and solve for y:

$$y^2 = e^1 - 1 + 15$$

$$= e + 14$$

$$y = \pm\sqrt{e + 14}$$

The initial condition (or a quick glance at the answer choices) tells you that you want the positive square root. So **(A)** is correct.

2.   B

Use the separation of variables technique.

$$\frac{dy}{dx} = y\sec^2 x$$

$$\frac{1}{y}\,dy = \sec^2 x\,dx$$

$$\int \frac{1}{y}\,dy = \int \sec^2 x\,dx$$

Now antidifferentiate:

$$\ln|y| = \tan x + C$$

$$|y| = e^{\tan x + C}$$

$$= e^C \cdot e^{\tan x}$$

Because $y = 8$ when $x = 0$, $e^C = 8$, and the answer is $y = 8e^{\tan x}$. That's **(B)**.

3.   C

To solve this differential equation, use the technique of separation of variables:

$$\frac{dy}{dx} = (1 - 2x) \cdot y \Rightarrow \frac{1}{y}\,dy = (1 - 2x)\,dx$$

Notice that after separating the variables, the equation is of the form $\frac{dy}{y} = f(x)\,dx$, which means the answer will involve an exponential function and a multiplicative constant. Integrate both sides of the equation to find:

$$\int \frac{1}{y}\,dy = \int (1 - 2x)\,dx$$

$$\ln|y| = x - x^2 + C$$

$$y = e^{x - x^2 + C} = e^{x - x^2} \cdot e^C = A \cdot e^{x - x^2}$$

Now use the initial condition $y(1) = 10$ to find A:

$$y = A \cdot e^{x - x^2}$$

$$10 = A \cdot e^{1 - 1^2} = A \cdot e^0 = A$$

Because $A = 10$, the equation is $y = 10 \cdot e^{x - x2}$. That's **(C)**.

4.   C

Use the technique of separation of variables to solve this differential equation:

$$\frac{dy}{dx} = \frac{1}{4}\,y \Rightarrow \frac{1}{y}\,dy = \frac{1}{4}\,dx$$

Notice that after separating the variables, the equation is of the form $\frac{1}{y}\,dy = f(x)\,dx$, so the solution will involve an exponential function and a multiplicative constant.

$$\int \frac{1}{y}\,dy = \int \frac{1}{4}\,dx \Rightarrow \ln|y| = \frac{1}{4}x + C \Rightarrow y = Ae^{\frac{1}{4}x}$$

Use the initial condition that when $x = 0$, $y = 5$ to find A:

$$5 = Ae^{\frac{1}{4} \cdot 0} = A$$

Now evaluate the function when $x = 4$:

$$y(4) = 5e^{\frac{1}{4} \cdot 4} = 5e$$

So **(C)** is correct.

14.5 EXPONENTIAL GROWTH AND DECAY

To answer a question like this:

A bandana falls off a tall building. At time $t \geq 0$, the velocity of the bandana satisfies the differential equation $\frac{dv}{dt} = -10v - 32$ with initial condition $v(0) = 0$. Use separation of variables to find an expression for v in terms of t when t is measured in seconds.

(A) $v = e^{-3.2t} - 32$

(B) $v = e^{-10t} - 3.2$

(C) $v = 10e^{-3.2t} - 3.2$

(D) $v = 3.2e^{-10t} - 3.2$

You need to know this:

Exponential growth

- The Calculus AB syllabus specifically mentions equations of the type $y' = ky$, where the rate of change of a quantity is proportional to the amount present.

- This type of equation models phenomena such as radioactive growth and decay, absorption of medicine in the bloodstream, and investments.

- This type of problem is an example of a separable differential equation.

Identifying exponential-growth type problems

- Sometimes it's easy to identify an exponential-growth or decay problem.

- If given a differential equation of the form $\frac{dy}{dx} = ky$, it's clear, but things aren't always this simple.

- Any question that has the form $\frac{dy}{y} = f(x) \, dx$ after variables are separated is likely to have an exponential solution—i.e., the solution will be of the form $y = Ae^{g(x)}$.

Integrals

You need to do this:

- Write the equation as $\dfrac{dy}{dx} = ky$. Separate the variables and integrate both sides to find:

$$\frac{dy}{y} = k\,dx \Rightarrow \int \frac{dy}{y} = \int k\,dx \Rightarrow \ln|y| = kx + C.$$

- To solve this equation for y, you have to be careful. Because $\ln|y| = kx + C$, $e^{\ln|y|} = e^{kx+C}$.

- That is $|y| = e^{kx+C}$. Because y is always positive, drop the absolute value signs. Use the laws of exponents to write:

$$y = e^{kx+C} \Rightarrow y = e^{kx} \cdot e^{C}.$$

- Rewrite e^c as some other constant A and find the solution $y = Ae^{kx}$.

Answer and Explanation:

D

Start by separating the variables: $\dfrac{dv}{-10v - 32} = dt$. Integrate both sides of the equation, $\displaystyle\int \frac{dv}{-10v - 32} = \int dt$. Integrating the left side requires a u-substitution, where $u = -10v - 32$ and $du = -10\,dv$:

$$\int \underbrace{-\frac{1}{10}(-10)}_{\text{multiply by 1}} \frac{dv}{-10v - 32} = -\frac{1}{10} \int \frac{\overbrace{-10\,dv}^{du}}{\underbrace{-10v - 32}_{u}} = -\frac{1}{10} \int \frac{1}{u}\,du = \int dt$$

It's easier to bring the $-\dfrac{1}{10}$ to the other side of the equation before integrating, mostly to help keep track of the constant:

$$\int \frac{1}{u}\,du = -10 \int dt \Rightarrow \ln|u| = -10t + C \Rightarrow u = e^{-10t} + C = Ae^{-10t}$$

Substitute back to get: $-10v - 32 = Ae^{-10t}$. Keeping in mind the rule to find the constant before solving for v: use the information $v(0) = 0$: $-10 \cdot 0 - 32 = Ae^{-10 \cdot 0} \Rightarrow -32 = A$. Fill this information into the equation and solve for v:

$$-10v - 32 = -32e^{-10t} \Rightarrow -10v = -32e^{-10t} + 32 \Rightarrow v = 3.2e^{-10t} - 3.2$$

This matches **(D)**.

Integrals

PRACTICE SET

1. If $A(t)$ is the amount of money in an account at time t, which of the following differential equations describes linear growth in the amount of money in the account?

   (A) $\dfrac{dA}{dt} = 50A$

   (B) $\dfrac{dA}{dt} = 50t^2$

   (C) $\dfrac{dA}{dt} = 25t$

   (D) $\dfrac{dA}{dt} = 100$

2. The population of a country doubles every 20 years. If the population (in millions) of the country was 100 in 2002, what will the population be in 2014 to the nearest million?

   (A) 132 million

   (B) 151 million

   (C) 167 million

   (D) 180 million

3. Zoologists tracking a population of rhinos noted a severe decline in the local population starting in 2015 after a sudden spike in poaching. If the population of rhinos was 1,272 in January 2015 and the rate of decline of the population is modeled by $r(t) = -200e^{-0.16t}$ rhinos per year, in what year will the population fall below 100 ?

   (A) 2019

   (B) 2027

   (C) 2032

   (D) 2040

4. During its growth phase, the population P of a bacterial colony grows according to the differential equation $\dfrac{dP}{dt} = kP$, where k is a constant and t is measured in minutes. If the population of the colony doubles every 23 minutes, what is the value of k ?

   (A) 0.030 per minute

   (B) 0.079 per minute

   (C) 0.230 per minute

   (D) 0.816 per minute

Integrals

ANSWERS AND EXPLANATIONS

1. D

Linear growth means that the rate of change is constant. Only (D) has a rate of change, $\frac{dA}{dt}$, which is constant, so **(D)** is correct.

2. B

Interpret the given information as "The rate of growth of the population is proportional to its size and $y(20) = 2y(0)$."

That is $\frac{dy}{dt} = ky$. The information $y(20) = 2y(0)$ will enable us to find k—the growth constant. Solve this differential equation using the technique of separation of variables:

$$\frac{1}{y}\,dy = k\,dt \Rightarrow \int \frac{1}{y}\,dy = \int k\,dt$$

$$\ln|y| = kt + C \Rightarrow e^{\ln|y|} = e^{kt+C}$$

$$|y| = e^{kt} \cdot e^{C} = Ae^{kt}$$

When $t = 0$, the population of the town is $y(0) = Ae^{k \cdot 0} = A$, so $y(20) = 2y(0) = 2A$. Use this information to find k:

$$y(20) = Ae^{k \cdot 20} = 2A$$

$$e^{20k} = 2$$

$$20k = \ln 2$$

$$k = \frac{\ln 2}{20}$$

That is, $y = Ae^{\frac{\ln 2}{20}t} = A\left(e^{\ln 2}\right)^{\frac{t}{20}} = A \cdot 2^{\frac{t}{20}}$.

Set $t = 0$ to be the year 2002, and use the given information $y(0) = 100$ to find A. $100 = A2^{\frac{0}{20}} = A \Rightarrow A = 100$. The population at time t is therefore given by $y(t) = 100 \cdot 2^{\frac{t}{20}}$ (where $t = 0$ is the year 2002). Find the population in 2014; that is $y(12)$: $y(12) = 100 \cdot 2^{\frac{12}{20}} \approx 151.172$. The population in the year 2014 will be approximately 151 million people, so **(B)** is correct.

3. C

Unlike the other questions, this question has provided the complete exponential equation. The negative constant (-0.16) indicates it is a formula for exponential decay, which makes sense because the rate of decline of the rhinos will decrease with their decreasing population. This question also does not involve differential equations; however, it will still require integration. The exponential decay formula only provides how much the population will decrease in a given year. It must undergo antidifferentiation to obtain the population after an interval of t years.

$$R(t) = \int r(t)\,dt = \int -200e^{-0.16t}\,dt$$

$$= \frac{-200e^{-0.16t}}{-0.16} = 1250e^{-0.16t} + C$$

Next, solve for C given the initial condition $R(0) = 1272$:

$$R(0) = 1250e^{0} + C$$

$$1272 = 1250 + C \Rightarrow C = 22$$

This completes the equation modeling the population of rhinos as $R(t) = 1250e^{-0.16t} + 22$. Setting this equation equal to 99 (when the population falls below 100) yields:

$$1250e^{-0.16t} + 22 = 99 \Rightarrow e^{-0.16t} = \frac{77}{1250}$$

$$-0.16t = \ln\left(\frac{77}{1250}\right) \Rightarrow t = \frac{\ln\left(\frac{77}{1250}\right)}{-0.16} \approx 17.4$$

The passage of 17.4 years from January 2015 would land midway through 2032, matching **(C)**.

4. A

The generic differential equation $\frac{dP}{dt} = kP$ will give the generic formula for exponential growth, $P = Ae^{kt}$, when integrated. While there seems to be a relative lack of information in this question compared to the prior examples, it is sufficient to solve for k. First, we know that when $t = 0$, P will be equal to A. Second, we know that the colony doubles every 23 minutes, so when $t = 23$, P will equal $2A$. Putting this into the equation yields:

$$2A = Ae^{23k} \Rightarrow 2 = e^{23k} \Rightarrow$$

$$23k = \ln 2 \Rightarrow k = \frac{\ln 2}{23} = 0.030$$

Therefore, the growth constant is 0.030 per minute, matching **(A)**.

14.6 SLOPE FIELDS

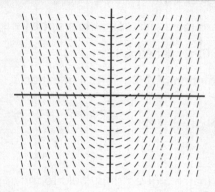

The above slope field describes which of the following differential equations?

(A) $\dfrac{dy}{dx} = \dfrac{x}{y}$

(B) $\dfrac{dy}{dx} = \dfrac{x}{3}$

(C) $\dfrac{dy}{dx} = 3x$

(D) $\dfrac{dy}{dx} = 3xy$

Slope fields

- One way to visualize solutions of differential equations of the type $\dfrac{dy}{dx} = f(x, y)$ is by drawing a short line segment with slope $f(x, y)$ through each point (x, y) in the plane.

- While this can't be done actually for *every* point, a sampling of many points will suffice.

- Following this procedure we see a pattern of lines.

- This pattern is called a *slope field* (or *direction field* or *vector field*) of the differential equation $\dfrac{dy}{dx} = f(x, y)$.

✔ **AP Expert Note**

In general, finding solutions to differential equations requires using the tools for integration, but using slope fields allows for the visualization of solutions to equations of the type discussed here.

Integrals

You need to do this:

Slope fields appear on the AP exam in a few different ways and may ask you to:

- Sketch a slope field for a given differential equation

- Sketch a solution curve through a given point in a slope field

- Match a slope field to a differential equation

- Match a slope field to a solution of a differential equation

Answer and Explanation:

B

This question requires that we match a given slope field to the appropriate differential equation. A solution of a differential equation is a function whose graph follows the slope field. Sketch a solution by starting at one point on the graph and connecting line segments that flow into each other—a kind of calculus "connect-the-dots."

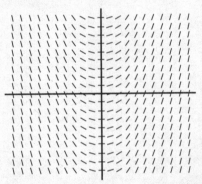

Note the wide parabolic shape of the connected lines. The derivative of a parabolic should be in terms of x to the first power, eliminating (A) and (D). (C) would be parabolic, but narrower than the standard parabolic $y = x^2$. **(B)** appropriately matches the slope field, being the derivative of a parabolic that is wider than the standard parabolic $y = x^2$.

Integrals

PRACTICE SET

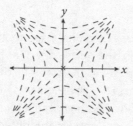

1. A slope field for which of the following differential equations is shown above?

(A) $\dfrac{dy}{dx} = \dfrac{x}{y}$

(B) $\dfrac{dy}{dx} = y^2 - x^2$

(C) $\dfrac{dy}{dx} = xy$

(D) $\dfrac{dy}{dx} = \dfrac{y}{x^2}$

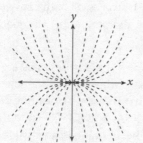

2. The above slope field describes which of the following differential equations?

(A) $\dfrac{dy}{dx} = \dfrac{2y}{x}$

(B) $\dfrac{dy}{dx} = x^2$

(C) $\dfrac{dy}{dx} = \dfrac{x^2}{y}$

(D) $\dfrac{dy}{dx} = 2xy$

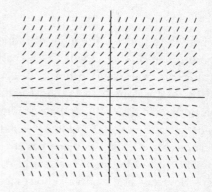

3. A slope field for which of the following differential equations is shown above?

(A) $\dfrac{dy}{dx} = \ln|x|$

(B) $\dfrac{dy}{dx} = 2xy$

(C) $\dfrac{dy}{dx} = \dfrac{y}{4}$

(D) $\dfrac{dy}{dx} = e^{x+y}$

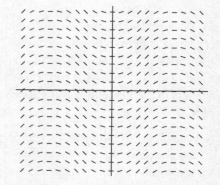

4. The above slope field describes which of the following differential equations?

(A) $\dfrac{dy}{dx} = 3\cos y$

(B) $\dfrac{dy}{dx} = \cos\left(x^2\right)$

(C) $\dfrac{dy}{dx} = \sin\left(\dfrac{y}{x}\right)$

(D) $\dfrac{dy}{dx} = 3\sin x$

ANSWERS AND EXPLANATIONS

1. A

The task here is to find a differential equation whose solutions are curves on the slope field. The easiest way to identify the correct answer is to look for distinguishing characteristics of the slope field. In this case, the slopes are always positive in the first and third quadrants, and always negative in the second and fourth quadrants. In addition, slopes are 0 when $x = 0$ and y is any nonzero value, but when $y = 0$ the slopes appear to be vertical (undefined). Now look at the answer choices for the differential equation that has those characteristics: **(A)** meets the description perfectly.

Alternatively, because this is a multiple choice test, you can also use the process of elimination to find the right answer. Choose "easy" values of x and y to plug into the answer choices. For instance, plugging in $x = 1$, $y = 1$ would rule out (B), because $1^2 - 1^2$ is 0 and the slope is clearly not 0 at the point (1, 1). Similarly, plugging in $x = 1$, $y = -1$ allows you to eliminate (D), and plugging in $x = 0$, $y = 0$ would eliminate (C). After eliminating three answer choices, **(A)** must be correct.

2. A

In this slope field, notice that for line segments close to the x-axis (where $y = 0$), the slope is 0, while for line segments close to the y-axis (where $x = 0$), the slope is vertical (undefined). This means the correct answer is (A), where y appears in the numerator and x appears in the denominator.

Another way to solve is to eliminate the choices you can rule out quickly and solve the remaining answer choices to see what the graphs of the solutions would look like. If you start with (A), you'd solve:

$$\frac{dy}{dx} = \frac{2y}{x}$$
$$\int \frac{1}{y}\, dy = 2 \cdot \int \frac{1}{x}\, dx$$
$$\ln y = 2 \ln x + C$$
$$y = e^{2 \ln x + C}$$
$$y = Ax^2$$

Solutions are in the form of parabolas with varying coefficients—exactly what is shown in the slope field. So **(A)** is correct.

3. C

When given a slope field and asked to match it to a differential equation, the quickest way to find the answer is to look for aspects of the slope field that stand out. Here, the slope field has line segments that are the same in horizontal bands. This means that for any y-value, the slope is constant, so look for the differential equation that does not depend on x. Only **(C)** fits this criterion, so it is correct.

Alternatively, you could plug in some "easy" numbers for x and y into each of the answer choices to find the slope, then look at the graphs to see whether the slope field has approximately that value at that x and y value. If you use this method, it may take a bit of time to eliminate the answer choices until you find the correct one.

4. D

In this case, looking at the characteristics of the slope field should suggest that you are looking for a trigonometric function. However, all of the answer choices have trig functions in them! So you will need to go a little farther.

One thing to notice is that the slopes are constant for a given x, no matter what y-value the line segment is at. This suggests that the slope does not depend on y, so you can rule out (A) and (C). Check the remaining two answer choices by plugging in an x-value, like $x = 0$.

(B) yields $\cos(0^2)$, which equals 1, while (D) produces $3 \sin 0$, which equals 0. Because the slope at all points where $x = 0$ appears to be 0, **(D)** is the right answer.

RAPID REVIEW

If you take away only five things from this chapter:

1. The average value integral: $\dfrac{1}{b-a}\displaystyle\int_{a}^{b} f(x)\, dx$.

2. Accumulation functions are defined by $G(x) = \displaystyle\int_{0}^{x} f(t)\, dt$, where the function $G(x)$ is the area under the curve $f(t)$ from some specified starting time to x.

3. If given a velocity or acceleration function, take the integrals $v(t) = \displaystyle\int_{a}^{t} a(t)\, dt$ and $x(t) = \displaystyle\int_{a}^{t} v(t)\, dt$ to work backwards to a position function.

4. The most basic type of differential equation tells us that the rate of change of a function is $f(x)$, $\dfrac{dy}{dx} = f(x)$.

5. A differential equation of the form $\dfrac{dy}{dx} = ky$ will be an exponential equation and the solution will be of the form $y = Ae^{g(x)}$.

TEST WHAT YOU LEARNED

1. If the graph of $y = f(x)$ contains the point $(0, 2)$, $\dfrac{dy}{dx} = \dfrac{-x}{ye^{\frac{x^2}{2}}}$, and $f(x) > 0$ for all x, then $f(x) =$

    (A) $1 + e^{\frac{-x^2}{2}}$

    (B) $3 - e^{\frac{-x^2}{2}}$

    (C) $\sqrt{2e^{\frac{-x^2}{2}} + 2}$

    (D) $\sqrt{e^{\frac{-x^2}{2}} + 3}$

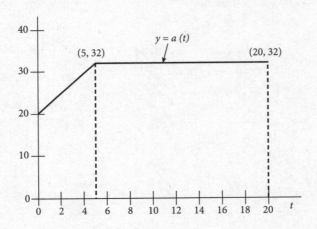

2. The acceleration of an airplane from the moment of liftoff $(t = 0)$ to 20 minutes into the flight is shown above. If the speed at liftoff is 900 ft/min, what is the speed of the plane after 20 minutes?

    (A) 610 ft/min   (B) 900 ft/min   (C) 1,510 ft/min   (D) 1,800 ft/min

Integrals

3. A wilderness preserve has been working to increase the population of tigers in their region. When they began their work in 2007, the tigers only numbered 34. As the population has increased, it continues to go up at an increasing rate. If in 2017 there are 212 tigers, which of the following is the appropriate formula for the tiger population in terms of time t in years?

    (A) $P = e^{0.183t} + 34$

    (B) $P = 34e^{0.183t}$

    (C) $P = e^{0.624t} + 34$

    (D) $P = 34e^{0.624t}$

4. A particle is moving along the x-axis with a velocity given by the function $v(t) = 4t^3 - 6t + 7$. When $t = 1$ the particle is at $x = 2$. Where is the particle when $t = 4$?

    (A) 233   (B) 236   (C) 248   (D) 251

Integrals

5. Which of the following is a slope field for the differential equation $\dfrac{dy}{dx} = \dfrac{2x^3}{y}$?

(A)

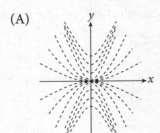

(B)

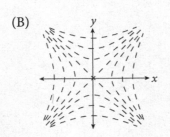

(C)

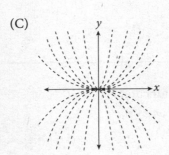

(D)

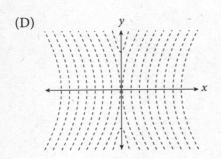

6. What is the average value of the function $f(x) = 4 - \sin x$ from $x = 0$ to $x = 3\pi$?

    (A) $4 - \dfrac{2}{3\pi}$

    (B) 4

    (C) $12\pi - 1$

    (D) $12\pi - \dfrac{1}{3\pi}$

Integrals

Answer Key

Test What You Already Know	Test What You Learned
1. **D** **Learning Objective:** 14.1	1. **C** **Learning Objective:** 14.4
2. **D** **Learning Objective:** 14.2	2. **C** **Learning Objective:** 14.2
3. **C** **Learning Objective:** 14.3	3. **B** **Learning Objective:** 14.5
4. **A** **Learning Objective:** 14.4	4. **A** **Learning Objective:** 14.3
5. **B** **Learning Objective:** 14.5	5. **A** **Learning Objective:** 14.6
6. **C** **Learning Objective:** 14.6	6. **A** **Learning Objective:** 14.1

REFLECTION

Test What You Already Know score: _____

Test What You Learned score: _____

Use this section to evaluate your progress. After working through the pre-quiz, check off the boxes in the "Pre" column to indicate which Learning Objectives you feel confident about. Then, after completing the chapter, including the post-quiz, do the same to the boxes in the "Post" column. Keep working on unchecked Objectives until you're confident about them all!

Pre	Post		
☐	☐	**14.1**	Find the average value of a function
☐	☐	**14.2**	Determine the net change of a function over an interval
☐	☐	**14.3**	Use integration to solve problems involving motion along a line
☐	☐	**14.4**	Solve differential equations with given initial conditions
☐	☐	**14.5**	Solve problems involving exponential growth and decay
☐	☐	**14.6**	Match a slope field with its corresponding differential equation

FOR MORE PRACTICE

Complete more practice online at kaptest.com. Haven't registered your book yet? Go to kaptest.com/booksonline to begin.

ANSWERS AND EXPLANATIONS

Test What You Already Know

1. D Learning Objective: 14.1

All of the statements are true except **(D)**. During the 24-hour period, the average rate of rainfall is $\frac{1}{24}\int_0^{24} r(t)\,dt$ in/hr. The quantity $\frac{r(24)-r(0)}{24}$ is the average rate of change in the rate of rainfall, a different concept.

2. D Learning Objective: 14.2

To find $f(6)$, compute $f(6) = f(0) +$ total change in f on $0 \le x \le 6$. The area under the graph of f' from $x = 0$ to $x = 6$ gives the total change in f on $0 \le x \le 6$. Express this as the definite integral:

$$f(6) = f(0) + \int_0^6 f'(x)\,dx$$

The definite integral $\int_0^6 f'(x)\,dx$ is the area under the graph of f'. To compute this area, break it into three regions, as shown below:

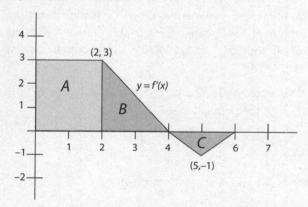

The area under the curve is therefore *Area of rectangle A + Area of triangle B − Area of triangle C*. Note that the area of triangle C is *subtracted* because the definite integral considers the area beneath the x-axis to be negative. This gives:

$$\int_0^6 f'(x)\,dx = 3\cdot2 + \frac{1}{2}\cdot2\cdot3 - \frac{1}{2}\cdot2\cdot1 = 8$$

Adding this to $f(0) = 7$ yields **(D)**.

$$f(6) = f(0) + \int_0^6 f'(x)\,dx = 7 + 8 = 15$$

3. C Learning Objective: 14.3

Recall that velocity is the antiderivative of acceleration, so begin by antidifferentiating the function.

$$v(t) = \int a(t)\,dt = \int 9t^2 - 12t\,dt$$
$$= 3t^3 - 6t^2 + C$$

Next, solve for C based on the initial condition $v(0) = 15$, giving $C = 15$. Last, solve for $v(3)$ with the complete equation: $v(3) = 3\cdot3^3 - 6\cdot3^2 + 15 = 42$, which is **(C)**.

4. A Learning Objective: 14.4

Because the slope of the curve is $4x^3y$, it follows that $\frac{dy}{dx} = 4x^3y$.

$$\frac{1}{y}\,dy = 4x^3\,dx$$
$$\int \frac{1}{y}\,dy = \int 4x^3\,dx$$
$$\ln|y| = x^4 + C$$
$$|y| = e^{x^4+C}$$
$$= Ae^{x^4}$$

Given the point $(0, 4)$, then its equation is $y = 4e^{x^4}$, which is **(A)**.

5. B Learning Objective: 14.5

Exponential growth functions follow the general form $y = Ae^{kx}$. In this case A will be the amount of money that initially begins in an account (200 for account 1, 150 for account 2). The variable k represents the growth constant, provided for each account. The easiest approach to solving this question is to graph the two functions: $y = 200e^{0.005x}$ for account 1 and $y = 150e^{0.008x}$ for account 2. On the CALC (2nd TRACE) menu, scroll down to **5:inter**sect and press ENTER. The intersection is at (95.894, 323.044), indicating that account 2 would surpass account 1 in approximately 96 days, which is **(B)**.

Integrals

6. **C** **Learning Objective:** 14.6

To start,

$$\frac{dy}{dx} = \sec^2(0.5x)$$

$$dy = \sec^2(0.5x)\, dx$$

$$\int dy = \int \sec^2(0.5x)\, dx$$

Now antidifferentiate to get $y = 2\tan(0.5x) + C$. Only **(C)** agrees.

Test What You Learned

1. **C** **Learning Objective**: 14.4

$$\frac{dy}{dx} = \frac{-x}{ye^{\frac{x^2}{2}}}$$

$$y\, dy = -xe^{\frac{-x^2}{2}}\, dx$$

$$\int y\, dy = \int -xe^{\frac{-x^2}{2}}\, dx$$

It follows that

$$\frac{1}{2}y^2 = e^{\frac{-x^2}{2}} + C_1$$

$$y^2 = 2e^{\frac{-x^2}{2}} + C_2$$

$$y = \pm\sqrt{2e^{\frac{-x^2}{2}} + C_2}$$

Given $f(x) > 0$ and the point $(0, 2)$, $2 = \sqrt{2 + C_2} \Rightarrow C_2 = 2$, which means $f(x) = \sqrt{2e^{\frac{-x^2}{2}} + 2}$. This matches **(C)**.

2. **C** **Learning Objective:** 14.2

Think of this as an accumulation problem. The speed of the airplane 20 minutes after liftoff is the initial speed plus the change in speed on $0 \le t \le 20$. The change in speed is given by the definite integral $\int_0^{20} a(t)\, dt$.

Therefore: $s(20) = s(0) + \int_0^{20} a(t)\, dt = 900 + \int_0^{20} a(t)\, dt$

The definite integral $\int_0^{20} a(t)\, dt$ is the area under the acceleration curve on $0 \le t \le 20$.

Break this area into two regions: trapezoid A and rectangle B:

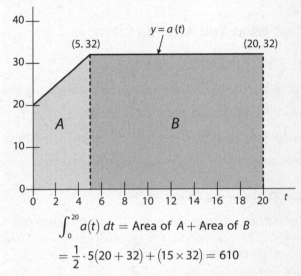

$$\int_0^{20} a(t)\, dt = \text{Area of } A + \text{Area of } B$$

$$= \frac{1}{2} \cdot 5(20 + 32) + (15 \times 32) = 610$$

Therefore, the speed at $t = 20$ is $s(20) = 900 + 610 = 1{,}510$ ft/min, which is **(C)**.

3. **B** **Learning Objective:** 14.5

The fact that the rate of growth increases with increasing population indicates that this is an exponential growth question. The exponential growth equation takes the form of $P = Ae^{kt}$. Begin filling in this equation with the initial condition, that in 2007 $(t = 0)$ the population was 34: $34 = Ae^{k(0)} \Rightarrow A = 34$.

Now input the second provided condition, that in 2017 $(t = 10)$ the population is 212:

$$212 = 34e^{k(10)} \Rightarrow 10k = \ln\frac{212}{34}$$

$$k = \frac{\ln\frac{212}{34}}{10} \approx 0.183$$

Putting it all together gives $P = 34e^{0.183t}$, matching **(B)**.

4. A Learning Objective: 14.3

The position of the particle $x(t)$ is an antiderivative of the velocity function, so start by computing the indefinite integral of the velocity:

$$v(t) = 4t^3 - 6t + 7$$

$$x(t) = \int v(t)\, dt = \int 4t^3 - 6t + 7\, dt$$

$$x(t) = t^4 - 3t^2 + 7t + C$$

Next, use the initial condition to determine C. Because $x(1) = 2$,

$$2 = 1^4 - 3 \cdot 1^2 + 7 \cdot 1 + C$$
$$2 = 5 + C$$
$$C = -3$$

So $x(t) = t^4 - 3t^2 + 7t - 3$. Finally, evaluate the position function at time $t = 4$:

$$x(4) = 4^4 - 3 \cdot 4^2 + 7 \cdot 4 - 3$$
$$= 256 - 48 + 28 - 3$$
$$= 233$$

That's a match for **(A)**.

5. A Learning Objective: 14.6

Cross-multiply and integrate to get $\int y\, dy = \int 2x^3\, dx$. Thus, $\dfrac{y^2}{2} = \dfrac{x^4}{2} + C$. Solving for y, we see that $y = \pm\sqrt{x^4 + C}$. For example, when $C = 0$, we have $y = \pm x^2$. These integral curves are given in **(A)**.

6. A Learning Objective: 14.1

Use the average value formula: $\dfrac{1}{b-a}\displaystyle\int_a^b f(x)\, dx$.

Then evaluate the integral and simplify the result:

$$\frac{1}{3\pi - 0}\int_0^{3\pi} 4 - \sin x\, dx = \frac{1}{3\pi}\left(4x + \cos x\right)\Big|_0^{3\pi}$$

$$\frac{1}{3\pi}\big((12\pi - 1) - 1\big) = 4 - \frac{2}{3\pi}$$

Distribute the $\dfrac{1}{3\pi}$ and that's a perfect match for **(A)**.

Integrals

PART 5

Calculus BC Topics

CHAPTER 15

Parametric, Polar, and Vector Functions

LEARNING OBJECTIVES

15.1 Define and graph parametric functions

15.2 Differentiate and integrate parametric functions

15.3 Define and graph polar functions

15.4 Differentiate and integrate polar functions

15.5 Define and graph vector functions

15.6 Differentiate and integrate vector functions

TEST WHAT YOU ALREADY KNOW

1. Eliminating the parameter in the equations $x = t^2 - t$, $y = t + 1$ yields which of the following equations?

    (A) $x = y^2 - 3y$

    (B) $x = y^2 - 3y - 2$

    (C) $x = y^2 - 3y + 2$

    (D) None of the above

2. Compute the derivative for the parametric equation $x = \sqrt{t} - t$ and $y = \frac{1}{2}t^4 + t$.

    (A) $\dfrac{4t^{\frac{7}{2}} + 2\sqrt{t}}{1 - 2\sqrt{t}}$

    (B) $\dfrac{4t^{\frac{5}{2}} + 2\sqrt{t}}{\sqrt{t}}$

    (C) $\dfrac{4t^{\frac{7}{2}} - 2}{1 - 2\sqrt{t}}$

    (D) $\dfrac{2t^{\frac{7}{2}} - 4\sqrt{t}}{2 + \sqrt{t}}$

3. Which of the following is a Cartesian equivalent to the polar equation: $2r = 1 + 3r \cos \theta$?

    (A) $2y^2 - 7x^2 - 1 = 0$

    (B) $4y^2 - 5x^2 - 1 = 0$

    (C) $4y^2 + 5x^2 - 6x - 1 = 0$

    (D) $4y^2 - 5x^2 - 6x - 1 = 0$

4. The figure below shows the portions of the graphs of the ray $\theta = \frac{\pi}{4}$ and the curve $r = 2 \cos \theta + \sin \theta$ that lie in the first quadrant. Which of the following integrals expresses the area of the shaded region R ?

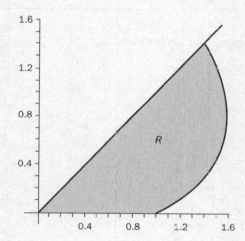

    (A) $\int_0^{1.6} \left(\frac{\pi}{4} - r \right) d\theta$

    (B) $\int_0^{\frac{\pi}{4}} r^2 \, d\theta$

    (C) $\int_0^{\frac{\pi}{4}} \frac{r^2}{2} \, d\theta$

    (D) $\int_0^{1.6} \left(\frac{\pi}{4} - \frac{r^2}{2} \right) d\theta$

5. State the component form and length of the vector v with initial point $A(2, -1)$ and terminal point $B(-1, 3)$.

   (A) $v\langle 3, -4 \rangle$, $|v| = 5$

   (B) $v\langle -3, 4 \rangle$, $|v| = 5$

   (C) $v\langle 3, -4 \rangle$, $|v| = \sqrt{7}$

   (D) $v\langle -3, 4 \rangle$, $|v| = \sqrt{7}$

6. Given $r(t) = \left\langle \dfrac{1}{t}, \ln(t) \right\rangle$ compute the derivative vector.

   (A) $r'(t) = \left\langle -\dfrac{1}{t^2}, \dfrac{1}{t} \right\rangle$

   (B) $r'(t) = \left\langle -\dfrac{1}{t}, \dfrac{1}{t} \right\rangle$

   (C) $r'(t) = \left\langle \dfrac{1}{t^2}, -\dfrac{1}{t} \right\rangle$

   (D) $r'(t) = \left\langle \dfrac{1}{t^2}, \dfrac{1}{t} \right\rangle$

15.1 DEFINE AND GRAPH PARAMETRIC FUNCTIONS

To answer a question like this:

Which curve below is represented by the parametric equations $x = 3 \sin t$ and $y = t^2$, where $-\pi \leq t \leq \pi$?

(A)

(B)

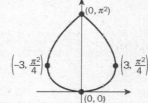

(C)

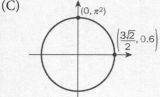

(D)

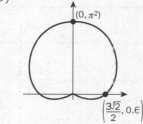

You need to know this:

Ways to represent a function

- Prior chapters have considered a function as a rule in which a dependent variable is controlled by an independent variable such as $f(x)$ or $h(t)$.

- However, when describing planar curves, other types of functions such as parametric and polar functions are better suited.

- Many of the questions on the BC exam take the same calculus concepts that are covered on the AB exam, but apply them to parametric, polar, or vector functions.

You need to do this:

- To write an equation to represent a planar curve, introduce a new variable t and write the x- and y-coordinates of the curve as functions of the new variable t. That is, write $x = f(t)$ and $y = g(t)$.

- Each value t determines a point $(x, y) = (f(t), g(t))$ on the curve and as t changes, so do $f(t)$ and $g(t)$.

Answer and Explanation:

B

t	$x = 3 \sin t$	$y = t^2$
$-\pi$	0	$\pi^2 \approx 9.8$
$\dfrac{-\pi}{2}$	-3	$\dfrac{\pi^2}{4} \approx 2.4$
$\dfrac{-\pi}{4}$	$\dfrac{-3\sqrt{2}}{2}$	$\dfrac{\pi^2}{16} \approx 0.6$
0	0	0
$\dfrac{\pi}{4}$	$\dfrac{3\sqrt{2}}{2}$	$\dfrac{\pi^2}{16} \approx 0.6$
$\dfrac{\pi}{2}$	3	$\dfrac{\pi^2}{4} \approx 2.4$
π	0	$\pi^2 \approx 9.8$

When you plot these points on the xy-plane you get the curve.

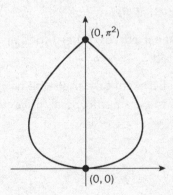

That's **(B)**.

PRACTICE SET

1. Eliminating the parameter in the equations $x = 4 - 2t$, $y = 3 + 6t - 4t^2$ yields which of the following equations?

    (A) $y = 8x^2 - 2x + 7$

    (B) $y = -x^2 - 11x + 31$

    (C) $y = 4x^2 + 8x - 5$

    (D) $y = -x^2 + 5x - 1$

2. What range for the parameter will result in the curve being traced exactly once for $x = \sin(2t)$ and $y = \cos(2t)$?

    (A) $0 \leq t \leq \dfrac{\pi}{2}$

    (B) $0 \leq t \leq \pi$

    (C) $0 \leq t \leq 2\pi$

    (D) $0 \leq t \leq 3\pi$

3. The graph of the parametric equations $x = 2 \sin t$, $y = 3 \cos t$ is

    (A) a circle, centered at the origin

    (B) an ellipse, centered at the origin

    (C) an ellipse, centered at $(2, 3)$

    (D) none of the above

4. Given that $x = 3 \sin 2t$ and $y = -2 \cos 2t$ and assuming that t increases from 0, which of the following statements is true?

    (A) The curve begins on the positive y-axis and travels in an anti-clockwise direction.

    (B) The curve begins on the negative y-axis and travels in a clockwise direction.

    (C) The curve begins on the negative y-axis and travels in an anti-clockwise direction.

    (D) None of the above

ANSWERS AND EXPLANATIONS

1. D

To eliminate the parameter, you must solve one equation for t, then plug that expression into the other equation and solve.

Because $x = 4 - 2t$ is simpler to solve for t, use this and plug into $y = 3 + 6t - 4t^2$.

$$x = 4 - 2t \Rightarrow t = -\frac{x}{2} + 2$$

$$y = 3 + 6t - 4t^2$$

$$= 3 + 6\left(-\frac{x}{2} + 2\right) - 4\left(-\frac{x}{2} + 2\right)^2$$

$$= 3 - 3x + 12 - 4\left(\frac{x^2}{4} - 2x + 4\right)$$

$$= -x^2 + 5x - 1$$

Hence, **(D)** is correct.

2. B

Because the period of both $x = \sin(2t)$ and $y = \cos(2t)$ is $\frac{2\pi}{2} = \pi$, you know that both the x- and y-coordinates will repeat every π radians. Thus, the desired range for t is: $0 \le t \le \pi$.

You can confirm this by making a table of values:

t	$x = \sin(2t)$	$y = -\cos(2t)$
0	0	−1
$\frac{\pi}{4}$	1	0
$\frac{2\pi}{4} = \frac{\pi}{2}$	0	1
$\frac{3\pi}{4}$	−1	0
$\frac{4\pi}{4} = \pi$	0	−1

That's **(B)**.

3. B

$$x = 2\sin t \Rightarrow \frac{x}{2} = \sin t$$

$$y = 3\cos t \Rightarrow \frac{y}{3} = \cos t$$

$$\left(\frac{x}{2}\right)^2 + \left(\frac{y}{3}\right)^2 = \sin^2 t + \cos^2 t = 1$$

$$\frac{x^2}{4} + \frac{y^2}{9} = 1$$

This is an ellipse because circles have coefficients of 1 for the x^2 and y^2. The center is (h, k) from the equation of an ellipse, $\frac{(x-h)^2}{a^2} + \frac{(y-k)^2}{b^2} = 1$, so the center is $(0, 0)$, or the origin. Thus, **(B)** is correct.

4. C

To see the starting point and direction of motion, set up a table of values and plot the points on a Cartesian plane.

t	$x = 3\sin(2t)$	$y = -2\cos(2t)$
0	0	−2
$\frac{\pi}{4}$	3	0
$\frac{2\pi}{4} = \frac{\pi}{2}$	0	2
$\frac{3\pi}{4}$	−3	0
$\frac{4\pi}{4} = \pi$	0	−2

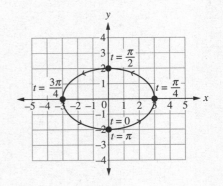

Therefore, **(C)** is correct.

15.2 DIFFERENTIATE AND INTEGRATE PARAMETRIC FUNCTIONS

To answer a question like this:

At time t the position of a particle on the xy-plane is given by $x(t) = e^t$, $y(t) = e^t \sin t$. Which of the following integrals gives the distance traveled by the particle from time $t = 0$ to time $t = 1$?

(A) $\displaystyle\int_0^1 \sqrt{e^t(1 + \sin t + \cos t)}\ dt$

(B) $\displaystyle\sqrt{2}\int_0^1 e^t \sqrt{1 + \sin t \cos t}\ dt$

(C) $\displaystyle\int_0^1 e^t \sqrt{(1 + \sin t + \cos t)}\ dt$

(D) $\displaystyle\int_0^1 e^{\frac{t}{2}} \sqrt{(1 + \sin t + \cos t)}\ dt$

You need to know this:

Differentiating parametric functions

- Suppose a function is defined with parametric equations $x = f(t)$ and $y = g(t)$. To calculate the tangent line of y (which is also a function of x), the chain rule must be used, giving $\dfrac{dy}{dt} = \dfrac{dy}{dx} \cdot \dfrac{dx}{dt}$.

- Provided $\dfrac{dx}{dt}$ is not equal to zero, this equality can be rearranged to get:

$$\frac{dy}{dx} = \frac{\frac{dy}{dt}}{\frac{dx}{dt}}, \text{ with } \left(\frac{dx}{dt}\right) \neq 0$$

- This is the equation used to calculate the slope of tangent lines to parametric functions.

You need to do this:

- The distance traveled by the particle from time $t = 0$ to $t = 1$ is the length of the curve traced by the particle over this interval, i.e., the arc length.

- Because the particle's position is given parametrically, use the parametric form of the equation for arc length: $\int_{t=a}^{t=b} \sqrt{\left(\dfrac{dx}{dt}\right)^2 + \left(\dfrac{dy}{dt}\right)^2}\ dt$

- Compute the derivatives and substitute them into the formula.

Answer and Explanation:

B

Compute the derivatives

$$\frac{dx}{dt} = e^t;\ \frac{dy}{dt} = e^t \sin t + e^t \cos t,$$

then substitute the derivatives and starting and ending times into the formula for arc length:

$$\int_{t=a}^{t=b} \sqrt{\left(\frac{dx}{dt}\right)^2 + \left(\frac{dy}{dt}\right)^2}\ dt$$

$$\int_{t=0}^{t=1} \sqrt{\left(e^t\right)^2 + \left(e^t \sin t + e^t \cos t\right)^2}\ dt$$

Simplify the integrand to find the answer

$$\sqrt{2} \int_0^1 e^t \sqrt{1 + \sin t \cos t}\ dt.$$

PRACTICE SET

1. The position of a particle moving in the xy-plane is determined by the curve $(x(t), y(t))$, $x'(t) = 3e^{3t}$, and $y'(t) = 2t - 4$. At $t = 0$, the particle is at the point $(0, 0)$. Compute the speed of the particle

   (A) 5

   (B) 3

   (C) $\sqrt{7}$

   (D) -5

2. The position of a particle moving in the xy-plane is determined by the curve $(x(t), y(t))$, $x'(t) = e^t$, and $y'(t) = 2t$. At $t = 0$, the particle is at the point $(0, 0)$. Compute the x- and y-coordinates of the particle at $t = 4$.

   (A) $(e, 4)$

   (B) $(4, e^2)$

   (C) $(8, 5)$

   (D) $(e^4, 16)$

3. Compute the horizontal and vertical tangents to the parametric curve: $x = t^3 - 3t^2$, $y = t^3 - 3t$.

   (A) $y = 2$ and $x = 0$

   (B) $y = -2, 1$ and $x = -2, 1$

   (C) $y = -2, 2$ and $x = -4, 0$

   (D) $y = 0$ and $x = -1, 1$

4. Compute $\dfrac{dy}{dx}$ given that $x = t^3 + t^2 - 1$ and $y = 1 - t^2$.

   (A) $\dfrac{6}{(3t + 2)^2}$

   (B) $-\dfrac{2}{3t + 2}$

   (C) $\dfrac{2}{3t + 2}$

   (D) $\dfrac{6}{(3t^2 + 2t)^3}$

ANSWERS AND EXPLANATIONS

1. A

The speed of the particle at time t is the length of the tangent vector of velocity. At $t = 0$ this is:

$$x\sqrt{x'(0)^2 + y'(0)^2} = \sqrt{\left(3e^{3(0)}\right)^2 + \left(2(0) - 4\right)^2}$$

$$= \sqrt{\left(3(1)\right)^2 + (-4)^2} = \sqrt{25} = 5$$

That's **(A)**.

2. D

The x- and y-coordinates are computed by integrating the velocity vector equations and taking into account the initial conditions of the position. Here, the initial condition is at $t = 0$ and position $(0, 0)$.

$$\int_0^t x'(t)\,dt = \int_0^t e^t\,dt = e^t - x(0) = x(t)$$

$$\int_0^t y'(t) = \int_0^t 2t = t^2 - y(0) = y(t)$$

At $t = 4$, compute the x- and y-coordinates as:

$$x(4) = e^4 - 0 = e^4$$

$$y(4) = 4^2 - 0 = 16$$

Hence, **(D)** is correct.

3. C

To find the horizontal tangents, set $\dfrac{dy}{dt} = 0$ and solve to determine at which values of t the tangents occur. Once the t values are found then you need to plug them into $y = t^3 - 3t$. To get the vertical tangents use the same process, but set $\dfrac{dx}{dt} = 0$, solve for t, and plug the t values into $x = t^3 - 3t^2$.

$$\frac{dy}{dt} = 3t^2 - 3 = 3\left(t^2 - 1\right) = 0$$

Hence, there are horizontal tangents at $t = -1$, 1 so $y(-1) = (-1)^3 - 3(-1) = 2$ and $y(1) = (1)^3 - 3(1) = -2$. The horizontal tangents are $y = -2, 2$. Now solve for the vertical tangents.

$$\frac{dx}{dt} = 3t^2 - 6t = 3t(t - 2) = 0$$

Solving for t, you get $t = 0$, 2 so $x(0) = (0)^3 - 3(0)^2 = 0$ and $x(2) = (2)^3 - 3(2)^2 = -4$. The vertical tangents are $x = -4, 0$. Therefore, **(C)** is correct.

4. B

The formula for the first derivative of a parametric equation is $y' = \dfrac{\frac{dy}{dt}}{\frac{dx}{dt}}$, therefore

$$\frac{dy}{dx} = \frac{\frac{dy}{dt}}{\frac{dx}{dt}} = \frac{-2t}{3t^2 + 2t} = -\frac{2}{3t + 2}$$

That's **(B)**.

15.3 DEFINE AND GRAPH POLAR FUNCTIONS

To answer a question like this:

Which of the following is the graph of $r = 2 - 2\sin\theta$?

(A)

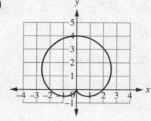

(B)

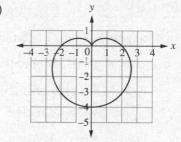

(C)

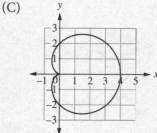

(D)

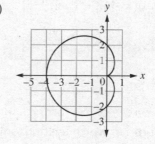

You need to know this:

Polar coordinate system

- Polar coordinates provide a different labeling system for the coordinate plane that is more useful in many situations.

- When using polar coordinates, use the variables r and θ to label points in the plane. The value of r denotes the distance from the origin, and the value of θ denotes the angle of rotation from the positive x-axis to the point.

- Any point in the plane can be written in either polar or Cartesian coordinates, and trigonometric functions can be used to convert between the systems as follows: $x = r\cos\theta$ and $y = r\sin\theta$.

Using your calculator in polar mode

- When graphing polar curves in polar mode on the calculator, be sure that the window settings for θ_{min} and θ_{max} will accommodate your curve's period. Either 0 to π or 0 to 2π will suffice for most curves (circles, cardioids, limaçons, and roses).

- Also, to minimize the distortion of polar graphs on the TI-83/84 calculator, choose your settings so that the ratio [Xmin, Xmax] : [Ymin, Ymax] $= 3 : 2$.

- Additionally, in polar mode, the default $\theta_{step} = 0.1308996\ldots$ represents the number $\frac{\pi}{24}$. This number works well in the TRACE mode because it allows you to trace to points on your polar curve that are multiples of $\frac{\pi}{24}$. These include those important and popular trig inputs such as $\frac{\pi}{6}$, $\frac{\pi}{4}$, $\frac{\pi}{3}$, and $\frac{\pi}{2}$.

You need to do this:

Learn to recognize the general patterns of common shapes

- Cardioids: $r = a \pm a \sin \theta$ (vertically oriented) or $r = a \pm a \cos \theta$ (horizontally oriented).
- Limacons: $r = a \pm b \sin \theta$ (vertically oriented) or $r = a \pm b \cos \theta$ (horizontally oriented). $\left|\frac{a}{b}\right| < 1$ will be looped, $1 < \left|\frac{a}{b}\right| < 2$ will be dimpled, and $\left|\frac{a}{b}\right| \geq 2$ will be convex.
- Roses: $r = a \sin n\theta$ or $r = a \cos n\theta$. When n is odd, the number of petals is odd. When n is even, the number of petals is even.

Answer and Explanation:

B

You recognize that all curves of the form $r = a \pm a \sin \theta$ are vertically oriented cardioids, which eliminates (C) and (D) (both of which are horizontally oriented). Plugging in a couple of values for θ allows you to choose between (A) and (B). For example, when

$$\theta = \frac{\pi}{6} \Rightarrow r = 2 - 2 \sin\left(\frac{\pi}{6}\right) = 2 - 2\left(\frac{1}{2}\right) = 1,$$

which rules out (A) and leaves **(B)** as the only remaining option.

Alternatively, you could plot a table of values for the curve and use this to identify the correct graph.

θ	$r = 2 - 2\sin\theta$	(θ, r)
0	2	$(0, 2)$
$\frac{\pi}{6}$	1	$\left(\frac{\pi}{6}, 1\right)$
$\frac{2\pi}{4} = \frac{\pi}{2}$	0	$\left(\frac{\pi}{2}, 0\right)$
$\frac{5\pi}{6}$	1	$\left(\frac{5\pi}{6}, 1\right)$
$\frac{4\pi}{4} = \pi$	2	$(\pi, 2)$
$\frac{3\pi}{2}$	4	$\left(\frac{3\pi}{2}, 4\right)$

PRACTICE SET

1. Which of the following Cartesian coordinates corresponds to the polar coordinate $\left(-2, \frac{4\tau}{3}\right)$?
   - (A) $\left(1, \sqrt{3}\right)$
   - (B) $\left(-1, \sqrt{3}\right)$
   - (C) $\left(1, -\sqrt{3}\right)$
   - (D) $\left(-1, -\sqrt{3}\right)$

2. Convert $x^2 = 4y$ to a polar equation.
   - (A) $r = \dfrac{\sin^2 \theta}{4 \cos^2 \theta}$
   - (B) $r = \dfrac{4 \sin^2 \theta}{\cos \theta}$
   - (C) $r = \dfrac{\sin \theta}{\cos \theta}$
   - (D) $r = \dfrac{4 \sin \theta}{\cos^2 \theta}$

3. Which of the following is a polar equation for the circle centered at $(4, 0)$ with radius 4 ?
   - (A) $r = 8 \cos \theta$
   - (B) $r = 4 \cos \theta$
   - (C) $r = 8 \sin \theta$
   - (D) $r = -4 \sin \theta$

4. Convert $y = 2x - 1$ to a polar equation.
   - (A) $\dfrac{2}{\cos \theta + \sin \theta}$
   - (B) $\dfrac{1}{2 \sin \theta - \cos \theta}$
   - (C) $\dfrac{-1}{\sin \theta - 2 \cos \theta}$
   - (D) $\dfrac{-4}{\sin^2 \theta + 2 \cos^2 \theta}$

ANSWERS AND EXPLANATIONS

1. A

You are given that $(r, \theta) = \left(-2, \frac{4\pi}{3}\right)$. Thus: $r = -2$ and $\theta = \frac{4\pi}{3}$. To convert from polar coordinates into Cartesian coordinates you use the conversion equations: $x = r \cos \theta$ and $y = r \sin \theta$.

$$x = r \cos \theta = -2 \cos \frac{4\pi}{3} = -2\left(-\frac{1}{2}\right) = 1$$

So:

$$y = r \sin \theta = -2 \sin \frac{4\pi}{3} = -2\left(-\frac{\sqrt{3}}{2}\right) = \sqrt{3}$$

and

$$(r, \theta) = \left(-2, \frac{4\pi}{3}\right) \rightarrow (x, y) = \left(1, \sqrt{3}\right)$$

Hence, **(A)** is correct.

2. D

To convert Cartesian equations to polar equations you will use the fact that $x = r \cos \theta$ and $y = r \sin \theta$. Plug these in and solve for r.

$$(r \cos \theta)^2 = 4(r \sin \theta)$$
$$r^2 \cos^2 \theta = 4r \sin \theta$$
$$r = \frac{4 \sin \theta}{\cos^2 \theta}$$

That's **(D)**.

3. A

You begin by recalling that the standard form of a circle centered at (h, k) with radius r is: $(x - h)^2 + (y - k)^2 = r^2$

Thus, the Cartesian equation for the given circle is: $(x - 4)^2 + y^2 = 16$. Expanding and simplifying yields:

$$x^2 - 8x + 16 + y^2 = 16$$
$$x^2 + y^2 = 8x$$

Remember that $x^2 + y^2 = r^2$ and that $x = r \cos \theta$ so

$$r^2 = 8r \cos \theta$$
$$r = 8 \cos \theta$$

Thus, **(A)** is correct.

4. C

To convert a Cartesian equation to a polar equation use the fact that $x = r \cos \theta$ and $y = r \sin \theta$. Plug these in and solve for r.

$$r \sin \theta = 2(r \cos \theta) - 1$$
$$r \sin \theta - 2r \cos \theta = -1$$
$$r(\sin \theta - 2 \cos \theta) = -1$$
$$r = \frac{-1}{\sin \theta - 2 \cos \theta}$$

Therefore, **(C)** is the appropriate match.

15.4 DIFFERENTIATE AND INTEGRATE POLAR FUNCTIONS

To answer a question like this:

Find the formula for the area in the first quadrant bordered by $r = 2 + \sin \theta$ but do not evaluate.

(A) $\int_{\frac{\pi}{2}}^{0} \frac{1}{2}(2 + \sin \theta) \, d\theta$

(B) $\int_{0}^{\frac{\pi}{2}} (2 + \sin \theta) \, d\theta$

(C) $\int_{0}^{\frac{\pi}{2}} (2 \sin \theta + 2)^2 \, d\theta$

(D) $\int_{0}^{\frac{\pi}{2}} \frac{1}{2}(2 + \sin \theta)^2 \, d\theta$

You need to know this:

Differentiating polar functions

- For polar curves, compute $\dfrac{dy}{dx}$ using the chain rule. If the polar curve is described by the equation $r = f(\theta)$, then θ plays the role of t in a parametric equation and $x = r \cos \theta = f(\theta) \cos \theta$ and $y = r \sin \theta = f(\theta) \sin \theta$. Therefore, calculate $\dfrac{dy}{dx}$ as follows (don't forget the product rule):

$$\frac{dy}{dx} = \frac{\frac{dy}{d\theta}}{\frac{dx}{d\theta}} = \frac{\frac{dr}{d\theta} \sin \theta + r \cos \theta}{\frac{dr}{d\theta} \cos \theta - r \sin \theta}.$$

- To find the slope of a line tangent to the curve, follow the same method as with parametric

  equations: $\dfrac{dy}{dx} = \dfrac{\frac{dy}{dt}}{\frac{dx}{dt}}$, with $\left(\dfrac{dx}{dt}\right) \neq 0$.

BC Topics

You need to do this:

- Finding an area will involve setting up an integral of the polar function.
- The region in the first quadrant is defined when $0 \leq \theta \leq \frac{\pi}{2}$.
- To find an area enclosed by a polar curve, use the formula $A = \int_{\alpha}^{\beta} \frac{1}{2} r^2 \, d\theta$.

Answer and Explanation:

D

Graphing $r = 2 + 2 \cos \theta$ in polar mode, you find the region in the first quadrant is defined when $0 \leq \theta \leq \frac{\pi}{2}$. The formula for the area of a polar curve is $\int_{\alpha}^{\beta} \frac{1}{2} r^2 \, d\theta$; hence,

$$\int_{0}^{\frac{\pi}{2}} \frac{1}{2} (2 + \sin \theta)^2 \, d\theta$$

That's a perfect match for **(D)**!

PRACTICE SET

1. A particle is moving along the curve $r = 2 + \cos\theta$ so that at time t, $\theta = t$. Find the velocity vector of the particle at time $t = \dfrac{\pi}{2}$.

    (A) $< -2, \; -1 >$

    (B) $< 1, 1 >$

    (C) $< 3, \; -2 >$

    (D) $< -3, 1 >$

2. Calculate the area of $r = \sqrt{\sin\theta}$ with $\dfrac{\pi}{2} \le \theta \le \pi$.

    (A) -1

    (B) -3

    (C) $\dfrac{1}{2}$

    (D) 1

3. Let r be a polar curve described by $r = 1 + 3\cos\theta$. What is the slope of the line tangent to the graph of r at $\theta = \dfrac{\pi}{2}$?

    (A) -3

    (B) $-\dfrac{1}{3}$

    (C) $\dfrac{1}{3}$

    (D) 3

4. Compute $\dfrac{dy}{dx}$ for $r = \cos^2\theta$.

    (A) $\dfrac{4\cos\theta}{\sin^2\theta}$

    (B) $\dfrac{\cos^2\theta - 2\sin^2\theta}{-3\sin\theta\cos\theta}$

    (C) $\dfrac{\cos^2\theta - 2\sin^2\theta}{-3\cot\theta\cos\theta}$

    (D) $2\sin\theta\cos^2\theta$

ANSWERS AND EXPLANATIONS

1. A

$x = r \cos \theta = (2 + \cos \theta)\cos \theta$ and

$y = r \sin \theta = (2 + \cos \theta)\sin \theta$.

The velocity vector is $\left\langle \dfrac{dx}{d\theta}, \dfrac{dy}{d\theta} \right\rangle$ evaluated at $\theta = t = \dfrac{\pi}{2}$.

$\dfrac{dx}{d\theta} = (2 + \cos \theta)\cdot(-\sin \theta) + \cos \theta \cdot (-\sin \theta) =$

$$-2 \sin \theta - 2 \sin \theta \cos \theta$$

$\dfrac{dy}{d\theta} = (2 + \cos \theta)\cdot \cos \theta + \sin \theta \cdot (-\sin \theta) =$

$$2 \cos \theta + \cos^2 \theta - \sin^2 \theta$$

Evaluating at $\theta = t = \dfrac{\pi}{2}$ gives $\dfrac{dx}{d\theta} = -2$ and $\dfrac{dy}{d\theta} = -1$. Therefore, the velocity vector is $<-2, -1>$, which means that **(A)** is the correct answer.

2. C

Use the formula $\dfrac{1}{2}\displaystyle\int_\alpha^\beta r^2 \, d\theta$ for calculating the area for polar curves:

$$\frac{1}{2}\int_{\frac{\pi}{2}}^{\pi} \left(\sqrt{\sin \theta}\right)^2 d\theta$$

$$\frac{1}{2}\int_{\frac{\pi}{2}}^{\pi} \sin \theta \, d\theta$$

$$-\frac{1}{2}(\cos \theta)\Big|_{\frac{\pi}{2}}^{\pi}$$

$$-\frac{1}{2}\left(\cos \pi - \cos \frac{\pi}{2}\right)$$

$$-\frac{1}{2}(-1 - 0) = \frac{1}{2}$$

That's **(C)**.

3. D

Tangent slopes in polar format are done parametrically.

$x = r \cos \theta \Rightarrow x = (1 + 3 \cos \theta)\cos \theta$

$\dfrac{dx}{d\theta} = (1 + 3 \cos \theta)(-\sin \theta) + \cos \theta(-3 \sin \theta)$

$y = r \sin \theta \Rightarrow y = (1 + 3 \cos \theta)\sin \theta$

$\dfrac{dy}{d\theta} = (1 + 3 \cos \theta)(\cos \theta) + \sin \theta(-3 \sin \theta)$

$$\frac{dy}{dx} = \frac{\dfrac{dy}{d\theta}}{\dfrac{dx}{d\theta}} = \frac{(1 + 3 \cos \theta)(\cos \theta) + \sin \theta(-3 \sin \theta)}{(1 + 3 \cos \theta)(-\sin \theta) + \cos \theta(-3 \sin \theta)}\Bigg|_{\theta = \frac{\pi}{2}} = 3$$

Thus, **(D)** is correct.

4. B

There are a few equations needed to start this problem. First you will need to convert the polar equation into parametric to be able to calculate the derivative. Use the fact that $x = r \cos \theta$ and $y = r \cos \theta$. Solve for r and plug in like so:

$$r = \frac{x}{\cos \theta} \Rightarrow \left(\frac{x}{\cos \theta}\right) = \cos^2 \theta \Rightarrow x = \cos^3 \theta$$

$$r = \frac{y}{\sin \theta} \Rightarrow \left(\frac{y}{\sin \theta}\right) = \cos^2 \theta \Rightarrow y = \sin \theta \cos^2 \theta$$

Now use the fact that $y' = \dfrac{\dfrac{dy}{d\theta}}{\dfrac{dx}{d\theta}}$ so:

$$\frac{dx}{d\theta} = 3 \cos^2 \theta(-\sin \theta) = -3 \cos^2 \theta \sin \theta$$

$$\frac{dy}{d\theta} = \cos \theta\left(\cos^2 \theta\right) + \sin \theta\left(2 \cos \theta(-\sin \theta)\right) =$$

$$\cos^3 \theta - 2 \sin^2 \theta \cos \theta$$

$$\frac{\frac{dy}{d\theta}}{\frac{dx}{d\theta}} = \frac{\cos^3 \theta - 2 \sin^2 \theta \cos \theta}{-3 \cos^2 \theta \sin \theta} = \frac{\cos^2 \theta - 2 \sin^2 \theta}{-3 \sin \theta \cos \theta}$$

Therefore, **(B)** is correct. Make sure you realize that most trigonometry answers can be simplified (in this case with the power simplifying formulas).

15.5 DEFINE AND GRAPH VECTOR FUNCTIONS

To answer a question like this:

Find the angle between the vectors $a = \left\langle -\sqrt{3}, 1 \right\rangle$ and $b = \left\langle -3, -\sqrt{3} \right\rangle$.

(A) $\dfrac{\pi}{4}$

(B) $\dfrac{\pi}{3}$

(C) $\dfrac{\pi}{2}$

(D) $\dfrac{2\pi}{3}$

You need to know this:

Vector coordinate system

- Vector functions take one real number and return three real numbers. These functions are usually written in the form $v(t) = (v_1(t), v_2(t), v_3(t))$.

- Think of a vector as a direction in 3-space. Each component tells you where the vector is pointing. The first component tells you how far to move in the x-direction, the second component tells you how far to move in the y-direction, and the last component tells you how far to move in the z-direction.

- The formula to find an angle between two vectors is $\theta = \cos^{-1}\left(\dfrac{a \cdot b}{|a||b|} \right)$

- The dot product $(a \cdot b)$ of the vectors $a = <x_1, y_1>$ and $b = <x_2, y_2>$ is $x_1 x_2 + y_1 y_2$

- The magnitude of a vector $a = <x_1, y_1>$ is $\sqrt{(x_1)^2 + (y_1)^2}$. This is considered the actual length of the vector.

You need to do this:

- You will need to break the vector angle formula into its components and solve them before finding the angle.

- Use the dot product formula to solve for the numerator.

- Calculate the magnitudes of both a and b and then multiply them to solve for the denominator.

- Lastly, combine those values and calculate the inverse cosine, or arccosine. Remember that the range of arccosine is $0 \leq y \leq \pi$ and that $\cos^{-1}\theta = x$ or $\cos^{-1}x = \theta$.

Answer and Explanation:

B

The angle θ between the vectors a and b is given by: $\theta = \cos^{-1}\left(\dfrac{a \cdot b}{|a||b|}\right)$

$$a \cdot b = \left\langle -\sqrt{3}, 1 \right\rangle \cdot \left\langle -3, -\sqrt{3} \right\rangle = \left(-\sqrt{3}\right)(-3) + (1)\left(-\sqrt{3}\right) = 3\sqrt{3} - \sqrt{3} = 2\sqrt{3}$$

$$|a| = \sqrt{\left(-\sqrt{3}\right)^2 + 1^2} = \sqrt{4} = 2$$

$$|b| = \sqrt{(-3)^2 + \left(-\sqrt{3}\right)^2} = \sqrt{12} = 2\sqrt{3}$$

$$\theta = \cos^{-1}\frac{a \cdot b}{|a||b|} = \cos^{-1}\left(\frac{2\sqrt{3}}{2\left(2\sqrt{3}\right)}\right) = \cos^{-1}\left(\frac{1}{2}\right) = \frac{\pi}{3}$$

Thus, **(B)** is the appropriate match.

PRACTICE SET

1. Consider the vectors $a = \langle -2, 3 \rangle$ and $b = \langle 1, -4 \rangle$. Find $|-3a - 2b|$.

   (A) $\sqrt{17}$

   (B) $\sqrt{65}$

   (C) $\sqrt{305}$

   (D) $\sqrt{353}$

2. Consider the vectors $a = \langle -2, 3 \rangle$ and $b = \langle 1, -4 \rangle$. Find $a \cdot b$.

   (A) -14

   (B) -10

   (C) 14

   (D) $\sqrt{221}$

3. Calculate the magnitude of $a = \langle 4, 7 \rangle$.

   (A) 5

   (B) $\sqrt{33}$

   (C) $\sqrt{65}$

   (D) 7

4. Compute the vector of a particle traveling from points a to b if $a = (-3, 2)$ and $b = (2, -5)$.

   (A) $\langle -1, -3 \rangle$

   (B) $\langle -6, -10 \rangle$

   (C) $\langle 4, 2 \rangle$

   (D) $\langle 5, -7 \rangle$

ANSWERS AND EXPLANATIONS

1. A

Begin by finding $-3a - 2b$:

$$-3a - 2b = -3\langle -2, 3 \rangle - 2\langle 1, -4 \rangle$$
$$= \langle 6, -9 \rangle - \langle 2, -8 \rangle$$
$$= \langle 4, -1 \rangle$$

To find the magnitude of this vector, use the Pythagorean theorem:

$$|\langle 4, -1 \rangle| = \sqrt{(4)^2 + (-1)^2} = \sqrt{17}$$

Therefore, **(A)** is correct.

2. A

$a \cdot b = \langle -2, 3 \rangle \cdot \langle 1, -4 \rangle = (-2)(1) + (3)(-4) = -2 - 12$
$= -14$

That's **(A)**.

3. C

The formula to calculate the magnitude is

$|x, y| = \sqrt{x^2 + y^2}$, so:

$$|a| = \sqrt{(4)^2 + (7)^2} = \sqrt{65}$$

Hence, **(C)** is correct.

4. D

To find the vector algebraically from points (x_1, y_1) and (x_2, y_2), subtract the starting point from the second point: $(x_2 - x_1, y_2 - y_1)$.

$$\langle 2 - (-3), (-5) - 2 \rangle = \langle 5, -7 \rangle$$

That's **(D)**.

15.6 DIFFERENTIATE AND INTEGRATE VECTOR FUNCTIONS

To answer a question like this:

Compute the position vector from the given velocity vector and initial condition:

$$v(t) = t^2 i + 4t^3 j, \; s(0) = j$$

(A) $\frac{1}{3}t^3 i + t^4 j$

(B) $t^4 + 1$

(C) t^4

(D) $\frac{1}{3}t^3 i + \left(t^4 + 1\right)j$

You need to know this:

- The different notations of vectors are $x_1 i + y_1 j + z_1 k = \langle x_1, y_1, z_1 \rangle$

- Just like normal derivatives and antiderivatives, or integration, vector derivatives are extremely similar with the exception that you do not combine the *ijk* terms, but rather keep them separate.

- This means that the velocity vector, (v, t), is the derivative of the position vector, (s, t), and the acceleration vector, (a, t), is the derivative of the velocity vector.

- Conversely, when you integrate the acceleration vector you will get the velocity vector, and when you integrate the velocity vector you will get the position vector.

- Remember that if some of the *ijk*'s are not mentioned or in the equation, then those are equal to zero.

You need to do this:

- First realize that you were given the velocity vector($v(t)$) with the initial condition $s(0) = j$. Because $s(t)$ is the position vector, you will need to integrate and plug in the initial condition, which will allow you to solve for the general term, C.

- Make sure to integrate the *ijk* terms separately and do not combine or add those terms together.

- Once everything is integrated and solved for, then rearrange all terms in the descending order of *i*, then *j*, then *k*.

Answer and Explanation:

D

To achieve the position vector from the velocity vector you need to take the integral of the velocity vector then use the initial condition to find the general term C.

$$\int v'(t)\, dt = \int t^2 i + 4t^3 j\, dt = \frac{1}{3}t^3 i + t^4 j + C$$

Now input the initial condition to find the general term C and get the final position vector.

$$j = \frac{1}{3}(0)^3 i + (0)^4 j + C, C = j$$

$$v(t) = \frac{1}{3}t^3 i + t^4 j + j$$

$$v(t) = \frac{1}{3}t^3 i + \left(t^4 + 1\right)j$$

PRACTICE SET

1. Compute the derivative vector of
   $r(t) = \cos(3t)i + tj$.

   (A) $\sin(3t)i + j$

   (B) $\tan(3t)i - tj$

   (C) $3\cos(t)i + j$

   (D) $-3\sin(3t)i + j$

2. Calculate the tangent vector at the specified
   value for $w(t) = t^3i + t^2j$, $t = 1$.

   (A) $2i - 3j$

   (B) $3i + 2j$

   (C) $6i - 4j$

   (D) $3t^2i + 2tj$

3. Calculate the tangent vector of $x(t) = \left\langle t - t^2, \sqrt{t} \right\rangle$
   at $t = 1$.

   (A) $\left\langle \dfrac{1}{2}, -1 \right\rangle$

   (B) $\left\langle \dfrac{1}{2}, -\dfrac{1}{4} \right\rangle$

   (C) $\langle 2, 3 \rangle$

   (D) $\left\langle -1, \dfrac{1}{2} \right\rangle$

4. Evaluate $\int_0^1 \left(\sqrt[3]{t}\,i + \dfrac{1}{t+1}j \right) dt$.

   (A) $\dfrac{3}{4}i + \ln(2)j$

   (B) $\dfrac{3}{4}t^{\frac{4}{3}}i + \ln|t+1|j + C$

   (C) $\dfrac{3}{4}i + \ln(2)j$

   (D) $\dfrac{4}{3}i + \ln(2)j$

ANSWERS AND EXPLANATIONS

1. D

With vectors in the $i + j$ notation the derivative vector will look like $i' + j'$, so take the derivatives separately and add/subtract appropriately. Make sure to use the chain rule when needed!

$$i' = (-3) \sin(3t) = -3 \sin(3t)$$
$$j' = (1) = 1$$
$$r'(t) = -3 \sin(3t) \, i + j$$

That's a match for **(D)**.

2. B

With vectors in the $i + j$ notation the derivative vector will look like $i' + j'$, so take the derivatives separately and add/subtract appropriately.

$$i' = 3t^2$$
$$j' = 2t$$
$$w'(t) = 3t^2 i + 2t j$$

Now that you have the derivative vector plug in the initial condition and evaluate $w'(1)$.

$$w'(1) = 3(1)^2 i + 2(1) j = 3i + 2j$$

Thus, **(B)** is correct.

3. D

Notice that the formula for a tangent vector at $t = 1$ is $\langle x'(1) \rangle$, so

$$x'(t) = \left\langle 1 - 2t, \frac{1}{2} t^{-\frac{1}{2}} \right\rangle = \left\langle 1 - 2t, \frac{1}{2\sqrt{t}} \right\rangle$$

$$x'(1) = \left\langle 1 - 2(1), \frac{1}{2\sqrt{(1)}} \right\rangle = \left\langle -1, \frac{1}{2} \right\rangle$$

That's **(D)**.

4. C

Treat this problem exactly like a normal definite integration problem and treat the ij's as separate terms like so:

$$\frac{3}{4} t^{\frac{4}{3}} i + \ln|t + 1| j \Big|_0^1$$

$$= \frac{3}{4} (1)^{\frac{4}{3}} i + \ln|(1) + 1| j - \left[\frac{3}{4} (0)^{\frac{4}{3}} i + \ln|(0) + 1| j \right]$$

$$= \frac{3}{4} i + \ln|2| j - \left[(0) i + (0) j \right]$$

$$= \frac{3}{4} i + \ln(2) j$$

Therefore, **(C)** is correct.

TEST WHAT YOU LEARNED

1. The graphs of the polar curves $r = 2 + \cos\theta$ and $r = -3\cos\theta$ are shown on the graph below. The curves intersect when $\theta = \frac{2}{3}\pi$ and $\theta = \frac{4}{3}\pi$. Region R is in the second quadrant, bordered by each curve and the y-axis.

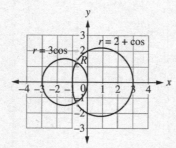

Set up but do not evaluate a formula to find the area of R.

(A) $\frac{1}{2}\displaystyle\int_{\frac{\pi}{2}}^{\frac{2}{3}\pi}\left((2 + \cos\theta)^2 - (-3\cos\theta)^2\right)d\theta$

(B) $\displaystyle\int_{\frac{\pi}{4}}^{\frac{\pi}{3}}\left((2 + \cos\theta)^2 - (-3\cos\theta)^2\right)d\theta$

(C) $\frac{1}{2}\displaystyle\int_{\frac{\pi}{2}}^{\frac{2}{3}\pi}\left((-3\cos\theta)^2 - (2 + \cos\theta)^2\right)d\theta$

(D) $\displaystyle\int_{\frac{\pi}{4}}^{\frac{\pi}{3}}\left((-3\cos\theta)^2 - (2 + \cos\theta)^2\right)d\theta$

2. Consider the curve given parametrically by $x = \sin\left(\frac{\pi}{2}t\right), y = 2t^3$. The equation of the line tangent to the curve at the point where $y = \frac{1}{4}$ is

(A) $y - \frac{1}{4} = \frac{3}{2}\left(x - \frac{\sqrt{2}}{2}\right)$

(B) $y - \frac{1}{4} = \frac{3\sqrt{2}}{\pi}x$

(C) $y - \frac{1}{4} = \frac{3\sqrt{2}}{\pi}\left(x - \frac{\sqrt{2}}{2}\right)$

(D) Not enough information to determine

3. Which of the following polar coordinates corresponds to the Cartesian coordinate $(-1, 1)$?

   (A) $\left(1, \dfrac{3\pi}{4}\right)$

   (B) $\left(-1, \dfrac{3\pi}{4}\right)$

   (C) $\left(-\sqrt{2}, \dfrac{3\pi}{4}\right)$

   (D) $\left(\sqrt{2}, \dfrac{3\pi}{4}\right)$

4. Compute the dot product for the vectors $\langle 4, -2 \rangle$, and $\langle 1, \sqrt{7} \rangle$.

   (A) $4 - 2\sqrt{7}$

   (B) $\dfrac{13}{10}$

   (C) 2

   (D) $\sqrt{7}$

5. Eliminating the parameter in the equations $x = 2 \sin t$, $y = 3 \cos t$ yields which of the following equations?

(A) $\dfrac{x^2}{3} - \dfrac{y^2}{2} = 1$

(B) $\dfrac{x^2}{9} + \dfrac{y^2}{4} = 1$

(C) $\dfrac{x^2}{4} + \dfrac{y^2}{9} = 1$

(D) $\dfrac{x^2}{2} - \dfrac{y^2}{3} = 1$

6. Compute the tangent vector at the specified value of $z(t) = \langle 2 \sin(t), -3 \cos(t) \rangle$, $t = \dfrac{\pi}{3}$.

(A) $\left\langle \sqrt{3}, \dfrac{3}{2} \right\rangle$

(B) $\left\langle 1, \dfrac{3\sqrt{3}}{2} \right\rangle$

(C) $\left\langle -1, -\dfrac{3\sqrt{3}}{2} \right\rangle$

(D) $\langle 2, 3 \rangle$

Answer Key

Test What You Already Know	Test What You Learned
1. C	1. C
2. A	2. C
3. D	3. D
4. C	4. A
5. B	5. A
6. A	6. B

REFLECTION

Test What You Already Know score: _____

Test What You Learned score: _____

Use this section to evaluate your progress. After working through the pre-quiz, check off the boxes in the "Pre" column to indicate which Learning Objectives you feel confident about. Then, after completing the chapter, including the post-quiz, do the same to the boxes in the "Post" column. Keep working on unchecked Objectives until you're confident about them all!

Pre	Post		
☐	☐	**15.1**	Define and graph parametric functions
☐	☐	**15.2**	Differentiate and integrate parametric functions
☐	☐	**15.3**	Define and graph polar functions
☐	☐	**15.4**	Differentiate and integrate polar functions
☐	☐	**15.5**	Define and graph vector functions
☐	☐	**15.6**	Differentiate and integrate vector functions

FOR MORE PRACTICE

Complete more practice online at kaptest.com. Haven't registered your book yet? Go to kaptest.com/booksonline to begin.

ANSWERS AND EXPLANATIONS

Test What You Already Know

1. C

Eliminating the parameter yields:

$$y = t + 1 \Rightarrow t = y - 1$$
$$x = t^2 - t$$
$$= (y-1)^2 - (y-1)$$
$$= y^2 - 2y + 1 - y + 1$$
$$= y^2 - 3y + 2$$

Therefore, **(C)** is correct.

2. A

The formula for the first derivative of a parametric equation is

$$y' = \frac{\frac{dy}{dt}}{\frac{dx}{dt}} = \frac{2t^3 + 1}{\frac{1}{2}t^{-\frac{1}{2}} - 1} = \frac{4t^{\frac{7}{2}} + 2\sqrt{t}}{1 - 2\sqrt{t}}$$

That's **(A)**.

3. D

You begin by making use of the two known facts: $x = r \cos \theta$ and $r = \sqrt{x^2 + y^2}$. Substituting these into the given equation yields:

$$2r = 1 + 3r \cos \theta$$
$$2\sqrt{x^2 + y^2} = 1 + 3x$$

Squaring both sides of this equation gives:

$$\left(2\sqrt{x^2 + y^2}\right)^2 = (1 + 3x)^2$$
$$4\left(x^2 + y^2\right) = 1 + 6x + 9x^2$$
$$4x^2 + 4y^2 = 1 + 6x + 9x^2$$
$$4y^2 - 5x^2 - 6x - 1 = 0$$

Thus, **(D)** is correct.

4. C

The area of a simple region enclosed by a polar curve r between two rays $\theta = \alpha$ and $\theta = \beta$ is $\int_{\alpha}^{\beta} \frac{r^2}{2} \, d\theta$. In our case, the simple polar region R is enclosed by r and the rays $\theta = 0$ and $\theta = \frac{\pi}{4}$; therefore, the integral in answer **(C)** expresses the area of R.

5. B

Consider the graphical representation of the vector, shown below:

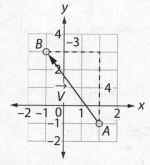

You can see that in going from A to B, you go -3 in the x (horizontal) direction and $+4$ in the y (vertical) direction. Thus, the vector representation of v is $\langle -3, 4 \rangle$. To find the length of the vector, use the Pythagorean Theorem:

$$|v| = \sqrt{(-3)^2 + 4^2} = \sqrt{25} = 5$$

So **(B)** is correct.

6. A

The derivative vector of a vector $\langle x, y \rangle$ is $\left\langle \dfrac{dx}{dt}, \dfrac{dy}{dt} \right\rangle$. For $\dfrac{dx}{dt}$ you can use the power rule or the quotient rule. The quotient rule will work out as follows:

$$\frac{dx}{dt} = \frac{(0)(t) - (1)(1)}{t^2} = -\frac{1}{t^2}$$

The power rule will work out as follows:

$$\frac{1}{t} = t^{-1}$$

$$\frac{dx}{dt} = (-1)t^{-2} = -\frac{1}{t^2}$$

For $\dfrac{dy}{dt}$ remember that the derivative of $\ln(x)$ is $\dfrac{x'}{x}$ hence,

$$\frac{dy}{dt} = \frac{1}{t}$$

Now put both derivatives into the vector form.

$$r'(t) = \left\langle -\frac{1}{t^2}, \frac{1}{t} \right\rangle$$

Therefore, **(A)** is the appropriate match.

Test What You Learned

1. C

When you see sine and cosine, most of the time it will involve the Pythagorean theorem, which states $\sin^2 x + \cos^2 x = 1$. You will want to solve for sine and cosine: $\dfrac{x}{2} = \sin t$ and $\dfrac{y}{3} = \cos t$. Now square both to get $\dfrac{x^2}{4} = \sin^2 t$ and $\dfrac{y^2}{9} = \cos^2 t$. Add the equations and simplify.

$$\frac{x^2}{4} + \frac{y^2}{9} = \sin^2 t + \cos^2 t = 1$$

$$\frac{x^2}{4} + \frac{y^2}{9} = 1$$

Hence, **(C)** is correct.

2. C

To determine the equation of the tangent line, you need to find the x- and y-coordinates of a point on the line and the slope, which is the derivative $\dfrac{dy}{dx}$ of the function at the point where $y = \dfrac{1}{4}$. You'll use this point as your point on the line.

Next, you need to find its x-coordinate. Substitute into the definition $y = 2t^3$ and solve for t to find $t = \dfrac{1}{2}$. Substitute this value for t into the definition $x = \sin\left(\dfrac{\pi}{2}t\right)$ and simplify to find $x = \dfrac{\sqrt{2}}{2}$. Now you've found the points you need: $\left(\dfrac{\sqrt{2}}{2}, \dfrac{1}{4}\right)$.

Next, find the slope. Because the curve is defined parametrically, use the formula $\dfrac{dy}{dx} = \dfrac{\frac{dy}{dt}}{\frac{dx}{dt}}$ and evaluate this derivative at the value of t corresponding to $x = \dfrac{\sqrt{2}}{2}$, $y = \dfrac{1}{4}$ which was computed above as $t = \dfrac{1}{2}$. Therefore, you need to find $\dfrac{dy}{dx}\bigg|_{\frac{1}{4}, \frac{\sqrt{2}}{2}} = \dfrac{\frac{dy}{dt}}{\frac{dx}{dt}}\bigg|_{t=\frac{1}{2}}$.

Compute $\dfrac{dy}{dt} = 6t^2$ and $\dfrac{dx}{dt} = \dfrac{\pi}{2}\cos\dfrac{\pi}{2}t$; now evaluate these derivatives at $t = \dfrac{1}{2}$:

$$\frac{dy}{dx}\bigg|_{\frac{1}{4}, \frac{\sqrt{2}}{2}} = \frac{\frac{dy}{dt}}{\frac{dx}{dt}}\bigg|_{t=\frac{1}{2}} = \frac{6t^2}{\frac{\pi}{2}\cos\frac{\pi}{2}t}\bigg|_{t=\frac{1}{2}} = \frac{3\sqrt{2}}{\pi}.$$

Now you have the point $(x_0, y_0) = \left(\dfrac{1}{4}, \dfrac{\sqrt{2}}{2}\right)$ and slope $m = \dfrac{3\sqrt{2}}{\pi}$ of the line and you can use the point-slope form of the line equation, $y - y_0 = m(x - x_0)$ to find the equation of the line tangent to the curve:

$$y - \frac{1}{4} = \frac{3\sqrt{2}}{\pi}\left(x - \frac{\sqrt{2}}{2}\right).$$

Thus, **(C)** is correct.

3. D

You are given that $(x, y) = (-1, 1)$. Thus: $x = -1$ and $y = 1$. To convert from Cartesian coordinates into polar coordinates, use the conversion equations: $r = \sqrt{x^2 + y^2}$ and $\theta = \tan^{-1}\left(\dfrac{y}{x}\right)$.

So:

$$r = \sqrt{x^2 + y^2} = \sqrt{(-1)^2 + 1^2} = \sqrt{2}$$

and

$$\theta = \tan^{-1}\left(\frac{1}{-1}\right) = -\frac{\pi}{4} + n\pi.$$

Because your Cartesian coordinate, $(-1, 1)$ is in the second quadrant, $\theta = -\dfrac{\pi}{4} + \pi = \dfrac{3\pi}{4}$. Thus, an equivalent polar coordinate is given by $\left(\sqrt{2}, \dfrac{3\pi}{4}\right)$. Therefore, **(D)** is correct.

4. A

First determine the angles that sweep out the area of region R. $r = 2 + \cos\theta$ intersects the positive y-axis at $\theta = \dfrac{\pi}{2}$ radians. $r = -3\cos\theta$ is at the origin at $\theta = \dfrac{\pi}{2}$ radians. The curves intersect each other in the second quadrant at $\theta = \dfrac{2}{3}\pi$ radians. Therefore, because $A = \dfrac{1}{2}\displaystyle\int_\alpha^\beta r^2\, d\theta$,

$$A = \frac{1}{2}\int_{\frac{\pi}{2}}^{\frac{2}{3}\pi}\left((2 + \cos\theta)^2 - (-3\cos\theta)^2\right) d\theta.$$

Therefore, **(A)** is the appropriate match.

5. A

The formula for the dot product of vectors $\langle x_1, y_1\rangle$ and $\langle x_2, y_2\rangle$ is $(x_1 \cdot x_2) + (y_1 \cdot y_2)$. So,

$$(4)(1) + (-2)\left(\sqrt{7}\right)$$

$$4 - 2\sqrt{7}$$

Hence, **(A)** is correct.

6. B

The tangent vector of the form $\langle x, y\rangle$ is $\langle x', y'\rangle$ so

$$x' = (1)2\cos(t) = 2\cos(t)$$
$$y' = (-1)(-3)\sin(t) = 3\sin(t)$$
$$z'(t) = \langle 2\cos(t),\ 3\sin(t)\rangle$$

Now calculate $z'\left(\dfrac{\pi}{3}\right)$.

$$z'\left(\frac{\pi}{3}\right) = \left\langle 2\cos\left(\frac{\pi}{3}\right),\ 3\sin\left(\frac{\pi}{3}\right)\right\rangle = \left\langle 2\left(\frac{1}{2}\right),\ 3\left(\frac{\sqrt{3}}{2}\right)\right\rangle = \left\langle 1,\ \frac{3\sqrt{3}}{2}\right\rangle$$

That's **(B)**.

CHAPTER 16

Additional Techniques of Differentiation and Integration

LEARNING OBJECTIVES

16.1 Solve differential equations via Euler's method

16.2 Apply integration by parts to integrands with two functions

16.3 Evaluate fractional integrands using simple partial fractions

16.4 Perform antidifferentiation of improper integrals

16.5 Solve logistic differential equations and use them in modeling

TEST WHAT YOU ALREADY KNOW

1. Let $y = f(x)$ be the solution to the differential equation $\frac{dy}{dx} = x - y - 2$ with the initial condition $f(-1) = 3$. What is the approximation for $f(1)$ using Euler's method with two steps of equal length beginning at $x = -1$?

   (A) -3 (B) -2 (C) 0 (D) 2

2. $\int x\cos(2x)\,dx =$

   (A) $\frac{1}{2}x\sin(2x) - \frac{1}{4}\cos(2x) + C$

   (B) $\frac{1}{2}x\sin(2x) + \frac{1}{4}\cos(2x) + C$

   (C) $\frac{1}{2}x\sin(2x) - \frac{1}{2}\cos(2x) + C$

   (D) $2x\sin(2x) - 4\cos(2x) + C$

3. Evaluate $\int \frac{7}{(x-3)(x+4)}\,dx$.

   (A) $\ln\left|\frac{x-3}{x+4}\right| + C$

   (B) $\ln|(x-3)(x+4)| + C$

   (C) $\ln\left|\frac{x+4}{x-3}\right| + C$

   (D) $7\ln\left|\frac{x-3}{x+4}\right| + C$

4. Determine if the improper integral $\int_1^\infty \frac{1}{x}\,dx$ converges or diverges; if it converges, calculate where.

   (A) $-\infty$, diverges

   (B) converges, 1

   (C) converges, 0

   (D) ∞, diverges

5. The population $P(t)$ of the island fox on Catalina Island satisfies the logistic differential equation $\frac{dP}{dt} = P\left(2 - \frac{P}{5000}\right)$. According to this model, the $\lim\limits_{t \to \infty} P(t)$ is equal to

   (A) 2,500

   (B) 3,000

   (C) 5,000

   (D) 10,000

BC Topics

16.1 EULER'S METHOD

To answer a question like this:

Let $y = f(x)$ be the solution to the differential equation $\dfrac{dy}{dx} = x + y$ with the initial condition $f(2) = 6$. What is the approximation for $f(3)$ using Euler's method with two steps of equal length beginning at $x = 2$?

(A) $\dfrac{13}{4}$

(B) $\dfrac{65}{4}$

(C) 17

(D) $\dfrac{61}{2}$

You need to know this:

When to apply Euler's method

- Sometimes it is impossible to obtain an explicit formula for the solution of a differential equation.

- However, important information can still be gathered about the shape of the solution. Euler's method is a very useful technique in this regard.

- Beginning with an initial value problem and using not much more than the definition of derivatives, numerically plot points approximating the solution of the given differential equation to get a good idea of what the solution looks like.

✔ AP Expert Note

On the Calc BC exam you may be asked to determine whether an approximation obtained by Euler's method is an overestimate or an underestimate of the exact value. To do this, evaluate the second derivative at the indicated point to determine the curve's concavity. If the solution curve is concave up (the second derivative is positive), Euler's approximation is an underestimate. If the solution curve is concave down, Euler's approximation is an overestimate.

BC Topics

You need to do this:

- Start with the limit definition of a derivative: $\frac{dy}{dt}(a) = y'(a) = \lim\limits_{h \to 0} \frac{y(a+h) - y(a)}{h}$.

- A limit really only approximates a function, so $\frac{dy}{dt}(a) \approx \frac{y(a+h) - y(a)}{h}$, which can be rearranged to get $y(a+h) \approx y(a) + h\frac{dy}{dt}(a)$.

- Thus, for a general first-order differential equation, $\frac{dy}{dt} = F(x,y)$, $y(x_0) = y_0$, let $a = x_0$ and write the approximation function as:

$$y(x_0 + h) \approx y(x_0) + h\frac{dy}{dt}(x_0)$$
$$\approx y_0 + h \cdot F(x_0, y_0)$$

- To use Euler's method, generalize this process and develop a large collection of points that approximate actual values of $y(t)$. The x-values will all differ by a prescribed step increment, h. The smaller h is, the more accurate the graph will be, but the more complicated the computations become. Pick an h that's small enough to give useful information, but still large enough to be manageable.

- To calculate more values of $y(t)$:

$$y_1 = y(x_0 + h) \approx y_0 + h \cdot F(x_0, y_0)$$
$$y_2 = y(x_0 + 2h) \approx y_1 + h \cdot F(x_1, y_1) \quad \text{where } x_k = x_{k-1} + h$$
$$y_3 = y(x_0 + 3h) \approx y_2 + h \cdot F(x_2, y_2)$$

Answer and Explanation:

B

Original x	Original y	$dy = (x+y)dx$	Next x	Next y
2	6	$(2+6)\frac{1}{2} = 4$	2.5	10
2.5	10	$\left(\frac{5}{2}+10\right)\frac{1}{2} = \frac{25}{4}$	3	$\frac{65}{4}$

The two steps of equal length between will be at $x = 2$ and $x = 2.5$ to approximate $x = 3$. As the table above shows, the next y-value to follow will be $\frac{65}{4}$, matching **(B)**.

PRACTICE SET

1. Use Euler's method with step increment $h = 0.1$ to approximate $y(0.3)$, where $y(x)$ is the solution to the initial value problem $y' = y + xy$, $y(0) = 1$.

   (A) 0.1

   (B) 1.1

   (C) 1.221

   (D) 1.37

2. $\dfrac{dy}{dx} = 2x + 5y - 3$ and $y(1) = 0$. Estimate $y(2)$ with $\Delta x = \dfrac{1}{2}$.

   (A) -1.75

   (B) -0.5

   (C) 1.5

   (D) 2.25

3. $\dfrac{dy}{dx} = \sin(x) - \cos(y)$ and $y\left(\dfrac{\pi}{2}\right) = 0$. Estimate $y(\pi)$ with step-size $\dfrac{\pi}{4}$.

   (A) $-\dfrac{\sqrt{3}}{2}$

   (B) $\dfrac{\left(\sqrt{2} - 2\right)\pi}{8}$

   (C) $\dfrac{1}{2}$

   (D) $\dfrac{\pi}{4}$

4. $\dfrac{dy}{dx} = 3x + 2y + 1$ and $y(0) = k$. Using Euler's method, starting at $x = 0$ with a step-size of 2, gives the approximation $y(2) \approx 1$. What is k?

   (A) $-\dfrac{2}{7}$

   (B) $-\dfrac{1}{4}$

   (C) $-\dfrac{1}{5}$

   (D) $\dfrac{1}{10}$

ANSWERS AND EXPLANATIONS

1. D

In this question you are given that $h = 0.1$, $x_0 = 0$, $y_0 = 1$, and $F(x, y) = y + xy$. Using Euler's method for approximating solutions, you have:

$$y_0 = y(x_0) = y(0) = 1$$

$$y_1 = y_0 + hF(x_0, y_0) = 1 + (0.1)F(0, 1) = 1 + 0.1 = 1.1$$

$$y_2 = y_1 + hF(x_1, y_1) = 1.1 + (0.1)F(0.1, 1.1) = 1.1 + (0.1)1.21 = 1.221$$

$$y_3 = y_2 + hF(x_2, y_2) = 1.221 + (0.1)F(0.2, 1.221) = 1.37$$

Therefore, because $y_3 = y(x_0 + 0.3) = y(0.3)$, you have the approximation $y(0.3) \approx 1.37$. The answer is **(D)**.

2. A

Create a table. Start at $x = 1$ because the point $(1, 0)$ was given to you. Plug in the x- and y-values with the delta x as dx and evaluate to determine dy and your next y-value.

x	y	$dy = (2x + 5y - 3)dx$
1	0	$[2(1) + 5(0) - 3]0.5 = -0.5$
1.5	-0.5	$[2(1.5) + 5(-0.5) - 3]0.5 = -1.25$
2	-1.75	

Therefore, **(A)** is correct.

3. B

Create a table. Start at $x = \frac{\pi}{2}$ because the point was given to you. Plug in the x- and y-values with the step-size as dx and evaluate to determine dy and your next y-value.

x	y	$dy = (\sin(x) - \cos(y))dx \, \frac{\pi}{4}$
$\frac{\pi}{2}$	0	$\left[\sin\left(\frac{\pi}{2}\right) - \cos(0)\right]\frac{\pi}{4} = (1 - 1) = 0$
$\frac{3\pi}{4}$	0	$\left[\sin\left(\frac{3\pi}{4}\right) - \cos(0)\right]\frac{\pi}{4} = \left(\frac{\sqrt{2}}{2} - 1\right)\frac{\pi}{4} = \frac{(\sqrt{2} - 2)\pi}{8}$
π	$\frac{(\sqrt{2} - 2)\pi}{8}$	

That's a match for **(B)**.

4. C

Create a table. Start at $x = 0$ because the point was given to you. Plug in the x- and y-values with the step-size as dx and evaluate to find dy. Solve for k.

x	y	$dy = (3x + 2y + 1)dx$
0	k	$[3(0) + 2(k) + 1]2 = 4k + 2$
2	1	

This means that $k + 4k + 2 = 1$ and $k = -\frac{1}{5}$. Thus, **(C)** is correct.

16.2 INTEGRATION BY PARTS

To answer a question like this:

Evaluate $\int_1^e (\ln x)^2 \, dx$. (Hint: use integration by parts twice).

(A) $e - 2$

(B) 1

(C) e

(D) $\frac{1}{3}(e^3 - 1)$

You need to know this:

Relationship between integration by parts and the product rule

- Integration by parts can be thought of as the antidifferentiation of the product rule. The formula is $\int u \, dv = uv - \int v \, du$.

- Let $u = f(x)$ and $v = g(x)$, to obtain $\int f(x) g'(x) \, dx = f(x) g(x) - \int g(x) f'(x) \, dx$.

- From there, differentiate both sides to get $f(x) g'(x) = \frac{d}{dx}[f(x) g(x)] - g(x) f'(x)$, which can be rearranged to the product rule: $\frac{d}{dx}[fg] = gf' + fg'$.

You need to do this:

- Decide which function is *u* and which function is *dv*.

- Much like the substitution method, this choice really just takes practice. If the formula doesn't work the way you think it should with your initial guess, then just guess again.

- In general, however, it helps to keep in mind that *u* should be simpler after *differentiating* and *dv* should be simpler after *integrating*.

- If neither function becomes simpler after differentiation or integration, make an arbitrary assignment and see what happens.

Answer and Explanation:

A

First you set up $u = (\ln x)^2$ and $dv = dx$. Using integration by parts, the integral becomes

$$\left[x(\ln x)^2\right]_1^e - 2\int_1^e \ln x \, dx .$$

You again use integration by parts to compute the integral in the resulting expression, this time setting $u = \ln x$ and $dv = dx$, to get

$$\int_1^e \ln x \, dx = \left[x \ln x - x\right]_1^e .$$

You combine the information to find

$$\int_1^e (\ln x)^2 \, dx = \left[x(\ln x)^2 - 2(x \ln x - x)\right]_1^e .$$

You substitute the limits of integration into the expression and simplify to get the answer $e - 2$, matching **(A)**.

PRACTICE SET

1. Evaluate $\int xe^x\,dx$.

   (A) $e^x + 1 + C$

   (B) $xe^x - e^x + C$

   (C) $e^x + C$

   (D) $xe^x + C$

2. Evaluate the indefinite integral $\int x^2 \sin x\,dx$.

   (A) $x^2 \cos x + 2x\,\sin x + \cos x + C$

   (B) $-x^2 \cos x - 2x\,\sin x - 2\,\cos x + C$

   (C) $-x^2 \cos x + 2x\,\sin x + 2\,\cos x + C$

   (D) $\cos x^2 + C$

3. $\int \sin^{-1}x\,dx =$

   (A) $\dfrac{\left(\sin^{-1}x\right)^2}{2} + C$

   (B) $x\cos^{-1}x - \displaystyle\int \dfrac{x}{\sqrt{1-x^2}}\,dx$

   (C) $\sin^{-1}x + \displaystyle\int \dfrac{1}{\sqrt{1-x^2}}\,dx$

   (D) $x\sin^{-1}x - \displaystyle\int \dfrac{x}{\sqrt{1-x^2}}\,dx$

4. Evaluate the integral $\int e^{2\theta} \cos 3\theta\,d\theta$.

   (A) $\dfrac{1}{13}\left(2e^{2\theta} \cos 3\theta + 3e^{2\theta} \sin 3\theta\right)$

   (B) $\left(2e^{2\theta} \cos 3\theta\right)$

   (C) $\dfrac{1}{13}\left(e^{2\theta} \cos 3\theta + e^{3\theta} \sin 3\theta\right)$

   (D) $\left(3e^{3\theta} \cos 2\theta + 2e^{3\theta} \sin 2\theta\right)$

ANSWERS AND EXPLANATIONS

1. B

When dealing with repeating derivatives multiplied by another variable, integration by parts will be the method to apply. Set the repeating derivative to be dv and the rest to be u.

$$u = x \qquad dv = e^x dx$$
$$du = dx \qquad v = e^x$$

Plug into the formula and evaluate $uv - \int v\,du$.

$$xe^x - \int e^x dx$$
$$xe^x - e^x + C$$

Therefore, **(B)** is correct.

2. C

This problem requires integration by parts to solve. Because there is an x^2 you will need to use this method twice to get the trigonometry function by itself. Make sure to always make the function with the repeating derivative your dv and the rest your u.

$$u = x^2 \qquad dv = \sin x\,dx$$
$$du = 2x\,dx \qquad v = -\cos x$$

Use the formula and plug in $uv - \int v\,du$.

$$x^2(-\cos x) - \int (-\cos x)2x\,dx$$
$$-x^2 \cos x + 2\int x\cos x\,dx$$

Repeat again by correctly selecting your dv and u.

$$u = x \qquad dv = \cos x$$
$$du = dx \qquad v = \sin x$$

Plug into the formula and simplify.

$$-x^2 \cos x + 2\left[x\sin x - \int \sin x\,dx \right]$$
$$-x^2 \cos x + 2x\sin x + 2\cos x + C$$

That's **(C)**.

3. D

Integration by parts

$$\int \sin^{-1} dx = \int u\,dv \text{ with}$$

$$u = \sin^{-1} x \qquad dv = dx$$
$$du = \frac{1}{\sqrt{1-x^2}}\,dx \qquad v = x$$

$\int u\,dv = uv - \int v\,du$ means

$$\int \sin^{-1} dx = \sin^{-1} x \cdot x - \int x \cdot \frac{1}{\sqrt{1-x^2}}\,dx$$

$$\int \sin^{-1} dx = x\sin^{-1} x - \int \frac{x}{\sqrt{1-x^2}}\,dx$$

Hence, **(D)** is correct.

4. A

As you might have guessed, you have to use integration by parts to solve this integral. This is a good question because it requires integration by parts twice and you have to use the trick that you learned in the section on integration by parts.

Given $\int e^{2\theta} \cos 3\theta \, d\theta$, you get

$$u = \cos 3\theta \qquad\quad dv = e^{2\theta} d\theta$$
$$du = -3 \sin 3\theta \, d\theta \quad v = \frac{1}{2} e^{2\theta}$$

$$\int e^{2\theta} \cos 3\theta \, d\theta = \frac{1}{2} e^{2\theta} \cos 3\theta - \int \frac{1}{2} e^{2\theta} (-3 \sin 3\theta) d\theta =$$
$$\frac{1}{2} e^{2\theta} \cos 3\theta + \frac{3}{2} \int e^{2\theta} \sin 3\theta \, d\theta$$

Now you have to use integration by parts again on $\int e^{2\theta} \sin 3\theta \, d\theta$ you get

$$u = \sin 3\theta \qquad\quad dv = e^{2\theta} d\theta$$
$$du = 3 \cos 3\theta \, d\theta \quad v = \frac{1}{2} e^{2\theta}$$

$$\int e^{2\theta} \sin 3\theta \, d\theta = \frac{1}{2} e^{2\theta} \sin 3\theta - \int \frac{1}{2} e^{2\theta} (3 \cos 3\theta) d\theta =$$
$$\frac{1}{2} e^{2\theta} \sin 3\theta - \frac{3}{2} \int e^{2\theta} \cos 3\theta \, d\theta$$

Now, just as it seems that you are going in circles, you rewrite your equations and make some grand simplifications. From the very beginning,

$$\int e^{2\theta} \cos 3\theta \, d\theta = \frac{1}{2} e^{2\theta} \cos 3\theta + \frac{3}{2} \int e^{2\theta} \sin 3\theta \, d\theta$$
$$= \frac{1}{2} e^{2\theta} \cos 3\theta + \frac{3}{2} \left[\frac{1}{2} e^{2\theta} \sin 3\theta - \frac{3}{2} \right.$$
$$\left. \int e^{2\theta} \cos 3\theta \, d\theta \right]$$
$$= \frac{1}{2} e^{2\theta} \cos 3\theta + \frac{3}{4} e^{2\theta} \sin 3\theta - \frac{9}{4}$$
$$\int e^{2\theta} \cos 3\theta \, d\theta$$

Now all you have to do is solve this equation for $\int e^{2\theta} \cos 3\theta \, d\theta$ and you are done. You get

$$\int e^{2\theta} \cos 3\theta \, d\theta = \frac{1}{2} e^{2\theta} \cos 3\theta + \frac{3}{4} e^{2\theta} \sin 3\theta - \frac{9}{4}$$
$$\int e^{2\theta} \cos 3\theta \, d\theta$$
$$\frac{13}{4} \int e^{2\theta} \cos 3\theta \, d\theta = \frac{1}{2} e^{2\theta} \cos 3\theta + e^{2\theta} \sin 3\theta \frac{3}{4}$$
$$\int e^{2\theta} \cos 3\theta \, d\theta = \frac{4}{13} \left(\frac{1}{2} e^{2\theta} \cos 3\theta + \frac{3}{4} e^{2\theta} \sin 3\theta \right)$$
$$\int e^{2\theta} \cos 3\theta \, d\theta = \frac{1}{13} \left(2 e^{2\theta} \cos 3\theta + 3 e^{2\theta} \sin 3\theta \right)$$

The same result can be achieved by initial choosing $u = e^{2\theta} d\theta$ and $dv = \cos 3\theta \, d\theta$. Thus, **(A)** is correct.

16.3 PARTIAL FRACTION DECOMPOSITION

To answer a question like this:

Evaluate the integral $\int_0^1 \dfrac{x-1}{x^2+3x+2}\,dx$.

(A) $3\ln 3$

(B) $-5\ln 2 + 3\ln 3$

(C) $-5\ln 3 + 2\ln 2$

(D) $\ln 2 + 3\ln 3$

You need to know this:

Take a fractional integrand that can't be readily integrated and break it up into the sum of two or more simpler integrands that can be integrated.

- Consider the integral $\displaystyle\int \frac{7x+39}{x^2+9x+14}\,dx$ Notice that the denominator can be factored to get

$$\frac{7x+39}{x^2+9x+14} = \frac{7x+39}{(x+2)(x+7)}.$$

- Now, the goal is to split this fraction into two smaller fractions. Let A and B be two constants such that $\dfrac{7x+39}{(x+2)(x+7)} = \dfrac{A}{x+2} + \dfrac{B}{x+7}$.

- To determine the value of the constants, simply cross multiply and create a system of equations that can be solved:

$$\frac{A}{x+2} + \frac{B}{x+7} = \frac{Ax+7A+Bx+2B}{(x+2)(x+7)} = \frac{(A+B)x+(7A+2B)}{(x+2)(x+7)}$$

- Comparing this to the original equation, it can be gathered that $A+B=7$ and $7A+2B=39$.

- Solving this system of two equations and two unknowns yields $A=5$ and $B=2$. Therefore, simplify the initial integral to get:

$$\int \frac{7x+39}{x^2+9x+14}\,dx = \int \frac{5}{x+2} + \frac{2}{x+7}\,dx$$

$$= 5\int \frac{1}{x+2}\,dx + 2\int \frac{1}{x+7}\,dx$$

$$= 5\ln(x+2) + 2\ln(x+7) + C$$

BC Topics

You need to do this:

- Look for ways to factor the denominator. In the previous example, the denominator was the product of two linear factors; as a result there were two partial fractions.

- If, instead, there were three factors in the denominator, the same procedure would be followed, but there would be a system of three equations and three unknowns.

- Consider all factors in establishing partial fractions.

Answer and Explanation:

B

To evaluate the integral, you need to first use partial fractions to break up the integrand. You write

$$\frac{x-1}{x^2+3x+2} = \frac{x-1}{(x+1)(x+2)} = \frac{A}{x+1} + \frac{B}{x+2}$$

and try to solve for A and B.

By cross multiplying and setting the numerators equal to one another you get
$A(x+2) + B(x+1) = x(A+B) + (2A+B) = x-1$.

Thus, to solve for A and B you have to solve the system of equations below:

$$A + B = 1$$
$$2A + B = -1$$

You get that $A = -2$, $B = 3$. Therefore, you can rewrite the integral to get:

$$\int_0^1 \frac{x-1}{x^2+3x+2}\,dx = -2\int_0^1 \frac{1}{x+1}\,dx + 3\int_0^1 \frac{1}{x+2}\,dx = -2\ln|x+1|\big|_0^1 + 3\ln|x+2|\big|_0^1$$
$$= -2(\ln 2 - \ln 1) + 3(\ln 3 - \ln 2) = -2\ln 2 + 3\ln 3 - 3\ln 2 = -5\ln 2 + 3\ln 3$$

and you are finished. Thus, **(B)** is correct.

PRACTICE SET

1. $\int \dfrac{1}{x^2 + x}\, dx =$

   (A) $\dfrac{1}{2}\arctan\left(x + \dfrac{1}{2}\right) + C$

   (B) $\ln|x^2 + x| + C$

   (C) $\ln\left|\dfrac{x + 1}{x}\right| + C$

   (D) $\ln\left|\dfrac{x}{x + 1}\right| + C$

2. $\int \dfrac{2}{x^2 - 3x - 4}\, dx =$

   (A) $\dfrac{2}{5} \ln\left|\dfrac{x + 1}{x - 4}\right| + C$

   (B) $\dfrac{2}{5} \ln|(x - 4)(x + 1)| + C$

   (C) $\dfrac{2}{5} \ln\left|\dfrac{x - 4}{x + 1}\right| + C$

   (D) $\dfrac{4}{5} \ln|x - 4| - \dfrac{2}{5} \ln|x + 1| + C$

3. Evaluate $\int \dfrac{x^2 + 1}{x^2 - x}\, dx$.

   (A) $\ln\left|\dfrac{x}{x - 1}\right| + C$

   (B) $-\ln|x| + 2 \ln|x - 1| + C$

   (C) $-\ln|x - 1| + 2 \ln|x| + C$

   (D) $\ln\left|\dfrac{x - 1}{x^2}\right| + C$

4. Compute $\int \dfrac{2}{x^2 - 9}\, dx$.

   (A) $-\dfrac{1}{3} \ln\left|\dfrac{x + 3}{x - 3}\right| + C$

   (B) $2 \ln|x^2 - 9| + C$

   (C) $\dfrac{1}{3} \ln\left|\dfrac{x - 3}{x + 3}\right| + C$

   (D) $-\ln|x + 3| + C$

ANSWERS AND EXPLANATIONS

1. D

Integration by partial fractions

$$\int \frac{1}{x^2 + x}\, dx = \int \frac{1}{x(x+1)}\, dx = \int \frac{A}{x} + \frac{B}{x+1}\, dx$$

$$\frac{1}{x(x+1)} = \frac{A}{x} + \frac{B}{x+1} \Rightarrow 1 = A(x+1) + Bx$$

$$0 \cdot x + 1 = (A+B)x + A$$

$$\begin{cases} A+B=0 \\ A=1 \end{cases} \Rightarrow B = -1$$

$$\int \frac{A}{x} + \frac{B}{x+1}\, dx = \int \frac{1}{x} - \frac{1}{x+1}\, dx = \ln|x| - \ln|x+1| + C$$

$$\therefore \ln\left|\frac{x}{x+1}\right| + C$$

Therefore, **(D)** is correct.

2. C

When u-substitution doesn't work and there is a quadratic expression in the denominator, try separation by partial fractions.

$$\frac{2}{x^2 - 3x - 4} = \frac{2}{(x-4)(x+1)} = \frac{A}{(x-4)} + \frac{B}{(x+1)}$$

Multiply both sides by the common denominator.

$$2 = A(x+1) + B(x-4)$$

When $x = -1$,

$$2 = A(0) + B(-5) \Rightarrow B = -\frac{2}{5}.$$

When $x = 4$,

$$2 = 5A + B(0) \Rightarrow A = \frac{2}{5}.$$

So,

$$\int \frac{2}{(x-4)(x+1)}\, dx = \int \left(\frac{\frac{2}{5}}{x-4} - \frac{\frac{2}{5}}{x+1} \right) dx$$

$$\Rightarrow \frac{2}{5} \int \frac{1}{x-4} - \frac{1}{x+1}\, dx$$

$$= \frac{2}{5}\left(\ln|x-4| - \ln|x+1| \right) + C = \frac{2}{5} \ln\left|\frac{x-4}{x+1}\right| + C$$

Thus, **(C)** is correct.

3. B

To integrate you will have to use partial fraction decomposition to separate into multiple terms. Completely factor the denominator and separate the factors into fractional terms.

$$\frac{x^2 + 1}{x^2 - x} = \frac{x^2 + 1}{x(x-1)} = \frac{A}{x} + \frac{B}{x-1}$$

Multiply every term by $x(x-1)$ and set the x to values that eliminate A and B by making the terms equal to zero. This will give you a system of equations to solve or give values for A and B directly.

$$x^2 + 1 = A(x-1) + Bx$$

$$\text{Let } x = 0 : 1 = -A, A = -1$$

$$\text{Let } x = 1 : 2 = B$$

Plug in the values and integrate.

$$\int -\frac{1}{x} + \frac{2}{x-1}\, dx$$
$$= -\ln|x| + 2\ln|x-1| + C$$

That's **(B)**.

4. C

Because this problem can not be solved with u-substitution you will need to use partial fraction decomposition. First, completely factor the denominator and separate each factor into fractions.

$$\frac{2}{x^2 - 9} = \frac{2}{(x + 3)(x - 3)} = \frac{A}{x + 3} + \frac{B}{x - 3}$$

Multiply each term by $(x + 3)(x - 3)$ and set x to values that will eliminate A and B by making the terms equal to zero. This will give you a system of equations to solve or give values for A and B directly.

$$\text{Let } x = 3 : 2 = 6B, B = \frac{1}{3}$$

$$\text{Let } x = -3 : 2 = -6A, A = -\frac{1}{3}$$

Plug the values in, integrate, and simplify.

$$\int -\frac{1}{3(x + 3)} + \frac{1}{3(x - 3)} \, dx$$

$$= -\frac{1}{3} \ln |x + 3| + \frac{1}{3} \ln |x - 3| + C$$

$$= \frac{1}{3} \ln \left| \frac{x - 3}{x + 3} \right| + C$$

Thus, **(C)** is correct.

16.4 IMPROPER INTEGRALS

To answer a question like this:

Evaluate the improper integral $\int_1^{+\infty} e^{-3x}\, dx$.

(A) $-\dfrac{1}{3}\left(1 - \dfrac{1}{e^3}\right)$

(B) $\dfrac{1}{e^3}$

(C) $\dfrac{1}{3e^3}$

(D) The integral diverges to ∞.

You need to know this:

Improper integrals

- Integrals where one or both of the endpoints are at $\pm$ infinity, i.e., $\int_{-\infty}^{b} f(x)\, dx$, $\int_{a}^{\infty} f(x)\, dx$, or $\int_{-\infty}^{\infty} f(x)\, dx$.

- Solve it by finding a definite integral and then taking a limit and letting one of the endpoints tend to infinity, thus calculating the integral of the function with an endpoint at infinity.

- More specifically, use one of the following formulas:

$\displaystyle\int_{a}^{\infty} f(x)\, dx = \lim_{t \to \infty} \int_{a}^{t} f(x)\, dx$ provided this limit exists and is finite.

$\displaystyle\int_{-\infty}^{b} f(x)\, dx = \lim_{t \to -\infty} \int_{t}^{b} f(x)\, dx$ provided this limit exists and is finite.

$\displaystyle\int_{-\infty}^{\infty} f(x)\, dx = \int_{-\infty}^{a} f(x)\, dx + \int_{a}^{\infty} f(x)\, dx$ provided both integrals on the right exist.

DC Topics

You need to do this:

- If the corresponding limit of the improper integral is finite, the integral is *convergent*. If, however, the limit that defines the improper integral approaches infinity, the integral is *divergent*.

- It is very useful to know some basic rules for distinguishing between convergent and divergent improper integrals. An important rule of this type is the comparison test for improper integrals.

- If $f(x)$ and $g(x)$ are continuous functions and $f(x) \geq g(x) \geq 0$ for $x \geq a$, then:

  - If $\int_a^\infty f(x)\,dx$ is convergent, $\int_a^\infty g(x)\,dx$ then is also convergent. Meaning, if a given function is a convergent (finite) integral, then any function less than the given function must also have a convergent integral.

  - If $\int_a^\infty g(x)\,dx$ is divergent, then $\int_a^\infty f(x)\,dx$ is also divergent. Thus, if a given function is a divergent integral, then any function greater than the given function will also be divergent.

Answer and Explanation:

C

To compute this improper integral, rewrite the integral as the limit of a definite integral and compute the resulting definite integral and its limit to find

$$\int_1^{+\infty} e^{-3x}\,dx = \lim_{c \to \infty} \int_1^c e^{-3x}\,dx = \lim_{c \to \infty}\left[-\frac{1}{3}e^{-3x}\right]_1^c = \frac{1}{3e^3}.$$

That matches **(C)**.

PRACTICE SET

1. Evaluate the improper integral $\int_{-\infty}^{0} \frac{1}{\sqrt{3-x}}\,dx$ and determine if it converges or diverges, and where.

   (A) Diverges, $-\infty$

   (B) Diverges, ∞

   (C) Converges, $-2\sqrt{3}$

   (D) Converges, 0

2. Evaluate the improper integral $\int_{0}^{\infty} \frac{1}{2x-5}\,dz$ and determine if it converges or diverges, and where.

   (A) Diverges, $-\infty$

   (B) Diverges, ∞

   (C) Converges, $\frac{1}{2}\ln 5$

   (D) Converges, $\ln 5$

3. Evaluate the improper integral $\int_{-2}^{\infty} \sin x\,dx$ and determine if it converges, and where, or if it diverges at positive/negative infinity.

   (A) Diverges, $-\infty$

   (B) Diverges, limit does not exist

   (C) Converges, $\cos(2)$

   (D) Converges, 1

4. Evaluate the improper integral $\int_{1}^{\infty} x^{-2}\,dx$ and determine if it converges, and where, or if it diverges at positive/negative infinity.

   (A) Diverges, ∞

   (B) Converges, 1

   (C) Diverges, $-\infty$

   (D) Converges, 0

ANSWERS AND EXPLANATIONS

1. B

Integrate like normal but do not forget about the chain rule inside of the square root.

$$\int_{-\infty}^{0} \frac{1}{\sqrt{3-x}}\, dx = \lim_{t\to\infty} \left. -2\sqrt{3-x}\, \right|_{t}^{0}$$

$$\lim_{t\to\infty}\left(-2\sqrt{3} + 2\sqrt{3-t}\right)$$

$$-2\sqrt{3} + \infty = \infty$$

Because the integral diverges at infinity, **(B)** is correct.

2. B

To integrate, start with u-substitution or just realize that because of the chain rule you need to multiply by $\frac{1}{2}$.

$$\int_{0}^{\infty} \frac{1}{2x-5}\, dx = \lim_{t\to\infty} \frac{1}{2}\, \ln|2x-5|\Big|_{0}^{t}$$

$$\frac{1}{2}\lim_{t\to\infty} \ln|2x-5|\Big|_{0}^{t} = \frac{1}{2}\lim_{t\to\infty}\left(\ln|2t-5| - \ln(5)\right)$$

$$\frac{1}{2}\left(\infty - \ln(5)\right) = \infty$$

This integral diverges towards infinity, so **(B)** is correct.

3. B

Integrate like normal and evaluate at the bounds. Make sure to keep your eye on any negatives when integrating with cosine or sine.

$$\int_{-2}^{\infty} \sin x\, dx = \lim_{t\to\infty} \int_{-2}^{t} \sin x\, dx$$

$$\lim_{t\to\infty}\left(-\cos x\right)\Big|_{-2}^{t} = \lim_{t\to\infty}\left(-\cos t + \cos(-2)\right)$$

Cosine is a wave function so it will not converge to anything but oscillates forever. In other words, this function's limit does not exist so it is divergent but does not go towards positive or negative infinity. Hence, **(B)** is correct.

4. B

Integrate as normal by adding 1 to the exponent and dividing by the result. Then evaluate by the given bounds.

$$\int_{1}^{\infty} x^{-2}\, dx = \lim_{t\to\infty} \int_{1}^{t} x^{-2}\, dx = \lim_{t\to\infty} \left. -\frac{1}{x}\, \right|_{1}^{t}$$

$$= \lim_{t\to\infty} -\frac{1}{t} - \left(-\frac{1}{(1)}\right) = 0 + 1 = 1$$

Thus, the integral converges to 1 and **(B)** is correct.

16.5 LOGISTICAL MODELING

To answer a question like this:

Solve the logistic differential equation:

$$\frac{dP}{dt} = 0.08\left(1 - \frac{P}{1{,}000}\right)P, \quad P(0) = 100$$

(A) $P(t) = \dfrac{1{,}000}{1 - 9e^{-0.8t}}$

(B) $P(t) = \dfrac{0.08}{1 + 9e^{-1008t}}$

(C) $P(t) = \dfrac{1{,}000}{1 + 9e^{-008t}}$

(D) $P(t) = \dfrac{900}{1 + 9e^{-0.08t}}$

You need to know this:

A more complex conception of population growth

- When a population is small, it grows quite steadily and the growth factor is almost constant. During this time, the exponential growth model is in effect.

- However, the population will not continue to grow undeterred in this way for all time. After a while, the population becomes too big for its own good and the relative rate of growth actually slows down.

- In fact, if the population exceeds its carrying capacity, then the population is likely to decrease and the growth rate will be negative.

- The logistic model is a mathematical equation for population growth that is more realistic than the previously discussed model of exponential growth and will better reflect this more complex behavior.

You need to do this:

- The equation below is called the *logistic differential equation* and it can do this more complicated job: $\dfrac{dP}{dt} = kP\left(1 - \dfrac{P}{K}\right)$, where P = population, K = carrying constant, and k = constant.

- Note that if P is very close to the carrying constant, K, then $\dfrac{P}{K}$ is close to 1 and so $\dfrac{dP}{dt}$ is close to 0, as would be expected. If P is instead small compared to the carrying constant, K, then $\dfrac{P}{K}$ is close to 0 and so the population P is growing exponentially with a growth factor of k.

- The logistic differential equation is a separable differential equation and thus can be solved to $P(t) = \dfrac{K}{1 + Ae^{-kt}}$, where $A = \dfrac{K - P_0}{P_0}$ and P_0 is the initial population.

Answer and Explanation:

C

The first thing to note is that the carrying capacity, K, is 1,000 and the proportionality constant, k, is 0.08. The solution to a logistic differential equation takes the form $P(t) = \dfrac{K}{1 + Ae^{-kt}}$.

Solve for A:

$$A = \frac{K - P_0}{P_0} = \frac{1{,}000 - 100}{100}$$

$$= \frac{900}{100} = 9$$

Plug the values into $P(t)$ to get $P(t) = \dfrac{1{,}000}{1 + 9e^{-0.08t}}$.

That's a match for **(C)**.

PRACTICE SET

1. A population is modeled by the logistic differential equation $\frac{dP}{dt} = \frac{P}{3}\left(1 - \frac{P}{8}\right)$.

   If $P(0) = 2$, for what value of P is the population growing the fastest?

   (A) 0

   (B) 3

   (C) 4

   (D) 8

2. If $P(t) = \frac{200}{1 + 3e^{-0.02t}}$, calculate $\frac{dP}{dt}$.

   (A) $P + 0.02P^2$

   (B) $0.02P - 0.0001P^2$

   (C) $0.02 - 0.01P^2$

   (D) $0.004P^2 - 0.02P$

3. If $\frac{dP}{dt} = 0.04P - 0.0001P^2$, when is $\frac{dP}{dt}$ the greatest?

   (A) 0.04

   (B) 100

   (C) 200

   (D) 400

4. $\frac{dP}{dt} = 0.2P - 0.0004P^2$

   Compute $P(t)$ with the initial condition $P_0 = 200$.

   (A) $\frac{500}{1 + 1.5e^{0.2t}}$

   (B) $\frac{500}{1 + 1.5e^{-0.2t}}$

   (C) $0.2P\left(1 - \frac{P}{500}\right)$

   (D) $\frac{500}{1 + e^{-0.2t}}$

ANSWERS AND EXPLANATIONS

1. C

The quickest way to solve this problem is to recognize that the population growth is modeled by a logistic differential equation, $\frac{dy}{dt} = rP\left(1 - \frac{P}{K}\right)$, and that the rate of growth is greatest when $P = \frac{K}{2}$. In this case $K = 8$, so the population is growing the fastest when $P = 4$. Thus, **(C)** is correct.

Note: A logistic differential equation can take many forms. If the form that shows up on the exam is not the same as the form you learned in class, you may not be able to apply quick rules to answer questions. All of the forms are separable differential equations, so you can always tackle questions about logistic growth by solving the differential equation and proceeding from there. It may not be elegant, but it will work.

2. B

The given formula is in the state of $P(t) = \frac{K}{1 + Ae^{-kt}}$, so $K = 200$ and $k = 0.02$. Plug into the formula $\frac{dP}{dt} = kP\left(1 - \frac{P}{K}\right)$ and simplify.

$$\frac{dP}{dt} = (0.02)P\left(1 - \frac{P}{(200)}\right)$$

$$\frac{dP}{dt} = 0.02P - 0.0001P^2$$

That's **(B)**.

3. C

First get the equation into the standard form of $\frac{dP}{dt} = kP\left(1 - \frac{P}{K}\right)$.

$$\frac{dP}{dt} = 0.04P - 0.0001P^2$$

$$= 0.04P(1 - 0.0025P)$$

$$= 0.04P\left(1 - \frac{P}{400}\right)$$

From here you can tell that the carrying capacity is 400, so use the rule that $\frac{dP}{dt}$ is at its greatest when $P = \frac{K}{2}$ so:

$$P = \frac{K}{2} = \frac{400}{2} = 200$$

When $P = 200$, $\frac{dP}{dt}$ is at its greatest, so **(C)** is correct.

4. B

First get the equation into the standard form of $\frac{dP}{dt} = kP\left(1 - \frac{P}{K}\right)$.

$$\frac{dP}{dt} = 0.2P - 0.0004P^2$$

$$= 0.2P(1 - 0.002P)$$

$$= 0.2P\left(1 - \frac{P}{500}\right)$$

Now that you have the carrying capacity, use the initial condition and $A = \frac{K - P_0}{P_0}$ to get A.

$$A = \frac{K - P_0}{P_0}$$

$$= \frac{(500) - (200)}{(200)}$$

$$= \frac{3}{2} = 1.5$$

Plug everything into the equation $P(t) = \frac{K}{1 + Ae^{-kt}}$.

$$P(t) = \frac{K}{1 + Ae^{-kt}}$$

$$= \frac{500}{1 + 1.5e^{-0.2t}}$$

Hence, **(B)** is correct.

TEST WHAT YOU LEARNED

1. $\frac{dP}{dt}$ is greatest when $P = 100$. If the rate of growth is 0.08, what is the equation for $\frac{dP}{dt}$?

   (A) $\frac{dP}{dt} = 0.08P\left(1 - \frac{P}{200}\right)$

   (B) $\frac{dP}{dt} = 0.08\left(1 - \frac{P}{200}\right)$

   (C) $\frac{dP}{dt} = 200P\left(1 - \frac{P}{0.08}\right)$

   (D) $\frac{dP}{dt} = 0.08P\left(1 - \frac{P}{100}\right)$

x	1.5	2.5	3.5	4.5
$f'(x)$	2	-3	3	1

2. You are given that $f(1.5) = 4$ and a table of values of the derivative f'. Use Euler's method with a step-size of 1 to approximate $f(4.5)$.

   (A) 1

   (B) 3

   (C) 6

   (D) 7

3. Evaluate the integral $\int_{-\infty}^{1} xe^{2x^2}\, dx$.

   (A) $-\infty$

   (B) $\frac{e}{4}$

   (C) $\frac{e^2}{3}$

   (D) ∞

4. Evaluate $\int_0^1 (x^2 + 1)e^{-x}\, dx$.

   (A) $-\dfrac{6}{e} + 3$

   (B) $3 - e$

   (C) 1

   (D) $2e$

---

5. Solve $\int \dfrac{x^2 + 2x - 1}{2x^3 + 3x^2 - 2x}\, dx$.

   (A) $\ln\left|\dfrac{x}{x+2}\right| + C$

   (B) $\dfrac{1}{20}\ln|2x - 1| + C$

   (C) $\dfrac{1}{2}\ln|x| + \dfrac{1}{10}\ln\left|\dfrac{2x-1}{x+2}\right| + C$

   (D) $\dfrac{1}{5}\ln\left|\dfrac{2x^2 - x}{x+2}\right| + C$

THIS PAGE LEFT INTENTIONALLY BLANK.
Answers and explanations to the practice set are located on the next page.

Answer Key

Test What You Already Know	Test What You Learned
1. B	1. A
2. B	2. C
3. A	3. A
4. D	4. A
5. D	5. C

 REFLECTION

Test What You Already Know score: _____

Test What You Learned score: _____

Use this section to evaluate your progress. After working through the pre-quiz, check off the boxes in the "Pre" column to indicate which Learning Objectives you feel confident about. Then, after completing the chapter, including the post-quiz, do the same to the boxes in the "Post" column. Keep working on unchecked Objectives until you're confident about them all!

Pre	Post		
☐	☐	**16.1**	Solve differential equations via Euler's method
☐	☐	**16.2**	Apply integration by parts to integrands with two functions
☐	☐	**16.3**	Evaluate fractional integrands using simple partial fractions
☐	☐	**16.4**	Perform antidifferentiation of improper integrals
☐	☐	**16.5**	Solve logistic differential equations and use them in modeling

 FOR MORE PRACTICE

Complete more practice online at kaptest.com. Haven't registered your book yet? Go to kaptest.com/booksonline to begin.

ANSWERS AND EXPLANATIONS

Test What You Already Know

1. B

The first step in Euler's method is to solve for dy. The easiest way to view what you're doing is to create a table, just like below, and fill out the x- and y-coordinates by starting at the given x-value and moving one step, which is dx, until you reach the value your trying to approximate. Plug in the x- and y-values and dx to find dy, which tells you where the next y-value will land. For example, the first y-coordinate is 3 and the first dy is -6, so add them to get the next y-value, which is -3. Repeat until you reach $x = 1$ for the approximation of $y = -2$.

x	y	$dy = (x - y - 2)dx$
-1	3	$[(-1) - (-3) - 2](1) = -6$
0	-3	$[(0) - (3) - 2](1) = 1$
1	-2	

Therefore, **(B)** is the correct answer.

2. B

Use integration by parts to solve this problems.

Let

$$u = x \Rightarrow du = dx$$
$$dv = \cos(2x)dx \Rightarrow v = \frac{1}{2}\sin(2x)$$
$$\int x \cos(2x)\,dx = \frac{1}{2}x \sin(2x) - \int \frac{1}{2}\sin(2x)\,dx$$
$$\Rightarrow \frac{1}{2}x \sin(2x) + \frac{1}{4}\cos(2x) + C$$

Thus, the correct answer is **(B)**.

3. A

To compute this integral, use the technique of partial fractions. Rewrite the integrand as

$$\frac{7}{(x - 3)(x + 4)} = \frac{A}{(x - 3)} + \frac{B}{(x + 4)}$$

Simplify and substitute values for x into the simplified expression to find $A = 1$ and $B = -1$. Rewrite the integrand as

$$\int \frac{7}{(x - 3)(x + 4)}\,dx = \int \frac{1}{(x - 3)} + \frac{-1}{(x + 4)}\,dx$$

and compute these basic antiderivatives to find

$$\int \frac{1}{(x - 3)} + \frac{-1}{(x + 4)}\,dx = \ln|x - 3| - \ln|x + 4| + C.$$

Apply properties of the logarithm to rewrite the answer as $\ln\left|\frac{x - 3}{x + 4}\right| + C$.

Thus, **(A)** is the correct answer.

4. D

Integrate like normal and evaluate at the bounds.

$$\int_1^\infty \frac{1}{x}\,dx = \lim_{t \to \infty} \ln|x|\Big|_1^t$$
$$\lim_{t \to \infty} (\ln|t| - \ln|1|)$$
$$\infty - 0 = \infty$$

Because the answer is infinity, this integral diverges and **(D)** is the correct answer.

5. D

The logistic differential equation $\frac{dP}{dt} = P\left(2 - \frac{P}{5,000}\right)$ achieves population equilibrium for those values of P where $\frac{dP}{dt} = 0$. You solve $\frac{dP}{dt} = 0$ and get $P = 0$ or $P = 10,000$. The carrying capacity is $\lim_{t \to \infty} P(t)$, which is equal to 10,000, so **(D)** is the answer.

Test What You Learned

1. A

You need to know that when $\frac{dP}{dt}$ is at its greatest, the carrying capacity is double the population at that point, or $P = \frac{K}{2}$. After finding the carrying capacity, plug everything into this equation: $\frac{dP}{dt} = kP\left(1 - \frac{P}{K}\right)$.

$$P = \frac{K}{2}$$
$$100 = \frac{K}{2}$$
$$K = 200$$

$K = 200$ and you were given that $k = 0.08$, so:

$$\frac{dP}{dt} = kP\left(1 - \frac{P}{K}\right) = 0.08P\left(1 - \frac{P}{200}\right)$$

Therefore, **(A)** is the correct answer.

2. C

Because you are given the derivative values, follow this formula for each step until you arrive at the approximation of $f(4.5)$: $f(x) + f(x) = f(x + 1)$.

$$f(2.5) = f(1.5) + f'(1.5) = (4) + (2) = 6$$
$$f(3.5) = f(2.5) + f'(2.5) = (6) + (-3) = 3$$
$$f(4.5) = f(3.5) + f'(3.5) = (3) + (3) = 6$$

Hence, **(C)** is the correct answer.

3. A

The integral $\int_{-\infty}^{1} xe^{2x^2}\, dx$ is an improper integral; therefore you must first calculate $\int_{-t}^{1} xe^{2x^2}\, dx$ and then take the limit as t approaches infinity. Begin by calculating the definite integral using the substitution $u = 2x^2$, so $du = 4x\, dx$. You get:

$$\int_{-t}^{1} xe^{2x^2}\, dx = \int_{-t}^{1} e^{2x^2} x\, dx = \int_{-t}^{1} e^{u} \frac{1}{4}\, du$$
$$= \frac{1}{4} \int_{2t^2}^{2} e^{u}\, du = \frac{1}{4}\left(e^2 - e^{2t^2}\right)$$

Therefore, you get

$$\int_{-\infty}^{1} xe^{2x^2}\, dx = \lim_{t \to \infty} \int_{-t}^{1} xe^{2x^2}\, dx = \lim_{t \to \infty} \frac{1}{4}\left(e^2 - e^{2t^2}\right)$$
$$= \lim_{x \to \infty} -e^{2t^2} = -\infty$$

Because there is a negative sign in front of e^{2t^2}, this number becomes increasingly negative as t approaches infinity. Thus, $\int_{-\infty}^{1} xe^{2x^2}\, dx = -\infty$.

Hence, **(A)** is the correct answer.

4. A

This problem requires integration by parts twice because the e repeats derivatives and the x^2 will go away after the two derivatives. For this reason you will want to make the $x^2 + 1$ the u and the $e^{-x} dx$ the dv.

$$u = x^2 + 1 \quad dv = e^{-x} dx$$
$$du = 2x\, dx \quad v = -e^{-x}$$

Now you will want to use the integration by parts formula $uv - \int v\, du$

$$\left(x^2 + 1\right)\left(-e^{-x}\right)\Big|_0^1 - \int_0^1 \left(-e^{-x}\right)(2x)\, dx$$
$$= -\left(x^2 + 1\right)\left(e^{-x}\right)\Big|_0^1 + 2\int_0^1 xe^{-x}\, dx$$

Set up the u and dv again.

$$u = x \quad dv = e^{-x} dx$$
$$du = dx \quad v = -e^{-x}$$

Use the integration by parts formula again.

$$-\left(x^2 + 1\right)\left(e^{-x}\right)\Big|_0^1 + 2(x)\left(-e^{-x}\right)\Big|_0^1 - 2\int_0^1 \left(-e^{-x}\right)dx$$
$$= -\left(x^2 + 1\right)\left(e^{-x}\right) - 2xe^{-x} - 2e^{-x}\Big|_0^1$$

Now evaluate at the bounds and simplify to get the final answer.

$$-\left((1)^2 + 1\right)e^{-(1)} - 2(1)e^{-(1)} - 2e^{-(1)}$$
$$-\left[-\left((0)^2 + 1\right)e^{-(0)} - 2(0)e^{-(0)} - 2e^{-(0)}\right]$$
$$= -\frac{2}{e} - \frac{2}{e} - \frac{2}{e} - (-1 - 0 - 2)$$
$$= -\frac{6}{e} + 3$$

Hence, **(A)** is the correct answer.

5.   C

Because you can't find a u and du that work for this problem, you will have to use partial fraction decomposition. First factor the denominator completely and separate each term into separate fractions.

$$\frac{x^2 + 2x - 1}{2x^3 + 3x^2 - 2x} = \frac{x^2 + 2x - 1}{x(2x - 1)(x + 2)}$$

$$= \frac{A}{x} + \frac{B}{2x - 1} + \frac{C}{x + 2}$$

Multiply every term by $x(2x - 1)(x + 2)$.

$$x^2 + 2x - 1 = A(2x - 1)(x + 2) + Bx(x + 2) + Cx(2x - 1)$$

You will want to set x equal to values that make the terms including A, B, or C equal to zero. This will set up a system of equations for you to solve by either substitution or elimination.

$$\text{Let } x = 0: -1 = -2A, \ A = \frac{1}{2}$$

$$\text{Let } x = \frac{1}{2}: \frac{1}{4} = \frac{5}{4}B, \ B = \frac{1}{5}$$

$$\text{Let } x = -2: -1 = 10C, \ C = -\frac{1}{10}$$

Lastly, plug in, integrate, and simplify.

$$\int \frac{1}{2x} + \frac{1}{5(2x - 1)} - \frac{1}{10(x + 2)} \, dx$$

$$= \frac{1}{2} \ln|x| + \frac{1}{10} \ln|2x - 1| - \frac{1}{10} \ln|x + 2| + C$$

$$= \frac{1}{2} \ln|x| + \frac{1}{10} \ln\left|\frac{2x - 1}{x + 2}\right| + C$$

Thus, **(C)** is the correct answer.

CHAPTER 17

Series

LEARNING OBJECTIVES

17.1 Evaluate the limits of series using partial sums

17.2 Define convergence in a series and identify geometric and harmonic series

17.3 Apply the integral test and determine convergence in *p*-series

17.4 Determine convergence with the comparison test and the ratio test

17.5 Evaluate alternating series for convergence and estimate their limit

TEST WHAT YOU ALREADY KNOW

1. Determine whether the series $\sum_{n=1}^{\infty} \frac{(2n+1)^2}{n(2n+4)}$ is convergent or divergent and evaluate the series.

   (A) The series is convergent and equals 42.

   (B) The series is convergent and equals 56.

   (C) The series is convergent and equals 31π.

   (D) The series is divergent.

2. What type of series is $\sum_{n=1}^{\infty} \frac{(-1)^{n-1}}{n}$?

   (A) Geometric Series

   (B) Alternating Series

   (C) Harmonic Series

   (D) Alternating Harmonic Series

3. Given $\sum_{n=1}^{\infty} \frac{1}{\sqrt{n}}$, is this series convergent or divergent? If convergent, find the value the series converges to.

   (A) Divergent

   (B) Convergent, 4

   (C) Convergent, 1

   (D) None of the above

4. Which choice below best describes a sequence with terms defined by $a_n = \dfrac{1}{2n+3}$?

   (A) Increasing, bounded

   (B) Decreasing, bounded

   (C) Increasing, not bounded

   (D) Decreasing, not bounded

$$\left\{-2, \frac{3}{4}, -\frac{4}{9}, \frac{5}{16}, -\frac{6}{25}, \ldots\right\}$$

5. Find a formula for the general term a_n for the sequence above.

   (A) $a_n = \dfrac{(n+1)}{n^2}$

   (B) $a_n = \dfrac{(-1)^n(n+2)}{n^2}$

   (C) $a_n = \dfrac{(-1)^n(n+1)}{n^2}$

   (D) $a_n = \dfrac{n}{(n+1)^2}$

17.1 LIMITS OF SERIES

To answer a question like this:

Which of the following formulas represent the sequence $-1, \frac{1}{2}, -\frac{1}{4}, \frac{1}{8} \ldots$?

(A) $\left\{ \frac{(-1)^n}{2^n} \right\}_{n=0}^{\infty}$

(B) $\left\{ \frac{(-1)^n}{2^n} \right\}_{n=1}^{\infty}$

(C) $\left\{ \frac{(-1)^n}{2n} \right\}_{n=0}^{\infty}$

(D) $\left\{ \frac{(-1)^{n+1}}{2^n} \right\}_{n=0}^{\infty}$

You need to know this:

Sequences

- A sequence is just a list of numbers that are written down in a particular order: $a_1, a_2, a_3, a_4, \ldots, a_n, \ldots$

- To avoid having to write down infinitely many terms to refer to a given sequence, use the notation $\{a_n\}$ or $\{a_n\}_{n=1}^{\infty}$ to represent the sequence above. While in many sequences n is indexed from 1 to infinity, n can begin at any number.

Partial sums

- Adding all the terms of an infinite sequence gives an infinite series. This infinite series is denoted:

$$\sum_{n=1}^{\infty} a_n = a_1 + a_2 + a_3 + \ldots + a_n + \ldots$$

- Adding together infinitely many numbers may result in an infinite number, such as with the series:

$$\sum_{n=1}^{\infty} n = 1 + 2 + 3 + 4 + 5 + \ldots + n + \ldots$$

- However, if each term of the series becomes progressively small enough, then an infinite series will yield a finite sum, such as with the series:

$$\sum_{n=1}^{\infty} \frac{1}{2^n} = \frac{1}{2} + \frac{1}{4} + \frac{1}{8} + \frac{1}{16} + \ldots + \frac{1}{2^n} + \ldots$$

- As n gets larger and larger the sum gets closer and closer to 1. Thus,

$$\sum_{n=1}^{\infty} \frac{1}{2^n} = \frac{1}{2} + \frac{1}{4} + \frac{1}{8} + \frac{1}{16} + \ldots + \frac{1}{2^n} + \ldots = 1.$$

BC Topics

You need to do this:

- To determine whether a given series is finite or infinite, examine the partial sums of the series. With these partial sums, create a new sequence $\{s_k\}$ which has terms defined by

$$s_1 = a_1$$
$$s_2 = a_1 + a_2$$
$$s_3 = a_1 + a_2 + a_3$$
$$\ldots$$
$$s_k = a_1 + a_2 + a_3 + \ldots + a_k$$
$$\ldots$$

- If this new sequence $\{s_k\}$ has a limit, then the series is finite and is equal to the limit of the sequence $\{s_k\}$. So, if $\lim\limits_{k \to \infty} s_k = S$ (and S is finite), then $\sum\limits_{n=1}^{\infty} a_n = S$.

- If instead $\lim\limits_{k \to \infty} s_k = \infty$, then the series cannot have a finite sum.

Answer and Explanation:

D

You can try each formula until you find the one that works:

(A) $\left\{ \dfrac{(-1)^n}{2^n} \right\}_{n=0}^{\infty} = \dfrac{(-1)^0}{2^0}$

$\qquad = \dfrac{1}{1}$

(B) $\left\{ \dfrac{(-1)^n}{2^n} \right\}_{n=1}^{\infty} = \dfrac{(-1)^1}{2^1}$

$\qquad = \dfrac{-1}{2}$

(C) $\left\{ \dfrac{(-1)^n}{2n} \right\}_{n=0}^{\infty} = \dfrac{(-1)^0}{2(0)}$

(D) $\left\{ \dfrac{(-1)^{n+1}}{2^n} \right\}_{n=0}^{\infty} = \dfrac{(-1)^1}{2^0}, \dfrac{(-1)^2}{2^1}, \dfrac{(-1)^3}{2^2}, \dfrac{(-1)^4}{2^3}$

$\qquad = -1, \dfrac{1}{2}, -\dfrac{1}{4}, \dfrac{1}{8}$

The very last sequence works, so **(D)** is the answer.

PRACTICE SET

1. Suppose $\sum_{n=1}^{\infty} a_n$ is a converging series with corresponding partial sums $\{s_k\}$. The graph below plots the values of $\{s_k\}$ for $k = 1, 2, 3\ldots$

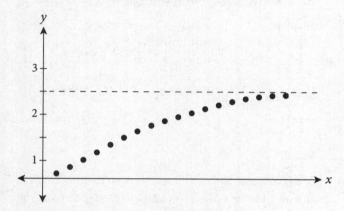

Which statement below is FALSE?

(A) $\lim_{n\to\infty} a_n = \dfrac{1}{2}$

(B) $\lim_{n\to\infty} a_n = 0$

(C) $\lim_{k\to\infty} s_k = 2\dfrac{1}{2}$

(D) The partial sums of $\sum_{n=1}^{\infty} a_n$ are increasing.

2. Determine the sum of the series generated by

$$\sum_{n=0}^{\infty} \frac{2n^3 + \cos(n)}{(2n+1)^3 + e^{-n}}.$$

(A) 0

(B) $\dfrac{1}{1 + \dfrac{1}{e}}$

(C) $\dfrac{4}{3}$

(D) The series diverges.

3. Which series below is divergent?

(A) $\displaystyle\sum_{n=1}^{\infty} \frac{n^3 + 1}{3e^n + 1}$

(B) $\displaystyle\sum_{n=1}^{\infty} \frac{n^2 + n}{3n^3 - 7}$

(C) $\displaystyle\sum_{n=1}^{\infty} \frac{n^3 + n^2}{3n^3 - 7}$

(D) $\displaystyle\sum_{n=1}^{\infty} \frac{(-1)^n (n+1)^2}{3n^3}$

4. Which (if any) of the three series listed below are equivalent?

I. $\displaystyle\sum_{n=1}^{\infty} a_n$

II. $\displaystyle\sum_{n=0}^{\infty} a_{n-1}$

III. $\displaystyle\sum_{n=0}^{\infty} a_{n+1}$

(A) I and II only

(B) I and III only

(C) II and III only

(D) None of these are equivalent.

ANSWERS AND EXPLANATIONS

1. A

Clearly this series is convergent because the sequence of partial sums is approaching the value $2\frac{1}{2}$. Therefore, $\lim_{n\to\infty} a_n \neq \frac{1}{2}$. In fact, if a series is convergent then $\lim_{n\to\infty} a_n = 0$, and you were told that the series was convergent. Therefore, **(A)** is correct.

2. D

Begin by considering the limit of the terms of the series as n goes to infinity (consider the dominant terms):

$$\lim_{n\to\infty} \frac{2n^3 + \cos(n)}{(2n+1)^3 + e^{-n}}$$

$$= \lim_{n\to\infty} \frac{2n^3}{(2n+1)^3}$$

$$= \lim_{n\to\infty} \frac{2n^3}{8n^3 + \dots}$$

$$= \frac{1}{4}$$

This means that the terms of the series get closer and closer to $\frac{1}{4}$ as n gets larger, which means you are, in effect, adding an infinite number of $\frac{1}{4}$'s.

Adding an infinite number of positive terms leads to infinity unless (possibly) the terms are going to zero. This is not the case here, and so the series diverges. Hence, **(D)** is the appropriate match.

3. C

You want to find which series is divergent. Recall that the divergence theorem says if the $\lim_{n\to\infty} a_n \neq 0$, then the series is divergent. Given the series $\sum_{n=1}^{\infty} \frac{n^3 + n^2}{3n^3 - 7}$, calculate

$$\lim_{n\to\infty} \frac{n^3 + n^2}{3n^3 - 7} = \frac{1}{3} \neq 0.$$

Thus, $\sum_{n=1}^{\infty} \frac{n^3 + n^2}{3n^3 - 7}$ is divergent, and **(C)** is the correct answer.

4. B

This is called an index shift, and is used to adjust indices of a given series. Whenever you increase (decrease) the initial value in the index, you decrease (increase) the value of each n in the series. In this case, in order to get from the first series, $\sum_{n=1}^{\infty} a_n$, to either of the other two, you decrease the initial value by one. Therefore, you need to increase the value of n in order to maintain equivalence. Thus:

$$\sum_{n=1}^{\infty} a_n = \sum_{n=1-1}^{\infty} a_{n+1} = \sum_{n=0}^{\infty} a_{n+1},$$

so you see that series I and III are equivalent, while series II is not. Thus, **(B)** is correct.

17.2 GEOMETRIC AND HARMONIC SERIES

To answer a question like this:

Evaluate the infinite series $\displaystyle\sum_{n=1}^{\infty}\frac{3^n + 2^n}{6^n}$.

(A) $1\frac{1}{2}$ 　　 (B) $\dfrac{\pi}{2}$ 　　 (C) $2\frac{1}{2}$ 　　 (D) $5\frac{1}{2}$

You need to know this:

Describing sequences and series

- A sequence $\{a_n\}$ has a limit L if the terms c_n become successively closer to L as n approaches infinity: $\lim_{n\to\infty} a_n = L$ or simply $a_n \to L$ as $n \to \infty$.

- Furthermore, if a sequence has a finite limit L, then we say the sequence is *convergent*. If, on the other hand, the sequence does not approach any limit L, then we say the sequence is *divergent*.

- A sequence is increasing if the terms of the sequence are increasing, meaning $a_n < a_{n+1}$ for each n. Similarly, a sequence is decreasing if the terms of the sequence are decreasing, meaning $a_n > a_{n+1}$.

- Lastly, a sequence is said to be bounded if there exists a real number M such that $a_n \leq M$, for all n.

- A series is said to converge (or is convergent) if $\lim_{k\to\infty} s_k = S$, where S is a real number less than infinity. This series is said to diverge (or is divergent) if $\lim_{k\to\infty} s_k = \infty$ or if the limit does not exist.

- However, some series will require special tests to either determine convergence or the value to which they converge.

Geometric series

- Each term in this series is obtained by multiplying the preceding one by a common factor,

$$r : 1 + r + r^2 + r^3 + \cdots + r^{n-1} + \cdots = \sum_{n=1}^{\infty} r^{n-1}$$

- Clearly if $r = 1$ then this series is divergent because then

$$s_k = \underbrace{1 + 1 + 1 + \cdots + 1}_{k-\text{times}} = k \text{ and as } k \text{ approaches infinity so does } s_k.$$

- Furthermore, the same must be true if r is any number bigger than 1 by the same reasoning. Thus, if $|r| \geq 1$, then the geometric series is divergent.

- However, when $|r| < 1$ the series is convergent and can be defined by:

$$\sum_{n=1}^{\infty} r^{n-1} = 1 + r + r^2 + r^3 + \cdots + r^n + \cdots = \frac{1}{1-r}.$$

Harmonic Series

- The harmonic series is given by $\sum_{n=1}^{\infty} \frac{1}{n} = 1 + \frac{1}{2} + \frac{1}{3} + \frac{1}{4} + \cdots + \frac{1}{n} + \cdots$

- This series diverges and, thus, if a given series can be written in this form, there is no need to try and determine its value.

You need to do this:

- Consider sequences as limits and note whether there is an obvious solution to that limit.

- Consider whether partial sums can be readily used to determine if a series converges to a finite number or goes to infinity.

- Look for common series types that are easily solvable. Recall that a geometric series diverges if $|r| \geq 1$, and converges to $\frac{1}{1-r}$ if $|r| < 1$. Also, a harmonic series always diverges.

- If none of these methods are working, consider the integral test and comparing to a known series, which will be discussed in the upcoming sections.

Answer and Explanation:

A

This question relies on the convergence of the geometric series. You know that $\sum_{n=1}^{\infty} r^{n-1} = \frac{1}{1-r}$ provided $|r| < 1$. To make this clear, simplify the equation and use the geometric series to compute:

$$\sum_{n=1}^{\infty} \frac{3^n + 2^n}{6^n} = \sum_{n=1}^{\infty} \frac{3^n}{6^n} + \frac{2^n}{6^n} = \sum_{n=1}^{\infty} \frac{3^n}{6^n} + \sum_{n=1}^{\infty} \frac{2^n}{6^n}$$

$$= \sum_{n=1}^{\infty} \left(\frac{1}{2}\right)^n + \sum_{n=1}^{\infty} \left(\frac{1}{3}\right)^n$$

$$= \frac{1}{2} \sum_{n=1}^{\infty} \left(\frac{1}{2}\right)^{n-1} + \frac{1}{3} \sum_{n=1}^{\infty} \left(\frac{1}{3}\right)^{n-1}$$

$$= \frac{1}{2} \left(\frac{1}{1 - \frac{1}{2}} \right) + \frac{1}{3} \left(\frac{1}{1 - \frac{1}{3}} \right)$$

$$= 1 + \frac{1}{2} = 1\frac{1}{2}$$

Thus, **(A)** is correct.

PRACTICE SET

1. Determine if the harmonic series $\sum\limits_{n=1}^{\infty} \dfrac{1}{n}$ diverges or converges. If it converges find the value.

   (A) Diverges

   (B) Converges, $\dfrac{1}{8}$

   (C) Converges, 3

   (D) None of the above

2. Which of the following series diverge?

   I. $24 + 18 + \dfrac{27}{2} + \dfrac{81}{8} + \ldots$

   II. $1 - 2 + 4 - 8 - 16\ldots$

   III. $6 - 6 + 6 - 6 + 6\ldots$

   (A) I only

   (B) II only

   (C) III only

   (D) II and III only

3. Evaluate the series $\sum\limits_{n=1}^{\infty} 2^{n+1} \cdot 3^{1-n}$.

   (A) $\dfrac{2}{3}$

   (B) 6

   (C) 12

   (D) The series diverges.

4. For what values of r is the sequence $\{r^n\}$ convergent?

   (A) $-1 < r \le 1$

   (B) $-1 \le r \le 1$

   (C) $-1 < r < 1$

   (D) The sequence is never convergent.

ANSWERS AND EXPLANATIONS

1. A

Write out a few of the first terms and see if there are sums that you notice that diverge.

$$\sum_{n=1}^{\infty}\frac{1}{n} = 1+\frac{1}{2}+\left(\frac{1}{3}+\frac{1}{4}\right)+\left(\frac{1}{5}+\frac{1}{6}+\frac{1}{7}+\frac{1}{8}\right)+\dots$$
$$1+\frac{1}{2}+\frac{1}{2}+\frac{1}{2}$$

As you can see, you can group multiple terms and eventually you will just be adding 0.5 forever. Because the infinite sum of 0.5 diverges, so does the harmonic series. Therefore, **(A)** is the correct answer.

2. D

Each of these is a geometric series. The convergence/divergence of a geometric series is based on the value of the common ratio.

I. $24+18+\frac{27}{2}+\frac{81}{8}+\dots r=\frac{3}{4} \Rightarrow$ series converges.

II. $1-2+4-8+16\dots r=-2 \Rightarrow$ series diverges.

III. $6-6+6-6+6\dots r=-1 \Rightarrow$ series diverges.

Therefore, **(D)** is the correct answer.

3. C

With some rearranging, you can recognize this as a geometric series:

$$\sum_{n=1}^{\infty}2^{n+1}\cdot3^{1-n} = \sum_{n=1}^{\infty}\frac{2^{n+1}}{3^{n-1}} = \sum_{n=1}^{\infty}\frac{2^2\cdot2^{n-1}}{3^{n-1}}$$
$$=4\sum_{n=1}^{\infty}\frac{2^{n-1}}{3^{n-1}} = 4\sum_{n=1}^{\infty}\left(\frac{2}{3}\right)^{n-1}$$

Therefore, the sum to infinity is given by:

$$S_{\infty} = \frac{a}{1-r} = \frac{4}{1-\frac{2}{3}} = \frac{4}{\frac{1}{3}} = 12$$

Hence, **(C)** is the correct answer.

4. A

While discussing limits you learned how to compute the limit of an exponential function. Using this information, you have

$$\lim_{n\to\infty} r^n = \begin{cases} \infty & \text{if } r>1 \\ 0 & \text{if } 0<r<1. \end{cases}$$

So you know the sequence converges for $0<r<1$ and diverges for $r>1$. You still have more values of r to check.

When $r=1$:

$$\lim_{n\to\infty} r^n = \lim_{n\to\infty} 1^n = 1$$

so the sequence converges.

When $r=0$: The limit is clearly 0, so the sequence converges.

When $-1<r<0$: Then $0<|r|<1$, so

$$\lim_{n\to\infty}|r^n| = \lim_{n\to\infty}|r|^n = 0$$

and this forces your original sequence to converge.

When $r=-1$: You can write out some terms of the sequence and see that it diverges; you get

$$\{-1, 1, -1, 1, -1, 1,\dots\}.$$

Compiling this information, your result is that the sequence $\{r^n\}_{n=1}^{\infty}$ converges for $-1<r\le1$.

Thus, **(A)** is the correct answer.

17.3 INTEGRAL TEST AND *p*-SERIES

To answer a question like this:

How many terms of the series do you need in order to show that $\sum_{n=1}^{\infty} \dfrac{3}{n^2}$ approximates its limit within an error of 0.001 ?

   (A) 1,800   (B) 2,525   (C) 2,650   (D) 3,000

You need to know this:

The integral test and remainder estimate

- Compare the area of an improper integral with the series itself. To do this, consider each element a_n of the series as a box of width 1 and height a_n; that is, a box of area a_n.

- If each of these boxes can fit *below* a function whose improper integral is *finite,* then clearly the series itself must also be finite (recall the integral of a function measures the area under the curve).

- If, however, each of these boxes can fit *above* a function whose improper integral is *infinite,* then clearly the series itself must also be infinite.

- The integral test says that a series is convergent (or divergent) if and only if its corresponding integral is convergent (or divergent).

- Recall the remainder of a series $\sum a_n$ is given by $R_n = s - s_n$, where s is the full sum and s_n is the nth partial sum. To control the remainder term we use the following inequality:
$$\int_{n+1}^{\infty} f(x)\,dx \le R_n \le \int_{n}^{\infty} f(x)\,dx.$$

- These upper and lower bounds give a way of determining how close a remainder is to the actual sum of the series.

The *p*-series

- The integral test can also be used to evaluate the convergence of another important series in mathematics, the *p*-series $\sum_{n=1}^{\infty} \dfrac{1}{n^p}$.

- According to the integral test, this series will converge exactly for those values of p when $\int_{1}^{\infty} \dfrac{1}{x^p}\,dx$ is finite. The series will diverge for all other values of p.

- Thus, only $\int_{1}^{\infty} \dfrac{1}{x^p}\,dx$ needs to be evaluated.

You need to do this:

- If f is a continuous, positive, decreasing function on $[1, \infty]$ and $a_n = f(n)$, then
  - If $\int_1^\infty f(x)\,dx$ is finite, then $\sum_{n=1}^\infty a_n$ is convergent.
  - If $\int_1^\infty f(x)\,dx$ is infinite, then $\sum_{n=1}^\infty a_n$ is divergent.

- There are three cases to consider for the p-series:
  - If $p < 0$, then $\lim_{n\to\infty} \dfrac{1}{n^p} = \lim_{n\to\infty} n^{|p|} = \infty$; the partial sums cannot converge to a finite number if each of the terms goes to infinity.
  - If $p = 0$, then $\sum_{n=1}^\infty \dfrac{1}{n^p} = \sum_{n=1}^\infty 1 = \infty$; the sum diverges.
  - If $p > 0$, the function $f(x) = \dfrac{1}{x^p}$ is continuous, positive, and decreasing on $[1, \infty]$; the integral test can be applied:

$$\int_1^\infty \frac{1}{x^p}\,dx = \left(\lim_{x\to\infty} \frac{x^{1-p}}{1-p}\right) - \frac{1}{1-p}$$

and thus is finite only if $1 - p < 0$; that is, if $p > 1$. So, $\int_1^\infty \dfrac{1}{x^p}\,dx$ is finite if $p > 1$ and $\int_1^\infty \dfrac{1}{x^p}\,dx$ is infinite if $p \le 1$.

Answer and Explanation:

D

An accuracy of 0.001 means that you must determine your value of n such that the remainder $R_n \le 0.001$. From the remainder estimate for the integral test, you have the following inequality:

$$R_n \le \int_n^\infty \frac{3}{x^2}\,dx = \lim_{t\to\infty} \int_n^t \frac{3}{x^2}\,dx$$

$$= \lim_{t\to\infty} \left[\frac{-3}{x}\right]_n^t$$

$$= \lim_{t\to\infty} \frac{-3}{t} + \frac{3}{n} = \frac{3}{n}.$$

Thus, you want $\dfrac{3}{n} < 0.001$, and solving this inequality for n, you get $n > \dfrac{3}{0.001} = 3{,}000$. So to get an accuracy of 0.001, you must sum up 3,000 terms of the series, matching **(D)**.

PRACTICE SET

1. Consider the series $\sum_{n=1}^{\infty} \dfrac{7}{5n^p}$. Which inequality below contains *all* the values of p where this series is divergent?

   (A) $p > \dfrac{7}{5}$

   (B) $p \geq 1$

   (C) $p \leq 1$

   (D) $p \leq \dfrac{7}{5}$

2. Given $1 + \dfrac{1}{2^{0.25}} + \dfrac{1}{3^{0.25}} + \dfrac{1}{4^{0.25}} + \cdots + \dfrac{1}{n^{0.25}} + \cdots$ is the series convergent or divergent? If convergent, find the value the series converges to.

   (A) Divergent

   (B) Convergent, 1

   (C) Convergent, 3

   (D) None of the above

3. A certain function $f(x)$ is known to be positive and decreasing on $(1, \infty)$, and $\int_{1}^{\infty} f(x)\,dx = 2$. Which of the following statements must be true if $f(n) = a_n$?

   I. $\sum_{n=1}^{\infty} a_n$ is convergent.

   II. $\sum_{n=1}^{\infty} a_n = 2$

   III. $\sum_{n=1}^{\infty} a_n > 2$

   (A) I only

   (B) III only

   (C) I and II only

   (D) I and III only

4. Which of the following statements are true for the two series below?

   I. $\sum_{n=1}^{\infty} \dfrac{3n^2}{e^{n^3}}$

   II. $\sum_{n=1}^{\infty} \dfrac{1}{n} (\ln(n))^4$

   (A) Series I and II both converge.

   (B) Series I converges and series II diverges.

   (C) Series I diverges and series II converges.

   (D) Series I and II both diverge.

ANSWERS AND EXPLANATIONS

1. C

The important point here is to find the inequality that contains ALL (not just some of) the values of p where the series is divergent.

Perhaps this series looks familiar. The material that you covered on the p-series, $\sum_{n=1}^{\infty} \frac{1}{n^p}$ should ring a bell for you.

You know that $\sum_{n=1}^{\infty} \frac{1}{n^p}$ converges for $p > 1$ and diverges for $p \leq 1$.

Clearly, $\sum_{n=1}^{\infty} \frac{7}{5n^p}$ diverges precisely when $\sum_{n=1}^{\infty} \frac{1}{n^p}$ diverges; the factor $\frac{7}{5}$ doesn't change anything. Thus, $\sum_{n=1}^{\infty} \frac{7}{5n^p}$ diverges if $p \leq 1$; this is the largest such interval for p.

Therefore, **(C)** is the correct answer.

2. A

Notice that this is a p-series so if the exponent of n, p, is less than or equal to 1 the series is divergent and if p is greater than 1 the series converges. Because p is 0.25 this series is divergent. Hence, **(A)** is the appropriate match.

3. D

The integral test tells you that if $\int_{1}^{\infty} f(x)\,dx = 2$ then $\sum_{n=1}^{\infty} a_n$ is convergent. It does not tell you what the series converges to, but the series gives you an upper bound. In this case, $\int_{1}^{\infty} f(x)\,dx = 2$ implies that $\sum_{n=1}^{\infty} a_n > 2$. Therefore, statements I and III are true, and **(D)** is the correct answer.

4. B

Use the integral test to determine the convergence of both series.

Series I

$$\int_{1}^{\infty} \frac{3x^2}{e^{x^3}}\,dx = \lim_{t \to \infty} \int_{1}^{t} \frac{3x^2}{e^{x^3}}\,dx$$

Let $u = x^3$
$\quad du = 3x^2\,dx$

$$\int \frac{3x^2}{e^{x^3}}\,dx = \int e^{-u}\,du = -e^{-u} + C = -e^{-x^3} + C$$

$$\lim_{t \to \infty} \int_{1}^{t} \frac{3x^2}{e^{x^3}}\,dx = \lim_{t \to \infty} -e^{-x^3}\Big|_{1}^{t}$$

$$= \lim_{t \to \infty}\left(-e^{-t^3} - \left(-\frac{1}{e}\right)\right) = \frac{1}{e}$$

Series II

$$\int_{1}^{\infty} \frac{1}{x}\left(\ln(x)\right)^4 dx = \lim_{t \to \infty} \int_{1}^{t} \frac{1}{x}\left(\ln(x)\right)^4 dx$$

Let $u = \ln(x)$
$\quad du = \frac{1}{x}\,dx$

$$\int \frac{1}{x}\left(\ln(x)\right)^4 dx = \int u^4\,du = \frac{1}{5}u^5 + C = \frac{1}{5}\left(\ln(x)\right)^5 + C$$

$$\lim_{t \to \infty} \int_{1}^{t} \frac{1}{x}\left(\ln(x)\right)^4 dx = \lim_{t \to \infty} \frac{1}{5}\left(\ln(x)\right)^5\Big|_{1}^{t}$$

$$= \frac{1}{5}\lim_{t \to \infty}\left(\left(\ln(t)\right)^5 - 0\right) = \infty$$

The first series converges and the second series diverges by the integral test. Thus, **(B)** is correct.

17.4 COMPARISON AND RATIO TESTS

To answer a question like this:

Determine whether the series $\sum_{n=1}^{\infty} \dfrac{6n+1}{n^2}$ is convergent or divergent. If it is convergent, find its sum.

(A) The series is convergent and equals $6\dfrac{3}{4}$.

(B) The series is convergent and equals $\dfrac{2\pi}{3}$.

(C) The series is divergent and equals $\pi + \dfrac{1}{2}$.

(D) The series is divergent.

You need to know this:

The comparison test

- Suppose $\sum a_n$ and $\sum b_n$ are series with positive terms.

  - If $\sum b_n$ is convergent and $a_n \leq b_n$ for each n, then $\sum a_n$ must also be convergent.

  - If $\sum b_n$ is divergent and $a_n \geq b_n$ for each n, then $\sum a_n$ must also be divergent.

The ratio test

- Let $\sum_{n=1}^{\infty} a_n$ be any given series.

- If $\lim_{n \to \infty} \left| \dfrac{a_{n+1}}{a_n} \right| < 1$, then the series is absolutely convergent.

- If $\lim_{n \to \infty} \left| \dfrac{a_{n+1}}{a_n} \right| > 1$ or $\lim_{n \to \infty} \left| \dfrac{a_{n+1}}{a_n} \right| = \infty$, then the series is divergent.

You need to do this:

- Keep in mind that there are two criteria to use the comparison test: 1) a series $\sum b_n$ whose convergence/divergence is known, and 2) a relation between the elements a_n and b_n.

- Most of the time $\sum b_n$ can be the geometric series, the harmonic series, or the p-series, and these known series can be used as the comparison to the unknown series.

- Clearly, the ratio test is a very valuable too. Using only the terms of the series itself, the ratio test can check the series' convergence or divergence.

- However, if $\lim_{n \to \infty} \left| \dfrac{a_{n+1}}{a_n} \right| = 1$, the ratio test is inconclusive and does not give any information.

  When this happens another means must be used to evaluate the convergence of the series.

Answer and Explanation:

D

Your first check for convergence should always be to look at the limit $\lim\limits_{n\to\infty} \dfrac{6n+1}{n^2}$. If this limit is not zero, then, by the divergence test, the series MUST be divergent. If it is zero, then the series *could* be convergent. You get $\lim\limits_{n\to\infty} \dfrac{6n+1}{n^2} = 0$. So it could be convergent; take a closer look.

Simplify the series to get:

$$\sum_{n=1}^{\infty} \frac{6n+1}{n^2} = \sum_{n=1}^{\infty} \frac{6}{n} + \frac{1}{n^2} = 6\sum_{n=1}^{\infty} \frac{1}{n} + \sum_{n=1}^{\infty} \frac{1}{n^2}$$

What do you know about these series? The series $\sum\limits_{n=1}^{\infty} \dfrac{1}{n^2}$ is convergent; however, the series $\sum\limits_{n=1}^{\infty} \dfrac{1}{n}$ is the harmonic series and you know that this diverges. Thus, the entire series $\sum\limits_{n=1}^{\infty} \dfrac{6n+1}{n^2}$ must diverge as well, making **(D)** correct.

PRACTICE SET

1. Suppose that $\sum_{n=1}^{\infty} b_n$ and $\sum_{n=1}^{\infty} c_n$ are two series with positive terms. Which statement below is definitely TRUE?

   (A) If $\sum_{n=1}^{\infty} b_n$ converges and $c_n \geq b_n$ for all n,

   then $\sum_{n=1}^{\infty} c_n$ diverges.

   (B) If $\sum_{n=1}^{\infty} b_n$ diverges and $b_n \geq c_n$ for all n, then

   $\sum_{n=1}^{\infty} c_n$ converges.

   (C) If $\sum_{n=1}^{\infty} b_n$ diverges and $c_n \geq \frac{1}{2} b_n$ for all n,

   then $\sum_{n=1}^{\infty} c_n$ also diverges.

   (D) None of the above

2. Consider the two positive series $\sum_{n=1}^{\infty} u(n)$ and $\sum_{n=1}^{\infty} c(n)$. It is known that $\sum_{n=1}^{\infty} c(n)$ converges, but the convergence/divergence of $\sum_{n=1}^{\infty} u(n)$ is unknown. A student correctly uses the comparison test to prove that $\sum_{n=1}^{\infty} u(n)$ converges. Which of the following statements must be true?

   I. $u_n \leq c_n$ for all n.
   II. Both $u_n > 0$ and $c_n > 0$ for all n.
   III. $\lim_{n \to \infty} c_n = 0$.

   (A) I and II only

   (B) II and III only

   (C) I, II, and III

   (D) None of the above

3. Which of the following statements is true about the series below?

   $$\sum_{n=1}^{\infty} \frac{(\sin n)(n+1)}{(2n+3)^4}$$

   (A) The series converges by the alternating series test.

   (B) You can conclude by the comparison test—comparing the series of absolute values with the series $\sum_{n=1}^{\infty} \frac{1}{n^3}$ —that the given series converges absolutely.

   (C) You can conclude by the limit comparison test—comparing the series of absolute values with the series $\sum_{n=1}^{\infty} \frac{1}{n^4}$ —that the series converges absolutely.

   (D) You can conclude by the limit comparison test—comparing the series of absolute values with the series $\sum_{n=1}^{\infty} \frac{1}{n^4}$ —that the given series does *not* converge absolutely.

4. For which of the following series is the ratio test inconclusive?

   (A) $\sum_{n=1}^{\infty} \frac{n}{2^n}$

   (B) $\sum_{n=1}^{\infty} \frac{(-3)^{n-1}}{\sqrt{n}}$

   (C) $\sum_{n=1}^{\infty} \frac{\sqrt{n}}{1+n^2}$

   (D) $\sum_{n=1}^{\infty} \frac{n^n}{n!}$

ANSWERS AND EXPLANATIONS

1.   C

If you guessed (A) or (B), then go back and read the rules for the comparison test. Each of these answers is *similar* to the statement, but none of them is the same. In fact, you can even find examples of series that contradict the statements in (A) and (B).

Now, suppose you know that $\sum_{n=1}^{\infty} b_n$ is divergent. Then clearly, $\sum_{n=1}^{\infty} \frac{1}{2} b_n = \frac{1}{2} \sum_{n=1}^{\infty} b_n$ must also be divergent simply because $\sum_{n=1}^{\infty} b_n$ is. If you also know that $c_n \geq \frac{1}{2} b_n$ for all n, then by the comparison test, $\sum_{n=1}^{\infty} c_n$ must also be divergent. That's a match for **(C)**.

2.   C

In order to use the comparison test to prove that a series converges, the unknown series must be smaller than a known convergent series for all values of n. Therefore, if $\sum_{n=1}^{\infty} c(n)$ is to be used to prove that $\sum_{n=1}^{\infty} u(n)$ converges, then $u(n)$ must be less than $c(n)$ for all values of n, so statement I is true. Furthermore, the convergence test requires that both series are positive for all values of n and so statement II must be true. Statement III is a necessary condition for the convergence of $c(n)$ by the divergence test: if the terms of a series do not go to zero as n approaches infinity, the series cannot be finite and therefore must diverge. Thus, all three statements must be true and **(C)** is the correct answer.

3.   B

This is a sneaky series to work with because of the sin n in the numerator. The series does take on both positive and negative values, but it does not alternate in a regular "$+ - + + - \ldots$" pattern. It is thus not an alternating series and you can eliminate answer (A). Compare the series of absolute values $\sum_{n=1}^{\infty} \left| \frac{(\sin n)(n+1)}{(2n+3)^4} \right|$ to the *p*-series $\sum_{n=1}^{\infty} \frac{1}{n^3}$, because

$$\left| \frac{(\sin n)(n+1)}{(2n+3)^4} \right| \leq \left| \frac{1 \cdot (n+1)}{(2n+3)^4} \right| \leq \frac{1}{n^3}.$$

Because the general term of the series of absolute values is smaller than the general term $\frac{1}{n^3}$ (the series is smaller than a series we know converges), you can conclude that the series of absolute values $\sum_{n=1}^{\infty} \left| \frac{(\sin n)(n+1)}{(2n+3)^4} \right|$ converges.

This means the original series, $\sum_{n=1}^{\infty} \frac{(\sin n)(n+1)}{(2n+3)^4}$, converges absolutely.

Note: The limit comparison test, suggested in (C), will not work because the limit of the ratios of the general terms of the series does not exist (sin n in the numerator). The limit comparison test would work if you compared the series of absolute values with the convergent *p*-series $\sum_{n=1}^{\infty} \frac{1}{n^2}$. Hence, **(B)** is correct.

4.   C

Recall that the ratio test is inconclusive if $\lim_{x \to a} \left| \frac{a_{n+1}}{a_n} \right| = 1$.

This occurs for $\sum_{n=1}^{\infty} \frac{\sqrt{n}}{1+n^2}$:

$$\lim_{n \to \infty} \left| \frac{a_{n+1}}{a_n} \right| = \lim_{n \to \infty} \left| \frac{\dfrac{\sqrt{n+1}}{1+(n+1)^2}}{\dfrac{\sqrt{n}}{1+n^2}} \right|$$

$$= \lim_{n \to \infty} \left| \frac{\sqrt{n+1}}{\sqrt{n}} \cdot \frac{1+n^2}{1+(n+1)^2} \right|$$

$$= \lim_{n \to \infty} \frac{\sqrt{n+1}}{\sqrt{n}} \cdot \lim_{n \to \infty} \frac{1+n^2}{1+(n+1)^2}$$

$$= 1 \cdot 1 = 1$$

So, the ratio test is inconclusive and **(C)** is the correct answer.

You are encouraged to calculate the limits of (A), (B), and (D) to show that the ratio test does indeed determine absolute convergence (or divergence) in these cases.

17.5 ALTERNATING SERIES

To answer a question like this:

Given $\sum_{n=1}^{\infty}(-1)^{n+1}\dfrac{n^2}{e^n}$, does this series converge absolutely, converge conditionally, diverge, or not have enough information?

(A) Converge absolutely

(B) Converge conditionally

(C) Diverge

(D) None of the above

You need to know this:

Alternating series

- Consider the following series: $\sum_{n=1}^{\infty}(-1)^{n+1}\dfrac{1}{n}=1-\dfrac{1}{2}+\dfrac{1}{3}-\dfrac{1}{4}+\cdots$

- The terms in this series are alternately positive and negative; thus, a series like this one is called an alternating series. More specifically, this is an alternating harmonic series, and while harmonic series are divergent, this one is actually convergent!

- The idea is that the negative terms in the series are just enough to keep the series from diverging to infinity. To help determine the convergence and divergence of alternating series use the following:

The alternating series test

- Given an alternating series: $\sum_{n=1}^{\infty}(-1)^{n+1}a_n=a_1-a_2+a_3-a_4+\cdots$

  where each $a_n>0$, the alternating series must be convergent if

  1) $a_{n+1}\le a_n$

  2) $\lim_{n\to\infty}a_n=0$

You need to do this:

- If presented with an alternating series, then apply the alternating series test to determine convergence.

- The alternating series test does not necessarily have to hold for every n. As long as eventually $a_{n+1} \leq a_n$ for large enough n, then this test can be employed.

- In short, this test says that if the absolute value of the terms of the series is decreasing to zero, then the series is convergent.

- To determine the limit of a convergent alternating series, the following estimation can be applied:

  - If $S = \sum_{n=1}^{\infty} (-1)^n a_n$ is the finite value of the alternating series, then $|E_n| = |S - s_n| \leq a_{n+1}$.
  - This theorem says that given a convergent alternating series, the nth error term of the partial sums is always less than the $(n + 1)$th term, a_{n+1}.

Answer and Explanation:

A

You can go ahead and use the alternating series test, but that will only tell you if it converges. To figure out if the series absolutely or conditionally converges, use the ratio test.

$$\lim_{n \to \infty} \left| \frac{a^{n+1}}{a^n} \right| = \left| \frac{(-1)^{n+2} (n+1)^2}{e^{n+1}} \cdot \frac{e^n}{(-1)^{n+1} n^2} \right|$$

$$\lim_{n \to \infty} \frac{(n+1)^2}{en^2} = \frac{1}{e} \lim_{n \to \infty} \frac{(n+1)^2}{n^2}$$

$$\frac{1}{e}(1) = \frac{1}{e} < 1$$

The ratio test shows that this series converges absolutely, so **(A)** is correct.

PRACTICE SET

1. What is the tenth term in the following sequence
$$\left\{\frac{-1}{2}, \frac{2}{4}, \frac{-3}{8}, \frac{4}{16}, \frac{-5}{32}, \ldots\right\}?$$

    (A) $\dfrac{-9}{512}$

    (B) $\dfrac{9}{1024}$

    (C) $\dfrac{5}{512}$

    (D) $\dfrac{10}{512}$

2. Given $\displaystyle\sum_{n=1}^{\infty}(-1)^n \frac{2n+1}{3n+2}$, does the series absolutely converge, conditionally converge, diverge, or not have enough information?

    (A) Diverge

    (B) Absolutely converge

    (C) Conditionally converge

    (D) Not enough information

3. Calculate the max error at R_{15} for $\displaystyle\sum_{n=1}^{\infty}\frac{(-1)^n}{n^2}$.

    (A) $\dfrac{1}{256}$

    (B) $\dfrac{1}{225}$

    (C) $\dfrac{1}{169}$

    (D) $\dfrac{1}{100}$

4. Which (if any) of the series listed below converge?

    I. $\displaystyle\sum_{n=1}^{\infty}\frac{(-1)^{n+1}}{n+10}$

    II. $\displaystyle\sum_{n=1}^{\infty}\frac{(-1)^{n+1} n^3}{n^3+10}$

    III. $\displaystyle\sum_{n=1}^{\infty}\frac{(-1)^{n+1} n^2}{n^3+10}$

    (A) I and II only

    (B) I and III only

    (C) II and III only

    (D) I, II, and III only

ANSWERS AND EXPLANATIONS

1. C

This question really has two parts. Determine the general formula for the nth term of the sequence and then substitute $n = 10$ to find the tenth term. When you look at the sequence, you notice three things:

1. It is alternating, so you have a factor of $(-1)^n$.

2. The denominators are increasing in powers of 2^n.

3. The numerator is increasing with n.

Therefore, the general term for the sequence must be $a_n = \dfrac{(-1)^n n}{2^n}$, and, if you check a couple of terms, this is definitely correct. Now to find the tenth term in the sequence, let $n = 10$ to get

$$a_{10} = \frac{(-1)^{10} \, 10}{2^{10}} = \frac{10}{1024} = \frac{5}{512}$$

Hence, **(C)** is the appropriate match.

2. A

The easiest way to see if the series converges or diverges is to see if the limit from n to infinity is equal to zero or not. Remember from the alternating series test that the series diverges if the limit does not equal 0. Use L' Hopital's rule where needed.

$$\lim_{n \to \infty}(a_n) = \lim_{n \to \infty} \frac{2n+1}{3n+2} = \frac{2}{3} \neq 0$$

This series diverges because the limit of n to infinity does not equal 0. Thus, **(A)** is correct.

3. A

The max error will be no greater than the next number in the series, so, if you want to calculate max error at $n = 15$, then plug in $n = 16$.

$$|R_{15}| = |S - S_{15}| \le b_{16} = \frac{1}{16^2} = \frac{1}{256}$$

That's a match for **(A)**.

4. B

You recognize all three of these as alternating series (because of the presence of the $(-1)^{n+1}$ in all three series) and can try the alternating series test to test for the convergence in all three cases. In order to do so, you need to know two things for $\displaystyle\sum_{n=1}^{\infty} (-1)^{n+1} \cdot a^n$:

- Is $\displaystyle\lim_{n \to \infty} a_n = 0$?

- Is $a_{n+1} \le a_n$ beyond some value of n?

Examine each of the three series individually using these criteria.

In series I, you see that the terms are:

$$\frac{1}{11}, -\frac{1}{12}, \frac{1}{13}, -\frac{1}{14} \cdots$$

Clearly, as n approaches infinity, these terms are approaching zero, and so the first condition is met. Also, if you consider the magnitudes of the terms:

$$\frac{1}{11}, \frac{1}{12}, \frac{1}{13}, \frac{1}{14} \cdots$$

you see that each term is smaller than the previous one, and so these are decreasing. Thus, the alternating series test tells you that this series converges.

In series II, you see that $\displaystyle\lim_{n \to \infty} \frac{n^3}{n^3 + 10} = 1 \neq 0$. Therefore, the second condition is not met and the alternating series test fails. The divergence test shows you that $\displaystyle\lim_{n \to \infty} \frac{(-1)^{n+1} n^3}{n^3 + 10}$ does not exist (as n grows to infinity, the terms bounce back and forth between values approaching $+1$ and -1, so the limit does not exist), and therefore the series diverges.

In series III, you see that $\displaystyle\lim_{n \to \infty} \frac{n^2}{n^3 + 10} = 0$ and so the first condition is met. In order to determine if the values of $\dfrac{n^2}{n^3 + 10}$ are decreasing, you can apply differential calculus:

$$\frac{d}{dn}\left(\frac{n^2}{n^3+10}\right) = \frac{2n(n^3+10)-n^2(3n^2)}{(n^3+10)^2}$$

$$= \frac{2n^4+20n-3n^4}{(n^3+10)^2}$$

$$= \frac{20n-n^4}{(n^3+10)^2}$$

$$= \frac{n(20-n^3)}{(n^3+10)^2}$$

$$\frac{d}{dn}\left(\frac{n^2}{n^3+10}\right) = 0 \Rightarrow n=0, n=\sqrt[3]{20}$$

$$\frac{d}{dn}\left(\frac{n^2}{n^3+10}\right) \text{ does not exist: } n=\sqrt[3]{-10}$$

You have critical values at $n=0$, $n=\sqrt[3]{20}$ and $n=\sqrt[3]{-10}$.

$(n^3+10)^2 \geq 0$ for all real numbers, so the sign of the derivative will not change at $n=\sqrt[3]{-10}$.

Checking the signs of the derivative around the other two critical values, $n=0$, $n=\sqrt[3]{20}$, you see that:

$$\frac{d}{dn}\left(\frac{n^2}{n^3+10}\right) > 0 \text{ on } \left(0, \sqrt[3]{20}\right)$$

$$\frac{d}{dn}\left(\frac{n^2}{n^3+10}\right) < 0 \text{ on } \left(\sqrt[3]{20}, \infty\right)$$

Thus, the series will decrease once n is greater than $\sqrt[3]{20}$. Because the alternating series test does not require the terms to be decreasing for all values of n, but only past some finite value of n, the condition is met and the alternating series test tells you that series III is convergent. Therefore, **(B)** is the correct answer.

TEST WHAT YOU LEARNED

1. Which series below is NOT absolutely convergent?

   (A) $\displaystyle\sum_{n=1}^{\infty}(-1)^n \frac{n^5}{5^n}$

   (B) $\displaystyle\sum_{n=1}^{\infty}(-5)^n \frac{1}{\sqrt[3]{n}}$

   (C) $\displaystyle\sum_{n=1}^{\infty}\frac{\cos 3n}{n^3}$

   (D) All the above series are absolutely convergent.

2. Which of the following sequences is generated by the formula $\left\{\dfrac{n-1}{n^2}\right\}_{n=1}^{\infty}$?

   (A) $\left\{2, \dfrac{3}{4}, \dfrac{4}{9}, \dfrac{5}{16}, \dfrac{6}{25} \cdots\right\}$

   (B) $\left\{\dfrac{1}{4}, \dfrac{2}{9}, \dfrac{3}{64}, \dfrac{4}{125} \cdots\right\}$

   (C) $\left\{0, \dfrac{1}{4}, \dfrac{2}{9}, \dfrac{3}{16}, \dfrac{4}{25} \cdots\right\}$

   (D) $\left\{0, \dfrac{1}{2}, \dfrac{2}{3}, \dfrac{3}{4}, \dfrac{4}{5} \cdots\right\}$

3. Calculate the value of $\displaystyle\sum_{n=1}^{\infty}3^{-n+2}(-7)^{n-1}$.

   (A) $\dfrac{\pi - 2}{3}$

   (B) $\dfrac{9}{10}$

   (C) $2\dfrac{3}{10}$

   (D) It is divergent.

4. Which sequence below follows the rule

$$a_n = \frac{(-1)^{n+1} 2^{n-1}}{n^2} ?$$

(A) $\left\{1, \dfrac{-1}{2}, \dfrac{4}{9}, \dfrac{-1}{2}, \ldots\right\}$

(B) $\left\{1, \dfrac{-1}{2}, \dfrac{4}{7}, \dfrac{-1}{3}, \ldots\right\}$

(C) $\left\{1, \dfrac{1}{2}, \dfrac{4}{9}, \dfrac{1}{2}, \ldots\right\}$

(D) $\left\{1, \dfrac{-3}{2}, \dfrac{2}{9}, \dfrac{-2}{3}, \ldots\right\}$

5. For what values of p will both series $\displaystyle\sum_{n=1}^{\infty} \frac{1}{n^{p+3}}$ and $\displaystyle\sum_{n=1}^{\infty} \left(\frac{p}{3}\right)^n$ converge?

(A) $-2 < p < 3$ only

(B) $-2 < p < 1$ only

(C) $2 < p < 3$ only

(D) $p < -2$ or $p > 3$ only

Answer Key

Test What You Already Know	Test What You Learned
1. D	1. B
2. D	2. C
3. A	3. D
4. B	4. A
5. C	5. A

REFLECTION

Test What You Already Know score: _____

Test What You Learned score: _____

Use this section to evaluate your progress. After working through the pre-quiz, check off the boxes in the "Pre" column to indicate which Learning Objectives you feel confident about. Then, after completing the chapter, including the post-quiz, do the same to the boxes in the "Post" column. Keep working on unchecked Objectives until you're confident about them all!

Pre	Post		
☐	☐	**17.1**	Evaluate the limits of series using partial sums
☐	☐	**17.2**	Define convergence in a series and identify geometric and harmonic series
☐	☐	**17.3**	Apply the integral test and determine convergence in p-series
☐	☐	**17.4**	Determine convergence with the comparison test and the ratio test
☐	☐	**17.5**	Evaluate alternating series for convergence and estimate their limit

FOR MORE PRACTICE

Complete more practice online at kaptest.com. Haven't registered your book yet? Go to kaptest.com/booksonline to begin.

ANSWERS AND EXPLANATIONS

Test What You Already Know

1. D

If the terms of a series don't go to zero as n approaches infinity, then the series cannot be finite; that is, it cannot possibly converge. Note that

$$\lim_{n \to \infty} \frac{(2n+1)^2}{n(2n+4)} = \lim_{n \to \infty} \frac{4n^2 + 4n + 1}{2n^2 + 4n} = \frac{4}{2} = 2 \neq 0.$$

Therefore, the series is divergent, and **(D)** is correct.

2. D

You can tell right away that this is an alternating series because the $(-1)^{n-1}$ changes the sign every term. The rest of the series is $\frac{1}{n}$, which is the harmonic series because the exponent of n is 1. Hence, **(D)** is the correct answer.

3. A

Notice that this is a p-series, so if the exponent of n, p, is less than or equal to 1, the series is divergent, and if p is greater than 1, the series converges.

$$\sqrt{n} = n^{\frac{1}{2}}$$
$$\frac{1}{2} < 1$$

Hence, **(A)** is the correct answer.

4. B

You are given $a_n = \frac{1}{2n+3}$ so

$$a_{n+1} = \frac{1}{2n+2+3} = \frac{1}{2n+5}.$$

Because $2n + 5 > 2n + 3$ for all $n = 1, 2, 3, \ldots$ we get that $a_n > a_{n+1}$, so the sequence must be decreasing. Furthermore, because the sequence is decreasing, a_1 is the largest term and so the series must also be bounded. Hence, **(B)** is the correct answer.

5. C

You have the terms

$$a_1 = -2, \; a_2 = \frac{3}{4}, \; a_3 = -\frac{4}{9},$$
$$a_4 = \frac{5}{16}, \; a_5 = -\frac{6}{25}, \ldots$$

and you notice three things: First, the denominators equal n^2; second, the numerators equal $n + 1$; third, the signs alternate according to $(-1)^n$. Combine all this information and you get $a_n = \frac{(-1)^n(n+1)}{n^2}$. Therefore, **(C)** is the correct answer.

Test What You Learned

1. B

To show that $\sum_{n=1}^{\infty} (-5)^n \frac{1}{\sqrt[3]{n}}$ is NOT absolutely convergent, you will use the ratio test. This means you must show that $\lim_{n \to \infty} \left| \frac{a_{n+1}}{a_n} \right| > 1$.

You get:

$$\lim_{n \to \infty} \left| \frac{a_{n+1}}{a_n} \right| = \lim_{n \to \infty} \left| \frac{(-5)^{n+1} \frac{1}{(n+1)^{\frac{1}{3}}}}{(-5)^n \frac{1}{n^{\frac{1}{3}}}} \right|$$

$$= \lim_{n \to \infty} \left| \frac{(-5) n^{\frac{1}{3}}}{(n+1)^{\frac{1}{3}}} \right|$$

$$= 5 \lim_{n \to \infty} \left| \frac{n^{\frac{1}{3}}}{(n+1)^{\frac{1}{3}}} \right|$$

$$= 5 \cdot 1 = 5 > 1$$

Hence, **(B)** is the correct answer.

2. C

$$\left\{ \underbrace{0}_{\frac{1-1}{1^2}}, \underbrace{\frac{1}{4}}_{\frac{2-1}{2^2}}, \underbrace{\frac{2}{9}}_{\frac{3-1}{3^2}}, \underbrace{\frac{3}{16}}_{\frac{4-1}{4^2}}, \underbrace{\frac{4}{25}}_{\frac{5-1}{5^2}} \cdots \right\}$$

Therefore, **(C)** is the correct answer.

3. D

The series $\sum_{n=1}^{\infty} 3^{-n+2}(-7)^{n-1}$ doesn't make much sense as it is. However, a discerning eye may have picked out the geometric series hidden in there somewhere. Keep in mind that you want to get a series of the form $\sum_{n=1}^{\infty} r^{n-1}$ where $|r| < 1$. You have:

$$\sum_{n=1}^{\infty} 3^{-n+2}(-7)^{n-1} = \sum_{n=1}^{\infty} \frac{(-7)^{n-1}}{3^{n-2}}$$

$$= \sum_{n=1}^{\infty} \frac{(-7)^{n-1}}{3^{n-1} \cdot 3^{-1}}$$

$$= \sum_{n=1}^{\infty} 3\left(\frac{-7}{3}\right)^{n-1}$$

$$= 3\sum_{n=1}^{\infty} \left(\frac{-7}{3}\right)^{n-1}$$

Because $\left|\frac{-7}{3}\right| > 1$, the series diverges. Hence, **(D)** is the correct answer.

4. A

To find the proper series you need to only evaluate a few terms of the series given by $\left\{ a_n = \frac{(-1)^{n+1} 2^{n-1}}{n^2} \right\}$. You see that

$$a_1 = 1, \; a_2 = \frac{-2}{4} = \frac{-1}{2}, \; a_3 = \frac{2^2}{9} = \frac{4}{9}, \; a_4 = \frac{-2^3}{16} = \frac{-1}{2}, \ldots$$

so the series must be $\left\{ 1, \frac{-1}{2}, \frac{4}{9}, \frac{-1}{2}, \ldots \right\}$.

Therefore, **(A)** is the correct answer.

5. A

Solve the first series using the p-series rule. The second series is a geometric series.

$$\sum_{n=1}^{\infty} \frac{1}{n^{p+3}} \Rightarrow p + 3 > 1 \rightarrow p > -2 \text{ and } \sum_{n=1}^{\infty} \left(\frac{p}{3}\right)^n$$

$$\Rightarrow -1 < \frac{p}{3} < 1 \rightarrow -2 < p < 3$$

Therefore, **(A)** is the correct answer.

CHAPTER 18

Power Series

LEARNING OBJECTIVES

18.1 Find the radius and interval of convergence of functions defined by power series

18.2 Use Taylor polynomial expansion to approximate non-polynomial functions

18.3 Apply the Maclaurin series of special functions to more complex functions

18.4 Find the Lagrange error bound for Taylor polynomials

TEST WHAT YOU ALREADY KNOW

1. Suppose $|x| < \frac{1}{2}$. Which of the following gives a power series representation for $\dfrac{1}{1 + 4x^2}$?

   (A) $\displaystyle\sum_{n=0}^{\infty} 4^n x^n$

   (B) $\displaystyle\sum_{n=0}^{\infty} 4^n x^{2n}$

   (C) $\displaystyle\sum_{n=0}^{\infty} (-1)^n 4^n x^{n+1}$

   (D) $\displaystyle\sum_{n=0}^{\infty} (-1)^n 4^n x^{2n}$

2. Which of the following functions can be represented by the power series

   $$1 - \frac{x^2}{3} + \frac{x^4}{5} - \frac{x^6}{7} + \ldots + (-1)^n \frac{x^{2n}}{2n+1} + \ldots$$

   for $|x| \leq 1$?

   (A) $\arctan x$

   (B) $\dfrac{\arctan x}{x}$

   (C) $\dfrac{\sin x}{x}$

   (D) $\sin x$

3. The Maclaurin series for the function f is given by $f(x) = \sum_{0}^{\infty} \left(\frac{x}{3}\right)^n$. What is the value of $f(-2)$?

(A) -3

(B) $-\dfrac{2}{3}$

(C) $-\dfrac{3}{5}$

(D) $\dfrac{3}{5}$

---

4. A third-order Taylor polynomial centered at $x = 3$ is used to approximate a differentiable function f at $x = 3.6$. It is known that $\left|f^{(4)}(x)\right| \leq 0.7$ for all $x \in [3, 3.6]$. What is the Lagrange Error Bound for the maximum error on the closed interval $[3, 3.6]$?

(A) 0.00378

(B) 0.00400

(C) 0.00412

(D) 0.00444

18.1 RADIUS AND INTERVAL OF CONVERGENCE

To answer a question like this:

What is the radius of convergence of the power series $\sum \dfrac{(-1)^n x^n}{\sqrt{n+3}}$?

(A) $R = 0$

(B) $R = 1$

(C) $R = 2\dfrac{1}{2}$

(D) $R = \infty$

You need to know this:

Convergence in a power series

- A power series is a series in which the elements of the series are not just real numbers but can actually be functions of x. In general, a power series is a series of the form:

$$\sum_{n=0}^{\infty} a_n x^n = a_0 + a_1 x^1 + a_2 x^2 + a_3 x^3 + \ldots + a_n x^n + \ldots$$

where x is a **variable** and the a_n's are called the *coefficients of the power series*.

- Even more generally, a power series can be written in the form:

$$\sum_{n=0}^{\infty} a_n (x - c)^n = a_0 + a_1 (x - c)^1 + a_2 (x - c)^2 + \ldots + a_n (x - c)^n + \ldots$$

A power series in this form is called a *power series centered at c*.

- As one might expect, however, a single power series may converge for one value of x and diverge for a different value of x. To study convergence of a power series, therefore, ask "For what values of x is a given power series convergent or divergent?"

Radius and interval of convergence in a power series

- Any given power series centered at $x = c$, written as $\sum_{n=0}^{\infty} a_n (x - c)^n$, clearly converges when $x = c$, because when $x = c$ we have $\sum_{n=0}^{\infty} a_a (x - c)^n = 0$.

- In general, for a given power series $\sum_{n=0}^{\infty} a_n (x - c)^n$, there is a positive integer R such that the series converges if $|x - c| < R$ and diverges if $|x - c| > R$. This number R is called the *radius of convergence of the power series*.

- The interval of convergence of a power series is the exact interval (with or without endpoints) where the power series converges.

You need to do this:

- Usually the ratio test can be used to determine the radius of convergence of a power series, but the ratio test is inconclusive at the endpoints of this interval.

- Another test should be used to calculate the convergence at the endpoints of the interval.

- Input each endpoint, in turn, examining the type of series it becomes and what test is needed to determine its convergence.

Answer and Explanation:

B

$$\sum \frac{(-1)^n x^n}{\sqrt{n+3}}$$

To calculate the radius of convergence, use the ratio test. Determine which values of x will give $\lim\limits_{n \to \infty} \left| \frac{a_{n+1}}{a_n} \right| < 1$ where $a_n = \frac{(-1)^n x^n}{\sqrt{n+3}}$. You get

$$\lim_{n \to \infty} \left| \frac{a_{n+1}}{a_n} \right| = \lim_{n \to \infty} \left| \frac{\frac{(-1)^{n+1} x^{n+1}}{\sqrt{n+1+3}}}{\frac{(-1)^n x^n}{\sqrt{n+3}}} \right| = \lim_{n \to \infty} \left| \frac{x^{n+1}}{x^n} \cdot \frac{\sqrt{n+3}}{\sqrt{n+4}} \right| = |x| \cdot \lim_{n \to \infty} \frac{\sqrt{n+3}}{\sqrt{n+4}} = |x|$$

Therefore, this series converges for all values of x such that $|x| < 1$. So, $R = 1$ and **(B)** is correct.

PRACTICE SET

1. Find the radius of convergence and interval of convergence of the power series $\sum_{n=1}^{\infty} \dfrac{4^n x^n}{(n+2)^2}$.

   (A) $R = \dfrac{1}{4}, I = \left(-\dfrac{1}{4}, \dfrac{1}{4}\right)$

   (B) $R = \dfrac{1}{4}, I = \left[-\dfrac{1}{4}, \dfrac{1}{4}\right]$

   (C) $R = \dfrac{1}{4}, I = \left(-\dfrac{1}{4}, \dfrac{1}{4}\right]$

   (D) $R = 4, I = \left[-\dfrac{1}{4}, \dfrac{1}{4}\right]$

2. Calculate the interval of convergence for $\sum_{n=1}^{\infty} \dfrac{x^n}{n4^n}$.

   (A) $-4 \leq x < 4$

   (B) $-4 \leq x \leq 4$

   (C) $-4 < x < 4$

   (D) $-4 < x \leq 4$

3. Compute the interval of convergence for $\sum_{n=1}^{\infty} \dfrac{(x+3)^n}{n}$.

   (A) $2 \leq x < 4$

   (B) $-4 < x < -2$

   (C) $2 < x < 4$

   (D) $-4 \leq x < -2$

4. Using the fact that $\sin x = \sum_{n=0}^{\infty} \dfrac{(-1)^n}{(2n+1)!} x^{2n+1}$, what is the power series representation of $\int x^2 \sin x^3 \, dx$?

   (A) $\sum_{n=0}^{\infty} \dfrac{(-1)^n}{6(2n+1)!} x^{6(n+1)}$

   (B) $\sum_{n=0}^{\infty} \dfrac{x^{6n}}{(2n+1)!(n+1)}$

   (C) $\sum_{n=0}^{\infty} \dfrac{(-1)^n}{6(2n+1)!(n+1)} x^{6(n+1)}$

   (D) $\sum_{n=0}^{\infty} \dfrac{(-1)^n}{(n+1)!(n+2)} x^{6n}$

ANSWERS AND EXPLANATIONS

1. B

To calculate the radius of convergence, determine which values of x ensure that

$$\lim_{n \to \infty} \left| \frac{a_{n+1}}{a_n} \right| < 1, \text{ where } a_n = \frac{4^n x^n}{(n+2)^2}.$$

You get

$$\lim_{n \to \infty} \left| \frac{a_{n+1}}{a_n} \right| = \lim_{n \to \infty} \left| \frac{\frac{4^{n+1} x^{n+1}}{(n+1+2)^2}}{\frac{4^n x^n}{(n+2)^2}} \right| = \lim_{n \to \infty} \left| \frac{4^{n+1} x^{n+1}}{4^n x^n} \right|$$

$$\cdot \frac{(n+2)^2}{(n+1+2)^2} \Bigg| = \lim_{n \to \infty} 4|x| \frac{(n+2)^2}{(n+3)^2} = 4|x|$$

Thus, $\lim_{n \to \infty} \left| \frac{a_{n+1}}{a_n} \right| < 1$ precisely when $4|x| < 1$; that is, when $|x| < \frac{1}{4}$. So, $R = \frac{1}{4}$.

Now you have to check the endpoints to determine the exact interval of convergence.

If $x = \frac{1}{4}$, you have

$$\sum_{n=1}^{\infty} \frac{4^n \left(\frac{1}{4}\right)^n}{(n+2)^2} = \sum_{n=1}^{\infty} \frac{1}{(n+2)^2} < \sum_{n=1}^{\infty} \frac{1}{n^2} < \infty.$$

If $x = -\frac{1}{4}$, you have

$$\sum_{n=1}^{\infty} \frac{4^n \left(-\frac{1}{4}\right)^n}{(n+2)^2} = \sum_{n=1}^{\infty} \frac{(-1)^n}{(n+2)^2} < \sum_{n=1}^{\infty} \frac{1}{(n+2)^2} < \infty.$$

Therefore, both the endpoints are included. So, $I = \left[-\frac{1}{4}, \frac{1}{4} \right]$ and **(B)** is the correct answer.

2. A

The best way to find the interval of convergence will be to use the ratio test, hence

$$\lim_{n \to \infty} \left| \frac{\frac{x^{(n+1)}}{(n+1) 4^{(n+1)}}}{\frac{x^n}{n 4^n}} \right| = \lim_{n \to \infty} \left| \frac{xn}{4n+4} \right|$$

$$\lim_{n \to \infty} \left| \frac{x}{4 + \frac{4}{n}} \right| = \left| \frac{x}{4} \right|$$

In the ratio test, you know that if the result is less than 1, then the series is convergent so

$$\left| \frac{x}{4} \right| < 1$$

$$-1 < \frac{x}{4} < 1$$

$$-4 < x < 4$$

From this you might think that (C) is the answer but you have to check the endpoints to see if they converge or diverge. When $x = -4$:

$$\sum_{n=1}^{\infty} \frac{(-4)^n}{n 4^n} = \sum_{n=1}^{\infty} \frac{(-1)^n 4^n}{n 4^n} = \sum_{n=1}^{\infty} \frac{(-1)^n}{n}$$

You should recognize this as the alternating harmonic series. You could use the alternating series test but this is a series known to converge so $x = -4$ is going to be included in the interval of convergence. When $x = 4$:

$$\sum_{n=1}^{\infty} \frac{(4)^n}{n 4^n} = \sum_{n=1}^{\infty} \frac{1}{n}$$

You should recognize this as the harmonic series. This series is known to diverge so $x = 4$ is not going to be included in the interval of convergence. The interval of convergence is $-4 \leq x < 4$, so **(A)** is the correct answer.

3. D

The best way to find the interval of convergence will be to use the ratio test, hence

$$\lim_{n\to\infty}\left|\frac{\dfrac{(x+3)^{n+1}}{n+1}}{\dfrac{(x+3)}{n}}\right| = \lim_{n\to\infty}\left|\frac{xn+3n}{n+1}\right| = \lim_{n\to\infty}\left|\frac{x+3}{1+\dfrac{1}{n}}\right| = |x+3|$$

In the ratio test you know that if the result is less than 1 then the series is convergent so

$$|x+3| < 1$$
$$-1 < x+3 < 1$$
$$-4 < x < -2$$

From this you might think that (B) is the answer but you have to check the endpoints to see if they converge or diverge. When $x = -4$:

$$\sum_{n=1}^{\infty}\frac{((-4)+3)^n}{n} = \sum_{n=1}^{\infty}\frac{(-1)^n}{n}$$

You should recognize this as the alternating harmonic series. You could use the alternating series test but this is a series known to converge so $x = -4$ is going to be included in the interval of convergence. When $x = -2$:

$$\sum_{n=1}^{\infty}\frac{((-2)+3)^n}{n} = \sum_{n=1}^{\infty}\frac{1}{n}$$

You should recognize this as the harmonic series. This series is known to diverge so $x = -2$ is not going to be included in the interval of convergence. The interval of convergence is $-4 \le x < -2$, so **(D)** is correct.

4. C

To compute $\int x^2 \sin x^3 \, dx$ you first need to simplify $x^2 \sin x^3$. To do this you use the given information to get

$$\sin x^3 = \sum_{n=0}^{\infty}\frac{(-1)^n}{(2n+1)!}(x^3)^{2n+1} = \sum_{n=0}^{\infty}\frac{(-1)^n}{(2n+1)!}x^{6n+3}$$

Therefore,

$$x^2 \sin x^3 = x^2\sum_{n=0}^{\infty}\frac{(-1)^n}{(2n+1)!}x^{6n+3} = \sum_{n=0}^{\infty}\frac{(-1)^n}{(2n+1)!}x^{6n+5}$$

Now, using the linearity of the integral, you get

$$\int x^2 \sin x^3\,dx = \int \sum_{n=0}^{\infty}\frac{(-1)^n}{(2n+1)!}x^{6n+5}\,dx = \sum_{n=0}^{\infty}\frac{(-1)^n}{(2n+1)!}\int x^{6n+5}\,dx$$
$$= \sum_{n=0}^{\infty}\frac{(-1)^n}{(2n+1)!}\frac{x^{6n+6}}{(6n+6)}$$
$$= \sum_{n=0}^{\infty}\frac{(-1)^n\, x^{6(n+1)}}{6(2n+1)!(n+1)}$$

That's a perfect match for **(C)**!

18.2 TAYLOR POLYNOMIAL EXPANSION

To answer a question like this:

Let f be a function with derivatives of all orders for $x > 0$ such that $f(4) = 8, f'(4) = -2, f''(4) = -6$, and $f'''(4) = 24$. Which of the following is the third-degree Taylor polynomial for f about $x = 4$?

(A) $8 - 2x - 6x^2 + 24x^3$

(B) $8 - 2x - 3x^2 + 4x^3$

(C) $8 - 2(x - 4) - 3(x - 4)^2 + 4(x - 4)^3$

(D) $8 - 2(x - 4) - 3(x - 4)^2 + 8(x - 4)^3$

You need to know this:

Taylor series expansion

- Non-polynomial functions can be approximated by a power series called the *Taylor series expansion*.

- Suppose that for a given function $f(x)$ there already exists a power series expansion, giving:

$$f(x) = \sum_{n=0}^{\infty} a_n (x - c)^n = a_0 + a_1(x - c) + a_2(x - c)^2 + \ldots + a_n(x - c)^n + \ldots$$

- When $x = c$, $f(c) = a_0$ because all the other terms in the sum would be zero. Furthermore, the derivative of the function would be:

$$f'(x) = \sum_{n=0}^{\infty} na_n(x - c)^{n-1} = a_1 + 2a_2(x - c)^1 + \ldots + na_n(x - c)^{n-1} + \ldots$$

- Therefore, $f'(c) = a_1$ for the exact same reasoning as before. Another derivative would give:

$$f''(x) = \sum_{n=0}^{\infty} n(n - 1)a_n(x - c)^{n-2} = 2a_2 + 3 \cdot 2a_3(x - c)^1 \ldots + n(n - 1)a_n(x - c)^{n-2} + \ldots$$

- Similarly $f''(c) = 2a_2$. Clearly, repeating this process would yield $f'''(c) = 3 \cdot 2a_3$, and even more generally, $f^{(n)}(c) = n!a_n$.

- Therefore, the formula for the coefficients of the power series expansion of the function centered at the point $x = c$ is $a_n = \dfrac{f^{(n)}(c)}{n!}$.

You need to do this:

- The Taylor series expansion of a function is the power series expansion of the function with coefficients discussed above and centered at the point $x = c$.

- The Taylor series of the function $f(x)$ at $x = c$ is given by:

$$f(x) = \sum_{n=0}^{\infty} \frac{f^{(n)}(c)}{n!}(x-c)^n = f(c) + \frac{f'(c)}{1!}(x-c) + \frac{f''(c)}{2!}(x-c)^2 + \frac{f'''(c)}{3!}(x-c)^3 + \dots$$

- For the case when $c = 0$, there is a special name for the Taylor series expansion; it is called the Maclaurin series.

- The Maclaurin series of the function $f(x)$ is given by:

$$f(x) = \sum_{n=0}^{\infty} \frac{f^{(n)}(0)}{n!}x^n = f(0) + \frac{f'(0)}{1!}x + \frac{f''(0)}{2!}x^2 + \frac{f'''(0)}{3!}x^3 + \dots$$

Answer and Explanation:

C

This deals with the basic formula for a Taylor series centered at $x = a$.

$$f(x) = \frac{f(a)}{0!}(x-a)^0 + \frac{f'(a)}{1!}(x-a)^1 + \frac{f''(a)}{2!}(x-a)^2 + \frac{f'''(a)}{3!}(x-a)^3 + \dots + \frac{f^{(n)}(a)}{n!}(x-a)^n + \dots$$

A Taylor polynomial has a finite number of terms. The series formula is truncated at that term.

$$T_3(4) = f(4) + \frac{f'(4)}{1!}(x-4)^1 + \frac{f''(4)}{2!}(x-4)^2 + \frac{f'''(4)}{3!}(x-4)^3$$

$$= 8 + \frac{-2}{1}(x-4) + \frac{-6}{2}(x-4)^2 + \frac{24}{6}(x-4)^3$$

$$= 8 - 2(x-4) - 3(x-4)^2 + 4(x-4)^3$$

This expression is a match for **(C)**.

PRACTICE SET

1. A function f has derivatives of all orders for all real numbers. It is known that $f(2) = 4$, $f'(2) = -3$, $f''(2) = 3$, and $f'''(2) = -1$. Which of the following statements are necessarily true?

   I. $f(1)$ can be determined exactly from the given information.

   II. The Taylor series for f at $x = 2$ converges to $f(x)$ for all values of x.

   III. The third-degree Taylor polynomial for f at $x = 2$ can be determined from the given information.

   (A) I only

   (B) II only

   (C) III only

   (D) I and II only

2. Let f be the function given by $f(x) = \cos x$. Which of the following is the second degree Taylor polynomial for f about $x = \frac{\pi}{3}$?

   (A) $\frac{\sqrt{3}}{2} - \frac{1}{2}\left(x + \frac{\pi}{3}\right) - \frac{\sqrt{3}}{4}\left(x + \frac{\pi}{3}\right)^2$

   (B) $\frac{1}{2} - \frac{\sqrt{3}}{2}\left(x + \frac{\pi}{3}\right) - \frac{1}{4}\left(x + \frac{\pi}{3}\right)^2$

   (C) $\frac{\sqrt{3}}{2} - \frac{1}{2}\left(x - \frac{\pi}{3}\right) - \frac{\sqrt{3}}{4}\left(x - \frac{\pi}{3}\right)^2$

   (D) $\frac{1}{2} - \frac{\sqrt{3}}{2}\left(x - \frac{\pi}{3}\right) - \frac{1}{4}\left(x - \frac{\pi}{3}\right)^2$

3. The third-order Taylor polynomial P for a function $f(x)$ about $x = 3$ is given by: $P(x) = 5 - 2(x - 3)^2 + 7(x - 3)^3$. Which of the following is true about $f(x)$?

   I. The point $(3, 5)$ is on the graph of $f(x)$.

   II. f is concave down at $x = 3$.

   III. f has a local maximum of 5 at $x = 3$.

   (A) I and III only

   (B) II and III only

   (C) I, II, and III

   (D) None of the above

4. The third-order Taylor polynomial for $f(x) = -\frac{2}{x}$ at $x = 2$ is given by

   (A) $-1 + \frac{1}{2}(x - 2) - \frac{1}{2}(x - 2)^2 + \frac{3}{4}(x - 2)^3$

   (B) $-1 - \frac{1}{2}(x - 2) - \frac{1}{2}(x - 2)^2 - \frac{3}{4}(x - 2)^3$

   (C) $1 - \frac{1}{2}(x - 2) + \frac{1}{4}(x - 2)^2 - \frac{1}{8}(x - 2)^3$

   (D) $-1 + \frac{1}{2}(x - 2) - \frac{1}{4}(x - 2)^2 + \frac{1}{8}(x - 2)^3$

ANSWERS AND EXPLANATIONS

1. C

Statement I is false because the information given is only concerning the values of the function and its derivatives at $x = 2$. Statement II is not necessarily true because no information about the remainder is given. Statement III is true because you have all the information needed to write the third-degree Taylor polynomial for f about $x = 2$.

Thus, **(C)** is the appropriate match.

2. D

The second-order Taylor polynomial for f at $x = \frac{\pi}{3}$ is given by:

$$P_2(x) = f\left(\frac{\pi}{3}\right) + f'\left(\frac{\pi}{3}\right)\left(x - \frac{\pi}{3}\right) + f''\left(\frac{\pi}{3}\right)\left(x - \frac{\pi}{3}\right)^2$$

$$= \cos\left(\frac{\pi}{3}\right) - \sin\left(\frac{\pi}{3}\right)\left(x - \frac{\pi}{3}\right) - \frac{\cos\left(\frac{\pi}{3}\right)}{2}\left(x - \frac{\pi}{3}\right)^2$$

$$= \frac{1}{2} - \frac{\sqrt{3}}{2}\left(x - \frac{\pi}{3}\right) - \frac{1}{4}\left(x - \frac{\pi}{3}\right)^2$$

That's **(D)**.

3. C

The third-order Taylor polynomial for f at $x = 3$ is given by:

$$f(3) + f'(3)(x - 3) + \frac{f''(3)}{2!}(x - 3)^2 + \frac{f'''(3)}{3!}(x - 3)^3$$

Thus:

$$P(x) = 5 - 2(x - 3)^2 + 7(x - 3)^3 = f(3) + f'(3)(x - 3) +$$

$$\frac{f''(3)}{2}(x - 3)^2 + \frac{f'''(3)}{6}(x - 3)^3$$

You know that $f(3) = 5$, and therefore you know that the function goes through the point $(3, 5)$.

There is no linear term in the Taylor polynomial, which tells you that $f'(3) = 0$ and therefore $(3, 5)$ must be a critical point.

$$\frac{f''(3)}{2} = -2 \Rightarrow f''(3) = -4,$$

which tells you that f is concave down at $x = 3$ and therefore $(3, 5)$ is a local maximum by the second derivative test.

Therefore, **(C)** is correct.

4. D

$$f(x) = -\frac{2}{x} \Rightarrow f(2) = -1$$

$$f'(x) = \frac{2}{x^2} \Rightarrow f'(2) = \frac{1}{2}$$

$$f''(x) = -\frac{4}{x^3} \Rightarrow f''(2) = -\frac{1}{2}, \frac{f''(2)}{2!} = -\frac{1}{4}$$

$$f'''(x) = \frac{12}{x^4} \Rightarrow f'''(2) = \frac{3}{4}, \frac{f'''(2)}{3!} = \frac{1}{8}$$

$$P_3(x) = -1 + \frac{1}{2}(x - 2) - \frac{1}{4}(x - 2)^2 + \frac{1}{8}(x - 2)^3$$

That's a perfect match for **(D)**!

18.3 MACLAURIN SERIES

To answer a question like this:

Which of the following is the correct Maclaurin series for $f(x) = \sin \pi x$?

(A) $\displaystyle\sum_{n=0}^{\infty} \frac{(-1)^{n+1} \pi^{2n} x^{2n+1}}{(2n+1)!}$

(B) $\displaystyle\sum_{n=0}^{\infty} \frac{(-1)^{n} \pi^{2n} x^{2n+1}}{(2n)!}$

(C) $\displaystyle\sum_{n=0}^{\infty} \frac{(-1)^{n} x^{2n+1}}{(2n+1)!}$

(D) $\displaystyle\sum_{n=0}^{\infty} \frac{(-1)^{n} \pi^{2n+1} x^{2n+1}}{(2n+1)!}$

You need to know this:

Maclaurin series for special functions

$\dfrac{1}{1-x}$	$\displaystyle\sum_{n=0}^{\infty} x^n$	$1 + x + x^2 + x^3 + x^4 + \cdots$
e^x	$\displaystyle\sum_{n=0}^{\infty} \frac{x^n}{n!}$	$1 + x + \dfrac{x}{2!} + \dfrac{x}{3!} + \dfrac{x}{4!} + \cdots$
$\sin x$	$\displaystyle\sum_{n=0}^{\infty} \frac{(-1)^n}{(2n+1)!} x^{2n+1}$	$x - \dfrac{x^3}{3!} - \dfrac{x^5}{5!} - \dfrac{x^7}{7!} + \cdots$
$\cos x$	$\displaystyle\sum_{n=0}^{\infty} \frac{(-1)^n}{(2n)!} x^{2n}$	$1 - \dfrac{x^2}{2!} + \dfrac{x^4}{4!} - \dfrac{x^6}{6!} + \cdots$

- The Maclaurin series of these special functions allow for a polynomial expression in place of trigonometric and exponential functions.

- Because polynomials are so easy to work with, substituting in the formal power series expansion for the actual function makes differentiating and integrating a lot simpler.

You need to do this:

- Recall the appropriate Maclaurin series for the special function given.

- New Maclaurin series for more complex trigonometric and exponential functions can be calculated by substituting for x in the special function's respective Maclaurin series.

- Larger functions that contain a special function within them can be calculated by replacing the special function with its respective Maclaurin series.

Answer and Explanation:

D

This question is probably easier than you think. It also illustrates the usefulness of power series. To make your life simple, recall that $\sin x = \displaystyle\sum_{n=0}^{\infty} \frac{(-1)^n x^{2n+1}}{(2n+1)!}$. Then to find $f(x) = \sin \pi x$, you simply plug πx in for x. You get

$$\sin \pi x = \sum_{n=0}^{\infty} \frac{(-1)^n (\pi x)^{2n+1}}{(2n+1)!} = \sum_{n=0}^{\infty} \frac{(-1)^n \pi^{2n+1} x^{2n+1}}{(2n+1)!}$$

That matches **(D)**.

PRACTICE SET

1. $f(0) = 4$, $f'(0) = 2$, $f''(0) = 1$, $f^3(0) = -3$
    What are the first four non-zero terms of the Maclaurin series of f?

    (A) $4 + 2x + \dfrac{x^2}{2} - \dfrac{x^3}{2}$

    (B) $4 - 2x + x^2 - x^3$

    (C) $2 + 4x + \dfrac{x^2}{2} - \dfrac{x^3}{2}$

    (D) $-3 + 2x - x^2 + 4x^3$

2. Using $\dfrac{1}{1+x} = 1 - x + x^2 - x^3 + \dots$ find the Maclaurin series for the function $\dfrac{1}{2+x}$.

    (A) $1 - \dfrac{x}{2} + \dfrac{x^2}{4} - \dfrac{x^3}{8} + \dots$

    (B) $\dfrac{1}{2} - \dfrac{x}{2} + \dfrac{x^2}{2} - \dfrac{x^3}{2} + \dots$

    (C) $\dfrac{1}{2} - \dfrac{x}{4} + \dfrac{x^2}{8} - \dfrac{x^3}{16} + \dots$

    (D) $\dfrac{1}{4} + \dfrac{x}{8} + \dfrac{x^2}{16} + \dfrac{x^3}{32} + \dots$

3. Calculate the first 3 non-zero terms in the Maclaurin series for xe^{-x}.

    (A) $2x + 3x^2 - \dfrac{x^3}{4}$

    (B) $x - x^2 + \dfrac{x^3}{2}$

    (C) $x - 2x + 3x^2$

    (D) $3 + 4x - \dfrac{2x^2}{3} + \dfrac{3x^3}{5}$

4. Calculate the first 3 non-zero terms in the Maclaurin series for $\sin^2 x$.

    (A) $2x^2 + 4x^3 - x^5$

    (B) $1 + x - 2x^2$

    (C) $x^2 - \dfrac{x^4}{3} + \dfrac{2x^6}{45}$

    (D) $-x^2 + \dfrac{x^4}{2} - \dfrac{x^6}{35}$

ANSWERS AND EXPLANATIONS

1. A

The formula for the Taylor series is $\frac{f(c)}{0!} + \frac{xf'(c)}{1!} + \frac{x^2 f''(c)}{2!} + \frac{x^3 f^3(c)}{3!}$. Notice that with Maclaurin series the center is at 0 so c will always be 0, but this will change sometimes if you're working with Taylor polynomials. Everything was given to you so plug into the formula to get

$$\frac{f(0)}{0!} + \frac{x(2)}{1!} + \frac{x^2(1)}{2!} + \frac{x^3(-3)}{3!}$$

$$4 + 2x + \frac{x^2}{2} - \frac{x^3}{2}$$

Therefore, **(A)** is correct.

2. C

First get your given function $\frac{1}{2+x}$ in the form of $\frac{1}{1+x}$.
Then substitute in to get your series.

$$\frac{1}{2+x} = \frac{1}{2}\left(\frac{1}{1+\frac{x}{2}}\right)$$

$$= \frac{1}{2}\left(1 - \frac{x}{2} + \frac{x^2}{4} - \frac{x^3}{8} + \dots\right)$$

$$= \frac{1}{2} - \frac{x}{4} + \frac{x^2}{8} - \frac{x^3}{16}$$

Thus, **(C)** is the correct answer.

3. B

For the Maclaurin series, derive the first 3 derivatives. Because the center is at 0 that will be your c.

$$f(x) = xe^{-x} \Rightarrow$$
$$f(0) = 0$$
$$f'(x) = (1)e^{-x} + x(-e^{-x}) = e^{-x} - xe^{-x} \Rightarrow$$
$$f'(0) = 1$$
$$f''(x) = -e^{-x} - \left[(1)e^{-x} + x(-e^{-x})\right] = -2e^{-x} + xe^{-x} \Rightarrow$$
$$f''(0) = -2$$
$$f^3(x) = 2e^{-x} + (1)e^{-x} + x(-e^{-x}) = 3e^{-x} - xe^{-x} \Rightarrow$$
$$f^3(0) = 3$$

Now you have everything you need, so plug into the equation

$\frac{f(0)}{0!} + \frac{xf'(0)}{1!} + \frac{x^2 f''(0)}{2!} + \frac{x^3 f^3(0)}{3!}$ and simplify.

$$\frac{(0)}{0!} + \frac{x(1)}{1!} + \frac{x^2(-2)}{2!} + \frac{x^3(3)}{3!}$$

$$x - x^2 + \frac{x^3}{2}$$

Therefore, **(B)** is the appropriate match.

4. C

Calculate derivatives until you get 3 non-zero terms; in this case it will be 6 derivatives. Because the center is at zero, find each derivative at that point also. $f(0)$ is the easiest to find, which is just zero.

$$f'(x) = 2\sin x \cos x = \sin 2x \quad \Rightarrow \quad f'(0) = 0$$
$$f''(x) = 2\cos 2x \quad \Rightarrow \quad f''(0) = 2$$
$$f^3(x) = -4\sin 2x \quad \Rightarrow \quad f^3(0) = 0$$
$$f^4(x) = -8\cos 2x \quad \Rightarrow \quad f^4(0) = -8$$
$$f^5(x) = 16\sin 2x \quad \Rightarrow \quad f^5(0) = 0$$
$$f^6(x) = 32\cos 2x \quad \Rightarrow \quad f^6(0) = 32$$

Plug everything into the Maclaurin formula and simplify. Notice that every derivative at 0 that equals 0 will cancel out the term, so there is no worry about those terms.

$$\frac{x^2(2)}{2!} + \frac{x^4(-8)}{4!} + \frac{x^6(32)}{6!}$$

$$x^2 - \frac{x^4}{3} + \frac{2x^6}{45}$$

That's **(C)**.

18.4 LAGRANGE ERROR BOUND

To answer a question like this:

Given that $\cos x = 1 - \dfrac{x^2}{2!} +, \dfrac{x^4}{4!}$ what is the accuracy for $\cos(2)$?

(A) $\dfrac{4 \sin(2)}{15}$

(B) $\dfrac{2 \cos(2)}{15}$

(C) $\dfrac{-\sin(2)}{20}$

(D) $\dfrac{-\cos(2)}{15}$

You need to know this:

Inherent error of a Taylor polynomial

- As stated at the beginning of this chapter, the Taylor series expansion provides a polynomial approximation of the given function.

- As more and more terms are added to a Taylor series, the resulting sum function becomes closer and closer to the actual function.

- Taking the first K elements of a power series gives a corresponding *finite* partial sum, which is called $f_K(x)$. The limit as K approaches infinity is the actual function.

- However, notice that when using a sum of a finite number of terms, there is always some degree of error with the actual function being approximated.

Remainder terms

- The remainder or error of the Kth finite Taylor series, $R_K(x)$, is equal to $f(x) - f_K(x)$.

- If the remainder of the finite Taylor series of a given function goes to zero as K gets larger and larger, then the function is equal to its full Taylor series expansion on the interval where it is defined:

$$\lim_{K \to \infty} R_K(x) = 0 \text{ for } |x - a| < R$$

$$\Rightarrow f(x) = \sum_{n=0}^{\infty} \frac{f^{(n)}(a)}{n!}(x - a)^n$$

for all x such that $|x - a| < R$

You need to do this:

- To get a bound on the actual remainder term, use a theorem much like the one for infinite series.

- Suppose $|f^{(K+1)}(x)| \leq M$ when $|x - a| \leq R$. Then the remainder, $R_K(x)$, satisfies

$$R_K(x) \leq \frac{M}{(K+1)!}|x - a|^{K+1} \text{ for all } x \text{ such that } |x - a| \leq R.$$

- In short, this theorem says that if there is an upper bound on the $(K+1)$th derivative of $f(x)$, then there is also an upper bound on the error term of the Kth finite Taylor series of $f(x)$.

Answer and Explanation:

A

First realize what the question is asking. Accuracy means how close your answer would be to the real answer so this is asking for the error. For the Lagrange error bound, you need to know how many derivatives you'll need or the order of the polynomial being used. With the given equation the last term has an x^4, which means the polynomial is to the 4th degree, so $n = 4$. You'll have to compute 5 derivatives and plug into the Lagrange error bound equation $|R_n x| \leq \left| \dfrac{f^{(n+1)}(x)x^{(n+1)}}{(n+1)!} \right|$.

Notice the absolute values; this means you don't need to worry about the negative signs when doing the derivatives of the cosine function. The derivatives will flip back and forth between sine and cosine so just count 5 times (derivatives), which lands on sine.

$$|R_4(2)| \leq \frac{\sin(2)2^5}{5!} = \frac{4\sin(2)}{15}$$

That's a match for **(A)**.

PRACTICE SET

1. Calculate the Lagrange error bound in approximating $e^{-0.5}$ using the Maclaurin polynomial of degree 4.

   (A) $\dfrac{(0.5)^5}{120}$

   (B) $\dfrac{(0.4)^3}{125}$

   (C) $\dfrac{(0.4)^4}{25}$

   (D) $\dfrac{(0.5)^3}{45}$

2. Calculate the Lagrange error bound to estimate the error in using a 4th degree Maclaurin polynomial to approximate $\sin\left(\dfrac{\pi}{4}\right)$.

   (A) $\dfrac{\left(\frac{\pi}{4}\right)^5}{120}$

   (B) $\dfrac{\left(\frac{\pi}{4}\right)^2}{32}$

   (C) $\left(\dfrac{\pi}{4}\right)^3$

   (D) $\cos\left(\dfrac{\pi}{4}\right)$

 3. Estimating $e^{1.45}$ using a Taylor polynomial about $x = 2$, what is the least degree of polynomial that assures an error smaller than 0.001 ?

   (A) $n = 3$

   (B) $n = 4$

   (C) $n = 5$

   (D) $n = 6$

4. Calculate the error bound for the 2nd degree Taylor polynomial centered around $x = 2$ for $(1.5)^4$.

   (A) 0.02

   (B) 0.4

   (C) 1

   (D) 3

BC Topics

ANSWERS AND EXPLANATIONS

1. A

Because you are approximating $e^{-0.5}$, $f(x) = e^x$. You were given that $n = 4$ so you will need to find $f^{(5)}x$. Luckily, e^x repeats infinitely for its derivatives so you already know that the 5th derivative is just e^x. Now just plug into the Lagrange error bound equation $\left|R_n(x)\right| \leq \left|\dfrac{Mx^{n+1}}{(n+1)!}\right|$. Because this is a Maclaurin polynomial it is centered around 0 so $M = e^{(0)}$. Plug everything into the formula and simplify.

$$\left|R_4(x)\right| \leq \left|\frac{Mx^{4+1}}{(4+1)!}\right| = \left|\frac{(e^0)(-0.5)^5}{5!}\right| = \frac{(0.5)^5}{120}$$

That's **(A)**.

2. A

It is very important to note that most sine and cosine error problems have M as the amplitude for maximum error; in this case it would be 1. You technically could have the $n + 1$ derivative at $\frac{\pi}{4}$ be equal to M, but this will not give you maximum error. Read the question very carefully; being that it is a 4th degree Maclaurin polynomial, $n = 4$ and $c = 0$. Now that you have everything needed, plug into the equation $\left|R_n(x)\right| \leq \left|\dfrac{Mx^{n+1}}{(n+1)!}\right|$.

$$\left|R_4(x)\right| \leq \left|\frac{Mx^{4+1}}{(4+1)!}\right| = \frac{(1)\left(\frac{\pi}{4}\right)^5}{5!} = \frac{\left(\frac{\pi}{4}\right)^5}{120}$$

Therefore, **(A)** is correct.

3. C

The Taylor polynomial is about $x = 2$, which means $c = 2$. Now figure out your M, which is an interval containing x and c. For this problem the best would be from $[0, 2]$, which means the max error would fall on 2, as the function is an increasing function ($M = e^2$). Finally, plug everything into the Lagrange error bound formula $\left|R_n(x)\right| \leq \left|\dfrac{M(x-c)^{n+1}}{(n+1)!}\right|$ and set this to less than 0.001.

$$\left|R_n(x)\right| \leq \left|\frac{M(x-c)^{n+1}}{(n+1)!}\right| = \left|\frac{e^2(1.45-2)^{n+1}}{(n+1)!}\right| < 0.001$$

To solve for n, get all terms including n on the left and everything else to the right of the inequality. Do not forget about the absolute value, so:

$$\frac{e^2(0.55)^{n+1}}{(n+1)!} \leq 0.001 \Rightarrow \frac{(0.55)^{n+1}}{(n+1)!} < \frac{0.001}{e^2}$$

Now plug in numbers for n and see if it is less than $\dfrac{0.001}{e^2}$. You will find that the smallest n is 5. Hence, **(C)** is the correct answer.

4. C

Because $f^{(n+1)}(x) \leq M$ and $n = 2$, $f^3(x) \leq M$. This means that you will need to compute the third derivative of x^4.

$$f(x) = x^4$$
$$f'(x) = 4x^3$$
$$f^{(2)}(x) = 12x^2$$
$$f^{(3)}(x) = 24x$$

Realize that the max error will be at 2 for M because the interval is $[1.5, 2]$ and that $c = 2$. Now plug into the formula $\left|R_n(x)\right| \leq \left|\dfrac{M(x-c)^{n+1}}{(n+1)!}\right|$ and simplify.

$$\left|R_n(x)\right| \leq \left|\frac{M(x-c)^{n+1}}{(n+1)!}\right| = \left|\frac{[24(2)](1.5-2)^3}{3!}\right| = \frac{48(0.5)^3}{6} = 1$$

That's **(C)**.

TEST WHAT YOU LEARNED

1. The Maclaurin series for $f(x) = x^2 e^{-2x^3}$ is

   (A) $\displaystyle\sum_{n=0}^{\infty} \frac{x^{3n+2}}{n!}$

   (B) $\displaystyle\sum_{n=0}^{\infty} \frac{(-2)^n \cdot x^{3n+2}}{n!}$

   (C) $\displaystyle\sum_{n=0}^{\infty} \frac{(2)^n \cdot x^{3n+2}}{n!}$

   (D) $\displaystyle\sum_{n=0}^{\infty} \frac{(-2)^n \cdot x^{3n-2}}{n!}$

2. Which of the following is the Taylor series for $f(x) = \ln x$ centered at $a = 3$?

   (A) $\ln 3 + \displaystyle\sum_{n=1}^{\infty} \frac{(-1)^{n+1}}{3^n n}(x-3)^n$

   (B) $\displaystyle\sum_{n=1}^{\infty} \frac{(x-3)^n}{3^n n!}$

   (C) $\displaystyle\sum_{n=1}^{\infty} \frac{(-1)^n}{n}(x-3)^n$

   (D) $\ln 3 + \displaystyle\sum_{n=1}^{\infty} \frac{(-1)^{n+1}}{3^n n!}(x-3)^n$

3. What is the error bound for the 3rd degree Taylor polynomial of $f(x) = 2\sin(x)$ centered at $x = 0$ on $[0, 2\pi]$?

(A) $-\dfrac{5\pi^4}{7}$

(B) $\dfrac{2\pi}{3}$

(C) $\dfrac{\pi^3}{6}$

(D) $\dfrac{4\pi^4}{3}$

4. The radius of convergence of the power series $\displaystyle\sum_{n=0}^{\infty} \dfrac{n \cdot x^{2n+3}}{x^2 \cdot n!}$ is

(A) 2

(B) 3

(C) 4

(D) ∞

THIS PAGE LEFT INTENTIONALLY BLANK.
Answers and explanations to the practice set are located on the next page.

Answer Key

Test What You Already Know	Test What You Learned
1. D	1. B
2. B	2. A
3. D	3. C
4. A	4. D

REFLECTION

Test What You Already Know score: _____

Test What You Learned score: _____

Use this section to evaluate your progress. After working through the pre-quiz, check off the boxes in the "Pre" column to indicate which Learning Objectives you feel confident about. Then, after completing the chapter, including the post-quiz, do the same to the boxes in the "Post" column. Keep working on unchecked Objectives until you're confident about them all!

Pre	Post		
☐	☐	**18.1**	Find the radius and interval of convergence of functions defined by power series
☐	☐	**18.2**	Use Taylor polynomial expansion to approximate non-polynomial functions
☐	☐	**18.3**	Apply the Maclaurin series of special functions to more complex functions
☐	☐	**18.4**	Find the Lagrange error bound for Taylor polynomials

FOR MORE PRACTICE

Complete more practice online at kaptest.com. Haven't registered your book yet? Go to kaptest.com/booksonline to begin.

ANSWERS AND EXPLANATIONS

Test What You Already Know

1. D

For this question you want to use a geometric series, which states that $\sum_{n=0}^{\infty} x^n = \frac{1}{1-x}$, provided $|x| < 1$. Rewrite $\frac{1}{1+4x^2}$ so that it resembles the right-hand side. Note that you are given $|x| < \frac{1}{2}$, so the geometric series is convergent. You have

$$\frac{1}{1+4x^2} = \frac{1}{1-(-4x^2)} = \sum_{n=0}^{\infty}(-4x^2)^n = \sum_{n=0}^{\infty}(-1)^n 4^n (x^2)^n$$

$$= \sum_{n=0}^{\infty}(-1)^n 4^n x^{2n}$$

That's **(D)**.

2. B

You are expected to know the Taylor series expansion formulas for seven basic functions: sine, cosine, e^x, $\frac{1}{1-x}$, $\frac{1}{1+x}$, ln x, and arctan x.

This is the power series expansion for $\frac{\arctan x}{x}$. Thus, **(B)** is the appropriate match.

3. D

When dealing with series, try to determine the type of series shown. Then you can apply the tools associated with that series type.

This is a geometric series with first term of 1 and a common ratio of $\frac{-2}{3}$.

$$f(2) = \sum_{n=0}^{\infty}\left(\frac{-2}{3}\right)^n$$

$$S = \frac{a_0}{1-r} = \frac{1}{1-\frac{-2}{3}} = \frac{3}{5}$$

That's **(D)**.

4. A

The Lagrange Error Bound in using a third-order polynomial centered at c (in this case 3) is given by: $\frac{M}{4!}|x-c|^4$, where M is the maximum possible value of the fourth derivative of f on the interval (which is given as 0.7), and x is the value at which you are trying to estimate (in this case 3.6).

Thus, the Lagrange Error Bound is: $\frac{0.7}{4!}|3.6-3|^4 = 0.00378$. Therefore, **(A)** is correct.

Test What You Learned

1. B

You could apply the formula for Taylor series to this polynomial and get the answer, but this would require a great deal of work. You can simplify the process considerably by recalling that the Maclaurin series for $f(x) = e^x$ is:

$$e^x = 1 + x + \frac{x^2}{2!} + \frac{x^3}{3!} + \cdots = \sum_{n=0}^{\infty}\frac{x^n}{n!}$$

Then:

$$x^2 e^{-2x^3} = x^2\left(\sum_{n=0}^{\infty}\frac{(-2x^3)^n}{n!}\right)$$

$$= x^2\left(\sum_{n=0}^{\infty}\frac{(-2)^n \cdot x^{3n}}{n!}\right)$$

$$= \sum_{n=0}^{\infty}\frac{(-2)^n \cdot x^{3n+2}}{n!}$$

That's a match for **(B)**.

2. A

To compute the Taylor series of $f(x)$ at the point $x = a$, use the formula $\sum_{n=0}^{\infty} \frac{f^{(n)}(a)}{n!}(x-a)^n$. So calculate some derivatives of ln x.

	so	
$f(x) = \ln x,$		$f(3) = \ln 3$
$f'(x) = \dfrac{1}{x},$		$f'(3) = \dfrac{1}{3}$
$f''(x) = \dfrac{-1}{x^2},$		$f''(3) = \dfrac{-1}{3^2}$
$f'''(x) = \dfrac{2}{x^3},$		$f'''(3) = \dfrac{2}{3^3}$
......		......
$f^{(n)}(x) = \dfrac{(-1)^{n+1}(n-1)!}{x^n},$		$f^{(n)}(3) = \dfrac{(-1)^{n+1}(n-1)!}{3^n}$

So you get

$$\sum_{n=0}^{\infty} \frac{f^{(n)}(3)}{n!}(x-3)^n = f(3) + \sum_{n=1}^{\infty} \frac{\frac{(-1)^{n+1}(n-1)!}{3^n}}{n!}(x-3)^n$$

$$= \ln 3 + \sum_{n=1}^{\infty} \frac{(-1)^{n+1}}{3^n n}(x-3)^n$$

for the Taylor series at $x = 3$. That's **(A)**.

3. C

When dealing with sine and cosine, M will almost always be the amplitude of the wave function; hence, $M = 2$. Use the Lagrange error bound equation $|R_n(x)| \leq \frac{Mx^{n+1}}{(n+1)!}$ $|R_n(x)| \leq \frac{Mx^{n+1}}{(n+1)!}$ with n being the degree of the polynomial, in this case $n = 3$. On the interval given 2π would be the max so use that as your x.

$$|R_3(2\pi)| \leq \frac{(2)(2\pi)^4}{4!} = \frac{4\pi^4}{3}$$

Thus, **(C)** is correct.

4. D

You can begin by simplifying the power series:

$$\sum_{n=0}^{\infty} \frac{n \cdot x^{2n+3}}{x^2 \cdot n!} = \sum_{n=0}^{\infty} \frac{n \cdot x^{2n+3-2}}{n \cdot (n-1)!} = \sum_{n=0}^{\infty} \frac{x^{2n+1}}{(n-1)!}$$

You can now check for convergence by using the ratio test:

$$\lim_{n \to \infty} \left| \frac{A_{n+1}}{A_n} \right| = \lim_{n \to \infty} \left| \frac{\frac{x^{2(n+1)+1}}{((n+1)-1)!}}{\frac{x^{2n+1}}{(n-1)!}} \right|$$

$$= \lim_{n \to \infty} \left| \frac{x^{2n+3}}{n!} \cdot \frac{(n-1)!}{x^{2n+1}} \right|$$

$$= \lim_{n \to \infty} \left| \frac{x^2}{n \cdot (n-1)!} \cdot \frac{(n-1)!}{1} \right|$$

$$= \lim_{n \to \infty} \frac{|x^2|}{n} = 0$$

$$\lim_{n \to \infty} \left| \frac{A_{n+1}}{A_n} \right| = 0$$

Therefore, the series converges for all values of x, and the radius of convergence is infinity, matching **(D)**.

Graphing Calculators and Free-Response Questions

CHAPTER 19

Problems that Require Graphing Calculators

LEARNING OBJECTIVES

19.1 Plot the graph of a function within a given viewing window and find the zeros and/or relative extrema

19.2 Calculate the derivative of a function numerically

19.3 Calculate the value of a definite integral numerically

19.1 GRAPHING FUNCTIONS AND FINDING CRITICAL POINTS

To answer a question like this:

Find the minimum value of the function $f(x) = \ln(x) + \sin(x)$ on the closed interval $\left[\dfrac{\pi}{4}, \pi\right]$.

(A) 0   (B) 0.466   (C) 0.682   (D) 1.145

You need to know this:

Using a graph utility to evaluate a function

- Often there is a need to determine the value of a function at a specific point while working a problem. For example, evaluate the function $f(x) = x^2 + 1$ at $x = 3$.

- One way to approach this task is to ignore the powerful capabilities of the calculator and use just the regular, non-algebraic computing power of your calculator (or, for that matter, your brain). Type $3^2 + 1$ [ENTER] on the main screen. The answer 10 appears.

- However, there are fancier techniques that involve entering the function into the [Y =] screen. **TI-83:** On the [Y =] screen enter the function $Y_1 = x^2 + 1$.

- **TI-83:** View the graph of the function in a reasonable window, say $[0, 4] \times [0, 20]$. On the [CALC] ([2nd] [TRACE]) menu, choose value by pressing 1 or [ENTER]. You will be prompted for an x-value. Type 3 [ENTER]. The answer $y = 10$ appears at the bottom of the screen.

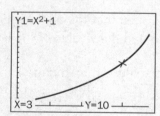

Other graphing calculator methods such as the Y-VARS technique, table technique, nderiv, and fnInt are covered in the PDF *More Advanced Calculator Techniques*, available in your Online Resources.

You need to do this:

- The intermediate value theorem states that a continuous function attains a maximum and a minimum value on a closed interval.

- Graph the function on the window of the given interval.

- Evaluate the graph's endpoints and any other potential critical points for the minimum.

- Evaluate the function for the associated *y*-value of the minimum.

> ✔ **AP Expert Note**
>
> All graphing calculators approved for use on the AP exam are able to evaluate a derivative at a point and compute a definite integral; make sure you are familiar with these utilities on your calculator.

Answer and Explanation:

B

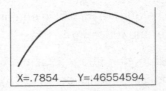

X=.7854 ___ Y=.46554594

TI-83: On the $\boxed{Y=}$ screen enter $Y_1 = \ln(x) + \sin(x)$ and view the graph in the window $\left[\frac{\pi}{4}, \pi\right] \times [0, 2]$. From looking at the graph, it can be seen that the minimum value of this function occurs at the left endpoint. Use the value utility to find the *y*-value of the left endpoint.

On the $\boxed{\text{CALC}}$ ($\boxed{2^{\text{nd}}}$ $\boxed{\text{TRACE}}$) menu, scroll down to **1:value** and press $\boxed{\text{ENTER}}$. When prompted for an *x*-value, enter $\frac{\pi}{4}$ or .7854 $\left(\frac{\pi}{4} \approx .7854\right)$. Press $\boxed{\text{ENTER}}$. The *y*-value .4655…appears at the bottom of the screen, corresponding to **(B)**.

PRACTICE SET

 1. Find the maximum value of the function $f(x) = \ln(x) + \sin(x)$ on the closed interval $\left[\frac{\pi}{4}, \pi\right]$.

   (A) 0.466

   (B) 1.145

   (C) 1.588

   (D) 1.605

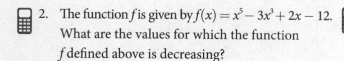 2. The function f is given by $f(x) = x^5 - 3x^3 + 2x - 12$. What are the values for which the function f defined above is decreasing?

   (A) $x < 1.782$

   (B) $-0.510 < x < 0.510$

   (C) $-1.241 < x < -0.510$ and $0.510 < x < 1.241$

   (D) $x < -1.241$, $-0.510 < x < 0.510$, and $x > 1.241$

 3. The derivative of the function f is given by $f'(x) = x + \sin(x^2) - 2$. At what value(s) of x does f have a relative maximum on the interval $0 < x < 4$?

   (A) 1.080

   (B) 1.677

   (C) 2.419

   (D) 1.080 and 2.419

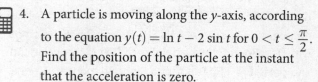 4. A particle is moving along the y-axis, according to the equation $y(t) = \ln t - 2\sin t$ for $0 < t \le \frac{\pi}{2}$. Find the position of the particle at the instant that the acceleration is zero.

   (A) -1.661

   (B) 1.277

   (C) 8.415

   (D) 9.817

ANSWERS AND EXPLANATIONS

1. D

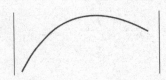

TI-83: Begin with the same setup as in the previous question. On the $\boxed{Y=}$ screen enter $Y_1 = \ln(x) + \sin(x)$ and view the graph in the window $\left[\frac{\pi}{4}, \pi\right] \times [0, 2]$. From looking at the graph it can be seen that the maximum value of this function occurs around the midpoint of the window. On the $\boxed{\text{CALC}}$ ($\boxed{2^{nd}}$ $\boxed{\text{TRACE}}$) menu, scroll down to **4:maximum** and press $\boxed{\text{ENTER}}$. The screen will prompt with "Left Bound?" Pick a point to the left of the maximum and press $\boxed{\text{ENTER}}$. Then set a Right Bound with a point to the right of the maximum and press $\boxed{\text{ENTER}}$. Last, the screen will prompt "Guess?" Select a point at approximately the maximum and press $\boxed{\text{ENTER}}$. The y-value 1.6055213 appears at the bottom of the screen, making **(D)** correct.

2. C

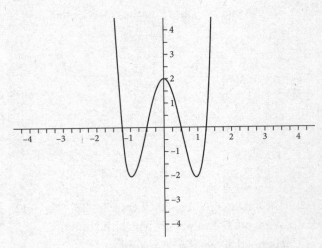

The function f is decreasing for all x such that $f'(x) < 0$. Use the power rule to take the derivative of each term. The result is $f'(x) = 5x^4 - 9x^2 + 2$. This equation is not easily factored, so use your calculator to find the zeros. After inputting the equation for the derivative on the $\boxed{Y=}$ screen, go to the $\boxed{\text{CALC}}$ ($\boxed{2^{nd}}$ $\boxed{\text{TRACE}}$) menu, scroll down to **2:zero** and press $\boxed{\text{ENTER}}$, set a left bound, right bound, and guess. Repeat this process for each of the four zeros and then examine the graph to determine where it is less than zero (below the x-axis). Based on the graph, $f'(x) < 0$ when $-1.241 < x < -0.510$ and when $0.510 < x < 1.241$. **(C)** is correct.

3. B

Relative extrema may occur when the first derivative of a function is equal to 0 or where it does not exist. Here, when given f' (which happens to be defined everywhere), the first task is to determine where $f' = 0$. Do this on your calculator by graphing the equation and using the Calculate Zero function. The interval is given from $0 < x < 4$, so restrict the window accordingly—this makes it much easier to see what's going on. The graph looks something like this:

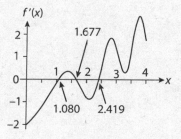

A relative maximum occurs where a function changes from increasing (when f' is positive, or above the x-axis) to decreasing (when f' is negative, or below the x-axis). Based on the graph, this happens at $x = 1.677$, so **(B)** is correct.

4. A

Start by graphing the original function. Sometimes just seeing the function is enough to get the answer, and this is one of those cases. A look at the graph shows that it resembles a straight line around -1.7. Because the slope of a straight line is constant, acceleration equals 0 at this point.

Alternatively, use the graphing utility to find the zero of the graph. Enter the function $Y_1 = -\dfrac{1}{t^2} + 2\sin t$ on the $\boxed{Y =}$ screen. View the graph in an appropriate window. Use the zero utility on the $\boxed{CALC}$ ($\boxed{2^{nd}}$ $\boxed{TRACE}$) menu to find the coordinates of the zero.

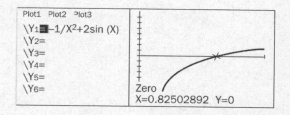

Now immediately press $\boxed{QUIT}$ ($\boxed{2^{nd}}$ $\boxed{MODE}$) to return to the main screen. Enter ln $(x) - 2\sin x$, press $\boxed{ENTER}$. The calculator evaluates this function at the x-value of the zero you just found, yielding -1.661. Thus, **(A)** is correct.

Graphing Calculator

19.2 FINDING A DERIVATIVE AT A POINT

To answer a question like this:

 If $f(x) = \sqrt{e^x + \tan x}$, then $f'(1) =$

(A) 1.486   (B) 1.498   (C) 2.068   (D) 2.168

You need to know this:

Evaluating a derivative at a point using nDeriv

- **TI-83:** The utility to evaluate a derivative at a point is called **nDeriv**, and it is found on the MATH menu. Press 8 or scroll with the arrow keys to find it. The syntax for the nDeriv command is nDeriv(*function, variable, value*).

- For example, to evaluate the derivative of $f(x) = x^2 + $ ˙ at $x = 3$ go to the nDeriv utility by pressing MATH 8. The prompt "nDeriv" appears.

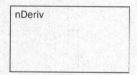

- Enter $(X^2 + 1,X,3)$ ENTER. The answer 6 appears.

Other graphing calculator methods for solving a derivative at a point such as the Y-VARS technique and graphing utility method are covered in the PDF *More Advanced Calculator Techniques*, available in your Online Resources.

You need to do this:

- All of the graphing calculators allowed on the exam will evaluate a derivative at a point.

- Check your owner's manual to see exactly where to find this utility on your calculator.

- To use this utility, tell the calculator the function, the variable, and the point at which the derivative is to be evaluated.

> ✔ **AP Expert Note**
>
> Before calculating the function, make sure your calculator is in radian mode.

Answer and Explanation:

A

Use the calculator's utility for evaluating a derivative at a point.

TI-83 (left)/TI-89 (right):

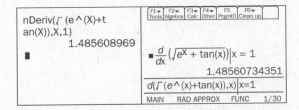

Notice that the syntax is different for the TI-83 and the TI-89. If you forget to include the vertical "when" bar, the TI-89 will return an expression rather than a number.

The provided answer is $f'(1) \approx 1.486$, matching **(A)**.

PRACTICE SET

 1. Let $f(x) = \cos(\ln x^2)$. What is the value of $f'(\pi)$, accurate to three decimal places?

    (A) -2.351

    (B) -0.479

    (C) 0.694

    (D) 1.863

 2. If $f(x) = \tan^{-1}(\cos(4x)) + 2$, then $f'\left(\frac{\pi}{3}\right) =$

    (A) -1.564

    (B) -0.428

    (C) 1.395

    (D) 2.771

 3. What is the slope of the curve $y = 4^{\cos x} - e$ at its first positive x-intercept?

    (A) -2.610

    (B) -1.884

    (C) 2.348

    (D) 3.195

 4. The position of a swinging pendulum t seconds after its release is defined by the function $f(t) = \dfrac{5\cos(t)}{\log(t+5)}$. What is the instantaneous velocity of the pendulum at $t = 5$?

    (A) -6.489

    (B) -2.519

    (C) 4.733

    (D) 8.298

ANSWERS AND EXPLANATIONS

1. B

This question is a prime example of when it is much faster to solve for a derivative at a point using the nDeriv utility. Press MATH 8, enter $(\cos(\ln(X^2)),X,\pi)$ and get -0.479, which is **(B)**.

2. D

This is another problem that the graphing calculator can solve quickly with the nDeriv utility. Press MATH 8, enter $(\tan^{-1}(\cos(4X)),X,\pi/3)$ and get 2.771, matching **(D)**.

3. A

This slope question has a slight twist, but it can be solved exactly as the previous two questions. You just don't know the value of x at which you need to evaluate the derivative. The equation is not easily solvable, so use the Calculate Zero function on your calculator to find the first positive x-intercept of the graph, which is $x = 0.76505$. To find the slope at this value of x, find the value of the derivative at $x = 0.76505$ using the nDeriv feature of your calculator. This gives y' at $x = 0.76505$ is -2.610, which is **(A)**. It is a good idea to use many decimal places when rounding in the middle steps of a problem because your final answer will need to be correct to 3 decimal places. Use the Store function on your calculator to save all nine or more decimal places.

4. C

Recall from Chapter 9.4 that the instantaneous rate of change of a function at a point is the slope of the tangent line at that point (or the derivative evaluated at that point). Because velocity is the derivative of position, this question is still just asking for the derivative at a point. This question can be answered with the same approach as those above, the nDeriv utility. Evaluating the function $f(5)$ using nDeriv yields 4.733, which is **(C)**.

19.3 EVALUATING A DEFINITE INTEGRAL

To answer a question like this:

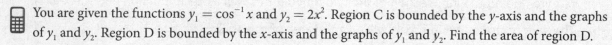 You are given the functions $y_1 = \cos^{-1} x$ and $y_2 = 2x^2$. Region C is bounded by the y-axis and the graphs of y_1 and y_2. Region D is bounded by the x-axis and the graphs of y_1 and y_2. Find the area of region D.

(A) 0.269   (B) 0.326   (C) 0.382   (D) 0.461

You need to know this:

Evaluating a definite integral using fnInt

- **TI-83:** The utility for computing a definite integral is called **fnInt**, and it is found on the MATH menu. Press 9 or scroll with the arrow keys to find it. The syntax for the fnInt command is fnint(*function,variable,lower,upper*).

- For example, to compute the definite integral of $\int_{0}^{4} x^2 + 1 \, dx$, go to the fnInt utility by pressing MATH 9. The prompt "fnInt" appears.

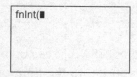

- Enter $(x^2 + 1, x,0,4)$ ENTER. The answer 25.33333 appears.

Other graphing calculator methods for computing a definite integral such as the Y-VARS technique and graphing utility method are covered in the PDF *More Advanced Calculator Techniques*, available in your Online Resources.

You need to do this:

- Graph the functions. Find the appropriate boundaries of region D.

- Integrate each function over the interval that it lies as a boundary of region D (Note: this will require finding the intersection of the two functions as well as each function and the x-axis).

- The sum of these two integrals will give the area of region D.

Answer and Explanation:

C

Your first task is to determine the graphical relationship between the two curves, $y_1 = \cos^{-1}x$ and $y_2 = 2x^2$. Graph these functions on the same set of axes, using the window $[-.2, 1.5] \times [-.2, 2]$:

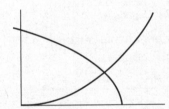

TI-83: After graphing the functions, press TRACE. The curve $y_1 = \cos^{-1}x$ lies on top to the left of the point of intersection; $y_2 = 2x^2$ lies on top to the right of the point of intersection. Therefore, the left curve bounding region D is $y_2 = 2x^2$ and the right curve bounding region D is $y_1 = \cos^{-1}x$.

The curve $y_2 = 2x^2$ intersects the x-axis at $x = 0$ and the curve $y_1 = \cos^{-1}x$ intersects the x-axis at $x = \cos 0 = 1$. Therefore, the area in question is found by computing the value of

$$Area_D = \int_0^A 2x^2 \, dx + \int_A^1 \cos^{-1} x \, dx,$$

where $A = 0.65463568$, the intersection point of the two curves (this can be found by pressing CALC (2nd TRACE) 5 (for the intersect option).

To evaluate the integral on the calculator, pull up the fnInt option on the MATH menu. Enter each integral with the appropriate syntax and sum the two values together. On the **TI-83** this should look like: fnInt($2X^2$,X,0,0.655) + fnInt($\cos^{-1}$X,X,0.655,1), giving the result, approximately 0.382, or **(C)**.

PRACTICE SET

1. Suppose $G(x)$ is an antiderivative of $\dfrac{\cos x}{e^x}$ and $G(-2) = -1$. Then, $G(5) =$

    (A) -3.485

    (B) -1.002

    (C) 0.371

    (D) 0.818

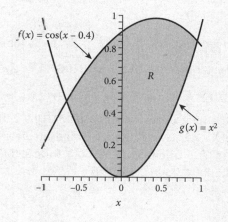

2. If the area under the curve $f(x) = \sqrt{x}$ over the closed interval $[k, 30]$ is 50, what is k ?

    (A) 19.609

    (B) 19.981

    (C) 20.013

    (D) 20.683

4. Consider the region R enclosed by the graphs of $f(x) = \cos(x - 0.4)$ and $g(x) = x^2$, shown above. Find the volume generated by revolving the region R about the line $y = 0$.

    (A) 3.339

    (B) 4.128

    (C) 12.285

    (D) 15.493

3. A particle is moving along the x-axis with a velocity of $v(t) = 2 \sin t + \cos t - 2$ for $t \geq 0$ in minutes. What is the total distance traveled after five minutes?

    (A) 8.024

    (B) 8.415

    (C) 9.526

    (D) 9.817

ANSWERS AND EXPLANATIONS

1. D

The antiderivative of $\frac{\cos x}{e^x}$ is $\int \frac{\cos x}{e^x}\,dx$. By the second fundamental theorem of calculus, if $G'(x) = g(x)$, then $\int_a^b g(x)\,dx = G(b) - G(a)$ so $\int_{-2}^{5} \frac{\cos x}{e^x}\,dx = G(5) - G(-2)$. Go to the fnInt utility by pressing MATH 9 and enter $(\cos X/e^x, X, -2, 5)$ ENTER. Evaluating the integral from -2 to 5 gives 1.818. Given that $G(-2) = -1$, solve for $G(5)$: $G(5) + 1 \approx 1.818 \Rightarrow G(5) \approx 0.818$. This matches **(D)**.

2. B

One approach here would be to use the calculator's solver utility to solve the equation $\int_k^{30} \sqrt{x}\,dx = 50$ for k. Go to the solver utility by pressing MATH B. The image below shows the appropriate syntax. This should yield $k = 19.981$.

EQUATION SOLVER	fnInt($\sqrt{}$(X),X,...=0
eqn:0=fnInt($\sqrt{}$(X)	X=0
,X,K,30)−50	K=19.98121979■...
	bound={⁻1ᴇ99,1...

F1▾	F2▾	F3▾	F4▾	F5	F6▾	
Tools	Algebra	Calc	Other	PrgmIO	Clean up	

■solve$\left(\int_k^{30} \sqrt{x}\,dx = 50, k\right)$

k = 19.9812198

solve($\int$($\sqrt{}$(x),x,k,30)=50,k...

| MAIN | RAD APPROX | FUNC | 1/30 |

If the above approach isn't working out or if you prefer to stick with using the fnInt method, this problem can also be solved by working backwards. Plug each of the answer choices in for k in the integral $\int_k^{30} \sqrt{x}\,dx$ and see which one gives 50. **(B)** yields the appropriate result.

3. D

The total distance traveled is given by the definite integral of the speed function (the absolute value of the velocity function.) That is, $\int_0^5 |v(t)|\,dt$. Use the fnInt method (your calculator will take the absolute value of a function; check your owner's manual about how to do this on your calculator). **TI-83:** The absolute value function is found on the MATH menu or in the catalog.

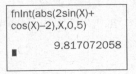

This matches **(D)**.

4. A

To find the volume of the solid of revolution, integrate the area of the typical cross section. First, draw the solid of revolution and the typical cross section.

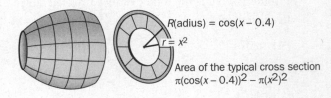

R(adius) $= \cos(x - 0.4)$

$r = x^2$

Area of the typical cross section
$\pi(\cos(x - 0.4))^2 - \pi(x^2)^2$

To compute the volume of the solid of revolution of the region R about the line $y = 0$, integrate the area of the typical cross section. In this case, the typical cross section is a washer. The boundaries of the definite integral are the intersection points of the curves; $x = -0.6839$ and $x = 0.9291$. Thus,

$$\int_{-0.6839}^{0.9291} \pi\left(\cos(x - 0.4)\right)^2 - \pi\left(x^2\right)^2\,dx$$

Use the utility of the graphing calculator to compute the value of the definite integral: 3.339. This matches **(A)**.

RAPID REVIEW

> ### If you take away only four things from this chapter:
>
> 1. Use a graph utility to evaluate a function by entering the function into the $\boxed{Y=}$ screen. From there use the $\boxed{CALC}$ ($\boxed{2^{nd}}$ $\boxed{TRACE}$) menu to find zeros, minimums, or maximums.
>
> 2. The utility to evaluate a derivative at a point is called **nDeriv**, and it is found on the $\boxed{MATH}$ menu. Press 8 or scroll with the arrow keys to find it.
>
> 3. The utility for computing a definite integral is called **fnInt**, and it is found on the $\boxed{MATH}$ menu. Press 9 or scroll with the arrow keys to find it.
>
> 4. The online PDF *More Advanced Calculator Techniques*, available in your Online Resources, reviews several additional methods for evaluating functions, derivatives, and integrals.

REFLECTION

Use this section to evaluate your progress. After working through each practice set, check off the boxes to indicate which Learning Objectives you feel confident about. Keep working on unchecked Objectives until you're confident about them all!

Practice Set

☐ **19.1** Plot the graph of a function within a given viewing window and find the zeros and/or relative extrema

☐ **19.2** Calculate the derivative of a function numerically

☐ **19.3** Calculate the value of a definite integral numerically

FOR MORE PRACTICE

Complete more practice online at kaptest.com. Haven't registered your book yet? Go to kaptest.com/booksonline to begin.

CHAPTER 20

Answering Free-Response Questions

LEARNING OBJECTIVES

20.1 Show appropriate work that is organized, includes proper notation, and justifies your answers

20.2 Clearly identify your final answers

20.3 Use a scoring rubric to self-score your answers

THE FREE-RESPONSE SECTION

Section II of the AP Calculus exam is worth 50% of your grade. The section consists of two parts, just like the multiple-choice section. Part A contains two free-response questions for which you are allowed (and in fact, will most likely need) to use your graphing calculator. Part B contains four free-response questions for which you may not use your calculator.

How Free-Response Questions Are Structured

Each free-response question typically has three or four parts. This means you won't get one broad question, but rather an initial setup, which may include a table of values or a graph of some sort, followed by questions labeled (a), (b), (c), and so on.

There are six free-response questions in all, and you're allowed a total of 90 minutes (30 minutes for the two calculator questions and 60 minutes for the four no-calculator questions). Do the math—that's 15 minutes per question. The timing alone should tell you that these questions are not meant to be quick, easy-to-answer questions. They require critical thinking, the use of proper notation, organization, and often justification for your answers.

Sometimes the question parts are related to each other, and sometimes they are independent. If you can't answer part (a), that doesn't necessarily mean that you can't answer any of the other parts, so don't give up!

Using a Scoring Rubric to Self-Score Your Free-Response Questions

Each of the six free-response questions is worth the same amount. Based on past released tests, that amount is nine points. Those nine points can be distributed evenly across the parts of the question, or not. For example, if there are three parts to a question, each part may be worth three points. Or, one part may be worth two points, one worth three points, and one worth four points. It all depends on the level of complexity associated with each part of the question.

Each year, The College Board releases the official free-response questions from the previous year's test, along with the scoring rubrics that were used to score the questions. It is a good idea to visit The College Board's website and download several free-response questions. The more scoring rubrics you read through, the better an idea you will have about exactly what earns points. For example, the readers are looking for proper notation, organization, numerical answers accurate to three decimal places, clearly identified final answers, and justifications where requested.

After each practice question in this chapter, you will find general notes and strategies for how to approach the question, followed by a scoring rubric that includes detailed solutions to each part of the question with corresponding points earned. Be sure to read through the entire scoring rubric—your goal is to not only master the content and arrive at the correct answer, but also to learn how to document your solution in such a way that you earn maximum points on Test Day. Remember, you get 15 minutes per question, so make sure you're using that time to clearly explain to a question reader how you arrived at your final answer.

SAMPLE QUESTION 1

A GRAPHING CALCULATOR IS REQUIRED TO ANSWER THIS QUESTION.

A particle moves along the x-axis so that its velocity at any time $t \geq 0$ is given by $v(t) = 5t^2 - 4t + 7$. The position of the particle, $x(t)$, is 8 for $t = 3$.

(a) Write an equation for the position, $x(t)$, of the particle at any time $t \geq 0$.

_____

(b) Find the total distance traveled by the particle from time $t = 0$ until time $t = 2$.

_____

(c) Does the particle achieve a minimum velocity? Justify your answer. And if so, what is the position of the particle at this time?

Part 6
Graphing Calculators and Free-Response Questions

ANSWERS AND EXPLANATIONS

1. (a)

Solutions	Scoring Rubric
$\int v(t)\,dt = \int 5t^2 - 4t + 7\,dt$	1: $x(t) = \int v(t)$
$x(t) = \dfrac{5}{3}t^3 - 2t^2 + 7t + C$	1: correctly integrating
$8 = \dfrac{5}{3}(27) - 2(9) + 21 + C$	1: finding C and stating final $x(t)$
$8 = 45 - 18 + 21 + C$	
$8 = 48 + C \rightarrow C = -40$	
$\therefore$	
$x(t) = \dfrac{5}{3}t^3 - 2t^2 + 7t - 40$	

(b)

Solutions	Scoring Rubric
Total distance $= \displaystyle\int_0^2 \lvert v(t)\rvert\,dt$	1: $\displaystyle\int_0^2 \lvert v(t)\rvert\,dt$
$v(t)$ is positive everywhere	1: noting $v(t)$ is positive everywhere
$\displaystyle\int_0^2 v(t)\,dt = 19.333$ or $\dfrac{58}{3}$	1: correct to 3 decimal places or fraction form

(c)

Solutions	Scoring Rubric
$v'(t) = 10t - 4$	1: correctly solving $v' = 0$
$0 = 10t - 4$	1: justifying that the result is a minimum
$t = \dfrac{4}{10} = \dfrac{2}{5}$	
$v''(t) = 10 > 0$, so the velocity is at a minimum when $t = \dfrac{2}{5}$.	1: correctly evaluating $x(t)$ at $t = \dfrac{2}{5}$ to 3 decimal places
The position is :	
$x\left(\dfrac{2}{5}\right) = \dfrac{5}{3}\left(\dfrac{8}{125}\right) - 2\left(\dfrac{4}{25}\right) + \dfrac{14}{5} - 40$	
$= \dfrac{-2{,}806}{75} = -37.413$	

Supplemental Information:

Part (a): Integrate the velocity function to find position. Then solve for C by substituting 8 for $x(t)$ and 3 for t.

Part (b): Use the Integrate function on your calculator to evaluate $\int_0^2 \lvert v(t)\rvert\,dt$. (See chapter 15 for a detailed explanation on how to use this function.) Note that you don't need to indicate that you used your calculator. If you leave your answer in decimal form, be sure to include three decimal places for full credit.

Part (c): The function g has a horizontal tangent where the slope of the tangent is 0. Find the (potentially) minimum velocity by setting $v'(t) = 0$ and solving for t. Justify that this value of t is indeed a minimum by considering $v''(t)$ or evaluating $v''(t)$ for a value to the left and right of t. Note that it isn't enough to just compute the zero of v'. You also need to explain, however briefly, why v actually attains a minimum. That is what is meant by "justify."

SAMPLE QUESTION 2

NO CALCULATOR IS ALLOWED ON THIS QUESTION.

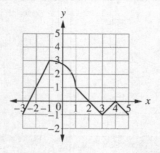

The graph of $y = f(x)$ is shown above. It consists of line segments and a quarter-circle centered at $(-1, 1)$. Let $g(x) = \int_1^x f(t)\, dt$.

(a) Find $g(-1)$ and $g(5)$.

_____

(b) Find an equation of the tangent line to g at the point where $x = -1$.

_____

(c) Find all values in $-3 < x < 5$ where g has a horizontal tangent line. Determine whether each value is a local maximum, local minimum, or neither. Justify your answers.

_____

(d) Find the x-coordinate of all points of inflection of g. Justify your answer.

ANSWERS AND EXPLANATIONS

2. **(a)**

Solutions	Scoring Rubric
$g(-1) = -$Area of $\left(2 \text{ unit sqs} + \dfrac{1}{4}\text{circle}\right)$ $= -\left(2 + \dfrac{1}{4}\pi \cdot 2^2\right) = -2 - \pi$ $g(5) = \dfrac{1}{2} - 1 - \dfrac{1}{2} = -1$	1: correct $g(-1)$ 1: correct $g(5)$

(b)

Solutions	Scoring Rubric
By the second fundamental theorem of calculus: $g'(x) = f(x) \Rightarrow m = g'(-1) = f(-1) = 3$ $y - (-2 - \pi) = 3(x + 1)$	1: finding $m = 3$ 1: writing equation in point-slope form

(c)

Solutions	Scoring Rubric
$g'(x) = f(x) = 0 \Rightarrow x = -2.5, x = 2, x = 4$ A local minimum occurs when $x = -2.5$ because g' (which is f) changes from negative (below the x-axis) to positive (above the x-axis). A local maximum occurs when $x = 2$ because g' changes from positive to negative. Neither a local maximum nor a local minimum occurs when $x = 4$ because g' does not change sign.	1: solving $g'(x) = 0$ by finding x-intercepts on the graph 1: identifying local min, local max, and x-value that is neither 1: providing justification for each of the above

(d)

Solutions	Scoring Rubric
Inflection points occur when $x = -1$ and $x = 4$ because f' changes from increasing to decreasing (relative maxima), which means g'' changes signs. An inflection point also occurs when $x = 3$ because f' changes from decreasing to increasing (relative minima), which means g'' also changes signs there.	1: identifying 3 inflection points 1: providing justification for each inflection point

Graphing Calculator

Supplemental Information:

Part (a): $g(-1)$ represents the area between the function and the x-axis between $x = 1$ and $x = -1$. It may help to shade this area on the graph. Notice that the limits of integration are not from least to greatest, so you'll need to multiply the area by -1. Similarly, $g(5)$ represents the area between the function and the x-axis between $x = 1$ and $x = 5$. Don't forget—the areas below the x-axis get a negative sign.

Part (b): Be sure to justify how you find the slope of the tangent line. This could be via a sentence or simply using the proper notation, as shown below. (If you use a sentence, be brief and don't use pronouns.) Then, use the point-slope form of a line and $y = g(-1)$ from part (a) to write the equation of the tangent line. It is important to note that the point-slope form of a line is a perfectly good form; you do not need to do extra work to write it in slope-intercept form.

PRACTICE SET

A GRAPHING CALCULATOR IS REQUIRED TO ANSWER QUESTIONS 1 AND 2.

1. Let R be the region bounded by the y-axis and the graphs of $f(x) = 3\ln(x+1)$ and $g(x) = \dfrac{1}{x+1} + 3$.

    (a) Sketch and shade the region R.

    (b) Find the area of R.

    (c) Find the volume if R is rotated around the x-axis.

    (d) The vertical line $x = k$ divides R into two regions of equal area. Set up, but do not solve, an integral equation that finds the value of k.

2. An online retailer has a warehouse that receives packages that are later shipped out to customers. The warehouse is open 18 hours per day. On one particular day, packages are received at the warehouse at a rate of $R(t) = 300\sqrt{t}$ packages per hour. Throughout the day, packages are shipped out at a rate of $S(t) = 60t + 300\sin\left(\frac{\pi}{6}t\right) + 300$ packages per hour. For both functions, $0 \le t \le 18$, where t is measured in hours. At the beginning of the workday, the warehouse already has 4,000 packages waiting to be shipped out.

(a) What is the rate of change of the number of packages in the warehouse at time $t = 4$?

(b) To the nearest whole number, evaluate $\int_3^{12} S(t)\,dt$. What is the meaning of $\int_3^{12} S(t)\,dt$?

(c) To the nearest whole number, how many packages are in the warehouse at the end of the 18-hour day?

(d) Over the 18-hour day, to the nearest whole number, what is the maximum number of packages in the warehouse and at what time did that occur?

NO CALCULATOR IS ALLOWED FOR THE REMAINING QUESTIONS.

3. The point $(4, 2)$ is on the graph of $y = f(x)$ and the derivative of $f(x)$ is $\dfrac{dy}{dx} = x(3 - y)$.

    (a) Find the equation of the tangent line to $y = f(x)$ at $(4, 2)$ and use it to approximate the value of $f(5)$.

    (b) Find an expression for $y = f(x)$ by solving the differential equation $\dfrac{dy}{dx} = x(3 - y)$ with the initial condition $f(4) = 2$.

    (c) A claim is made that $y = g(x)$ is a another solution to the differential equation $\dfrac{dy}{dx} = x(3 - y)$. The graph of $g(x)$ is shown below. Give a reason why $g(x)$ cannot be a solution as claimed.

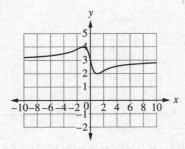

x	0	3	6	9	12
$f(x)$	17	16	14	11	6

4. Suppose a function f is continuous, decreasing, and concave down and has values shown in the table above.

   (a) Use the data to estimate $f'(9)$. Show the computations that lead to your answer.

   _____

   (b) Use the data to evaluate $\int_0^{12} f'(x)\, dx$. Show the computations that lead to your answer.

   _____

   (c) Use a right Riemann sum with four subintervals to find the average value of f over the interval $0 \le x \le 12$. Is your estimate an overestimate or underestimate? Explain.

   _____

   (d) The tangent line to f at $x = 6$ is used to approximate $f(7)$. The secant line from $x = 6$ to $x = 9$ is also used to approximate $f(7)$. Which calculation is an under-approximation of $f(7)$ and which is an over-approximation of $f(7)$? Justify your answer.

ANSWERS AND EXPLANATIONS

1. **(a)**

Solutions	Scoring Rubric
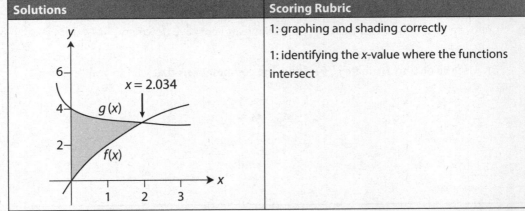	1: graphing and shading correctly 1: identifying the *x*-value where the functions intersect

(b)

Solutions	Scoring Rubric
$A = \int_0^{2.03395} \big(g(x) - f(x)\big)\, dx \approx 3.212 \text{ units}^2$	1: setting up integral correctly 1: evaluating correctly to 3 decimal places

(c)

Solutions	Scoring Rubric
Volume of a representative slice is: $\pi\Big(\big(g(x)\big)^2 - \big(f(x)\big)^2\Big) \cdot \Delta x$ Volume when *R* is revolved about the *x*-axis is: $V = \pi \int_0^{2.0339565} \Big(\big(g(x)\big)^2 - \big(f(x)\big)^2\Big)\, dx$ $= 16\pi \approx 50.265 \text{ units}^3$	1: representing volume of a single slice 1: representing volume of solid of revolution 1: evaluating correctly to 3 decimal places

(d)

Solutions	Scoring Rubric
$\int_0^k \big(g(x) - f(x)\big)\, dx = \frac{1}{2} \cdot 3.212$ OR $\int_0^k \big(g(x) - f(x)\big)\, dx = \int_k^{2.034} \big(g(x) - f(x)\big)\, dx$	1: setting up integral correctly (either one) 1: including correct limits of integration in terms of *k*

Supplemental Information:

Part (a): The question states that the graph is bounded by 3 things. Make sure to find a region whose only sides are those 3 things. Use your graphing calculator to graph the equations and then sketch a quick graph on your paper.

Part (b): Solving $f(x) = g(x)$, or using the graphing calculator's Intersect function, yields a point of intersection at $x \approx 2.0339565$. Include this on your sketch so it's clear why you're integrating from 0 to 2.0339565. Also, only round your <u>final</u> answers to 3 decimal places. Always use more than 3 decimal places (5 or more) when computing the answers. Rounding early will often affect the accuracy of your final answer.

Part (c): Focus on using correct notation here. For example, be sure to include the limits of integration and the *dx*.

Part (d): Two possible answers are given. You need only give one of these answers.

2. **(a)**

Solutions	Scoring Rubric
$R(4) - S(4) = 600 - 799.808 \approx -199.808$ packages per hour	1: writing rate as $R(4) - S(4)$ 1: evaluating correctly to 3 decimal places

(b)

Solutions	Scoring Rubric
$\int_3^{12} S(t)\,dt \approx 6{,}177.042 \rightarrow 6{,}177$ Between 3 and 12 hours after the warehouse opened, approximately 6,177 packages were shipped out.	1: evaluating integral and rounding correctly 1: adequately describing the meaning of the integral

(c)

Solutions	Scoring Rubric
$4000 + \int_0^{18}\big(R(t) - S(t)\big)\,dt \approx 3{,}008$	1: representing the number of packages using correct notation 1: evaluating and rounding correctly

(d)

Solutions	Scoring Rubric
Let $P(x)$ represent the number of packages. $P(x) = 4000 + \int_0^x \big(R(t) - S(t)\big)\,dt$ $P'(t) = R(t) - S(t) = 0$ $t \approx 5.526$ or $t \approx 12.111$ <table><tr><td>t</td><td>$P(t)$</td></tr><tr><td>0</td><td>4,000</td></tr><tr><td>5.526</td><td>2,895.783</td></tr><tr><td>12.111</td><td>4,394.908</td></tr><tr><td>18</td><td>3,007.591</td></tr></table> The maximum number of packages is approximately 4,395 at time $t \approx 12.111$ hours.	1: solving $P' = 0$ and identifying the two critical values 1: evaluating P at the endpoints and at the critical values 1: stating the time and number of packages that represent the maximum

Supplemental Information:

Part (a): The units should be "packages per hour." Both $S(t)$ and $R(t)$ are measured in packages per hour. Therefore, the rate of change of the number of packages is $R(4) - S(4)$.

Part (b): Use the Integrate function on your graphing calculator.

Part (c): The number of packages in the warehouse at the end of the day is the initial amount (4,000) plus the total change in the number of packages. Be sure to use proper notation, including limits of integration, as you set up a representation of this sum. Use your calculator to evaluate the integral.

Part (d): 1) In the calculator section, you do not need to show any work for how you solved the P' equation; you just need to set up the equation. 2) You need to evaluate P at the endpoints (0 hours and 18 hours) and at the critical values. Organize the information neatly so the grader can see that you evaluated each one. You could also show where P is increasing and decreasing and just evaluate P at $t = 0$ and 12.111.

Graphing Calculator

3. **(a)**

Solutions	Scoring Rubric
$m = 4(3-2) = 4$ so: $y - 2 = 4(x - 4) \Rightarrow y = 4x - 14$ $f(5) \approx 4(5) - 14 = 6$	1: finding the correct slope 1: writing the equation 1: approximating the value of $f(5)$

(b)

Solutions	Scoring Rubric
$\dfrac{dy}{dx} = x(3 - y) \Rightarrow \dfrac{1}{3 - y}\, dy = x \cdot dx$ $\displaystyle\int \dfrac{1}{3 - y}\, dy = \int x\, dx$ $-\ln\|3 - y\| = \dfrac{1}{2}x^2 + C$ $-\ln\|3 - 2\| = \dfrac{1}{2} \cdot 4^2 + C$ $\qquad 0 = 8 + C \Rightarrow C = -8$ $-\ln\|3 - y\| = \dfrac{1}{2}x^2 - 8$ $\ln\|3 - y\| = -\dfrac{1}{2}x^2 + 8$ $e^{\ln\|3 - y\|} = e^{\left(-\frac{1}{2}x^2 + 8\right)}$ $3 - y = e^{\left(-\frac{1}{2}x^2 + 8\right)}$ $y = 3 - e^{\left(-\frac{1}{2}x^2 + 8\right)}$	1: correctly separating the variables 1: finding the indefinite integral 1: finding the correct value of C using the initial condition 1: writing the equation in the form $y = f(x)$

(c)

Solutions	Scoring Rubric
Based on the given equation, when $x = 0$, $\dfrac{dy}{dx} = 0$. The graph of $y = g(x)$ has a slope that is negative at $x = 0$. Hence, $g(x)$ cannot be a solution to the differential equation.	1: stating g cannot be a solution 1: justifying

Supplemental Information:

Part (a): The equation of the derivative is given $\left(\dfrac{dy}{dx}\right)$. Find the slope of the tangent line by plugging the point (4, 2) into this equation. Then, use point-slope form to write the equation of the tangent line.

Part (b): Solve the differential equation by separating the variables. Don't forget to find the value of the constant C. To write the equation in the form $y = f(x)$, solve for y by making both sides of the equation powers of e. This will eliminate the natural log.

4. **(a)**

Solutions	Scoring Rubric
$f'(9) \approx \dfrac{6-14}{12-6} = -\dfrac{8}{6} = -\dfrac{4}{3}$ OR $f'(9) \approx \dfrac{6-11}{12-9} = -\dfrac{5}{3}$ OR $f'(9) \approx \dfrac{11-14}{9-6} = -\dfrac{3}{3} = -1$	2: using (12, 6) and (6, 14) OR 1: using (12, 6) and (9, 11) OR (9, 11) and (6, 14)

(b)

Solutions	Scoring Rubric
Using the fundamental theorem of calculus: $\displaystyle\int_0^{12} f'(x)\,dx = f(x)\Big|_0^{12}$ $= f(12) - f(0)$ $= 6 - 17 = -11$	1: setting up integral correctly 1: evaluating integral correctly

(c)

Solutions	Scoring Rubric
$\dfrac{1}{12-0} \displaystyle\int_0^{12} f(x)\,dx$ $\approx \dfrac{1}{12}\left[16\cdot3 + 14\cdot3 + 11\cdot3 + 6\cdot3\right]$ $= \dfrac{141}{12} = \dfrac{47}{4}$ For a decreasing function, a right Riemann sum is an underestimate of the integral. Therefore, $\dfrac{141}{12}$ underestimates the average value of f.	1: setting up integral correctly 1: evaluating integral correctly 1: stating and explaining the nature of the estimate

(d)

Solutions	Scoring Rubric
Because f is concave down and decreasing, the tangent line approximation of $f(7)$ will be an over-approximation, and the secant line approximation of $f(7)$ will be an under-approximation.	1: correctly stating which is over and which is under 1: justifying

Graphing Calculator

Supplemental Information:

Part (a): Use the slope formula. Be sure to show all numbers used. The first estimate given in the solution is the best of the three options shown because the points used are closest to $x = 9$.

Part (b): The integral of f' is f, so you can use the values in the table to evaluate the definite integral.

Part (c): You could also draw an estimate of the graph and the rectangles to explain your answer.

Part (d): You could also draw an estimate of the graph and the two lines to show the y-values at $x = 7$ for the actual value and the two approximations of $f(7)$ to explain your answer.

PART 7

Practice Exams

HOW TO TAKE THE PRACTICE EXAMS

The next section of this book consists of six practice exams. Taking a practice AP exam gives you an idea of what it's like to answer these exams questions for a longer period of time, one that approximates the real exam. You'll find out which areas you're strong in, and where additional review may be required. Any mistakes you make now are ones you won't make on the actual exam, as long as you take the time to learn where you went wrong.

The practice exams in this book each include 45 multiple choice questions (30 with calculator and 15 without) and 6 free-response questions (2 with calculator and 4 without). You will have 105 minutes for the multiple choice questions, a 10-minute break, and 90 minutes to answer the free-response questions. Before taking a practice exam, find a quiet place where you can work uninterrupted for over three hours and make sure you have your calculator available. Time yourself according to the time limit at the beginning of each section. It's okay to take a short break between sections, but for the most accurate results you should approximate real exam conditions as much as possible.

As you take the practice exams, remember to pace yourself. Train yourself to be aware of the time you are spending on each question. Skim each section before you start, so you can quickly tackle the questions you feel the most comfortable with. Then, you can tackle the more difficult problems. Try to be aware of the general types of questions you encounter, as well as being alert to certain strategies or approaches that help you to handle the various question types more effectively.

After taking a practice exam, be sure to access your online resources and read the detailed answer explanations. These will help you identify areas that could use additional review. Even when you've answered a question correctly, you can learn additional information by looking at the answer explanation.

Finally, it's important to approach the exam with the right attitude. You're going to get a great score because you've reviewed the material and learned the strategies in this book.

HOW TO COMPUTE YOUR SCORE

The practice exams are composed of multiple choice questions and free-response questions. The multiple choice questions are scored by an electronic scanner, while a team of trained reviewers scores the free-response questions by hand. Questions are evaluated during the month of June, and exam scores are sent out to students and colleges in July. If you haven't received your score by September, contact the College Board.

Scoring the Multiple Choice Questions

To compute your score on the multiple choice portion of the practice exams, calculate the number of questions you got right on each exam, then divide by 45 to get the percentage score for the multiple choice portion of that exam.

Scoring the Free-Response Questions

The readers have specific points that they want to see in each free-response question. Show your work in an organized fashion, using proper notation. Make sure that you present the various components of your answer in the right order, and in an order that the readers can understand. In addition to these basic structural concerns, readers will be seeking specific pieces of information (such as correct units) in your answer. Each piece of information that they are able to find and check off in your answer helps you earn a better score, so make sure your answer to each part of the question is easy to locate.

To figure out your approximate score for the free-response questions, use the scoring rubric provided in the Answers and Explanations section that is available as part of your online resources. Look at the key points found in the sample response for each question. For each key point you include, add a point. After you've scored all six questions, divide the sum of all the points you earned by the total number of points available for all the free-response questions to get the percentage score for the free-response portion of that exam.

Calculating Your Composite Score

Your score on the AP exam is a combination of your scores on the multiple choice section of the exam and the free-response section. Each of these two sections is worth one-half of the exam score, so average your two scores. If your score is a decimal, then round up to a whole number.

Remember, however, that your final score depends on how well all students taking the AP exam perform. If you do better than average, your score would be higher. The numbers here are just approximations.

The approximate score range is as follows:

5 = 65–100% Extremely well qualified

4 = 50–64% Very well qualified

3 = 40–49% Qualified

2 = 25–39% Possibly qualified

1 = 0–24% No recommendation

If your score falls between 50 and 100, you're doing great. Keep up the good work! If your score is lower than 49, there's still hope. Keep studying and you will be able to obtain a better score on the exam before you know it.

Good luck on the exam!

BC Practice Exam 1

Section I, Part A

1. (A) (B) (C) (D)
2. (A) (B) (C) (D)
3. (A) (B) (C) (D)
4. (A) (B) (C) (D)
5. (A) (B) (C) (D)
6. (A) (B) (C) (D)
7. (A) (B) (C) (D)
8. (A) (B) (C) (D)
9. (A) (B) (C) (D)
10. (A) (B) (C) (D)
11. (A) (B) (C) (D)
12. (A) (B) (C) (D)
13. (A) (B) (C) (D)
14. (A) (B) (C) (D)
15. (A) (B) (C) (D)
16. (A) (B) (C) (D)
17. (A) (B) (C) (D)
18. (A) (B) (C) (D)
19. (A) (B) (C) (D)
20. (A) (B) (C) (D)
21. (A) (B) (C) (D)
22. (A) (B) (C) (D)
23. (A) (B) (C) (D)
24. (A) (B) (C) (D)
25. (A) (B) (C) (D)
26. (A) (B) (C) (D)
27. (A) (B) (C) (D)
28. (A) (B) (C) (D)
29. (A) (B) (C) (D)
30. (A) (B) (C) (D)

Section I, Part B

76. (A) (B) (C) (D)
77. (A) (B) (C) (D)
78. (A) (B) (C) (D)
79. (A) (B) (C) (D)
80. (A) (B) (C) (D)
81. (A) (B) (C) (D)
82. (A) (B) (C) (D)
83. (A) (B) (C) (D)
84. (A) (B) (C) (D)
85. (A) (B) (C) (D)
86. (A) (B) (C) (D)
87. (A) (B) (C) (D)
88. (A) (B) (C) (D)
89. (A) (B) (C) (D)
90. (A) (B) (C) (D)

A A A A A A A A A A A A A A A A A A A A A A A A A A A A A A A A A A A

CALCULUS BC
SECTION I, Part A
Time—60 minutes
Number of questions—30

A CALCULATOR MAY NOT BE USED ON THIS PART OF THE EXAM.

Directions: Use the space provided to solve each of the following problems. After examining the given choices, decide which of the choices is the best and fill in the corresponding circle on the answer sheet. Credit will only be given for answers appropriately marked on the answer sheet, and not for anything written in the exam book. Do not spend too much time on any one problem.

In this exam:

(1) The domain of a function f should be assumed to be the set of all real numbers x for which $f(x)$ is a real number, unless the question indicates otherwise.

(2) The inverse function notation f^{-1} or the prefix "arc" (e.g., $\sin^{-1} x = \arcsin x$) may be used to indicate the inverse of a trigonometric function.

GO ON TO THE NEXT PAGE.

AAAAAAAAAAAAAAAAAAAAAAAAAAAAAAAAAAAAA

1. $\lim_{x \to 0} \left(\dfrac{3x^2 + 5 \cos x - 5}{2x} \right)$

   (A) 0   (B) $\dfrac{5}{2}$ (C) 5   (D) nonexistent

---

2. For what values of p does the series $\sum_{n=1}^{\infty} \dfrac{1}{n^{2p}}$ converge?

   (A) $p > 0$

   (B) $p \geq 1$

   (C) $p > \dfrac{1}{2}$

   (D) The series converges for all p.

GO ON TO THE NEXT PAGE.

AAAAAAAAAAAAAAAAAAAAAAAAAAAAAAAA

3. The Taylor series for $\sin x$ centered at $x = 0$ begins

$$x - \frac{x^3}{3!} + \frac{x^5}{5!} - \frac{x^7}{7!} + \cdots.$$

What is the coefficient of x^3 in the Taylor series for $x \sin x^2$ centered at $x = 0$?

(A) 1   (B) $\frac{2}{3}$ (C) $\frac{1}{2}$ (D) $\frac{1}{3}$

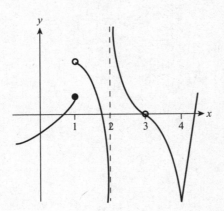

4. The graph of a function f is shown above. Which of the following statements about f is true?

(A) An infinite discontinuity occurs at $x = 4$.

(B) A jump discontinuity occurs at $x = 3$.

(C) A removable discontinuity occurs at $x = 2$.

(D) A jump discontinuity occurs at $x = 1$.

AAAAAAAAAAAAAAAAAAAAAAAAAAAAAAAAA

5. The cost of producing x units of a certain item is $C(x) = 2000 + 8.6x + 0.5x^2$. What is the instantaneous rate of change of C with respect to x when $x = 300$?

(A) 297.2   (B) 300.0   (C) 308.6   (D) 313.6

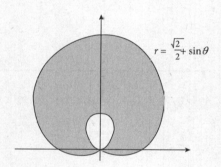

$r = \frac{\sqrt{2}}{2} + \sin\theta$

6. Which of the following integrals gives the area of the shaded region?

(A) $\displaystyle\int_{\frac{-\pi}{4}}^{\frac{5\pi}{4}} \frac{1}{2}\left(\frac{\sqrt{2}}{2} + \sin\theta\right)^2 d\theta$

(B) $\displaystyle\int_{0}^{2\pi} \frac{1}{2}\left(\frac{\sqrt{2}}{2} + \sin\theta\right)^2 d\theta$

(C) $\displaystyle\int_{\frac{-\pi}{4}}^{\frac{5\pi}{4}} \frac{1}{2}\left(\frac{\sqrt{2}}{2} + \sin\theta\right)^2 d\theta - \int_{\frac{5\pi}{4}}^{\frac{7\pi}{4}} \frac{1}{2}\left(\frac{\sqrt{2}}{2} + \sin\theta\right)^2 d\theta$

(D) $\displaystyle\int_{\frac{5\pi}{4}}^{\frac{7\pi}{4}} \frac{1}{2}\left(\frac{\sqrt{2}}{2} + \sin\theta\right)^2 d\theta - \int_{\frac{-\pi}{4}}^{\frac{5\pi}{4}} \frac{1}{2}\left(\frac{\sqrt{2}}{2} + \sin\theta\right)^2 d\theta$

GO ON TO THE NEXT PAGE.

AAAAAAAAAAAAAAAAAAAAAAAAAAAAAAAAAAAA

7. Which of the following gives the derivative of the function $f(x) = x^2$ at the point $(2, 4)$?

(A) $\displaystyle\lim_{h \to 0} \frac{(x + 2)^2 - x^2}{4}$

(B) $\displaystyle\lim_{h \to \infty} \frac{(2 + h)^2 - 2^2}{h}$

(C) $\displaystyle\lim_{h \to 0} \frac{(2 + h)^2 - 2^2}{h}$

(D) $\displaystyle\lim_{h \to 0} \frac{(4 + h)^2 - 4^2}{h}$

---

8. If $f(x) = 5x \sec x + x^3 \cos x + 17\pi$, then $f'(x)$ is which of the following?

(A) $5 \sec x \tan x - 3x^2 \sin x$

(B) $5 \sec x \tan x + 3x^2 \cos x + 17\pi$

(C) $5 \sec x + 5x \sec x \tan x + 3x^2 \cos x - x^3 \sin x$

(D) $5 \sec x + 5x \sec x \tan x - 3x^2 \cos x + x^3 \sin x + 17\pi$

GO ON TO THE NEXT PAGE.

AAAAAAAAAAAAAAAAAAAAAAAAAAAAAAAAAAAA

$$g(x) = \begin{cases} -x^2 + 4 & \text{for } x < 2 \\ 3x + k & \text{for } x \geq 2 \end{cases}$$

9. Let g be the function given above. If g is continuous for all real numbers, what is the value of k?

    (A) -6 (B) -3 (C) 0   (D) 2

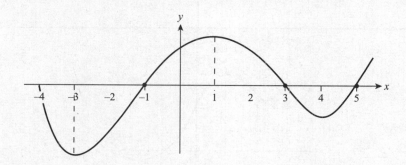

10. The graph of $f(x)$ appears above. Which of the following could be the zeros of the derivative function $f'(x)$?

    (A) $x = -2, 2.5$

    (B) $x = -3, 1, 4$

    (C) $x = -2, -1, 2.5$

    (D) $x = -4, -1, 3, 5$

GO ON TO THE NEXT PAGE.

11. Ford is water-skiing around Cartesian Lake, towed by a motorboat that is following a path parameterized by the equations:

$$y(t) = t^3 + 3t^2 + 4$$
$$x(t) = 2t^2 + 2$$

At time $t = 2$ Ford lets go and continues in a straight line. What is the slope of that line?

(A) 3   (B) 2   (C) $\frac{1}{2}$ (D) $\frac{1}{3}$

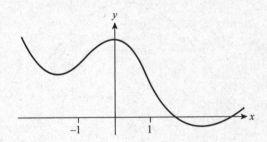

12. The graph of $f(x)$ is given above. Which of the following is true about $f''(x)$?

(A) $f''(x) < 0$ for all values of x

(B) $f''(x) > 0$ for all values of x

(C) $f''(x) > 0$ for $x < 1$ and $f''(x) < 0$ for $x > 1$

(D) $f''(x) > 0$ for $x > 1$ and $f''(x) < 0$ for $-1 < x < 1$

GO ON TO THE NEXT PAGE.

AAAAAAAAAAAAAAAAAAAAAAAAAAAAAAAAAA

13. What is the slope of the curve $y = 4x^{-2} + \frac{1}{2}x^2 + 3$ when $x = 2$?

(A) -1 (B) 1 (C) $\dfrac{127}{64}$ (D) 6

14. Compute $\displaystyle\int \frac{x+3}{(x+1)x}\,dx$.

(A) $-2\ln|x+1| + 3\ln|x| + C$

(B) $-\dfrac{1}{2(x+1)^2} + C$

(C) $-2\ln|x+1| + 3\ln|x| + C$

(D) $3\ln|x+1| + \ln|x| + C$

GO ON TO THE NEXT PAGE.

AAAAAAAAAAAAAAAAAAAAAAAAAAAAAAAAA

15. A 13-foot ladder is leaning against a 20-foot vertical wall when it begins to slide down the wall. During this sliding process, the bottom of the ladder is sliding away from the bottom of the wall at a rate of $\frac{1}{2}$ foot per second. At what rate is the top of the ladder sliding down the vertical wall when the top of the ladder is exactly 5 feet above the ground?

(A) -2 feet per second

(B) $-\frac{6}{5}$ feet per second

(C) $-\frac{12}{13}$ feet per second

(D) $\frac{5}{6}$ feet per second

16. What is the solution to the differential equation $\frac{dy}{dx} = \frac{\sin x}{e^y}$, where $y\left(\frac{\pi}{4}\right) = 0$?

(A) $y = \ln\left(\cos x - \frac{\sqrt{2}}{2}\right)$

(B) $y = \ln\left(-\cos x + \frac{\sqrt{2}}{2} + 1\right)$

(C) $y = -\frac{\cos x}{e^y}$

(D) $y = \ln(-\cos x)$

GO ON TO THE NEXT PAGE.

AAAAAAAAAAAAAAAAAAAAAAAAAAAAAAAAAAA

17. $\int_0^6 \left(2 - |4 - x|\right) dx =$

    (A) -4 (B) 2   (C) 6   (D) 12

---

18. Which of the following sequences converge?

    I. $\sum_{n=0}^{\infty} \dfrac{n+1}{e^n}$

    II. $\sum_{n=1}^{\infty} \dfrac{e^n}{4^n}$

    III. $\sum_{n=0}^{\infty} \sin\left(\dfrac{\pi}{2} - \dfrac{1}{n}\right)$

    (A) I only

    (B) II only

    (C) I and II

    (D) I, II, and III

GO ON TO THE NEXT PAGE.

AAAAAAAAAAAAAAAAAAAAAAAAAAAAAAAAAA

19. $\displaystyle\int \frac{\left(7+x^{\frac{2}{3}}\right)^5}{x^{\frac{1}{3}}}\,dx =$

   (A) $\dfrac{1}{6}\left(7x^{\frac{2}{3}} + x^{\frac{5}{3}}\right)^6 + C$

   (B) $\dfrac{1}{4}\left(7 + x^{\frac{1}{3}}\right)^6 + C$

   (C) $\dfrac{1}{4}\left(7 + x^{\frac{2}{3}}\right)^6 + C$

   (D) $\dfrac{1}{6}\left(7x^{-\frac{1}{3}} + x^{\frac{2}{3}}\right)^6 + C$

---

20. Compute $\displaystyle\int_1^\infty \frac{3}{x^2}\,dx$.

   (A) -1 (B) 1 (C) 3 (D) undefined

---

21. $\displaystyle\frac{d}{dt}\int_2^{t^4} e^{x^2}\,dx =$

   (A) $e^{t^8} - e^4$ (B) $4t^3 e^{t^8} - e^4$ (C) e^{t^8} (D) $4t^3 e^{t^8}$

GO ON TO THE NEXT PAGE.

A A A A A A A A A A A A A A A A A A A A A A A A A A A A A A A A

22. Let $g(x) = (\arccos(x^2))^5$. Then $g'(x) =$

(A) $\dfrac{10\left(\arccos\left(x^2\right)\right)^4}{\sqrt{1-x^4}}$

(B) $\dfrac{10x\left(\arccos\left(x^2\right)\right)^4}{\sqrt{1-x^2}}$

(C) $\dfrac{-10x\left(\arccos\left(x^2\right)\right)^4}{\sqrt{1-x^4}}$

(D) $\dfrac{-10x\left(\arcsin\left(x^2\right)\right)^4}{\sqrt{1-x^2}}$

---

23. $\dfrac{d}{dx}\left(\ln\left(3x\right)5^{2x}\right) =$

(A) $5^{2x}\left(\dfrac{1}{x} + 2\ln\left(5\right)\ln\left(3x\right)\right)$

(B) $5^{2x}\left(\dfrac{1}{3x} - 2x\ln\left(3x\right)\right)$

(C) $5^{2x}\left(\dfrac{1}{x} - \ln\left(5\right)\ln\left(3x\right)\right)$

(D) $5^{2x}\left(\dfrac{1}{3x} + 2\ln\left(3x\right)\right)$

GO ON TO THE NEXT PAGE.

AAAAAAAAAAAAAAAAAAAAAAAAAAAAAAAA

24. What is the slope of the normal line to the curve $x^3 + xy^2 = 10y$ at the point $(2, 1)$?

(A) $-\frac{7}{3}$ (B) $-\frac{6}{13}$ (C) $\frac{1}{2}$ (D) 2

---

25. Compute $\int xe^{-3x}\, dx$.

(A) $-\frac{1}{3}e^{-3x^2} + C$

(B) $-\frac{1}{3}xe^{-3x} + C$

(C) $-\frac{1}{3}xe^{-3x} - \frac{1}{9}e^{-3x} + C$

(D) $\frac{1}{9}xe^{-3x} - \frac{1}{3}e^{-3x} + C$

---

26. What is the volume of the solid of revolution determined by rotating the region bounded by the graphs of $f(x) = x^2$ and $g(x) = 3x$ about the x-axis?

(A) 9π (B) $\frac{27\pi}{2}$ (C) 28π (D) $\frac{162\pi}{5}$

GO ON TO THE NEXT PAGE.

AAAAAAAAAAAAAAAAAAAAAAAAAAAAAAAA

27. Which of the following slope fields describes the differential equation $\frac{dy}{dx} = \frac{x}{y}$?

(A)
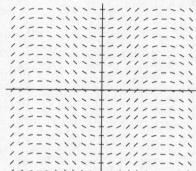

(B)
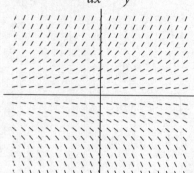

(C)
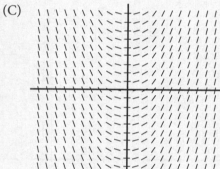

(D)
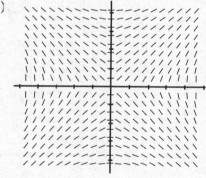

GO ON TO THE NEXT PAGE.

AAAAAAAAAAAAAAAAAAAAAAAAAAAAAAAAAAA

28. $\int_{-1}^{4} f(3x - 1) \, dx$ is equivalent to:

(A) $\frac{1}{3} \int_{-4}^{11} f(u) \, du$

(B) $3 \int_{-4}^{11} f(u) \, du$

(C) $\int_{-4}^{11} f(u) \, du$

(D) $\frac{1}{3} \int_{-1}^{4} f(u) \, du$

29. A population of voles is described by the differential equation $\frac{dP}{dt} = 0.6P\left(1 - \frac{P}{400}\right)$ with initial condition $P(0) = a$. For which values of a will $\lim_{x \to \infty} P(x) = 400$?

(A) $a = 0$ (B) $a \geq 400$ (C) $0 \leq a \leq 400$ (D) $a > 0$

AAAAAAAAAAAAAAAAAAAAAAAAAAAAAAAAA

30. Suppose $f(x) = x \sec x$ for $\dfrac{-\pi}{2} < x < \dfrac{\pi}{2}$ and $g(x) = f^{-1}(x)$. Given that the point $\left(\dfrac{\pi}{3}, \dfrac{2\pi}{3}\right)$ is on the graph of $f(x)$, what is the value of $g'\left(\dfrac{\pi}{3}\right)$?

(A) $\dfrac{3}{6 + 2\pi\sqrt{3}}$

(B) $2 + \dfrac{2\pi\sqrt{3}}{3}$

(C) $\dfrac{2\pi}{3}$

(D) $2 + \dfrac{2\pi}{3}$

END OF PART A OF SECTION I

**IF YOU FINISH BEFORE TIME IS CALLED, YOU MAY
CHECK YOUR WORK ON PART A ONLY.**

CALCULUS BC
SECTION I, Part B
Time—45 minutes
Number of questions—15

A GRAPHING CALCULATOR IS REQUIRED FOR SOME QUESTIONS ON
THIS PART OF THE EXAM.

Directions: Use the space provided to solve each of the following problems. After examining the given choices, decide which of the choices is the best and fill in the corresponding circle on the answer sheet. Credit will only be given for answers appropriately marked on the answer sheet, and not for anything written in the exam book. Do not spend too much time on any one problem.

In this exam:

(1) When the exact numerical value of the correct answer does not appear among the choices given, select the number that best approximates the numerical value.

(2) The domain of a function f should be assumed to be the set of all real numbers x for which $f(x)$ is a real number, unless the question indicates otherwise.

(3) The inverse function notation f^{-1} or the prefix "arc" (e.g., $\sin^{-1} x = \arcsin x$) may be used to indicate the inverse of a trigonometric function.

GO ON TO THE NEXT PAGE.

B B B B B B B B B

$$f(x) = x^3 + \frac{5}{4}x^2 - \frac{59}{4}x - 15$$

76. What are the values for which the function f defined above is increasing?

    (A) $x < 3.75$

    (B) $-2.673 < x < 1.839$

    (C) $x < -2.673$ and $x > 1.839$

    (D) $-4 < x < -1$ and $x > 3.75$

---

77. Let $f(x) = \sin(\sin x)$. What is the value of $f'\left(\frac{\pi}{4}\right)$, accurate to three decimal places?

    (A) 0.009   (B) 0.538   (C) 0.866   (D) 1.000

---

78. Consider the differential equation given by $y' = xy$, with initial condition $y(0) = 1$. Using Euler's method starting at $x = 0$ with step size 1, what is the approximate value of $y(2)$?

    (A) 0   (B) 1   (C) 2   (D) 4

GO ON TO THE NEXT PAGE.

79. The Maclaurin series for f is given by $x + x^3 + x^5 + x^7 \cdots$. Which of the following is an expression for f?

    (A) $e^{x^2} - 1$

    (B) $\dfrac{x}{1 - x^2}$

    (C) $\sin(x) - \cos(x^2)$

    (D) $\dfrac{1}{1 + x^2} - 1$

80. A spherical balloon is being inflated at a rate of 3 cubic inches per second. What is the rate of change of the radius of the balloon when the balloon's radius is 5 inches? The volume of a sphere is given by $V = \dfrac{4}{3}\pi r^3$.

    (A) 0.010 inches per second

    (B) 0.120 inches per second

    (C) 1.667 inches per second

    (D) 3.000 inches per second

81. The initial population of a colony of flies is 12. Ten days later, the population is 60. The population, P, of flies grows at a rate $\dfrac{dP}{dt} = kP$. What is the value of k?

    (A) $k \approx 0.161$ (B) $k \approx 0.843$ (C) $k \approx 1.208$ (D) $k \approx 1.609$

GO ON TO THE NEXT PAGE.

B B B B B B B B B

x	0	2	4	6	8	10
$g(x)$	9	25	30	16	25	32

82. The table above provides data points for a continuous function g. Use a right Riemann sum with 5 subdivisions to approximate the area under the curve of $y = g(x)$ on the closed interval $[0, 10]$.

(A) 206   (B) 210   (C) 235   (D) 256

$$s(t) = \frac{1}{3}t^3 - 4t^2 + 12t + 1$$

83. The position of a particle, s, is given above for $0 \le t \le 10$. On what interval(s) is the particle slowing down?

(A) $2 \le t \le 6$

(B) $0 \le t \le 4$

(C) $0 \le t \le 2 \cup 4 \le t \le 6$

(D) $2 \le t \le 6 \cup 6 \le t \le 10$

84. The area bounded by the curves $y = x^2 + 4$ and $y = -2x + 1$ between $x = -2$ and $x = 5$ equals

(A) 86.000   (B) 86.125   (C) 86.333   (D) 86.500

GO ON TO THE NEXT PAGE.

Practice Exams

85. Evaluate $\sum_{n=1}^{\infty} \frac{(-3)^{n+1}}{4^n}$.

    (A) $\frac{1}{7}$ (B) $\frac{4}{7}$ (C) 1   (D) $\frac{9}{7}$

86. What is the approximate average value of the function $f(x) = \frac{2x}{x^2 - 4}$ on the closed interval $[5, 8]$?

    (A) 0.350   (B) 0.456   (C) 0.743   (D) 1.050

87. The movement of a particle in the xy-plane is given by $(x(t), y(t))$ where $x(t) = e^t + 1$ and $y(t) = t^2 + 2$. What is the speed of the particle at time $t = 2$?

    (A) 3.375   (B) 8.402   (C) 14.389   (D) 33.314

B B B B B B B B B

88. The acceleration in m/s^2 of an object at any time $t \geq 0$ is given by $a(t) = t^2 + \sqrt{t+9} + e^{-t}$.
The velocity of the object at time $t = 0$ is 4 m/s and the position of the object at time $t = 0$ is 5 m.
What is the approximate position of the object at time $t = 7$?

(A) 15.507 m   (B) 103.668 m   (C) 246.752 m   (D) 321.351 m

---

89. The degree four Taylor polynomial of f centered at $x = 3$ is given by

$P(x) = 2 + (x - 3) + 3(x - 3)^2 + \frac{4}{3}(x - 3)^3 - \frac{1}{4}(x - 3)^4$. What is $f''(3)$?

(A) $\frac{1}{2}$ (B) $\frac{4}{3}$ (C) 3   (D) 6

---

90. Which of the following is a solution of $\dfrac{dy}{dx} = \dfrac{5}{y(x+4)}$ given that $y = 0$ at $x = (e - 4)$?
(A) $y = (t + 4)^2 - e^2$
(B) $y^2 = 5\ln|x + 4| - 5$
(C) $y = 2\sqrt{t + 4} - 2\sqrt{e}$
(D) $y^2 = 10\ln|x + 4| - 10$

END OF SECTION I

**IF YOU FINISH BEFORE TIME IS CALLED, YOU MAY
CHECK YOUR WORK ON PART B ONLY.**

CALCULUS BC
SECTION II, Part A
Time—30 minutes
Number of problems—2

A GRAPHING CALCULATOR IS REQUIRED FOR THESE PROBLEMS.

Directions: Write your solution to each part of the following questions in the space provided. Write clearly and legibly. Cross out any errors you make; erased or crossed-out work will not be scored.

You may wish to look over the problems before starting to work on them; on the actual test, it is not expected that everyone will be able to complete all parts of all problems. All problems are given equal weight, but the individual parts of a particular problem are not necessarily given equal weight. You should not spend too much time on any one problem.

- Show all of your work. Clearly label functions, graphs, tables, or anything else that you use to arrive at your final solution. Your work will be scored on the correctness and completeness of your methods as well as your final answers. Answers without supporting work will usually not receive credit.

- Justifications (i.e., the request that you "justify your answer") require that you give mathematical (non-calculator) reasons.

- Do not use calculator syntax. Your work must be expressed in standard mathematical notation.

- Numeric or algebraic answers do not need to be simplified, unless the question indicates otherwise.

- If you use decimal approximations in calculations, your final answers should be accurate to three places after the decimal point, unless the question indicates otherwise.

- The domain of a function f should be assumed to be the set of all real numbers x for which $f(x)$ is a real number, unless the question indicates otherwise.

1 1 1 1 1 1 1 1 1 1

1. The rate of change of charge passing into a battery is modeled by the function $C(t) = 10 + 6 \sin\left(\dfrac{t^2}{3}\right)$, for $t \geq 0$. $C(t)$ has units in coulombs per hour and t has units in hours. At $t = 0$, the battery is empty of any charge.

(a) Is the amount of charge in the battery increasing or decreasing at $t = 4$?
Give a reason for your answer.

_____

(b) Is the rate of change of charge in the battery increasing or decreasing at $t = 4$?
Give a reason for your answer.

GO ON TO THE NEXT PAGE.

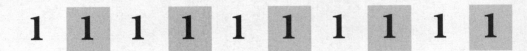

(c) What is the average rate of change of the charge between $t = 2$ and $t = 10$, specific to 3 decimals?

(d) The battery can hold a charge of 160 coulombs. How long, in hours, does it take to charge the battery?

GO ON TO THE NEXT PAGE.

2. A city was founded in 1900 with an initial population of 4,000. The rate of change in population is modeled by the equation $P'(t) = \ln(1 + 5t^2)$, where P is population in thousands and t is measured in years since 1900.

(a) Using the model, find and solve an integral formula to estimate the population of the city in the year 1950.

(b) Using the model, find and solve an integral formula to determine during which year the population reaches 1 million.

(c) Use Euler's method, starting at $t = 0$ and with a step size of 25 years, to estimate the population of the city in the year 2000.

END OF PART A OF SECTION II

**IF YOU FINISH BEFORE TIME IS CALLED, YOU MAY
CHECK YOUR WORK ON PART A ONLY.**

CALCULUS BC
SECTION II, Part B
Time—60 minutes
Number of problems—4

NO CALCULATOR IS ALLOWED FOR THESE PROBLEMS.

Note: If you have extra time, you can go back and work on Part A of Section II, but you cannot use a calculator to complete your work at this time.

GO ON TO THE NEXT PAGE.

3 3 3 3 3 3 3 3 3 3

NO CALCULATOR ALLOWED

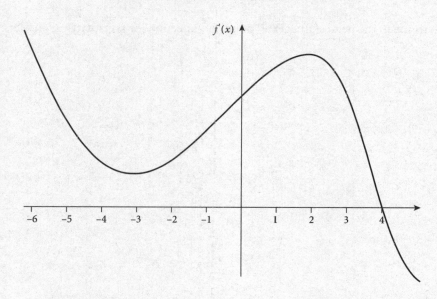

3. The graph of the function f' appears above, where f' is the derivative of some function f. The domain of f' here is the set of real numbers x satisfying the inequalities $-6 < x < 6$.

(a) Determine to the nearest integer all points at which f has a relative extremum (if any exist) and state whether each is a maximum or a minimum. Justify your answers.

GO ON TO THE NEXT PAGE.

3 **3** **3** **3** **3** **3** **3** **3** **3** **3**

NO CALCULATOR ALLOWED

(b) Determine to the nearest integer all points of inflection for f. Justify your answers.

_____

(c) Determine whether the function f is concave upward or concave downward on the closed intervals $[-6, -3]$ and $[0, 2]$. Justify your answers.

NO CALCULATOR ALLOWED

4. Let C be the curve defined by the equation $xy^2 + y^3 = 5$.

   (a) Find $\dfrac{dy}{dx}$ as a function of x and y.

   _____

   (b) Define when a horizontal tangent occurs. Determine all points (if any exist) where a horizontal tangent occurs. Justify your answer.

   _____

   (c) Define when a vertical tangent occurs. Determine the point on C where the tangent line is vertical. Justify your answer.

GO ON TO THE NEXT PAGE.

NO CALCULATOR ALLOWED

5. Let R be the region enclosed by the graphs of $f(x) = ax(2 - x)$ and $g(x) = ax$ for some positive real number a.

 (a) Find the area of the region R.

 _____

 (b) Find the volume of the solid of revolution generated when R is rotated about the x-axis. What is the shape of the typical cross-section of the solid?

 _____

 (c) Assume a solid exists with a cross-sectional area of R and uniform thickness π. Find the value of a for which this solid has the same volume as the solid in (b).

GO ON TO THE NEXT PAGE.

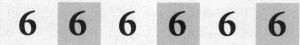

6. The derivatives of f at $x = 0$ are given by $f^{(n)}(0) = \dfrac{(n+1)!}{3^n}$.

   (a) Show that the Taylor series for f at $x = 0$ is $\displaystyle\sum_{n=0}^{\infty} \dfrac{(n+1)x^n}{3^n}$.

   (b) What is the radius of convergence of the series?

GO ON TO THE NEXT PAGE.

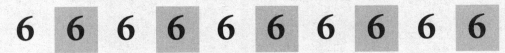

(c) Integrate the series term by term to get a power series for $\int f(x)\,dx$.

(d) Use the series from part (c) to show that $\dfrac{3}{1-\dfrac{x}{3}}$ is the antiderivative for $f(x)$ with $C = 3$.

(Note: The power series for $\dfrac{1}{1-x} = \displaystyle\sum_{n=0}^{\infty} x^n$.)

STOP

END OF EXAM

ANSWER KEY

Section I, Part A

1.	A	16.	B
2.	C	17.	B
3.	A	18.	C
4.	D	19.	C
5.	C	20.	C
6.	C	21.	D
7.	C	22.	C
8.	C	23.	A
9.	A	24.	B
10.	B	25.	C
11.	A	26.	D
12.	D	27.	D
13.	B	28.	A
14.	C	29.	D
15.	B	30.	A

Section I, Part B

76.	C
77.	B
78.	C
79.	B
80.	A
81.	A
82.	D
83.	C
84.	C
85.	D
86.	A
87.	B
88.	D
89.	D
90.	D

SCORING

Section II, Parts A & B

Use the scoring rubrics (found after the Section I answers and explanations) to self-score your free-response questions.

Section I, Part A Number Correct: _____

Section I, Part B Number Correct: _____

Section II, Part A Points Earned: _____

Section II, Part B Points Earned: _____

Enter your results to your Practice Exam 1 assignment to see your 1–5 score by logging in at kaptest.com.

Haven't registered your book yet? Go to kaptest.com/booksonline to begin.

BC PRACTICE EXAM 1 ANSWERS AND EXPLANATIONS

SECTION I, Part A

1. A

Plugging 0 in for x yields $\frac{0}{0}$, so manipulate the limit in some way. You could break it into pieces and use a special trig rule, but it's much easier here to use L'Hôpital's rule.

$$\lim_{x \to 0}\left(\frac{3x^2 + 5\cos x - 5}{2x}\right) = \lim_{x \to 0}\frac{6x - 5\sin x}{2}$$
$$= \frac{6(0) - 5\sin(0)}{2}$$
$$= \frac{0}{2} = 0$$

(A) is correct.

2. C

In general, the p-series $\sum_{n=1}^{\infty}\frac{1}{n^p}$ converges for $p > 1$.

(Remember, $\sum_{n=1}^{\infty}\frac{1}{n}$ does *not* converge, but $\sum_{n=1}^{\infty}\frac{1}{n^2}$ does.) You can use this to determine when $\sum_{n=1}^{\infty}\frac{1}{n^{2p}}$ converges.

In other words, you know that $\sum_{n=1}^{\infty}\frac{1}{n^{2p}}$ converges for $2p > 1$ or $p > \frac{1}{2}$. Thus, **(C)** is correct.

3. A

Though the function $x\sin(x^2)$ is itself fairly complicated, you don't need to compute all the derivatives in order to determine the power series. You know that a power series for $\sin x$ is given by:

$$\left(x - \frac{x^3}{3!} + \frac{x^5}{5!} - \frac{x^7}{7!} + \cdots\right)$$

This will give you a power series for $\sin(x^2)$ by composition:

$$\left((x^2) - \frac{(x^2)^3}{3!} + \frac{(x^2)^5}{5!} - \frac{(x^2)^7}{7!} + \cdots\right)$$
$$= \left(x^2 - \frac{x^6}{3!} + \frac{x^{10}}{5!} - \frac{x^{14}}{7!} + \cdots\right)$$

Finally, to get a power series for $x\sin(x^2)$, you just multiply the power series for x (which is just the series $(x + 0 + 0 + \cdots + 0 + \cdots)$ and $\sin(x^2)$, namely:

$$x\left(x^2 - \frac{x^6}{3!} + \frac{x^{10}}{5!} - \frac{x^{14}}{7!} + \cdots\right) = x^3 - \frac{x^7}{3!} + \frac{x^{11}}{5!} - \frac{x^{15}}{7!} + \cdots$$

The coefficient of x^3 is 1, which matches **(A)**.

4. D

Examine the discontinuity at each value of x. The graph is continuous at $x = 4$, so (A) is not correct. The discontinuity at $x = 3$ is a removable discontinuity (a hole), so (B) is not correct. The discontinuity at $x = 2$ is an infinite discontinuity (a vertical asymptote), so (C) is not correct. That leaves **(D)** as the correct choice. The discontinuity at $x = 1$ is a jump discontinuity because there is a gap in the graph there.

5. C

One interpretation of the first derivative of a function is that it represents the instantaneous rate of change, so calculate the derivative of the function and find the rate of change at the point specified. Because this question is in the no-calculator section of the test, the calculations should be fairly straightforward.

$$C'(x) = 8.6 + 2(0.5x) = 8.6 + x$$
$$C'(300) = 8.6 + 2(0.5 \times 300) = 308.6$$

That's **(C)**.

Note: The word "rate" in calculus nearly always means the derivative. The only time that it doesn't is when you are asked for "average rate."

6. C

To start with, how would you compute the total area enclosed by the outer curve?

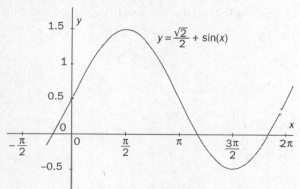

graph of $y = 1/2 + \sin(x)$

Here's a graph of $y = \dfrac{\sqrt{2}}{2} + \sin x$. Notice that it's positive except on the interval $\left(\dfrac{5\pi}{4}, \dfrac{7\pi}{4}\right)$. Plotting a few points if necessary, you see that the outer curve is given by $r = \dfrac{\sqrt{2}}{2} + \sin\theta$ on the interval $\left(-\dfrac{\pi}{4}, \dfrac{5\pi}{4}\right)$; i.e., exactly where the radius is positive. (Note: you can see by inspecting the graph that because the positive values of $y = \dfrac{\sqrt{2}}{2} + \sin x$ are much larger than the negative values, the points on the graph with positive radius correspond to the *outer curve*.) To compute the area of this region you need to compute

$$\int_{-\frac{\pi}{4}}^{\frac{5\pi}{4}} \frac{1}{2}[f(\theta)]^2\, d\theta = \int_{-\frac{\pi}{4}}^{\frac{5\pi}{4}} \frac{1}{2}\left(\frac{\sqrt{2}}{2} + \sin\theta\right)^2 d\theta$$

Just as above, the area of the inner shell is given by the integral

$$\int_{\frac{5\pi}{4}}^{\frac{7\pi}{4}} \frac{1}{2}[f(\theta)]^2\, d\theta = \int_{\frac{5\pi}{4}}^{\frac{7\pi}{4}} \frac{1}{2}\left(\frac{\sqrt{2}}{2} + \sin\theta\right)^2 d\theta$$

Finally, to compute the area you just take the difference. You need the polar integral over $\left(-\dfrac{\pi}{4}, \dfrac{5\pi}{4}\right)$ minus the polar integral over $\left(\dfrac{5\pi}{4}, \dfrac{7\pi}{4}\right)$.

$$\text{Area} = \int_{-\frac{\pi}{4}}^{\frac{5\pi}{4}} \frac{1}{2}\left(\frac{\sqrt{2}}{2} + \sin\theta\right)^2 d\theta - \int_{\frac{5\pi}{4}}^{\frac{7\pi}{4}} \frac{1}{2}\left(\frac{\sqrt{2}}{2} + \sin\theta\right)^2 d\theta$$

Thus, **(C)** is correct.

Note: If you are in a hurry you might think to compute the area in a polar equation just as you would for a standard function. That is, you might want to just integrate $\int r(\theta)\, d\theta$. Remember, the area formula in polar coordinates is:

$$A = \int_{\theta_1}^{\theta_2} \frac{1}{2}[f(\theta)]^2\, d\theta$$

7. C

The derivative at a point is defined as $\lim\limits_{h \to 0} \dfrac{f(x+h) - f(x)}{h}$. You can eliminate (B) right away because h is not approaching 0. The point $(2, 4)$ tells you that x is 2, so replace each x in the definition with a 2. This makes **(C)** the correct answer.

8. C

The derivative of a constant term (such as 17π) is 0, so you can eliminate (B) and (D) right away. For the other two terms, use the product rule, which states that $(fg)' = f'g + fg'$:

$$f(x) = 5x\sec x + x^3\cos x + 17\pi$$
$$f'(x) = (5)\sec x + 5x(\sec x \tan x)$$
$$+ (3x^2)\cos x + x^3(-\sin x) + 0$$
$$= 5\sec x + 5x\sec x \tan x + 3x^2\cos x - x^3\sin x$$

That's a perfect match for **(C)**.

9. A

You are given a piecewise function g and told that it is continuous for all real numbers. Both individual functions, or "pieces," are continuous on their own, so all you need to check is where the function potentially splits, which is at $x = 2$.

To be continuous at $x = 2$, $\lim\limits_{x \to 2^-} f(x)$ must equal $\lim\limits_{x \to 2^+} f(x)$, so substitute 2 for the variable in each piece of the function, set them equal, and solve for k. This gives $-(2)^2 + 4 = 3(2) + k$, or $0 = 6 + k$. Solving this equation yields $k = -6$, so **(A)** is correct.

10. B

The zeros of the derivative function $f'(x)$ are the values of x at which $f(x)$ has a horizontal tangent, in this case at the relative minimums and maximums. Here, this occurs at $x = -3$ (a relative minimum), $x = 1$ (a relative maximum), and $x = 4$ (a relative minimum). That's **(B)**.

11. A

At the heart of this word problem is a question about tangent lines to parametric curves. You are given a parametric curve in the xy-plane and asked to compute the slope of the tangent line at $t = 2$. In general, the trajectory of a point traveling along a curve given by a parametric equation is the vector $(x'(t), y'(t))$, which points along a line with slope $\frac{y'(t)}{x'(t)}$.

The equations for the curve are $\begin{matrix} y(t) = t^3 + 3t^2 + 4 \\ x(t) = 2t^2 + 2 \end{matrix}$. The general formula for the slope of the tangent line to a parametric curve is $\frac{y'(t)}{x'(t)}$ (this is just the infinitesimal version of "rise over run"). First you compute derivatives $\begin{matrix} y'(t) = 3t^2 + 6t \\ x'(t) = 4t \end{matrix}$. Now, you want the slope at $t = 2$:

$$\frac{y'(2)}{x'(2)} = \frac{3 \cdot 2^2 + 6 \cdot 2}{4 \cdot 2} = \frac{12 + 12}{8} = 3$$

Thus, **(A)** is correct.

12. D

Think about what the second derivative tells you about a graph: A graph is concave up (a cup that "holds water") if $f''(x) > 0$ (the second derivative is positive). A graph is concave down (a cup that "spills water") if $f''(x) < 0$. Start by identifying where the graph is concave up and where it is concave down. Put the x-values where it changes in a chart or on a number line like the one shown below. Then write down where f'' is positive and where it is negative:

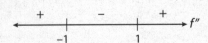

The graph is concave up ($f'' > 0$) to the left of $x = -1$ and to the right of $x = 1$, and concave down ($f'' < 0$) between -1 and 1. **(D)** is correct.

13. B

The slope of a curve at a given point means the slope of the tangent at that point. To find the slope of the tangent at $x = 2$, take the first derivative of the equation and plug in 2 for each x:

$$y = 4x^{-2} + \frac{1}{2}x^2 + 3$$

$$y' = -2 \cdot 4x^{-3} + 2 \cdot \frac{1}{2}x + 0$$

$$= \frac{-8}{x^3} + x$$

$$y'(2) = \frac{-8}{(2)^3} + 2$$

$$= \frac{-8}{8} + 2 = -1 + 2 = 1$$

(B) is correct.

14. C

In order to integrate this rational function you need to first compute the partial fraction decomposition. You start with your function $\frac{x+3}{(x+1)x}$ and try to write it as a sum: $\frac{A}{x+1} + \frac{B}{x}$. To solve this you add the two fractions by first forming their common denominator and then comparing that with your original equation.

$$\frac{x+3}{(x+1)x} = \frac{A}{x+1} + \frac{B}{x} = \frac{Ax}{(x+1)x} + \frac{B(x+1)}{x(x+1)}$$

$$\frac{x+3}{(x+1)x} = \frac{x(A+B) + 1(B)}{(x+1)x}$$

By comparing the coefficients of the numerator of both sides you have $\begin{matrix} A + B = 1 \\ B = 3 \end{matrix}$, and so $A = -2$. Finally you see that $\frac{x+3}{(x+1)x} = \frac{-2}{x+1} + \frac{3}{x}$.

Now integrate:

$$\int \frac{x+3}{(x+1)x}\,dx = \int \frac{-2}{x+1} + \frac{3}{x}\,dx = \int \frac{-2}{x+1}\,dx + \int \frac{3}{x}\,dx$$

$$= -2\int \frac{1}{x+1}\,dx + 3\int \frac{1}{x}\,dx = -2\ln|x+1| + 3\ln|x| + C$$

Thus, **(C)** is correct.

15. B

Let the horizontal distance from the bottom of the wall to the bottom of the ladder be x and let the vertical distance from the top of the ladder to the ground be y. Then, by the Pythagorean theorem, $x^2 + y^2 = 13^2$. You're given that $\frac{dx}{dt} = \frac{1}{2}$ (because the horizontal slide is $\frac{1}{2}$ foot per second), and you're looking for $\frac{dy}{dt}$ (the vertical slide). Use implicit differentiation to get:

$$2x\frac{dx}{dt} + 2y\frac{dy}{dt} = 0$$

At the time when $y = 5$, you know that $x = 12$ (again thanks to the Pythagorean theorem and the fact that the ladder is always 13 feet long). Substitute all this information into the derivative equation and solve for $\frac{dy}{dt}$:

$$2(12)\left(\frac{1}{2}\right) + 2(5)\frac{dy}{dt} = 0$$

$$12 + 10\frac{dy}{dt} = 0$$

$$10\frac{dy}{dt} = -12$$

$$\frac{dy}{dt} = -\frac{6}{5}$$

(B) is correct.

16. B

Cross-multiply so that all the x's are on one side and all the y's are on the other: $e^y\, dy = \sin x\, dx$. Next, integrate both sides:

$$\int e^y\, dy = \int \sin x\, dx$$

$$e^y = -\cos x + C$$

Plug in $y = 0$ and $x = \frac{\pi}{4}$ (because that was the other given information) to get:

$$e^0 = -\cos\frac{\pi}{4} + C$$

$$1 = -\frac{\sqrt{2}}{2} + C$$

$$\frac{\sqrt{2}}{2} + 1 = C$$

Now plug the value of C into the function. Then you need to write the equation in terms of y (not e^y), so take the natural log of both sides.

$$e^y = -\cos x + \frac{\sqrt{2}}{2} + 1$$

$$\ln(e^y) = \ln\left(-\cos x + \frac{\sqrt{2}}{2} + 1\right)$$

$$y = \ln\left(-\cos x + \frac{\sqrt{2}}{2} + 1\right)$$

That's (B).

17. B

You could try to split the given equation into a piecewise function (because there is an absolute value involved), but it's quicker to draw the graph of the integrand function, $y = 2 - |4 - x|$, then calculate areas. Transform the graph of $y = |x|$ by shifting right 4 units, flipping over the x-axis, and moving the graph up 2 units. The graph is shown below.

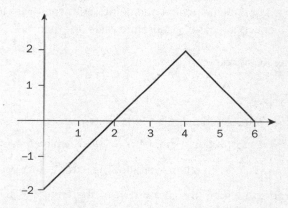

A definite integral is the sum of the areas bounded by the curve and the x-axis. Areas above the x-axis are given a positive sign, and areas below the x-axis are given a negative sign. Note that the triangle between $x = 0$ and $x = 2$ has an area of $\frac{1}{2}(2)(2) = 2$, but its contribution to the integral is -2 because the graph is below the x-axis. The area of the other triangle is $\frac{1}{2}(4)(2) = 4$, so the value of the definite integral is $-2 + 4 = 2$. (B) is correct.

18. C

First notice that $\sum_{n=0}^{\infty} \sin\left(\frac{\pi}{2}-\frac{1}{n}\right)$ diverges. To see this, observe that

$$\lim_{t \to \infty} \sin\left(\frac{\pi}{2}-\frac{1}{t}\right) = \sin\left(\frac{\pi}{2}-0\right) = 1$$

and so the series cannot converge (always remember that the first condition to check for series convergence is whether the limit of the summands is 0. $\lim_{n \to \infty} a_n = 0$ in order for $\sum_{n=0}^{\infty} a_n$ to converge). Notice that this immediately rules out (D). For statement I, you need to remember the order of growth of functions. Because you're comparing polynomials and exponential functions, this is relatively simple; exponentials dominate *any polynomial*, so $\sum_{n=1}^{\infty} \frac{n+1}{e^n}$ converges. For statement II, notice that $\sum_{n=1}^{\infty} \frac{e^n}{4^n} = \sum_{n=1}^{\infty} \left(\frac{e}{4}\right)^n$, so this is a geometric series with base $\frac{e}{4}$. Because $e \approx 2.71$, $\frac{e}{4} < 1$ and the series converges. (Remember, a geometric series converges if and only if the base has absolute value less than 1.)

Thus, **(C)** is correct.

19. C

When you see a complicated expression raised to a power, think *u*-substitution. Here, let $u = 7 + x^{\frac{2}{3}}$, which means $du = \frac{2}{3}x^{-\frac{1}{3}} dx$. To compensate for the extra $\frac{2}{3}$ that you need to multiply the dx portion of the integrand by, multiply the entire integral by $\frac{3}{2}$ and rearrange to get:

$$\frac{3}{2}\int \left(7 + x^{\frac{2}{3}}\right)^5 \times \frac{2}{3}x^{-\frac{1}{3}} dx$$

Written in terms of u, this is:

$$\frac{3}{2}\int u^5 \, du = \frac{3}{2} \times \frac{u^6}{6} + C = \frac{1}{4}u^6 + C$$

Converted back in terms of x, your final answer is $\frac{1}{4}\left(7 + x^{\frac{2}{3}}\right)^6 + C$, which is **(C)**.

20. C

This question asks you to compute an improper integral, which means that you will eventually need to compute more limits. To start, though, just find an antiderivative:

$$\int_1^{\infty} \frac{3}{x^2} \, dx = 3\int_1^{\infty} x^{-2} \, dx = -3x^{-1} \Big|_1^{\infty}$$

$$= \left(\lim_{x \to \infty} -3x^{-1}\right) - (-3 \cdot 1)$$

$$= -3\left(\lim_{x \to \infty} \frac{1}{x}\right) + 3 = 0 + 3 = 3$$

In order to evaluate, take the limit of the antiderivative as x goes to infinity (in place of evaluating the function at infinity), as above.

Thus, **(C)** is correct.

21. D

When you see the derivative of an integral, use the second fundamental theorem of calculus to evaluate the integral. Remember to use the chain rule and to take the derivative of both limits of integration. The derivative of 2 is 0 since it is a constant.

$$\left(e^{(t^4)^2} \cdot \frac{d}{dt}(t^4)\right) - \left(e^{2^2} \cdot \frac{d}{dt}(2)\right)$$
$$= e^{t^8} \cdot \left(4t^3\right) - e^4 \cdot 0$$

Simplify and rearrange to arrive at the correct answer, which is **(D)**.

22. C

This problem involves a repeated chain rule. The derivative of $(\text{something})^5 = 5(\text{something})^4 (\text{derivative of something})$. So in this case, the first step is:

$$5\left(\arccos\left(x^2\right)\right)^4 \cdot \frac{d}{dx}\arccos\left(x^2\right) \quad (*)$$

The derivative of arccos x is $\dfrac{-1}{\sqrt{1-x^2}}$. So:

$$\frac{d}{dx}\arccos\left(x^2\right) = \frac{-1}{\sqrt{1-\left(x^2\right)^2}} \cdot \frac{d}{dx}\left(x^2\right) \quad (**)$$

Apply the chain rule one more time and differentiate x^2 to get $2x$ (***). Combine all of the starred parts and simplify to get **(C)**.

23. A

Use the product rule along with the chain rule (in two places):

$$\frac{d}{dx}\left(\ln(3x)\cdot 5^{2x}\right) = \frac{d}{dx}\left(\ln(3x)\right)\cdot 5^{2x} + \ln(3x)\cdot \frac{d}{dx}\left(5^{2x}\right)$$

$$= \frac{1}{3x}\cdot \frac{d}{dx}(3x)\cdot 5^{2x} + \ln(3x)\cdot 5^{2x}$$

$$\ln(5)\cdot \frac{d}{dx}(2x)$$

$$= \frac{1}{3x}(3)\cdot 5^{2x} + \ln(3x)\cdot 5^{2x}\ln(5)\cdot 2$$

$$= \frac{5^{2x}}{x} + 2\ln(5)\ln(3x)5^{2x}$$

Factor out 5^{2x} to make your final answer look just like **(A)**.

24. B

The normal line to a curve is perpendicular to the tangent line. This means the slope of the normal line is the negative reciprocal of the slope of the line tangent at the given point. To find the slope of the tangent, use implicit differentiation. Don't forget to multiply by $\dfrac{dy}{dx}$ when you take the derivative of y. Also, use the product rule when differentiating the second term. The derivative is:

$$3x^2 + 1\cdot y^2 + x\cdot 2y\frac{dy}{dx} = 10\frac{dy}{dx}$$

You could solve for $\dfrac{dy}{dx}$ now, but because you are given values for x and y, plug those in first and then get $\dfrac{dy}{dx}$ by itself:

$$3(2)^2 + 1(1)^2 + (2)(2)(1)\frac{dy}{dx} = 10\frac{dy}{dx}$$

$$12 + 1 + 4\frac{dy}{dx} = 10\frac{dy}{dx}$$

$$13 = 6\frac{dy}{dx}$$

$$\frac{dy}{dx} = \frac{13}{6}$$

Thus, the slope of the tangent line at the point $(2, 1)$ is $\dfrac{dy}{dx} = \dfrac{13}{6}$. The negative reciprocal of this is $\dfrac{-6}{13}$, which is **(B)**.

25. C

Here you are given a relatively straightforward indefinite integral to compute. First look for a possible u-substitution. However, in this case no such substitution exists. That leads you next to integration by parts. (This is usually the next step for functions involving e^x, a popular element in integration-by-parts questions.)

First, recall:

$$\int u\, dv = uv - \int v\, du$$

The tactic involves simplifying one part of the function, u, (by differentiating) at the expense of complicating the other, dv, (by integrating). Because e^x integrated relatively easily, try $\begin{aligned}u &= x \\ dv &= e^{-3x}dx\end{aligned}$, which then makes:

$$du = dx$$

$$v = \int e^{-3x}\, dx = \frac{-1}{3}e^{-3x}$$

Plugging this into the formula above gives

$$\int xe^{-3x}\, dx = x\frac{-1}{3}e^{-3x} - \int \frac{-1}{3}e^{-3x}\, dx$$

$$= \frac{-xe^{-3x}}{3} + \frac{1}{3}\int e^{-3x}\, dx$$

$$= \frac{-xe^{-3x}}{3} + \frac{1}{3}\cdot \frac{-1}{3}e^{-3x} + C$$

$$= -\frac{1}{3}xe^{-3x} - \frac{1}{9}e^{-3x} + C$$

Thus, **(C)** is correct.

26. D

First, draw the region. The graph of $f(x)$ is a parabola in standard position. It is below the graph of $g(x)$, which is a line through the origin with a slope of 3. The graphs intersect at $x = 0$ and $x = 3$ (set the equations equal to each other to find where they intersect). The graph looks like the following:

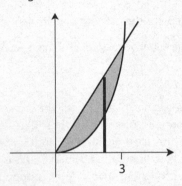

Use the washer method to find the volume. The vertical line between the curves represents the radius, $3x - x^2$. Each washer will have an area of $\pi((3x)^2 - (x^2)^2)$. Therefore, the volume can be found using:

$$\int_0^3 \pi\left((3x)^2 - \left(x^2\right)^2\right) dx$$

Simplify, then integrate to arrive at:

$$\int_0^3 \pi\left((3x)^2 - \left(x^2\right)^2\right) dx = \pi\int_0^3 \left(9x^2 - x^4\right) dx$$

$$= \pi\left[3x^3 - \frac{1}{5}x^5\right]_0^3$$

Now, evaluate from $x = 0$ to $x = 3$ to get $\pi\left(81 - \frac{243}{5} - 0\right) = \frac{162\pi}{5}$. **(D)** is correct.

27. D

There are at least two different approaches to this problem. The quickest way to determine which graph is the slope field is to plug in some "easy" numbers for x and/or y to find the slope. Then look at the graphs to see which ones have approximately that slope at those (x, y) points. First, notice that y cannot equal 0 (because it is in the denominator of the right side of the equation), so there cannot be any slopes drawn along the x-axis. Eliminate (A) and (C). At $x = 0$, the derivative is 0, so you must have short horizontal segments up the entire y-axis. This makes (D) correct. Also, the line segments corresponding to the points $(1, 1)$, $(2, 2)$, $(3, 3)$, and so on must all have a slope of 1.

Another way is to solve the differential equation given in the question. Because it is a first order, separable differential equation, this is straightforward. First, rewrite the original differential equation as: $y\,dy = x\,dx$. Then integrate both sides to obtain $\frac{1}{2}y^2 = \frac{1}{2}x^2 + C$ or $\frac{y^2}{2} - \frac{x^2}{2} = C$ for some constant C. Next, consider what the graph of such an equation looks like—it is a hyperbola. **(D)** most closely resembles a family of hyperbolas.

28. A

The problem requires you to change the variables from x to u. Make sure you change all the x's, including the dx and the limits of integration. Let:

$$u = 3x - 1 \Rightarrow du = 3\,dx \Rightarrow dx = \frac{1}{3}\,du$$

The limits of integration become $u = 3(-1) - 1 = -4$ and $u = 3(4) - 1 = 11$.

So:

$$\int_{-1}^{4} f(3x - 1)\,dx = \frac{1}{3}\int_{-4}^{11} f(u)\,du$$

(A) is correct.

29. D

First off, you need to recognize that you are modeling a population and $P(0) = a$ is the number of creatures you are starting with. Common sense should tell you that if you start with a population of 0, it will always stay 0. Looking at the model $\frac{dP}{dt} = 0.6P\left(1 - \frac{P}{400}\right)$, if $P = 0$, then $\frac{dP}{dt} = 0$, and so, as you'd expect, there is no population change and the population stays at 0. This should immediately rule out (A). Looking closer, $\frac{dP}{dt} = 0.6P\left(1 - \frac{P}{400}\right)$ is well behaved everywhere, and if $P > 400$, then the population is decreasing, because $\frac{dP}{dt} < 0$. If $P < 400$, then the population is increasing because $\frac{dP}{dt} > 0$. Finally, if $P = 400$, then $\frac{dP}{dt} = 0$, and again you have another constant solution. These clues all point to **(D)**.

In fact, the differential equation given is a *logistic differential equation* and limits to 400 for *every* nonzero initial condition. Below is a graph of the slope field given by

$$\frac{dP}{dt} = 0.6P\left(1 - \frac{P}{400}\right).$$

Thus, **(D)** is correct.

30. A

The function $g(x)$ is the inverse of $y = f(x)$. To find the inverse of a function, switch all the x's and y's, including the ones in the given point. The equation becomes $x = y \sec y$ and the point becomes $\left(\frac{2\pi}{3}, \frac{\pi}{3}\right)$.

Since x is now by itself, take the derivative of x with respect to y to make differentiating easier. Then flip both sides of the equation so that $\frac{dx}{dy}$ becomes $\frac{dy}{dx}$:

$$\frac{dx}{dy} = 1 \cdot \sec y + y \cdot \sec y \cdot \tan y$$

$$\frac{dy}{dx} = \frac{1}{1 \cdot \sec y + y \cdot \sec y \cdot \tan y}$$

Plug in the value of y from the new point and simplify. In the last step, multiply the numerator and the whole denominator by 3 to get rid of the fraction inside of the fraction:

$$\frac{dy}{dx} = \frac{1}{1 \cdot \sec\frac{\pi}{3} + \frac{\pi}{3} \cdot \sec\frac{\pi}{3} \cdot \tan\frac{\pi}{3}}$$

$$= \frac{1}{1 \cdot 2 + \frac{\pi}{3} \cdot 2 \cdot \sqrt{3}}$$

$$= \frac{1}{2 + \frac{\pi}{3} \cdot 2 \cdot \sqrt{3}} \times \frac{3}{3}$$

$$= \frac{3}{6 + 2\pi\sqrt{3}}$$

That's **(A)**.

SECTION I, Part B

76. C

The function f is increasing for all x such that $f'(x) > 0$. Use the power rule to take the derivative of each term.

The result is $f'(x) = 3x^2 + \frac{5}{2}x - \frac{59}{4}$. This equation is not easily factored, so use your calculator to find the zeros and then examine the graph to determine where it is greater than zero (above the x-axis):

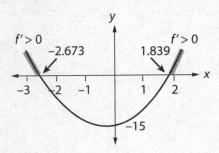

Based on the graph, $f'(x) > 0$ when $x < -2.673$ and when $x > 1.839$. **(C)** is correct.

77. B

You can do this problem in a number of ways. One way is to take the derivative by hand and then plug the information into your calculator. Use the chain rule to take the derivative. Start by taking the derivative of what is on the outside and leave the inside. Then multiply this by the derivative of the inside. The result is $\cos(\sin x) \cdot \cos x$. Now replace x with $\frac{\pi}{4}$ and plug into your calculator. This gives you an answer of 0.538, which is **(B)**. Note: Make sure your calculator is set to radian mode before you start the AP exam.

Alternatively, you could graph the equation in your calculator. On the $\boxed{Y=}$ screen enter the function and adjust the viewing window as needed. Press $\boxed{\text{CALC}}$ ($\boxed{\text{2nd}}$ $\boxed{\text{TRACE}}$) 6 (for the dy/dx option). Both methods yield the same answer.

78. C

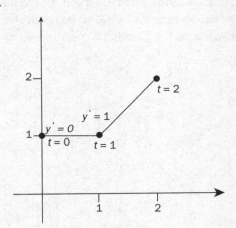

You are given the differential equation $y' = xy$ with the initial condition $y(0) = 1$ and asked to use Euler's method starting at $x = 0$ with step size 1, to approximate the value of $y(2)$. Euler's method works by approximating a function with its derivative. You start at the point $(0, 1)$ (remember your initial condition of $y(0) = 1$) and move in the direction given by $y' = xy$ for one step. At the point $(0, 1)$, $y' = 0 \cdot 1 = 0$. The slope is 0 so we move one step to the right along the line through $(0, 1)$ with slope 0. You're now at the point $(1, 1)$. Here $y' = 1 \cdot 1 = 1$, so you move one step to the right along the line through $(1, 1)$ with slope 1 and end up at the point $(2, 2)$. This is the final x-value and so your approximation of $y(2)$ is 2.

Thus, **(C)** is correct.

79. B

You're given a series, told that it's the Maclaurin series for a function, and asked to find that function. The most effective way to answer this question is through the process of elimination, using guess and check techniques. To start, you're told that this is the Maclaurin series, i.e., the Taylor series about 0. Thus you know the values for the function and all its derivatives at $x = 0$. The series is $x + x^3 + x^5 + x^7 + x^9 + \ldots$ so we know that

$$f(x) = 0$$
$$f'(x) = 1$$
$$f''(x) = 0$$

because these are the coefficients of 1 (x^0), x, and x^2, respectively, in the series. Start by looking at the values of the functions at $x = 0$.

(A) $e^{0^2} - 1 = 1 - 1 = 0$

(B) $\dfrac{0}{1 - 0^2} = 0$

(C) $\sin 0 - \cos(0^2) = -1$

(D) $\dfrac{1}{1 + 0^2} - 1 = 0$

That rules out (C). Now try the first derivative:

$f(x)$	$f'(x)$	$f'(0)$
$e^{x^2} - 1$	$e^{x^2} \cdot 2x$	$e^{0^2} \cdot 2 \cdot 0 = 0$
$\dfrac{x}{1 - x^2}$	$\dfrac{(1 - x^2)1 + x(2x)}{(1 - x^2)}$	$\dfrac{(1 - 0^2)1 - 0(2 \cdot 0)}{(1 - 0^2)^2}$ $= \dfrac{1}{1} = 1$
~~$\sin(x) - \cos(x^2)$~~		
$\dfrac{1}{1 + x^2} - 1$ $= (1 + x^2)^{-1} - 1$	$-1(1 + x^2)^{-2} \cdot 2x$	$-1(1 + 0^2)^{-2} \cdot 2 \cdot 0 = 0$

This rules out all choices other than $\dfrac{x}{1 - x^2}$. Computing the derivatives of $\dfrac{x}{1 - x^2}$ and $\dfrac{1}{1 + x^2} - 1$ can be time consuming sometimes, so look at these a little closer to see why you get the answer you do. Now, you should know the formula $\dfrac{1}{1 - x} = \sum\limits_{n=0}^{\infty} x^n$. This is just the formula for summing a geometric series. This will give you a formula for $\dfrac{1}{1 + x^2}$:

$$\frac{1}{1 - (-x^2)} = \sum_{n=0}^{\infty} (-x^2)^n = \sum_{n=0}^{\infty} (-1)^n x^{2n}$$

Even subtracting off 1, you don't get anything that even remotely resembles $x + x^3 + x^5 + x^7 + \cdots$. However, using this technique for $\dfrac{1}{1 - x^2}$ gives:

$$\frac{1}{1 - x^2} = \sum_{n=0}^{\infty} (x^2)^n = \sum_{n=0}^{\infty} x^{2n}$$

This gives you the power series:

$$\frac{x}{1 - x^2} = x \frac{1}{1 - x^2} = x \left(\sum_{n=0}^{\infty} x^{2n} \right) = \sum_{n=0}^{\infty} x^{2n+1}$$

This is looking better. In fact, you've got it. In a different notation you have the Maclaurin series for $\dfrac{1}{1 - x} = 1 + x + x^2 + x^3 + x^4 + \cdots$, so a power series for

$$\frac{1}{1 - x^2} = 1 + x^2 + x^4 + x^6 + x^8 + \cdots, \text{ and}$$

$$\frac{x}{1 - x^2} = x \left(\frac{1}{1 - x^2} \right) = x(1 + x^2 + x^4 + x^6 + x^8 + \cdots)$$

$$= x + x^3 + x^5 + x^7 + x^9 \cdots.$$

If you are worried that all you have is *some* power series for your function, and that you might get a different one by constructing the Taylor series using all of the derivatives, just remember that the Taylor series for a power series is the power series itself (just take the derivatives); so as soon as you have any power series, you have the Taylor (and Maclaurin) series.

Thus, **(B)** is correct.

80. A

This is a related rates question because you're being asked to relate the rate of change in the volume (how fast the balloon is being inflated) to the rate of change in the radius. This means you take the derivative of the equation that relates volume and radius, which is $V = \dfrac{4}{3}\pi r^3$.

The rates of change are given by the derivatives, so differentiate both sides of the equation:

$$\frac{dV}{dt} = 3 \times \frac{4}{3}\pi r^2 \frac{dr}{dt} = 4\pi r^2 \frac{dr}{dt}$$

The problem is asking for the rate of change of the radius, or $\dfrac{dr}{dt}$, when $r = 5$, given that the rate of change of the volume is $\dfrac{dV}{dt} = 3$. (Pay attention to the units for each number given. Cubic inches is a measurement of volume.)

Substitute this information into the differentiated equation and solve for $\frac{dr}{dt}$:

$$\frac{dV}{dt} = 4\pi r^2 \frac{dr}{dt}$$

$$3 = 4\pi(5)^2 \frac{dr}{dt}$$

$$\frac{3}{100\pi} = \frac{dr}{dt}$$

Use your calculator to find that this is approximately 0.010 inches per second, or **(A)**.

81. A

The only function whose derivative is a constant times itself is $y = Ce^{kx}$, or using the variables given, $P = Ce^{kt}$. If you don't notice that, you could solve the rate equation for P and then integrate. Either way, you arrive at $P = Ce^{kt}$. You are given additional information that will allow you to find the values of C and k. $P = 12$ when $t = 0$ (initial population) and $P = 60$ when $t = 10$. Plug these in to solve for C and k, one at a time.

To find C, plug in $P = 12$ and $t = 0$. The result is $12 = Ce^0$, or $12 = C$. To find k, plug in $P = 60$, $t = 10$, and the newly found $C = 12$. Remember, to ultimately get rid of the e, you can take the natural log of both sides and simplify:

$$p = Ce^{kt}$$

$$60 = 12e^{k \cdot 10}$$

$$5 = e^{10k}$$

$$\ln 5 = \ln e^{10k}$$

$$\ln 5 = 10k$$

$$k = \frac{\ln 5}{10} \approx 0.161$$

(A) is correct.

82. D

First, note that the width of all the rectangles in this problem is 2 (because that is the distance between all the x-values). The height of each of the rectangles is the y-value on the right-hand side of each subinterval (because the question instructs you to use a right Riemann sum). To find the area of a rectangle, multiply its width times its height. Then, add the areas to find the total area under the curve. The quickest way to do this is to add the required heights (the right sides of the rectangles only) together first and then multiply by 2 (the width):

$$g(2) + g(4) + g(6) + g(8) + g(10)$$

$$= 25 + 30 + 16 + 25 + 32$$

$$= 128$$

Therefore, the approximate area under the curve on the closed interval [0, 10] is $2 \times 128 = 256$, which is **(D)**.

Note that you could also multiply each width and height together first and then add the products, but this may take just a bit longer.

83. C

A particle is slowing down when the velocity and acceleration have opposite signs. This means you need to take the derivative of s to find velocity, v, and the second derivative to find acceleration, a. The result is $v(t) = s'(t) = t^2 - 8t + 12$ and $a(t) = v'(t) = 2t - 8$. The easiest way to determine where v and a have opposite signs is to graph their equations on your calculator. You're only interested in the interval from 0 to 10, so adjust your window accordingly. The graph should look like this:

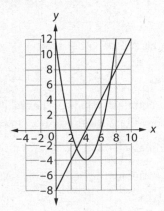

Examine the graph to find that the velocity and acceleration have opposite signs (one is above the x-axis while the other is below) on the intervals $0 \leq t \leq 2$ and $4 \leq t \leq 6$, which is **(C)**.

Note that you could also find where v and a are equal to zero and where they are positive and negative by hand, but using your calculator is the more efficient method.

84. C

Your first task is to determine the graphical relationship between the two curves, $y = x^2 + 4$ and $y = -2x + 1$. Using a graphing calculator, you can quickly see that $y = x^2 + 4$ is always above $y = -2x + 1$ on the closed interval $[-2, 5]$.

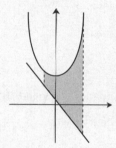

Therefore, the area in question is found by computing the value of

$$\int_{-2}^{5} \left((x^2 + 4) - (-2x + 1) \right) dx = \int_{-2}^{5} \left(x^2 + 2x + 3 \right) dx$$

This can be done by hand or by calculator. To evaluate the integral on the calculator, enter the simplified integrand into $\boxed{Y=}$ and adjust the viewing window as needed. Press $\boxed{CALC}$ ($\boxed{2^{nd}}$ $\boxed{TRACE}$) 7 (for the $\int f(x)\, dx$ option).

You'll also need to enter the limits of integration (here, the lower limit is -2 and the upper limit is 5). The result is approximately 86.333, or **(C)**.

85. D

For this question you modify your given series until you have a geometric series that you know how to solve. For example,

$$\sum_{n=1}^{\infty} \frac{(-3)^{n+1}}{4^n} = -3 \sum_{n=1}^{\infty} \frac{(-3)^n}{4^n}$$

You know how to compute this one:

$$-3 \sum_{n=1}^{\infty} \frac{(-3)^n}{4^n} = -3 \sum_{n=1}^{\infty} \left(\frac{-3}{4} \right)^n = -3 \times \frac{-\frac{3}{4}}{1 - \left(\frac{-3}{4} \right)}$$

$$= -3 \times \frac{-\frac{3}{4}}{\frac{7}{4}} = -3 \times \left(-\frac{3}{4} \cdot \frac{4}{7} \right) = \frac{9}{7}$$

Thus, **(D)** is correct.

86. A

The average value of a function f on the closed interval $[a, b]$ is given by:

$$\frac{1}{b-a} \int_{a}^{b} f(x)\, dx$$

Here, the average value is:

$$\frac{1}{8-5} \int_{5}^{8} \frac{2x}{x^2 - 4}\, dx = \frac{1}{3} \int_{5}^{8} \frac{2x}{x^2 - 4}\, dx$$

Because this question is in the calculator section, you can save yourself some time by using the integration function on your calculator. Plug the integrand $\left(\frac{2x}{x^2 - 4} \right)$ into the $\boxed{Y=}$ window, adjust the viewing window as needed, and press $\boxed{CALC}$ ($\boxed{2^{nd}}$ $\boxed{TRACE}$) 7. Use a lower limit of 5 and an upper limit of 8. The definite integral is approximately 1.050—but don't pick (D)! You need to divide this by 3 (because of the $\frac{1}{3}$ multiplier). The result is 0.350, which is **(A)**.

Alternatively, you could integrate by hand: Using u-substitution with $u = x^2 - 4$ and $du = 2x$, you have:

$$\frac{1}{3} \int \frac{2x}{x^2 - 4}\, dx = \frac{1}{3} \int \frac{1}{u}\, du$$

$$= \frac{1}{3} \ln u + C$$

$$= \frac{1}{3} \ln(x^2 - 4) + C$$

On the closed interval $[5, 8]$, the average value is:

$$\frac{1}{3} \left(\ln(8^2 - 4) - \ln(5^2 - 4) \right) \approx 0.350$$

87. B

Remember, for a particle traveling along a parametric curve, the speed of the particle is given by the equation $\sqrt{x'(t)^2 + y'(t)^2}$ (this is just the length of the tangent vector). For the parametric equation you were given, $x'(t) = e^t$ and $y'(t) = 2t$. To compute the speed at $t = 2$, you just plug in to get:

$$\sqrt{x'(2)^2 + y'(2)^2} = \sqrt{(e^2)^2 + (2 \cdot 2)^2} = 8.402$$

Thus, **(B)** is correct.

88. D

Acceleration is the derivative of velocity, so integrate $a(t)$ to find $v(t)$. Before you integrate, rewrite the radical term using a fractional exponent:

$$\int a(t) = \int \left[t^2 + (t+9)^{\frac{1}{2}} + e^{-t} \right] dt$$

$$v(t) = \frac{t^3}{3} + \frac{(t+9)^{\frac{3}{2}}}{\frac{3}{2}} + (-1)e^{-t} + C$$

$$v(t) = \frac{t^3}{3} + \frac{2}{3}(t+9)^{\frac{3}{2}} - e^{-t} + C$$

You're given that $v(0) = 4$, so:

$$4 = 0 + \frac{2}{3}(9)^{\frac{3}{2}} - e^0 + C$$

$$4 = 18 - 1 + C \rightarrow C = -13, \text{ so:}$$

$$v(t) = \frac{t^3}{3} + \frac{2}{3}(t+9)^{\frac{3}{2}} - e^{-t} - 13$$

Next, velocity is the derivative of position, so integrate $v(t)$ to find $s(t)$ and then use $s(0) = 5$ to find the position function:

$$s(t) = \frac{t^4}{12} + \frac{4}{15}(t+9)^{\frac{5}{2}} + e^{-t} - 13t + C$$

$$5 = 0 + \frac{4}{15}(9)^{\frac{5}{2}} + 1 + 0 + C$$

$$C = -\frac{304}{5}, \text{ so:}$$

$$s(t) = \frac{t^4}{12} + \frac{4}{15}(t+9)^{\frac{5}{2}} + e^{-t} - 13t - \frac{304}{5}$$

Evaluate this equation at $t = 7$ to get the final answer, which is **(D)**.

$$s(7) = \frac{7^4}{12} + \frac{4}{15}(7+9)^{\frac{5}{2}} + e^{-7} - 13(7) - \frac{304}{5}$$

$$= 321.351$$

89. D

The coefficient of $(x-3)^2$ in the degree four Taylor polynomial (in fact, in any degree Taylor polynomial) $P(x)$ is equal to $\frac{f''(3)}{2!}$. In this case the coefficient is 3, so

$$\frac{f''(3)}{2} = 3, \text{ or } f''(3) = 2 \cdot 3 = 6.$$

Thus, **(D)** is the answer.

90. D

The differential equation supplied in the question is separable, so rewrite it as $y \, dy = \frac{5}{x+4} dx$.

Integrating both sides yields:

$$\frac{1}{2}y^2 = 5\ln|x+4| + C$$

or

$$y^2 = 10\ln|x+4| + C$$

for some constant C. You're given that $y = 0$ at $x = (e-4)$, so:

$$0^2 = 10\ln|e-4+4| + C$$

$$0 = 10\ln|e| + C$$

$$0 = 10(1) + C \rightarrow C = -10$$

Hence, $y^2 = 10\ln|x+4| - 10$, which is **(D)**.

SECTION II, Part A

1. For calculator-based problems, it is enough to write down the integral you put into the calculator followed by the answer it yields. If you make any algebraic simplifications, though, or separate your integral into more than one piece, make sure to illustrate that in your answer. You will lose points if you lose significant digits by rounding two integrals and then summing them. Similarly, do not write more than is needed. If you had your calculator return $C'(t) = 9.309$ for part (b), it is not necessary to do the explicit derivation yourself. Finally, in part (d), make the calculator do all the work. Be familiar with your calculator's solve features and know how to use them for problems like this. Never forget your units!

(a)

Solutions	Scoring Rubric
At $t = 4$, $C(t) = 5.12 > 0$, so the amount of charge is increasing.	1: stating increasing 1: valid reasoning

(b)

Solutions	Scoring Rubric
$C'(t) = 0 + 6\cos\left(\dfrac{t^2}{3}\right)\left(\dfrac{2t}{3}\right) = 4t\cos\left(\dfrac{t^2}{3}\right)$ At $t = 4$, $C'(t) = 9.309 > 0$, so the rate of change of the charge in the battery is increasing at $t = 4$.	1: stating increasing 1: valid reasoning

(c)

Solutions	Scoring Rubric
$\dfrac{1}{10-2}\displaystyle\int_2^{10} C(t)\, dt$ $= \dfrac{1}{10-2}\displaystyle\int_2^{10} 10 + 6\sin\left(\dfrac{t^2}{3}\right) dt$ $= 10.264$ C/hr	1: correct formula for average 1: 10.264 1: correct units

(d)

Solutions	Scoring Rubric
$160 = \displaystyle\int_0^s C(t)\, dt = \displaystyle\int_0^s 10 + 6\sin\left(\dfrac{t^2}{3}\right) dt$ This happens at $s = 15.298$ hours.	1: correct set up of integral 1: 15.298 hours

2. Once you know what the integral needs to be for parts (a) and (b), make the calculator do all the work. The less you do by hand, the less likely you are to make a silly mistake. For part (b), use your calculator's solve feature or plot the integral. Remember, for Euler's method, you estimate the graph of the integral by a sequence of line segments where the slope is determined by the derivative of the function at the starting point of the line segment.

(a)

Solutions	Scoring Rubric
$P(t) = \int_0^t P'(s)\,ds + P(0) = \int_0^t \ln(1 + 5s^2)\,ds + P(0)$ Calculating at $t = 50$ gives $P(50) = 373.075 + P(0)$. So the population in 1950 is approximately 377,075.	1: appropriately set up integral 1: 373.075 1: 377,075

(b)

Solutions	Scoring Rubric
$1{,}000 = P(t) = \int_0^t P'(s)\,ds + P(0) = \int_0^t \ln(1 + 5s^2)\,ds + P(0)$ Calculating, this occurs at $t \approx 110.314$. Further, $\int_0^{110} \ln(1 + 5t^2)\,dt + P(0) \approx 997$ $\int_0^{111} \ln(1 + 5t^2)\,dt + P(0) \approx 1008$ So the population reaches 1 million between $t = 110$ and $t = 111$, that is, in the year 2010.	1: correct set up of integral 1: 110.314 1: 2010

(c)

Solutions	Scoring Rubric
The starting point is the initial population $t = 0$, $P_0 = 4$. $t = 0 \qquad\quad P_1 = 4 + 25 \cdot P'(0)$ $t = 25 \qquad\; P_2 = P_1 + 25 \cdot P'(25)$ $t = 50 \qquad\; P_3 = P_2 + 25 \cdot P'(50)$ $t = 75 \qquad\; P_4 = P_3 + 25 \cdot P'(75)$ $P = 4 + 25 \cdot (P'(0) + P'(25) + P'(50) + P'(75)) = 697{,}138$ So using Euler's method, the population in 2000 is 697,138.	1: correctly state $t = 0$, $P_0 = 4$ 1: correct set up of Euler's formula 1: 697,138

SECTION II, Part B

3. (a)

Solutions	Scoring Rubric
f' changes sign only at $x = 4$, so f has a relative extremum only there. Because f' changes from positive to negative there, it is a local max.	1: $x = 4$ 1: maximum 1: valid reasoning

(b)

Solutions	Scoring Rubric
The inflection points of f are the relative extrema of f'. These occur at $x = -3, 2$.	1: $x = -3$ 1: $x = 2$ 1: valid reasoning

(c)

Solutions	Scoring Rubric
Because f' is decreasing on $[-6, -3]$, f is concave down on that interval. Because f' is increasing on $[0, 2]$, f is concave up on that interval.	1: concave down on $[-6, -3]$ 1: concave up on $[0, 2]$ 1: valid reasoning

4. (a)

Solutions	Scoring Rubric
$y^2 + x\left(2y\dfrac{dy}{dx}\right) + 3y^2\dfrac{dy}{dx} = 0$ $x\left(2y\dfrac{dy}{dx}\right) + 3y^2\dfrac{dy}{dx} = -y^2$ $\dfrac{dy}{dx}\left(2xy + 3y^2\right) = -y^2$ $\dfrac{dy}{dx} = \dfrac{-y^2}{\left(2xy + 3y^2\right)}$	1: correct implicit differentiation of function 1: correct solution for $\dfrac{dy}{dx}$

(b)

Solutions	Scoring Rubric
Horizontal tangents occur when $\dfrac{dy}{dx} = 0$. By (a), this could only occur if $y = 0$. Plugging $y = 0$ into the equation for C gives $0 = 5$, so there are no points on the curve with $y = 0$ and no horizontal tangents.	1: correctly defined horizontal tangent 1: no tangents 1: valid reasoning

(c)

Solutions	Scoring Rubric
Vertical tangencies can only occur if $\frac{dy}{dx}$ is undefined. $\left(2xy + 3y^2\right) = 0$ $y\left(2x + 3y\right) = 0$ By (b), $y \neq 0$. So, $(2x + 3y) = 0$ or $x = -\frac{3}{2}y$. Plugging this back into the equation for C gives: $-\frac{3}{2}y \cdot y^2 + y^3 = 5$ $-\frac{3}{2}y^3 + y^3 = 5$ $-\frac{1}{2}y^3 = 5$ $y = \sqrt[3]{-10}$ So, $x = -\frac{3}{2}y = -\frac{3}{2}\sqrt[3]{-10} = \frac{3}{2}\sqrt[3]{10}$ $(x, y) = \left(\frac{3}{2}\sqrt[3]{10}, \sqrt[3]{-10}\right)$ The point $\left(\frac{3}{2}\sqrt[3]{10}, \sqrt[3]{-10}\right)$ is where the function is undefined because it makes the denominator of $\frac{dy}{dx}$ equal to 0.	1: correctly defined vertical tangent 1: $y = \sqrt[3]{-10}$ 1: $x = \frac{3}{2}\sqrt[3]{10}$ 1: valid reasoning

5. (a)

Solutions	Scoring Rubric
$f(x) = g(x)$ $ax(2 - x) = ax$ $x = 0$ or $2 - x = 1$ $x = 1$ $\int_0^1 f(x) - g(x)\, dx = \int_0^1 \left(2ax - ax^2\right) - ax\, dx$ $\int_0^1 ax - ax^2\, dx = \frac{ax^2}{2} - \frac{ax^3}{3}\Big|_0^1 = \frac{a}{2} - \frac{a}{3} = \frac{a}{6}$	1: limits 0, 1 1: correctly set up integral 1: $\frac{a}{6}$

(b)

Solutions	Scoring Rubric
$$\int_0^1 \pi\big(f(x)\big)^2 - \pi\big(g(x)\big)^2 \, dx$$ $$= \int_0^1 \pi\big(2ax - ax^2\big)^2 - \pi(ax)^2 \, dx$$ $$= \pi\int_0^1 4a^2x^2 - 4a^2x^3 + a^2x^4 - a^2x^2 \, dx$$ $$= \pi\left(a^2x^3 - a^2x^4 + \frac{a^2x^5}{5}\right)\Bigg|_0^1$$ $$= \pi\left(a^2 - a^2 + \frac{a^2}{5}\right) - 0 = \frac{\pi a^2}{5}$$ The cross section is a washer.	1: limits 0, 1 1: correctly set up integral 1: $\dfrac{\pi a^2}{5}$ 1: washer

(c)

Solutions	Scoring Rubric
The volume created by the cross-section R and thickness π is $\left(\dfrac{a}{6}\right)\pi = \dfrac{\pi a}{6}$. You need to find the positive value of a for which $\dfrac{\pi a}{6} = \dfrac{\pi a^2}{5}$. Therefore, for $a = \dfrac{5}{6}$ the volumes are equal.	1: $V = \dfrac{\pi a}{6}$ 1: $a = \dfrac{5}{6}$

6. (a)

Solutions	Scoring Rubric
The formula for the Taylor series for f about $x = 0$ is $$\sum_{n=0}^{\infty} \frac{f^{(n)}(0)}{n!}x^n = \sum_{n=0}^{\infty} \frac{(n+1)!}{n!3^n}x^n = \sum_{n=0}^{\infty} \frac{(n+1)}{3^n}x^n$$ You just plug the formula $f^{(n)}(0) = \dfrac{(n+1)!}{3^n}$ into the formula for the general term of the Taylor series: $\dfrac{f^{(n)}(0)}{n!}x^n$.	1: correctly set up Taylor series 1: reduce properly to get $$\sum_{n=0}^{\infty} \frac{(n+1)x^n}{3^n}$$

(b)

Solutions	Scoring Rubric
You can determine the radius of convergence for this series using the ratio test. $$\lim_{x \to \infty} \left| \frac{a_{n+1}}{a_n} \right| = \lim_{x \to \infty} \left| \frac{\frac{(n+2)}{3^{n+1}} x^{n+1}}{\frac{(n+1)}{3^n} x^n} \right| =$$ $$\lim_{x \to \infty} \left| \frac{3^n (n+2)}{3^{n+1}(n+1)} x \right| = \lim_{x \to \infty} \left| \left(\frac{n+2}{n+1} \right) \left(\frac{x}{3} \right) \right|$$ $$= \lim_{x \to \infty} \frac{(n+2)}{(n+1)} \left| \frac{x}{3} \right| = \left| \frac{x}{3} \right|$$ The series will diverge when the above limit is greater than 1 and will converge when the limit is less than 1. Therefore the series will converge for all x such that $\left| \frac{x}{3} \right| < 1 \Rightarrow |x| < 3$. The radius of convergence for this series is 3. Because you are not asked for the interval of convergence, you do not have to test the endpoints.	1: correctly set up ratio test. 1: $\left| \frac{x}{3} \right|$ 1: radius of convergence is 3.

(c)

Solutions	Scoring Rubric
Integrating the general term of the Taylor series of f gives $$\int \frac{(n+1)}{3^n} x^n \, dx = \frac{n+1}{3^n} \frac{x^{n+1}}{n+1} + C = \frac{x^{n+1}}{3^n} + C$$ A term-by-term integration of the Taylor series of f then gives the series $$\sum_{n=0}^{\infty} \int \frac{(n+1)x^n}{3^n} \, dx = \sum_{n=0}^{\infty} \frac{x^{n+1}}{3^n} + C$$	1: $\frac{x^{n+1}}{3^n} + C$ 1: $\sum_{n=0}^{\infty} \frac{x^{n+1}}{3^n} + C$

(d)

Solutions	Scoring Rubric
$3\dfrac{1}{1-\dfrac{x}{3}} = 3\displaystyle\sum_{n=0}^{\infty}\left(\dfrac{x}{3}\right)^n = \sum_{n=0}^{\infty}\dfrac{x^n}{3^{n-1}} = \sum_{n=0}^{\infty}\dfrac{x^{n+1}}{3^n} + 3$	1: properly sub in information given $3\displaystyle\sum_{n=0}^{\infty}\left(\dfrac{x}{3}\right)^n$
In part (c), Taylor's theorem tells you that on the interval of convergence, you can interchange a function and its Taylor series in almost any situation. In particular, suppose that $g'(x) = f(x)$, and so $g^{(n+1)}(x) = f^{(n)}(x)$. Then	1: prove through ending $\displaystyle\sum_{n=0}^{\infty}\dfrac{x^{r+1}}{3^n} + 3$
$\displaystyle\int\dfrac{f^{(n)}(0)}{n!}x^n = \dfrac{f^{(n)}(0)}{n!}\cdot\dfrac{x^{n+1}}{n+1} = \dfrac{g^{(n+1)}}{(n+1)!}x^{n+1}.$	
So integrating the Taylor series for f gives you the Taylor series for g, up to the constant. In fact, this tells you some information for part (d) as well. This means that the Taylor series for any power series is just the power series itself. You can also answer part (d) by going the other direction. Start with the antiderivative and show that it is the same as the power series for $\dfrac{3}{1-\dfrac{x}{3}}$, basically working backwards.	
$\displaystyle\sum_{n=0}^{\infty}\dfrac{x^{n+1}}{3^n} + 3 = 3\sum_{n=0}^{\infty}\dfrac{x^{n+1}}{3^{n+1}} + 3 = 3\left(\sum_{n=1}^{\infty}\dfrac{x^n}{3^n} + 1\right)$	
$= 3\left(\displaystyle\sum_{n=0}^{\infty}\dfrac{x^n}{3^n}\right) = 3\left(\dfrac{1}{1-\left(\dfrac{x}{3}\right)}\right) = \dfrac{3}{1-\dfrac{x}{3}}$	

BC Practice Exam 2

Section I, Part A

1. (A)(B)(C)(D)
2. (A)(B)(C)(D)
3. (A)(B)(C)(D)
4. (A)(B)(C)(D)
5. (A)(B)(C)(D)
6. (A)(B)(C)(D)
7. (A)(B)(C)(D)
8. (A)(B)(C)(D)
9. (A)(B)(C)(D)
10. (A)(B)(C)(D)
11. (A)(B)(C)(D)
12. (A)(B)(C)(D)
13. (A)(B)(C)(D)
14. (A)(B)(C)(D)
15. (A)(B)(C)(D)
16. (A)(B)(C)(D)
17. (A)(B)(C)(D)
18. (A)(B)(C)(D)
19. (A)(B)(C)(D)
20. (A)(B)(C)(D)
21. (A)(B)(C)(D)
22. (A)(B)(C)(D)
23. (A)(B)(C)(D)
24. (A)(B)(C)(D)
25. (A)(B)(C)(D)
26. (A)(B)(C)(D)
27. (A)(B)(C)(D)
28. (A)(B)(C)(D)
29. (A)(B)(C)(D)
30. (A)(B)(C)(D)

Section I, Part B

76. (A)(B)(C)(D)
77. (A)(B)(C)(D)
78. (A)(B)(C)(D)
79. (A)(B)(C)(D)
80. (A)(B)(C)(D)
81. (A)(B)(C)(D)
82. (A)(B)(C)(D)
83. (A)(B)(C)(D)
84. (A)(B)(C)(D)
85. (A)(B)(C)(D)
86. (A)(B)(C)(D)
87. (A)(B)(C)(D)
88. (A)(B)(C)(D)
89. (A)(B)(C)(D)
90. (A)(B)(C)(D)

A A A A A A A A A A A A A A A A A A A A A A A A A A A A A A A A A

CALCULUS BC
SECTION I, Part A
Time—60 minutes
Number of questions—30

A CALCULATOR MAY NOT BE USED ON THIS PART OF THE EXAM.

Directions: Use the space provided to solve each of the following problems. After examining the given choices, decide which of the choices is the best and fill in the corresponding circle on the answer sheet. Credit will only be given for answers appropriately marked on the answer sheet, and not for anything written in the exam book. Do not spend too much time on any one problem.

In this exam:

(1) The domain of a function f should be assumed to be the set of all real numbers x for which $f(x)$ is a real number, unless the question indicates otherwise.

(2) The inverse function notation f^{-1} or the prefix "arc" (e.g., $\sin^{-1} x = \arcsin x$) may be used to indicate the inverse of a trigonometric function.

GO ON TO THE NEXT PAGE.

AAAAAAAAAAAAAAAAAAAAAAAAAAAAAAAAAAA

1. Suppose that the function f satisfies $f(x) = 3x^2 - \sin \pi x$. Then the slope of the line tangent to the graph of f at the point $x = 2$ is

   (A) $8 - \dfrac{1}{\pi}$

   (B) $12 - \pi$

   (C) 12

   (D) 24

2. Which of the following is a necessary and sufficient condition for a function f to be continuous at $x = 4$?

   (A) $\lim\limits_{x \to 4} f(x)$ exists.

   (B) f is differentiable at $x = 4$.

   (C) $f(4)$ exists.

   (D) $\lim\limits_{x \to 4} f(x) = f(4)$.

3. Consider a continuous function f with the properties that f is concave up on the closed interval $[-1, 3]$ and concave down on the closed interval $[3, 5]$. Which of the following statements is true?

   (A) $f''(2) > 0$ and $f''(4) < 0$

   (B) $f''(2) < 0$ and $f''(4) > 0$

   (C) $f''(3) > 0$ and $x = 3$ is a point of inflection of f

   (D) $f''(3) < 0$ and $x = 3$ is a point of inflection of f

GO ON TO THE NEXT PAGE.

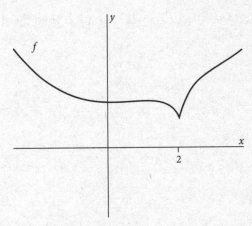

4. The graph of a function f is given above. Based on the graph, which of the following statements is true?

    (A) f is continuous and differentiable everywhere and $f'(2) = 0$.

    (B) f is continuous everywhere and differentiable everywhere except at $x = 2$.

    (C) f is discontinuous at $x = 2$ and f is differentiable everywhere except at $x = 2$.

    (D) f is discontinuous at $x = 2$, f is differentiable everywhere, and $f'(2) = 0$.

5. The position vector for a particle moving in the xy-plane is given by $p(t) = \langle 3t^2, 2\sqrt{t} \rangle$. What is the velocity vector of the particle at time $t = 4$?

    (A) $\left\langle 24, \dfrac{1}{2} \right\rangle$

    (B) $\langle 48, 4 \rangle$

    (C) $\left\langle 64, \dfrac{32}{3} \right\rangle$

    (D) $\left\langle 4, \sqrt{145} \right\rangle$

GO ON TO THE NEXT PAGE.

A A A A A A A A A A A A A A A A A A A A A A A A A A A A A A A A A A A A A A A

6. The derivative of $f(x) = \dfrac{x^3 \sin x}{x^2 + 3}$ is given by which of the following?

(A) $f'(x) = \dfrac{3x \cos x}{2}$

(B) $f'(x) = \dfrac{x^3 \cos x + 3x^2 \sin x}{x^2 + 3} - \dfrac{2x^4 \sin x}{\left(x^2 + 3\right)^2}$

(C) $f'(x) = \dfrac{x^2 \cos x + 3x \sin x}{2}$

(D) $f'(x) = \dfrac{\left(3x^2 \cos x\right)\left(x^2 + 3\right) - 2x^4 \sin x}{\left(x^2 + 3\right)^2}$

7. Which of the following series are convergent only for $-1 \leq x < 1$?

I. $\displaystyle\sum_{k=1}^{\infty} \dfrac{x^k}{\sqrt{k}}$

II. $\displaystyle\sum_{k=1}^{\infty} \dfrac{x^k}{k}$

III. $\displaystyle\sum_{k=1}^{\infty} \dfrac{x^k}{k^2}$

(A) I only

(B) II only

(C) I and II only

(D) I and III only

GO ON TO THE NEXT PAGE.

AAAAAAAAAAAAAAAAAAAAAAAAAAAAAAAAA

8. If $\displaystyle\sum_{k=1}^{\infty} b_k$ converges and $0 < a_k < b_k$ for all k, then which of the following statements is false?

(A) $\displaystyle\lim_{k \to \infty} a_k = 0$

(B) $\displaystyle\sum_{k=1}^{\infty} (-1)^k a_k$ converges

(C) $\displaystyle\sum_{k=1}^{\infty} (a_k + b_k)$ converges

(D) $\displaystyle\sum_{k=1}^{\infty} a_k$ diverges

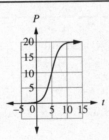

9. Which of the differential equations below could be used to model the logistic growth shown in the graph?

(A) $\dfrac{dP}{dt} = 20P - 20P^2$

(B) $\dfrac{dP}{dt} = 0.2P^2 + 0.1P$

(C) $\dfrac{dP}{dt} = 0.2P + 0.01P^2$

(D) $\dfrac{dP}{dt} = 0.2P - 0.01P^2$

GO ON TO THE NEXT PAGE.

A A A A A A A A A A A A A A A A A A A A A A A A A A A A A A A A A A

10. If $x^2 - xy + y^3 = 8$, then $\dfrac{dy}{dx} =$

    (A) $\dfrac{2x}{y + 3y^2}$

    (B) $\dfrac{-x^2 - 2xy^2 - 8}{\left(y^2 - x\right)^2}$

    (C) $\dfrac{8 - x^2 + y - 4y^2}{y^2 - x}$

    (D) $\dfrac{y - 2x}{3y^2 - x}$

11. Consider the function $f(x) = x^2 \sin \pi x - 3x$. Which of the following is a linear approximation to f at $x = 1$?

    (A) $f(x) = (2x \sin \pi x + x^2 \pi \cos \pi x - 3)x - 1$

    (B) $f(x) = -x - 1$

    (C) $f(x) = -\pi x + \pi$

    (D) $f(x) = (-\pi - 3)x + \pi$

GO ON TO THE NEXT PAGE.

AAAAAAAAAAAAAAAAAAAAAAAAAAAAAAAAAA

12. The function $f(x) = \dfrac{1}{\sqrt{x}} + 3$, $x > 0$ has asymptotes at which of the following?

(A) A horizontal asymptote at $y = 3$ and a vertical asymptote at $x = 0$

(B) A horizontal asymptote at $y = 0$ and a vertical asymptote at $x = 3$

(C) A horizontal asymptote at $y = 0$ and a vertical asymptote at $x = 0$

(D) A horizontal asymptote at $x = 3$ and a vertical asymptote at $y = 0$

13. Which of the following series converge?

I. $\displaystyle\sum_{n=1}^{\infty} \dfrac{n}{n^3 + 1}$

II. $\displaystyle\sum_{n=1}^{\infty} \dfrac{2^n}{e^n}$

III. $\displaystyle\sum_{n=1}^{\infty} \dfrac{n}{n^2 + 1}$

(A) II only

(B) III only

(C) I and II only

(D) I and III only

GO ON TO THE NEXT PAGE.

$$f(x) = x \sin\left(\frac{\pi}{2}\sqrt{x}\right)$$

14. Given the function f above, which of the following statements is a conclusion of the mean value theorem?

    (A) There exists c in $(1, 4)$ such that $f'(c) = -\frac{1}{3}$.

    (B) There exists c in $(0, 1)$ such that $f'(c) = -\frac{1}{3}$.

    (C) There exists c in $(0, 1)$ such that $f'(c) = 3$.

    (D) There exists c in $(1, 4)$ such that $f'(c) = 3$.

---

15. If $f(x) = \ln\left(\frac{2x - 3}{x + 8}\right)$, then $f'(x) =$

    (A) $\frac{1}{2}\left(\frac{x + 8}{2x - 3}\right)$

    (B) $\frac{2}{2x - 3} - \frac{1}{x + 8}$

    (C) $2\left(\frac{1}{2x - 3} + \frac{1}{x + 8}\right)$

    (D) $\frac{2}{2x - 3} \cdot \frac{1}{x + 8}$

GO ON TO THE NEXT PAGE.

AAAAAAAAAAAAAAAAAAAAAAAAAAAAAAAAA

16. A self-storage company wants to construct a rectangular storage unit with a base length 3 times the base width. The material used to build the top and bottom of the unit costs $10 per square meter and the material used to build the sides costs $6 per square meter. If the storage unit must have a volume of 50 cubic meters, what width will minimize the cost to build the unit?

(A) $\sqrt[3]{\dfrac{10}{3}}$ (B) $\sqrt[3]{\dfrac{20}{3}}$ (C) $\dfrac{10}{3}$ (D) $\dfrac{20}{3}$

$$f(x) = \begin{cases} 3 - x & \text{for } x < 3 \\ 10 & \text{for } x \geq 3 \end{cases}$$

17. For the piecewise function f defined above, what is the value of $\int_{1}^{7} f(x)\, dx$?

(A) 10 (B) 42 (C) 47 (D) 60

---

18. An equivalent representation of the definite integral $\int_{1}^{3} 2x \cos\left(x^2\right) dx$ is

(A) $\int_{1}^{\sqrt{3}} \cos u\, du$

(B) $\int_{1}^{3} \cos u\, du$

(C) $\int_{1}^{9} \cos u\, du$

(D) $\int_{1}^{9} 2\sqrt{u} \cos u\, du$

GO ON TO THE NEXT PAGE.

AAAAAAAAAAAAAAAAAAAAAAAAAAAAAAAAAAAAA

19. Let $f(x) = \int_{1}^{x^2} \left(3t^2 - 5t + 7\right) dt$.

Let $P(x)$ be a second degree Taylor polynomial for f centered at $x_0 = -1$. Find a formula for $P(x)$.

(A) $P(x) = 5(x+1) + 5(x+1)^2$

(B) $P(x) = -10(x+1) + 7(x+1)^2$

(C) $P(x) = -10(x+1) + 14(x+1)^2$

(D) $P(x) = -11(x+1) + 14(x+1)^2$

20. $\displaystyle\sum_{k=1}^{\infty} \frac{2^{k+3}}{5^k} =$

(A) $\dfrac{2}{5}$ (B) $\dfrac{16}{5}$ (C) $\dfrac{16}{3}$ (D) 8

21. The radius of a circle is increasing at a rate of $\dfrac{5}{2}$ centimeters per minute. At the instant when the area of the circle is 16π square centimeters, what is the rate of increase in the area of the circle, in square centimeters per minute?

(A) 4 (B) 8π (C) 16π (D) 20π

GO ON TO THE NEXT PAGE.

AAAAAAAAAAAAAAAAAAAAAAAAAAAAAAAA

22. If $f(x) = e^{\sin(3x)}$, then $f'(0) =$

(A) 0   (B) 1   (C) 3   (D) e^3

23. $\lim\limits_{h \to 0} \dfrac{(x+h)^9 \sin(x+h)^2 - x^9 \sin x^2}{h} =$

(A) 0

(B) ∞

(C) $x^8(9 \sin x^2 + 2x^2 \cos x^2)$

(D) $x^8(9 \sin x^2 + 2x \sin x \cos x)$

24. $\lim\limits_{h \to 0} \dfrac{(4+h)^3 - 4^3}{h} =$

(A) 0   (B) 48   (C) 64   (D) nonexistent

GO ON TO THE NEXT PAGE.

AAAAAAAAAAAAAAAAAAAAAAAAAAAAAAAAA

25. If the rate of change of a quantity over the closed interval $[-1, 5]$ is given by $f'(x) = xe^{x^2}$, then the net change of the quantity over the interval $[-1, 5]$ is

(A) $5e^{25} + e$

(B) $\frac{5}{2}e^{25} + \frac{1}{2}e$

(C) $51e^{25} - 3e$

(D) $\frac{1}{2}e^{25} - \frac{1}{2}e$

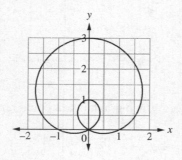

26. The graph of $r = 1 + 2\sin\theta$ is shown above. Which of the following integrals represents the area of the inside loop?

(A) $\frac{1}{2}\int_0^{2\pi}(1 + 2\sin\theta)^2\,d\theta$

(B) $\frac{1}{2}\int_{\frac{7}{6}\pi}^{\frac{3}{2}\pi}(1 + 2\sin\theta)^2\,d\theta$

(C) $\frac{1}{2}\int_0^{\pi}(1 + 2\sin\theta)^2\,d\theta$

(D) $\frac{1}{2}\int_{\frac{7}{6}\pi}^{\frac{11}{6}\pi}(1 + 2\sin\theta)^2\,d\theta$

GO ON TO THE NEXT PAGE.

AAAAAAAAAAAAAAAAAAAAAAAAAAAAAAAAA

27. Suppose $y(4) = 2$ and $\dfrac{dy}{dt} = t - y$. Use Euler's method with two steps to approximate the value of $y(3)$.

(A) -1 (B) -0.25 (C) 0.25 (D) 1.25

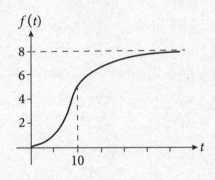

28. The graph of f is given above. Which graph below could represent the graph of f' ?

(A) $f'(t)$

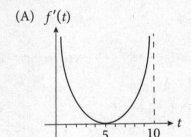

(B) $f'(t)$

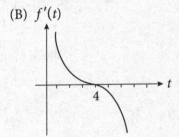

(C) $f'(t)$

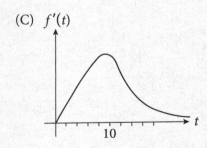

(D) $f'(t)$

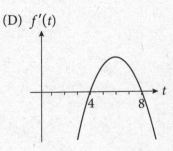

GO ON TO THE NEXT PAGE.

AAAAAAAAAAAAAAAAAAAAAAAAAAAAAAA

29. The series $\displaystyle\sum_{k=1}^{\infty}\frac{(-1)^{k-1}}{k!}$ is to be approximated by $\displaystyle\sum_{k=1}^{n}\frac{(-1)^{k-1}}{k!}$ for some positive integer n. What is the lowest possible value of n that is guaranteed to approximate $\displaystyle\sum_{k=1}^{\infty}\frac{(-1)^{k-1}}{k!}$ with error less than $\frac{1}{1000}$?

(A) $n = 4$ (B) $n = 5$ (C) $n = 6$ (D) $n = 7$

30. $\displaystyle\int x \cdot \sqrt{\frac{1}{6}x^2 + 5}\, dx =$

(A) $2\left(\frac{1}{6}x^2 + 5\right)^{\frac{3}{2}} + C$

(B) $\left(\frac{1}{9}x^2 + 5x\right)^{\frac{3}{2}} + C$

(C) $\frac{2}{3}\left(\frac{1}{6}x^2 + 5\right)^{\frac{3}{2}} + C$

(D) $\frac{x^2}{2}\left(\frac{1}{18}x^3 + 5x\right)^{\frac{3}{2}} + C$

END OF PART A OF SECTION I

**IF YOU FINISH BEFORE TIME IS CALLED, YOU MAY
CHECK YOUR WORK ON PART A ONLY.**

CALCULUS BC
SECTION I, Part B
Time—45 minutes
Number of questions—15

A GRAPHING CALCULATOR IS REQUIRED FOR SOME QUESTIONS ON THIS PART OF THE EXAM.

Directions: Use the space provided to solve each of the following problems. After examining the given choices, decide which of the choices is the best and fill in the corresponding circle on the answer sheet. Credit will only be given for answers appropriately marked on the answer sheet, and not for anything written in the exam book. Do not spend too much time on any one problem.

In this exam:

(1) When the exact numerical value of the correct answer does not appear among the choices given, select the number that best approximates the numerical value.

(2) The domain of a function f should be assumed to be the set of all real numbers x for which $f(x)$ is a real number, unless the question indicates otherwise.

(3) The inverse function notation f^{-1} or the prefix "arc" (e.g., $\sin^{-1} x = \arcsin x$) may be used to indicate the inverse of a trigonometric function.

GO ON TO THE NEXT PAGE.

76. The derivative of the function f is given by $f'(x) = x + \sin(x^2) - 2$. At what value(s) of x does f have a relative minimum on the interval $0 < x < 4$?

    (A) 1.080 only   (B) 1.677 only   (C) 2.419 only   (D) 1.080 and 2.419

x	-4	-1	0	3	7
$f(x)$	6	-5	-3	6	3

77. Suppose f is a continuous, differentiable function. Selected values of f are shown in the table above. Which of the following is the value of $\int_{-4}^{7} f(x)\ dx$ if it is approximated by a left Riemann sum using four subintervals?

    (A) 28   (B) 31   (C) 44   (D) 56

78. What is the domain of the function $f(x) = \sum_{k=1}^{\infty} \dfrac{(x-3)^k}{k}$?

    (A) $2 \le x < 4$

    (B) $2 \le x \le 4$

    (C) $-1 < x < 1$

    (D) $-1 \le x < 1$

GO ON TO THE NEXT PAGE.

79. If $\cos\left(\dfrac{1}{x^2+1}\right)$ is an antiderivative of $f(x)$, then $\displaystyle\int_0^2 f(x)\,dx =$

   (A) 0.032   (B) 0.440   (C) 0.540   (D) 1.643

---

80. The infinite series for $f(x) = \dfrac{e^x - 1}{x}$ can be written as

   I. $\displaystyle\sum_{k=0}^{\infty} \frac{x^{k+1}}{k!}$

   II. $\displaystyle\sum_{k=0}^{\infty} \frac{x^k}{(k+1)!}$

   III. $\displaystyle\sum_{k=1}^{\infty} \frac{x^k}{(k+1)!}$

   (A) I only

   (B) II only

   (C) III only

   (D) I and II only

---

81. Find the length of the parametric curve given by $x(t) = 2\cos(3t)$ and $y(t) = 4 - e^t$ for $-1 \le t \le 3$.

   (A) 19.560   (B) 28.155   (C) 87.651   (D) 184.098

82. The base of a solid region is bordered in the first quadrant by the x-axis, the y-axis, and the line $y = 3 - \frac{1}{2}x$. Cross-sections perpendicular to the x-axis are squares. What is the volume of the solid?

    (A) 6   (B) 9   (C) 15   (D) 18

83. If $f(x) = 3x^2 - 7x + \frac{1}{x}$, what is $f''(-2)$?

    (A) $-\frac{77}{4}$ (B) $\frac{19}{4}$ (C) $\frac{23}{4}$ (D) $\frac{25}{4}$

84. Let $y' = \frac{1}{3}y$ be the differential equation that approximately models the growth of a population of rabbits. At time $t = 0$ years, there are 50 rabbits. Approximately how many rabbits are there after 9 years?

    (A) A whole number of rabbits close to $50e^3$

    (B) A whole number of rabbits close to 150

    (C) A whole number of rabbits close to $e^3 + 49$

    (D) A whole number of rabbits close to $150e^{\frac{1}{3}}$

GO ON TO THE NEXT PAGE.

B B B B B B B B B

85. $\int x \sec^2 x \, dx =$

(A) $\frac{1}{2}x^2 \tan x + C$

(B) $x \tan x + \ln|\cos x| + C$

(C) $x \tan x - \ln|\cos x| + C$

(D) $x \tan x + \frac{1}{2}x^2 \sec^2 x + C$

---

86. $\int x \sec^2(x^2) \, dx =$

(A) $\frac{1}{2} \tan(x^2) + C$

(B) $2 \tan(x^2) + C$

(C) $\frac{1}{3} \sec^3(x^2) + C$

(D) $\frac{1}{3}(2x) \cdot \sec^3(x^2) + C$

GO ON TO THE NEXT PAGE.

B B B B B B B B B

87. What is the area in square units of the region enclosed by the graph of $y = 2x^2 + 1$ and the line $y = 2x + 5$?

(A) 3　　(B) $\dfrac{7}{2}$　　(C) 5　　(D) 9

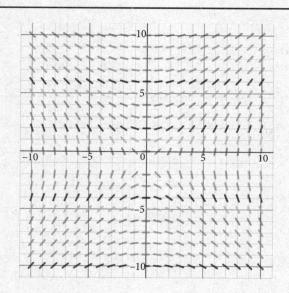

88. A slope field for which of the following differential equations is shown above?

(A) $\dfrac{dy}{dx} = \dfrac{x^2}{y}$

(B) $\dfrac{dy}{dx} = \dfrac{x^2}{y} - 1$

(C) $\dfrac{dy}{dx} = \dfrac{x}{y+1}$

(D) $\dfrac{dy}{dx} = \dfrac{x}{y-1}$

GO ON TO THE NEXT PAGE.

B B B B B B B B B

89. $\displaystyle\int \frac{5}{x^2 + x - 6}\, dx =$

(A) $5 \ln|x^2 + x - 6| + C$

(B) $\dfrac{5x}{x^3 + x^2 - 6x} + C$

(C) $\ln|x + 3| - \ln|x - 2| + C$

(D) $\ln|x - 2| - \ln|x + 3| + C$

90. Which of the following is the solution to the differential equation $\dfrac{dy}{dx} = \dfrac{x^2}{y^3}$, where $y(3) = 3$?

(A) $y = \sqrt[4]{\dfrac{3}{4}x^3 - 45}$

(B) $y = \sqrt[4]{\dfrac{4}{3}x^3 + 45}$

(C) $y = \sqrt[4]{\dfrac{4}{3}x^4 + 5}$

(D) $y = \sqrt[4]{\dfrac{4}{3}x^4 + 45}$

END OF SECTION I

**IF YOU FINISH BEFORE TIME IS CALLED, YOU MAY
CHECK YOUR WORK ON PART B ONLY.**

CALCULUS BC
SECTION II, Part A
Time—30 minutes
Number of problems—2

A GRAPHING CALCULATOR IS REQUIRED FOR THESE PROBLEMS.

Directions: Write your solution to each part of the following questions in the space provided. Write clearly and legibly. Cross out any errors you make; erased or crossed-out work will not be scored.

You may wish to look over the problems before starting to work on them; on the actual test, it is not expected that everyone will be able to complete all parts of all problems. All problems are given equal weight, but the individual parts of a particular problem are not necessarily given equal weight. You should not spend too much time on any one problem.

- Show all of your work. Clearly label functions, graphs, tables, or anything else that you use to arrive at your final solution. Your work will be scored on the correctness and completeness of your methods as well as your final answers. Answers without supporting work will usually not receive credit.

- Justifications (i.e., the request that you "justify your answer") require that you give mathematical (non-calculator) reasons.

- Do not use calculator syntax. Your work must be expressed in standard mathematical notation.

- Numeric or algebraic answers do not need to be simplified, unless the question indicates otherwise.

- If you use decimal approximations in calculations, your final answers should be accurate to three places after the decimal point, unless the question indicates otherwise.

- The domain of a function f should be assumed to be the set of all real numbers x for which $f(x)$ is a real number, unless the question indicates otherwise.

GO ON TO THE NEXT PAGE.

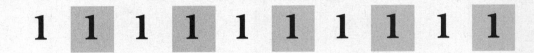

1. A particle moves along the x-axis so that its velocity v at time t, for $0 \le t \le 6$, is given by:

$$v(t) = 4 \, \sin\!\left(\frac{t^2}{2} - 2t + 2\right)$$

At time $t = 0$, the particle is at $x = 1$.

(a) Find the velocity and acceleration of the particle at time $t = 4$.

(b) Is the speed of the particle increasing or decreasing at time $t = 4$? Give a reason for your answer.

Practice Exams

(c) Find the time when the particle first changes direction. Justify your reasoning.

(d) Describe the movement of the particle before, at, and after $t = 2$.

2. A ski resort uses a snow machine to control the snow level on a ski slope. Over a 24-hour period the volume of snow added to the slope per hour is modeled by the equation:

$$S(t) = 24 - t \sin^2\left(\frac{t}{14}\right)$$

The rate that the snow melts is modeled by the equation:

$$M(t) = 10 + 8 \cos\left(\frac{t}{3}\right)$$

Both $S(t)$ and $M(t)$ have units of cubic yards per hour, and t is measured in hours for $0 \leq t \leq 24$. At time $t = 0$, the slope holds 50 cubic yards of snow.

(a) Set up the equation for and compute the total volume of snow added to the mountain over the first 6-hour period.

_____

(b) Determine the first time that the instantaneous rate of change of the volume of snow is 0. Justify your answer.

(c) Is the volume of snow increasing or decreasing at time $t = 4$? Justify your answer.

_____

(d) Suppose the snow machine is turned off at time $t = 6$. At what time will all the snow be melted?

END OF PART A OF SECTION II

**IF YOU FINISH BEFORE TIME IS CALLED, YOU MAY
CHECK YOUR WORK ON PART A ONLY.**

**CALCULUS BC
SECTION II, Part B
Time—60 minutes
Number of problems—4**

NO CALCULATOR IS ALLOWED FOR THESE PROBLEMS.

Note: If you have extra time, you can go back and work on Part A of Section II, but you cannot use a calculator to complete your work at this time.

3 3 3 3 3 3 3 3 3 3

NO CALCULATOR ALLOWED

3. Let $\dfrac{dy}{dt} = 2t\left(y^2 + 1\right)$. Also let $y = f(t)$ be the particular solution to the differential equation with initial condition $f(0) = \sqrt{3}$.

(a) Starting at $t = 0$, use Euler's method with two steps of equal size to approximate $f(1)$.

(b) Find $y = f(t)$, the solution to the differential equation with initial condition $f(0) = \sqrt{3}$.

(c) Find the second degree Taylor polynomial for $y = f(t)$, centered at $t = 0$.

GO ON TO THE NEXT PAGE.

4 **4** **4** **4** **4** **4** **4** **4** **4** **4**

NO CALCULATOR ALLOWED

4. Consider the differential equation $\dfrac{dy}{dx} = \dfrac{x^2 - 1}{y}$.

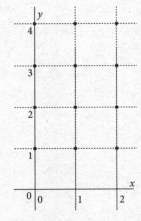

(a) On the axes provided, sketch a slope field for the given differential equation at the 12 points indicated.

_____

(b) Let $y = f(x)$ be the particular solution to the given differential equation with the initial condition $f(3) = 4$. Write an equation for the tangent line to the graph of f at $x = 3$.

_____

(c) Find the particular solution to the given differential equation with the initial condition $f(3) = 4$.

GO ON TO THE NEXT PAGE.

5 5 5 5 5 5 5 5 5 5

NO CALCULATOR ALLOWED

5. A circular oil slick of uniform thickness contains 100 cm^3 of oil.

(a) The volume of the oil remains constant. Use the equation for the volume of a cylinder to relate the thickness to the radius.

_____

(b) As the oil spreads, the thickness is decreasing at a rate of 0.01 cm/min. At what rate is the radius of the slick increasing when the diameter is 20 cm?

_____

(c) Suppose that after 1 minute, the diameter of the oil slick is 20 cm. How long will it take for the diameter of the slick to reach 40 cm?

GO ON TO THE NEXT PAGE.

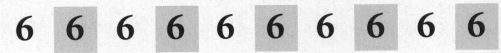

6. Let $f(x) = \dfrac{e^x - e^{-x}}{x}$.

   (a) Find the first three nonzero terms and the general term of the Taylor series representation for f centered at $x = 0$.

   _____

   (b) Use the Taylor series found in part (a) to evaluate $\displaystyle\lim_{x \to 0} \dfrac{e^x - e^{-x}}{x}$.

6 6 6 6 6 6 6 6 6 6

NO CALCULATOR ALLOWED

(c) Let $g(x) = \dfrac{1 - e^{-x}}{x}$. Find the first three nonzero terms and the general term of the Taylor series representation for g centered at $x = 0$.

(d) The first four terms of the series for g are used to approximate $g\!\left(\dfrac{1}{10}\right)$. Show the error is less than $\dfrac{1}{1,000,000}$. Justify your answer.

STOP

END OF EXAM

ANSWER KEY

Section I, Part A

1.	B	16.	B
2.	D	17.	B
3.	A	18.	C
4.	B	19.	B
5.	A	20.	C
6.	B	21.	D
7.	C	22.	C
8.	D	23.	C
9.	D	24.	B
10.	D	25.	D
11.	D	26.	D
12.	A	27.	B
13.	C	28.	C
14.	A	29.	C
15.	B	30.	A

Section I, Part B

76.	D
77.	A
78.	A
79.	B
80.	B
81.	B
82.	D
83.	C
84.	A
85.	B
86.	A
87.	D
88.	C
89.	D
90.	B

Scoring

Section II, Parts A & B

Use the scoring rubrics (found after the Section I answers and explanations) to self-score your free-response questions.

Section I, Part A Number Correct: _____

Section I, Part B Number Correct: _____

Section II, Part A Points Earned: _____

Section II, Part B Points Earned: _____

Enter your results to your Practice Exam 2 assignment to see your 1–5 score by logging in at kaptest.com.

Haven't registered your book yet? Go to kaptest.com/booksonline to begin.

BC PRACTICE EXAM 2 ANSWERS AND EXPLANATIONS

SECTION I, Part A

1. B

The slope of a tangent to the graph at a point is the derivative of the function at that point, so take the derivative of the given function and plug in 2 for each x. Don't forget to use the chain rule when taking the derivative of the second term:

$$f(x) = 3x^2 - \sin \pi x$$
$$f'(x) = 6x - \cos \pi x \cdot \pi$$
$$f'(2) = 6(2) - \pi \cos(2\pi)$$
$$= 12 - \pi(1) = 12 - \pi$$

Thus, **(B)** is correct.

2. D

This question is testing your understanding of what it means for a function to be continuous at a point, in terms of limits. The words "necessary and sufficient" mean that the correct answer has to both imply and be implied by the continuity of f at $x = 4$. In other words, the correct answer is equivalent to the continuity of f at $x = 4$. Because differentiability at a point is a stronger condition than continuity at a point, (B) can be eliminated. (C) is certainly implied by the continuity of f at $x = 4$, but it is not enough to imply continuity, so it may also be eliminated. Of the remaining two choices, you might recognize (D) as the definition of continuity of a function at $x = 4$. Hence, the correct answer is **(D)**.

3. A

A positive second derivative ($f'' > 0$) on an interval corresponds to the function being concave up on that interval, while a negative second derivative ($f'' < 0$) on an interval corresponds to the function being concave down on that interval. Careful—don't choose (A) just yet because (D) could still be the answer. You can, however,

eliminate (B). To finish answering the question, consider what is happening at the point $x = 3$. Based on the information given, the concavity of the function changes at that point, which means $x = 3$ is a point of inflection. The second derivative at a point of inflection is either equal to 0 or does not exist, so (C) is not correct. This means you can also eliminate (D), leaving **(A)** as the correct answer.

4. B

This question is testing your knowledge of what continuity and differentiability mean graphically. Continuity is the more intuitive concept: You don't need to take your pencil off the page to trace the graph where it is continuous. This means you can immediately eliminate (C) and (D). Next, recall that the derivative of a function at a point corresponds to the slope of the tangent line to the graph at that point. There is no tangent line to the function at the point where $x = 2$ (because there is a sharp corner), but there is a tangent line at every other point. From this, you can eliminate (A), leaving **(B)** as the correct answer. Careful—don't be fooled into thinking that $f'(2) = 0$ just because the graph has a minimum there. It would have to be differentiable there too, and it is not.

5. A

Velocity is the derivative of position, so you differentiate each part of $p(t)$ and evaluate each of them at $t = 4$.

$$v(t) = \frac{d}{dt}(p(t)) = \left\langle 6t, \frac{1}{\sqrt{t}} \right\rangle \Rightarrow v(4) = \left\langle 24, \frac{1}{2} \right\rangle$$

Hence, **(A)** is the correct answer.

6. B

The function is a quotient that can't be simplified, so use the quotient rule: $\left(\dfrac{t}{b}\right)' = \dfrac{t'b - tb'}{b^2}$ (t stands for "top" and b stands for "bottom"). The top is a product, so you'll also need to use the product rule. For $f(x) = \dfrac{x^3 \sin x}{x^2 + 3}$, the derivative is:

$$f'(x) = \frac{\left(x^3 \cos x + 3x^2 \sin x\right)\left(x^2 + 3\right) - \left(x^3 \sin x\right)(2x)}{\left(x^2 + 3\right)^2}$$

$$= \frac{\left(x^3 \cos x + 3x^2 \sin x\right)\left(x^2 + 3\right)}{\left(x^2 + 3\right)^2} - \frac{\left(x^3 \sin x\right)(2x)}{\left(x^2 + 3\right)^2}$$

$$= \frac{x^3 \cos x + 3x^2 \sin x}{x^2 + 3} - \frac{2x^4 \sin x}{\left(x^2 + 3\right)^2}$$

That's a perfect match for **(B)**.

7. C

Using the ratio test, you can find that all three series converge for $-1 < x < 1$. Test the endpoints of the interval on each series:

I. $x = 1$: $\displaystyle\sum_{k=1}^{\infty} \frac{1}{k^{\frac{1}{2}}}$ diverges because it is a

p-series, $p = \dfrac{1}{2} < 1$.

$x = -1$: $\displaystyle\sum_{k=1}^{\infty} \frac{(-1)^k}{k^{\frac{1}{2}}}$ converges because of the alternating series test.

Therefore, $-1 \leq x < 1$.

II. $x = 1$: $\displaystyle\sum_{k=1}^{\infty} \frac{1}{k}$ diverges because it is the harmonic series.

$x = -1$: $\displaystyle\sum_{k=1}^{\infty} \frac{(-1)^k}{k}$ converges because it is the alternating harmonic series.

Therefore, $-1 \leq x < 1$.

III. $x = 1$: $\displaystyle\sum_{k=1}^{\infty} \frac{1}{k^2}$ converges because it is a p-series,

$p = 2 > 1$.

$x = -1$: $\displaystyle\sum_{k=1}^{\infty} \frac{(-1)^k}{k^2}$ converges because it is absolutely convergent.

Therefore, $-1 \leq x \leq 1$. Because the series converges at the endpoint $x = 1$, III exceeds the range given in the problem and cannot be included as a correct answer. Hence, **(C)** is the correct answer.

8. D

$\displaystyle\sum_{k=1}^{\infty} a_k$ must converge by the basic comparison test. All the other statements are true. Hence, **(D)** is the correct answer.

9. D

Factoring each of the choices, the only one that results in a logistic differential equation with horizontal asymptotes at $P = 0$ and $P = 20$ is

$$\frac{dP}{dt} = 0.2P - 0.01P^2 = 0.01P(20 - P).$$

Remember, your zeros here correspond to regions where the slope of the given curve is zero, or flat. Thus, **(D)** is correct.

10. D

Because the equation is not already written in terms of y (or easily solvable for y), use implicit differentiation. Don't forget to multiply by $\dfrac{dy}{dx}$ when you take the derivative of y. Also, use the product rule when differentiating the second term. The derivative is:

$$2x - \left(1 \cdot y + x \cdot \frac{dy}{dx}\right) + 3y^2 \frac{dy}{dx} = 0$$

$$2x - y - x\frac{dy}{dx} + 3y^2 \frac{dy}{dx} = 0$$

Next, separate the terms that include $\dfrac{dy}{dx}$ from those that don't and then solve for $\dfrac{dy}{dx}$:

$$\left(3y^2 - x\right)\frac{dy}{dx} = (y - 2x)$$

$$\frac{dy}{dx} = \frac{y - 2x}{3y^2 - x}$$

That's **(D)**.

11. D

The tangent line to the function at the given point provides a linear approximation to the function at that point, so start by finding the first derivative and evaluating it at $x = 1$. This gives you the slope of the tangent line:

$$f(x) = x^2 \sin \pi x - 3x$$

$$f'(x) = x^2 \pi \cos \pi x + 2x \sin \pi x - 3$$

$$f'(1) = \pi(1)^2 \cos \pi + 2(1) \sin \pi - 3$$

$$f'(1) = \pi(-1) + 2(0) - 3 = -\pi - 3$$

Now, don't do more work than is needed to obtain the correct solution—because the answer choices all have different slopes (the coefficient of x), it suffices to know the slope of the tangent line at $x = 1$. In other words, finding $f'(1) = -\pi - 3$ allows you to find the correct answer, which is **(D)**.

12. A

Vertical asymptotes are typically easier to find, so start with those. Vertical asymptotes occur when the denominator of a rational function is zero: $\sqrt{x} = 0 \rightarrow x = 0$ is a vertical asymptote; eliminate (B) and (D). The horizontal asymptote, by definition, is the limit of the function as x tends to infinity, so imagine plugging in a huge value:

$y = \dfrac{1}{\sqrt{100,000}} + 3 = 0 + 3 = 3$, so $y = 3$ is the horizontal asymptote. Thus, **(A)** is correct.

Note that if this question were in the calculator section of the test, you could graph the function to find the asymptotes:

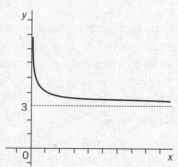

13. C

I. $\displaystyle\sum_{n=1}^{\infty} \dfrac{n}{n^3 + 1}$: $\qquad \dfrac{n}{n^3 + 1} < \dfrac{n}{n^3} = \dfrac{1}{n^2}$

$\displaystyle\sum_{n=1}^{\infty} \dfrac{1}{n^2}$ converges: p-series, $p = 2 > 1$. Therefore $\displaystyle\sum_{n=1}^{\infty} \dfrac{n}{n^3 + 1}$ converges by the basic comparison test.

II. $\displaystyle\sum_{n=1}^{\infty} \dfrac{2^n}{e^n}$: $\qquad \displaystyle\sum_{n=1}^{\infty} \dfrac{2^n}{e^n} = \sum_{n=1}^{\infty} \left(\dfrac{2}{e}\right)^n$

is a convergent geometric series because $r = \dfrac{2}{e} < 1$.

III. $\displaystyle\sum_{n=1}^{\infty} \dfrac{n}{n^2 + 1}$: Let $a_n = \dfrac{n}{n^2 + 1}$ and let $b_n = \dfrac{1}{n}$. $\displaystyle\sum_{n=1}^{\infty} b_n$ is the harmonic series and therefore diverges.

$\displaystyle\lim_{n \to \infty} \dfrac{a_n}{b_n} = \lim_{n \to \infty} \dfrac{n^2}{n^2 + 1} = 1$

Therefore $\displaystyle\sum_{n=1}^{\infty} \dfrac{n}{n^2 + 1}$ diverges by the limit comparison test. Hence, **(C)** is the correct answer.

14. A

The mean value theorem says, in essence, that in a given interval, there is some point such that the derivative of the function at that point equals the slope of the secant whose endpoints correspond to the endpoints of the interval. So you don't actually need to find the derivative of f. Rather, just start going down the list of answer choices. The slope of the secant in (A) and (D) is:

$$\dfrac{f(4) - f(1)}{4 - 1} = \dfrac{4 \cdot \sin \pi - 1 \cdot \sin \dfrac{\pi}{2}}{3} = \dfrac{0 - 1}{3} = -\dfrac{1}{3}$$

Thus, (D) can be eliminated and (A) looks like the right choice. For (B) and (C), the slope of the secant is:

$$\dfrac{f(1) - f(0)}{1 - 0} = \dfrac{1 \cdot \sin \dfrac{\pi}{2} - 0 \cdot \sin 0}{1} = \dfrac{1}{1} = 1$$

Neither (B) nor (C) gives 1 as a possible value of $f'(c)$, so you can eliminate them. The only viable answer is **(A)**.

15. B

Rather than having to use both the chain rule and the quotient rule, simplify the function using rules of logs first:

$$\ln \dfrac{a}{b} = \ln a - \ln b$$

Here, the function can be written as:

$$f(x) = \ln(2x - 3) - \ln(x + 8)$$

Now use rules of derivatives (don't forget the chain rule if it applies) to differentiate the function: $\dfrac{d}{du}(\ln u) = \dfrac{1}{u} \cdot u'$. The result is:

$$f'(x) = \dfrac{1}{2x - 3}(2) - \dfrac{1}{x + 8}(1)$$
$$= \dfrac{2}{2x - 3} - \dfrac{1}{x + 8}$$

Thus, **(B)** is correct.

16. B

Start with a quick sketch:

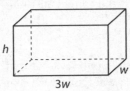

You want to minimize the cost of the materials subject to the constraint that the volume must be 50 cubic meters. The cost for each side is just the area of that side times the corresponding cost. The two functions are therefore:

$$C = 10(2lw) + 6(2wh + 2lh)$$
$$C = 10(2)(3w)(w) + 6(2wh + 2(3w)(h))$$
$$C = 60w^2 + 48wh$$
and
$$V = lwh = (3w)wh$$
$$50 = 3w^2h$$

Next, solve the constraint for one of the variables and plug this into the cost. It is easier to solve the constraint for h, so do that (besides, you're looking for the width, so it would be ideal if you can keep the w):

$$50 = 3w^2h \rightarrow h = \frac{50}{3w^2}$$
$$C = 60w^2 + 48w\left(\frac{50}{3w^2}\right) = 60w^2 + \frac{800}{w}$$

To minimize C, you need to find the values of w for which C' is either 0 or does not exist. Find C':

$$C' = 120w - \frac{800}{w^2} = \frac{120w^3 - 800}{w^2}$$

C' is undefined at $w = 0$, but this answer doesn't make sense in the given context, so set C' equal to 0 and solve for w:

$$0 = \frac{120w^3 - 800}{w^2}$$
$$0 = 120w^3 - 800$$
$$800 = 120w^3$$
$$w = \sqrt[3]{\frac{800}{120}} = \sqrt[3]{\frac{20}{3}}$$

That's **(B)**.

17. B

To evaluate a definite integral for a piecewise function, split the integral into parts using the "for $x = \dots$" pieces of the function's definition. Here, split the integral at 3 to get:

$$\int_1^7 f(x)\,dx = \int_1^3 f(x)\,dx + \int_3^7 f(x)\,dx$$
$$= \int_1^3 (3 - x)\,dx + \int_3^7 10\,dx$$

Next, evaluate each of the definite integrals and add the results:

$$\int_1^3 (3 - x)\,dx = \left(3x - \frac{x^2}{2}\right)\Big|_1^3 = \left(9 - \frac{9}{2}\right) - \left(3 - \frac{1}{2}\right) = 2$$
$$\int_3^7 10\,dx = 10x\Big|_3^7 = 70 - 30 = 40$$
$$\int_1^7 f(x)\,dx = 2 + 40 = 42$$

Thus, **(B)** is correct.

18. C

This is an exercise in the correct application of u-substitution in a definite integral. When doing a u-substitution, the most important thing is that the du part can be found within the integrand. Here, if $u = x^2$, then $du = 2x\,dx$, which is indeed part of the integrand. This means the integral can be written as $\int \cos u\,du$, and you can eliminate (D). Next, find the new limits of integration: The original limits are in terms of x. Now you are differentiating with respect to u, so the new limits should reflect this. Plug the original limits for x into the expression $u = x^2$ to get the new limits, 1 and 9, and the correct answer follows, which is **(C)**.

19. B

Because

$$f(x) = \int_1^{x^2} (3t^2 - 5t + 7)\,dt,$$

then $f(-1) = 0$.

$f'(x) = (3x^4 - 5x^2 + 7) \cdot 2x = 6x^5 - 10x^3 + 14x$, and $f(-1) = -10$.

$f''(x) = 30x^4 - 30x^2 + 14$, and $f''(-1) = 14$.

Therefore,

$$P(x) = -10(x+1) + \frac{14}{2!}(x-1)^2 = -10(x+1) + 7(x+1)^2$$

Therefore, **(B)** is the correct answer.

20. C

Look for a way to write both the numerator and denominator as a power of k, then use the formula for the sum of an infinite geometric series to evaluate.

$$\sum_{k=1}^{\infty} \frac{2^{k+3}}{5^k} = \sum_{k=1}^{\infty} \frac{2^3 \cdot 2^k}{5^k} = 8 \cdot \sum_{k=1}^{\infty} \left(\frac{2}{5}\right)^k = 8 \cdot \frac{\frac{2}{5}}{1 - \frac{2}{5}} = 8 \cdot \frac{2}{3} = \frac{16}{3}$$

Thus, **(C)** is the correct answer.

21. D

This is a typical related rates question, so jot down the variables involved and anything you already know about them: radius (r), area ($A = 16\pi$), change in radius $\left(\frac{dr}{dt} = \frac{5}{2}\right)$, and change in area $\left(\frac{dA}{dt}\right)$ From the given information, you already know the values of two of these and you can find a third, the radius, by solving $A = \pi r^2 = 16\pi$ for r, which gives $r = 4$. Thus, you're looking for $\frac{dA}{dt}$ at the instant when $r = 4$. To find $\frac{dA}{dt}$, differentiate the area equation and then substitute the values of the variables that you know:

$$A = \pi r^2$$
$$\frac{dA}{dt} = 2\pi r \frac{dr}{dt}$$
$$\frac{dA}{dt} = 2\pi(4)\left(\frac{5}{2}\right) = 20\pi$$

(D) is correct.

22. C

You have to apply the chain rule twice to compute the derivative here. You'll need to use the following: The derivative of e^u is e^u times du, and the derivative of $\sin(u)$ is $\cos(u)$ times du. Thus, $f'(x) = e^{\sin(3x)} \cdot \cos(3x) \cdot 3$. No need to simplify or rewrite anything—just evaluate the derivative at $x = 0$ to find that:

$$f'(0) = e^{\sin(0)} \cdot \cos(0) \cdot 3$$
$$= e^0 \cdot 1 \cdot 3 = 1 \cdot 1 \cdot 3 = 3$$

That's **(C)**.

23. C

This is a classic "definition of the derivative" question. Look at the term being subtracted; the limit represents the derivative of $x^9 \sin x^2$. Therefore, you just need to use the product rule $[(fg)' = f'g + fg']$ and the chain rule to find the derivative:

$$f(x) = x^9 \sin x^2$$
$$f'(x) = 9x^8 \sin x^2 + x^9 \cos x^2 \cdot 2x$$
$$= 9x^8 \sin x^2 + 2x^{10} \cos x^2$$
$$= x^8 \left(9 \sin x^2 + 2x^2 \cos x^2\right)$$

That's a perfect match for **(C)**.

24. B

The definition of the derivative is $\lim\limits_{h \to 0} \dfrac{f(x+h) - f(x)}{h}$. Therefore, $\lim\limits_{h \to 0} \dfrac{(4+h)^3 - 4^3}{h}$ is the derivative of the function $f(x) = x^3$ at the point where $x = 4$. For $f(x) = x^3$, the derivative is $f'(x) = 3x^2$, so the value of the given limit is $f'(4) = 3(4)^2 = 48$.

(B) is correct.

25. D

The definite integral of the rate of change of a quantity over an interval gives the net change of that quantity over that interval, so find the antiderivative of $f'(x)$ first. Using u-substitution with $u = x^2$ and $du = 2x\, dx$, the result is:

$$\int_{-1}^{5} xe^{x^2}\, dx = \frac{1}{2}e^{x^2} \Big|_{-1}^{5} = \frac{1}{2}e^{25} - \frac{1}{2}e$$

That's **(D)**. Note that various mistakes could lead you to the wrong answer, like plugging the interval into $f'(x)$ instead of $f(x)$, so be careful. A rate of change corresponds to the derivative of a function, while net change corresponds to the function itself.

26. D

The graph crosses through the origin when

$$1 + 2\sin\theta = 0 \Rightarrow \sin\theta = -\frac{1}{2} \Rightarrow \theta = \frac{7}{6}\pi \text{ and } \theta = \frac{11}{6}\pi.$$

Therefore, the area is represented by

$$\frac{1}{2}\int_{\frac{7}{6}\pi}^{\frac{11}{6}\pi}(1 + 2\sin\theta)^2 \, d\theta$$

Hence, **(D)** is the correct answer.

27. B

$$y(3.5) \approx y(4) + y'(4)(-0.5) \qquad y'(4) = 4 - 2 = 2$$
$$y(3.5) \approx 2 + 2(-0.5) = 1$$

$$y(3) \approx y(3.5) + y'(3.5)(-0.5) \qquad y'(3.5) = 3.5 - 1 =$$
$$y(3) \approx 1 + 2.5(-0.5) = -0.25 \qquad 2.5$$

Therefore, **(B)** is the correct answer.

28. C

Notice that the graph of f has a point of inflection at $x = 10$ (i.e., it changes concavity at $x = 10$). Points of inflection can only occur when $f'' = 0$ or is undefined. Because f'' is the rate of change of f', the slope of the tangent line to the graph of f' must be 0 at the point of inflection. You can therefore eliminate all except **(C)**, which is the correct answer.

29. C

For an alternating series, the error when summing the first n terms is less than the absolute value of the $(n + 1)st$ term. So $\frac{1}{(n+1)!} < \frac{1}{1000}$, or $(n + 1)! > 1000$.

$6! = 720$ and $7! = 5040$, so $n + 1 = 7$ and $n = 6$.

Hence, **(C)** is the correct answer.

30. A

This is a classic u-substitution question, because there is something quite messy inside a radical and an x outside, which means you can most likely create a nice, neat du.

If $u = \frac{1}{6}x^2 + 5$, then $du = \frac{1}{3}x \, dx$. You already have the x and the dx, but not the $\frac{1}{3}$, so multiply the integrand by a factor of $\frac{1}{3}$ and then compensate by multiplying the entire integral by 3:

$$\int x \cdot \sqrt{\frac{1}{6}x^2 + 5} \, dx = 3\int \sqrt{\frac{1}{6}x^2 + 5}\,\frac{1}{3}x \, dx$$

$$= 3\int \sqrt{u} \, du$$

$$= 3\int u^{\frac{1}{2}} \, du$$

$$= 3\left(\frac{2}{3}u^{\frac{3}{2}} + C\right)$$

$$= 2u^{\frac{3}{2}} + 3$$

Writing the final expression above back in terms of x gives you **(A)**.

SECTION I, Part B

76. D

Relative extrema may occur when the first derivative of a function is equal to 0 or where it does not exist. Here, you're given f' (which happens to be defined everywhere), so your first task is to determine where $f' = 0$. Do this on your calculator by graphing the equation and using the Calculate Zero function. You're only asked about the interval from $0 < x < 4$, so restrict your window accordingly—this makes it much easier to see what's going on. The graph looks something like this:

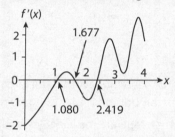

A relative minimum occurs where a function changes from decreasing (when f' is negative, or below the x-axis) to increasing (when f' is positive, or above the x-axis). Based on the graph, this happens at $x = 1.080$ and at 2.419, so **(D)** is correct.

77. A

First, note that the widths of the rectangles in this problem are not all the same (because the distance between the x-values varies). The height of each of the rectangles is the y-value on the left-hand side of each subinterval (because the question instructs you to use a left Riemann sum). To find the area of a rectangle, multiply its width times its height. Then, add the areas to find the total area under the curve:

$$\int_{-4}^{7} f(x)\,dx \approx 6 \cdot 3 + (-5) \cdot 1 + (-3) \cdot 3 + 6 \cdot 4 = 28$$

Thus, **(A)** is correct.

78. A

Let $a_k = \dfrac{(x-3)^k}{k}$, $a_{k+1} = \dfrac{(x-3)^{k+1}}{k+1}$

Then

$$\lim_{k \to \infty} \left| \frac{a_{k+1}}{a_k} \right| = \lim_{k \to \infty} \left| \frac{(x-3)^{k+1}}{k+1} \cdot \frac{k}{(x-3)^k} \right|$$

$$= |x-3| \cdot \lim_{k \to \infty} \frac{k}{k+1} = |x-3|$$

$|x-3| < 1 \Rightarrow 2 < x < 4.$

Testing the endpoints:

If $x = 4$, then $f(4) = \sum_{k=1}^{\infty} \dfrac{1}{k}$ which is divergent because this is the harmonic series.

If $x = 2$, then $f(2) = \sum_{k=1}^{\infty} \dfrac{(-1)^k}{k}$ which is convergent because this is the alternating harmonic series.

Therefore the domain of f is: $2 \le x < 4$, matching **(A)**.

79. B

First, note that you are told that the expression given is an antiderivative of f (call it F), so you don't need to integrate anything. This question is all about using the first fundamental theorem of calculus, so simply plug in the limits of integration and subtract (using your calculator):

$$\int_{0}^{2} f(x)\,dx = F(2) - F(0)$$

$$= \cos\left(\frac{1}{2^2 + 1}\right) - \cos\left(\frac{1}{0^2 + 1}\right)$$

$$= \cos\left(\frac{1}{5}\right) - \cos(1)$$

Note that it's easy to lose track of parentheses when entering expressions into your calculator, so the safest approach is to always mentally simplify the easy expressions and then plug those into your calculator. Here, the result is approximately 0.440, so **(B)** is correct.

80. B

$$e^x = 1 + x + \frac{x^2}{2!} + \frac{x^3}{3!} + \frac{x^4}{4!} + \cdots$$

$$\Rightarrow e^x - 1 = x + \frac{x^2}{2!} + \frac{x^3}{3!} + \frac{x^4}{4!} + \cdots$$

$$\Rightarrow \frac{e^x - 1}{x} = 1 + \frac{x}{2!} + \frac{x^2}{3!} + \frac{x^3}{4!} + \cdots$$

II is the only choice that fits this series. Hence, **(B)** is correct.

81. B

$x'(t) = -6\sin(3t)$ and $y'(t) = -e^t$

$$\text{arclength} = \int_{-1}^{3} \sqrt{\left(-6\sin(3t)\right)^2 + \left(-e^t\right)^2}\, dt \approx 28.155$$

Therefore, **(B)** is the correct answer.

82. D

In calculus, volume nearly always means integrate. Draw a quick sketch to get an idea of what you need to integrate.

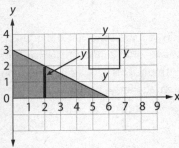

Each cross-section is a square with side length $y = 3 - \frac{1}{2}x$ and the area of a square is given by $A = s^2$, so the volume of the solid is:

$$\int_0^6 \left(3 - \frac{1}{2}x\right)^2 dx$$

Save yourself some time by using your calculator to evaluate the definite integral. To do this, enter the integrand into Y= and adjust the viewing window as needed. Press CALC (2nd TRACE) 7 (for the $\int f(x)\, dx$ option). You'll also need to enter the limits of integration (here, the lower limit is 0 and the upper limit is 6). The result is approximately 18, which is **(D)**.

83. C

This is a fairly straightforward question—you need to find the second derivative of the given function and then evaluate it at −2. Start by rewriting $\frac{1}{x}$ as x^{-1}, then differentiate each term twice, then plug in −2 for each x. The result is:

$$f(x) = 3x^2 - 7x + x^{-1}$$
$$f'(x) = 6x - 7 - x^{-2}$$
$$f''(x) = 6 + 2x^{-3}$$
$$f''(-2) = 6 + \frac{2}{(-2)^3} = 6 - \frac{1}{4} = \frac{23}{4}$$

That's **(C)**.

84. A

This differential equation says that the rate of growth of the rabbits is proportional to the number of rabbits. Rewrite y' as $\frac{dy}{dt}$, separate the variables, and use implicit differentiation and the given initial condition to find that:

$$\frac{dy}{dt} = \frac{1}{3}y$$
$$\frac{dy}{y} = \frac{1}{3}dt$$
$$\int \frac{dy}{y} = \int \frac{1}{3}dt$$
$$\ln y = \frac{1}{3}t + C$$
$$y = e^{\frac{1}{3}t + C} = e^{\frac{1}{3}t} \cdot e^C = Ae^{\frac{1}{3}t}$$
$$y(0) = Ae^{\frac{1}{3}(0)} \rightarrow A = 50$$

then, $y = 50e^{\frac{1}{3}t}$.

Now, plug $t = 9$ years into this equation to arrive at approximately $50e^3$ (or 1,004) rabbits after 9 years. That's **(A)**.

85. B

Using integration by parts,

$u = x$	$dv = \sec^2 x\, dx$
$du = dx$	$v = \tan x$

$$\int x\sec^2 x\, dx = x\tan x - \int \tan x\, dx$$
$$= x\tan x + \ln|\cos x| + C.$$

Hence, **(B)** is the correct answer.

86. A

You can't simplify the integrand and there is no "product rule" for integrals, so this must be a u-substitution question. Let $u = x^2$, which means $du = 2x\,dx$. You already have the x in the integrand, but not the 2, so multiply the integrand by a factor of 2 and multiply the entire integral by $\frac{1}{2}$ to compensate. Then integrate:

$$\int x \sec^2(x^2)\,dx = \frac{1}{2}\int \sec^2(x^2)\,2x\,dx$$

$$= \frac{1}{2}\int \sec^2(u)\,du$$

$$= \frac{1}{2}\tan(u) + C$$

$$= \frac{1}{2}\tan(x^2) + C$$

Thus, **(A)** is correct.

87. D

Use your calculator to graph the two equations and you'll find that the line $y = 2x + 5$ is above the parabola $y = 2x^2 + 1$, so you need to evaluate the integral:

$$\int_a^b (2x + 5) - (2x^2 + 1)\,dx$$

To determine the values of a and b, use the Intersect function on your calculator to find the x-values where the graphs intersect (or solve $2x^2 + 1 = 2x + 5$ by factoring) to get $x = 2$ and $x = -1$. The enclosed area is therefore given by the definite integral:

$$\int_{-1}^{2} (2x + 5) - (2x^2 + 1)\,dx$$

$$= \int_{-1}^{2} (-2x^2 + 2x + 4)\,dx$$

Save yourself some time by using the Integrate function on your calculator to find that the area is 9 square units, **(D)**.

88. C

When given a slope field and asked to match it to a differential equation, the quickest way to the correct answer is to look for aspects of the slope field that stand out. Here, the slope field has no line segments drawn along

the horizontal line $y = -1$. This means the derivative $\left(\frac{dy}{dx}\right)$ does not exist at $y = -1$, so look for the differential equation that is undefined when $y = -1$. Only (C) meets this criterion, so **(C)** is the correct answer. Alternatively, you could plug in some "easy" numbers for x and y in each of the answer choices to find the slope. Then look at the graph to see whether it has approximately that slope at those (x, y) points. If you use this method, it may take a bit of time to eliminate answer choices until you find the correct one.

89. D

Using partial fractions, $\dfrac{5}{x^2 + x - 6} = \dfrac{1}{x - 2} - \dfrac{1}{x + 3}$.

Therefore,

$$\int \frac{5}{x^2 + x - 6}\,dx = \int \left(\frac{1}{x - 2} - \frac{1}{x + 3}\right) dx$$

$$= \ln|x - 2| - \ln|x + 3| + C$$

Hence, **(D)** is the correct answer.

90. B

First, separate the variables: cross-multiply and then integrate both sides to get $\int y^3\,dy = \int x^2\,dx$. Thus:

$$\frac{1}{4}y^4 = \frac{1}{3}x^3 + C$$

$$y^4 = \frac{4}{3}x^3 + C$$

Now use the initial condition, $y(3) = 3$, to find C:

$$3^4 = \frac{4}{3} \cdot 27 + C$$

$$81 = 36 + C$$

$$45 = C$$

Plug in $C = 45$ and solve for y to get $y = \sqrt[4]{\frac{4}{3}x^3 + 45}$, which is **(B)**.

Note that when taking the 4th root, you typically have a $\pm$ before the root, but here, you are given the point $(3, 3)$, and because the y-value is positive when the x-value is plugged in, you can ignore the negative part of the root.

SECTION II, Part A

1. When answering questions that ask for a description of the system an equation is modeling, always write clearly and in complete sentences. (In fact, you should always write your answers as though you were explaining the answer to someone, and always in complete sentences.) You might get most of the credit for an answer such as "it stopped and then started again," but then again, you might not. For part (b), and similar problems, questions often come up where the velocity is negative *and* the acceleration is negative. In these cases, the *speed* is increasing, because negative velocity and acceleration mean that velocity is becoming more negative, and so the particle is moving faster and faster in the negative direction. If velocity were negative but acceleration positive, then the speed would be decreasing. For part (c), make sure you look at the graph of the function before you ask the calculator to find the zero. If not, you might end up with $t = 2$ as the first time the particle changes direction.

(a)

Solutions	Scoring Rubric
Velocity is given by $$v(t) = 4\sin\left(\frac{t^2}{2} - 2t + 2\right).$$ Acceleration is given by $$a(t) = v'(t) = 4\cos\left(\frac{t^2}{2} - 2t + 2\right)\left(\frac{2t}{2} - 2\right)$$ Thus, the velocity at $t = 4$ is $v(4) = 3.637$ unit/sec and the acceleration is $v'(4) = -3.329$ units/sec^2.	1: $v(4) = 3.637$ unit/sec 1: $a(4) = -3.329$ units/sec^2

(b)

Solutions	Scoring Rubric
At $t = 4$, $v(t)$ is positive and $v'(t)$ is negative, so the speed of the particle is decreasing. Speed is the absolute value of velocity, so because velocity is positive, speed is equal to velocity and the rate of change of speed is the rate of change of velocity.	1: speed is decreasing 1: valid reasoning

(c)

Solutions	Scoring Rubric
The particle changes direction when $v(t)$ changes sign; $v(t)$ first changes sign at $t = 4.507$.	1: $t = 4.507$ 1: valid reasoning

(d)

Solutions	Scoring Rubric
Near $t = 2$, the graph of the velocity of the particle decreases to 0 and then begins increasing again. Before $t = 2$, the particle is in the positive part of the x-axis, moving in the positive direction, but slowing down. As t approaches 2, the particle stops. At $t = 2$, the particle is stopped. After $t = 2$, the particle begins moving in the positive direction again. Informally, the particle is moving, slowing down, stopping for an instant, and then speeding back up again.	1: slowing down before $t = 2$ 1: stopping at $t = 2$ 1: speeding up after $t = 2$

2. For part (a), read the question carefully and make sure you understand what it's asking. (For example, it's not asking for the actual volume of snow on the mountain at $t = 6$.) Remember, because $M(t)$ measures the rate the snow melts, it contributes negatively to the rate of change of the volume of the snow. Again, don't forget your units!

(a)

Solutions	Scoring Rubric
The total volume of snow added to the mountain over the first 6-hour period is given by the integral $$\int_0^6 S(t)\, dt = \int_0^6 24 - t\sin^2\left(\frac{t}{14}\right) dt$$ $$= 142.413 \text{ yds}^3$$	1: limits 0, 6 1: appropriately set up integral 1: 142.413 yds^3

(b)

Solutions	Scoring Rubric
The first time that the instantaneous rate of change of the volume of snow is 0 occurs the first time $M(t) = S(t)$, at $t = 15.0404$ hours.	1: $t = 15.040$ hrs 1: valid reasoning

(c)

Solutions	Scoring Rubric
At time $t = 4$, the rate of change of the volume of snow is given by $S(4) - M(4) = 11.8004$ yds^3/hr. This is positive, so the volume of snow is increasing.	1: increasing 1: valid reasoning

(d)

Solutions	Scoring Rubric
At time $t = 6$, the amount of snow added to the mountain was 142.413 yds^3, bringing the total amount of snow needing to be melted to 192.413 yds^3. To determine the time when all the snow is melted we solve the equality $$\int_0^S M(t)\, dt = \int_0^S 10 + 8\cos\left(\frac{t}{14}\right) dt = 192.413$$ This occurs at $S = 11.204$ hours.	1: appropriately set up integral 1: valid reasoning

SECTION II, Part B

3. (a)

Solutions	Scoring Rubric
Remember, when using Euler's method, each value you calculate is based on the previous value. $f(0.5) \approx f(0) + f'(0) \cdot 0.5$ $\qquad$ $f'(0) = 2 \cdot 0(3+1) = 0$ $f(0.5) \approx \sqrt{3} + 0 \cdot 0.5 = \sqrt{3}$ $f(1) \approx f(0.5) + f'(0.5) \cdot 0.5$ $\qquad$ $f'(0.5) = 2 \cdot 0.5(3+1) = 4$ $f(1) \approx \sqrt{3} + 4 \cdot 0.5 = 2 + \sqrt{3}$	1: set up first step correctly 1: $\sqrt{3}$ 1: $2 + \sqrt{3}$

(b)

Solutions	Scoring Rubric
Separating the variables, $$\frac{1}{y^2 + 1} \, dy = 2t \, dt$$ $$\int \frac{1}{y^2 + 1} \, dy = \int 2t \, dt$$ $$\arctan y = t^2 + C$$ Substituting $\left(0, \sqrt{3}\right)$ gives $\arctan\sqrt{3} = 0^2 + C$, so $C = \frac{\pi}{3}$. Therefore, $\arctan y = t^2 + \frac{\pi}{3}$, so $y = \tan\left(t^2 + \frac{\pi}{3}\right)$.	1: separate variables correctly 1: $C = \frac{\pi}{3}$ 1: $y = \tan\left(t^2 + \frac{\pi}{3}\right)$

(c)

Solutions	Scoring Rubric
Because $\frac{dy}{dt} = 2t\left(y^2 + 1\right)$, $$\frac{d^2 y}{dt^2} = 2t \cdot 2y \frac{dy}{dt} + 2\left(y^2 + 1\right).$$ Evaluating $\frac{d^2 y}{dt^2}$ at $\left(0, \sqrt{3}\right)$ gives $f''(0) = 8$. Because you already know that $f'(0) = 0$: $$f(t) \approx \sqrt{3} + 0t + \frac{8}{2!}t^2 = \sqrt{3} + 4t^2.$$	1: $\frac{d^2 y}{dt^2} = 2t \cdot 2y \frac{dy}{dt} + 2\left(y^2 + 1\right)$ 1: $f''(0) = 8$ 1: $\sqrt{3} + 4t^2$

4.

x	y	$\dfrac{dy}{dx} = \dfrac{x^2-1}{y}$
0	1	-1
0	2	$-\dfrac{1}{2}$
0	3	$-\dfrac{1}{3}$
0	4	$-\dfrac{1}{4}$
1	1	0
1	2	0
1	3	0
1	4	0
2	1	3
2	2	$\dfrac{3}{2}$
2	3	1
2	4	$\dfrac{3}{4}$

For part (b), because you know the point you are looking at, the slope is just given by the slope field, namely

$$\frac{dy}{dx} = \frac{x^2-1}{y} = \frac{3^2-1}{4} = 2.$$

Use any form for the equation of the line and don't bother to simplify, because there's a small chance you could make an error and lose points. You won't lose points for leaving the equation for the tangent line unsimplified. For part (c), always remember to separate the variables before integrating. Even if you have the correct answer, you will lose points if you fail to separate the variables first. Finally, in the computation of C, use the simpler of the two equivalent formulations of the solution to ease calculations. Note also that if the initial condition was $f(3) = -4$, you would have ended up with the same solution only using the *negative* root.

(a)

Solutions	Scoring Rubric
	1: 1st column correct 1: 2nd column correct 1: 3rd column correct

(b)

Solutions	Scoring Rubric
Because $f(x)$ is a solution to the differential equation and $f(3) = 4$, the slope of the tangent line to the graph of f is $$\frac{dy}{dx} = \frac{x^2 - 1}{y} = \frac{9 - 1}{4} = 2.$$ The equation for the tangent line is $(y - 4) = 2(x - 3)$.	1: $\dfrac{dy}{dx} = 2$ 1: $y - 4$ 1: $x - 3$

(c)

Solutions	Scoring Rubric
Solving: $$\frac{dy}{dx} = \frac{x^2 - 1}{y}$$ $$y\,dy = (x^2 - 1)\,dx$$ $$\int y\,dy = \int x^2 - 1\,dx$$ $$\frac{1}{2}y^2 = \frac{1}{3}x^3 - x + C$$ $$y = \pm\sqrt{\frac{2}{3}x^3 - 2x + 2C}$$ To compute C, use $\frac{1}{2}y^2 = \frac{1}{3}x^3 - x + C$. Thus, $\frac{1}{2}(4^2) = \frac{1}{3}(3^3) - 3 + C$, and $C = 2$. Finally, the particular solution is given by the equation $f(x) = \sqrt{\frac{2}{3}x^3 - 2x + 4}$. (Because $f(3) = 4$, take the positive root.)	1: $y =$ $\pm\sqrt{\frac{2}{3}x^3 - 2x + 2C}$ 1: $C = 2$ 1: $f(x) = \sqrt{\frac{2}{3}x^3 - 2x + 4}$

5. (a)

Solutions	Scoring Rubric
For a cylinder, $\text{Vol} = \pi r^2 h$. Thus, $$100 = \pi r^2 h$$ $$h = \frac{100}{\pi r^2}$$	1: $\text{Vol} = \pi r^2 h$ 1: $h = \frac{100}{\pi r^2}$

(b)

Solutions	Scoring Rubric
You are given that $\frac{dh}{dt} = -0.01\,\text{cm/min}$. By (a), You have $$\frac{dh}{dt} = -\frac{200}{\pi} r^{-3} \frac{dr}{dt},$$ and we want $\frac{dr}{dt}$ at $r = 10$ cm. Solving: $$\frac{dr}{dt} = \frac{dh}{dt}\left(-\frac{\pi r^3}{200}\right) = -0.01\left(-\frac{\pi r^3}{200}\right) = 0.01\left(\frac{\pi r^3}{200}\right)$$ At $r = 10$, you have $$\frac{dr}{dt} = 0.01\left(\frac{\pi(10^3)}{200}\right) = 0.05\pi.$$ Thus, at $r = 10$ cm, the radius is increasing at 0.05π cm/min.	1: $\frac{dh}{dt} = -\frac{200}{\pi} r^{-3} \frac{dr}{dt}$ 1: $\frac{dr}{dt} = 0.01\left(\frac{\pi r^3}{200}\right)$ 1: 0.05π

(c)

Solutions	Scoring Rubric
At $r = 10$, $h(1) = \frac{100}{\pi r^2} = \frac{100}{\pi 100} = \frac{1}{\pi}$. At $r = 20$, $h(t) = \frac{100}{\pi r^2} = \frac{100}{\pi 400} = \frac{1}{4\pi}$. Applying the second fundamental theorem of calculus, $$h(t) = \int_0^t h'(s)\,ds = \int_0^1 h'(s)\,ds + \int_1^t h'(s)\,ds.$$ Because you know that $h'(t) = -0.01$ and $h(1) = \frac{1}{\pi}$. $$h(t) = \frac{1}{\pi} - 0.01\int_1^t ds = \frac{1}{\pi} - 0.01\,(t - 1)$$ Finally, when $r = 20$, $h(t) = \frac{1}{4\pi}$. So you need to find t so that $h(t) = \frac{1}{4\pi}$. $$\frac{1}{4\pi} = h(t) = \frac{1}{\pi} - 0.01\int_1^t ds = \frac{1}{\pi} - 0.01(t - 1)$$ $$\frac{1}{4\pi} - \frac{1}{\pi} = -0.01(t - 1)$$ $$t = \frac{1}{0.01}\left(\frac{1}{\pi} - \frac{1}{4\pi}\right) + 1 = 100\frac{3}{4\pi} + 1 = \frac{75 + \pi}{\pi}$$	1: $h(1) = \frac{1}{\pi}$ 1: $h(t) = \frac{1}{4\pi}$ when $r = 20$ 1: set up integral correctly 1: $t = \frac{(75 + \pi)}{\pi}$

6. (a)

Solutions	Scoring Rubric
Start with the series for e^x and build from there. $$e^x = 1 + x + \frac{x^2}{2!} + \frac{x^3}{3!} + \cdots$$ $$e^{-x} = 1 - x + \frac{x^2}{2!} - \frac{x^3}{3!} + \cdots$$ $$f(x) = \frac{e^x - e^{-x}}{x} = \frac{2x + \frac{2x^3}{3!} + \frac{2x^5}{5!} + \cdots}{x}$$ $$= 2 + \frac{2x^2}{3!} + \frac{2x^4}{5!} + \cdots + \frac{2x^{2k}}{(2k+1)!} + \cdots, \text{ for}$$ $$k = 0, 1, 2, \cdots$$	1: $e^x = 1 + x + \frac{x^2}{2!} + \frac{x^3}{3!} + \cdots$ 1: $e^{-x} = 1 - x + \frac{x^2}{2!} - \frac{x^3}{3!} + \cdots$ 1: $f(x) = \frac{e^x - e^{-x}}{x}$ $$= \frac{2x + \frac{2x^3}{3!} + \frac{2x^5}{5!} + \cdots}{x}$$ $$= 2 + \frac{2x^2}{3!} + \frac{2x^4}{5!} + \cdots$$ $$+ \frac{2x^{2k}}{(2k+1)!} + \cdots, \text{ for}$$ $$k = 0, 1, 2, \cdots$$

(b)

Solutions	Scoring Rubric
$$\lim_{x \to 0} \frac{e^x - e^{-x}}{x}$$ $$= \lim_{x \to 0} \left(2 + \frac{2x^2}{3!} + \frac{2x^4}{5!} + \cdots + \frac{2x^{2k}}{(2k+1)!} + \cdots \right)$$ $$= 2$$	1: set up the limit properly $$\lim_{x \to 0} \left(2 + \frac{2x^2}{3!} + \frac{2x^4}{5!} + \cdots + \frac{2x^{2k}}{(2k+1)!} + \cdots \right)$$ 1: 2

(c)

Solutions	Scoring Rubric
$$e^{-x} = 1 - x + \frac{x^2}{2!} - \frac{x^3}{3!} + \cdots$$ $$\frac{1 - e^{-x}}{x} = \frac{1 - \left(1 - x + \frac{x^2}{2!} - \frac{x^3}{3!} + \cdots \right)}{x}$$ $$= 1 - \frac{x}{2!} + \frac{x^2}{3!} - \cdots + (-1)^k \frac{x^k}{(k+1)!} + \cdots$$ for $k = 0, 1, 2, \ldots$	1: first 3 nonzero terms $1 - \frac{x}{2!} + \frac{x^2}{3!}$ 1: general term $(-1)^k \frac{x^k}{(k+1)!} + \cdots$

(d)

Solutions	Scoring Rubric
The error is less than the absolute value of the fifth term, evaluated at $x = \frac{1}{10}$. $$\left. \left| -\frac{x^4}{5!} \right| \right|_{x=\frac{1}{10}} = \frac{\left(\frac{1}{10}\right)^4}{120} = \frac{1}{1,200,000} < \frac{1}{1,000,000}$$	1: state that the error is less than the absolute value of the fifth term, evaluated at $x = \frac{1}{10}$ 1: $\left. \left| -\frac{x^4}{5!} \right| \right|_{x=\frac{1}{10}} = \frac{\left(\frac{1}{10}\right)^4}{120} = \frac{1}{1,200,000} < \frac{1}{1,000,000}$

BC Practice Exam 3

Section I, Part A

1 Ⓐ Ⓑ Ⓒ Ⓓ
2 Ⓐ Ⓑ Ⓒ Ⓓ
3 Ⓐ Ⓑ Ⓒ Ⓓ
4 Ⓐ Ⓑ Ⓒ Ⓓ
5 Ⓐ Ⓑ Ⓒ Ⓓ
6 Ⓐ Ⓑ Ⓒ Ⓓ
7 Ⓐ Ⓑ Ⓒ Ⓓ
8 Ⓐ Ⓑ Ⓒ Ⓓ
9 Ⓐ Ⓑ Ⓒ Ⓓ
10 Ⓐ Ⓑ Ⓒ Ⓓ
11 Ⓐ Ⓑ Ⓒ Ⓓ
12 Ⓐ Ⓑ Ⓒ Ⓓ
13 Ⓐ Ⓑ Ⓒ Ⓓ
14 Ⓐ Ⓑ Ⓒ Ⓓ
15 Ⓐ Ⓑ Ⓒ Ⓓ
16 Ⓐ Ⓑ Ⓒ Ⓓ
17 Ⓐ Ⓑ Ⓒ Ⓓ
18 Ⓐ Ⓑ Ⓒ Ⓓ
19 Ⓐ Ⓑ Ⓒ Ⓓ
20 Ⓐ Ⓑ Ⓒ Ⓓ
21 Ⓐ Ⓑ Ⓒ Ⓓ
22 Ⓐ Ⓑ Ⓒ Ⓓ
23 Ⓐ Ⓑ Ⓒ Ⓓ
24 Ⓐ Ⓑ Ⓒ Ⓓ
25 Ⓐ Ⓑ Ⓒ Ⓓ
26 Ⓐ Ⓑ Ⓒ Ⓓ
27 Ⓐ Ⓑ Ⓒ Ⓓ
28 Ⓐ Ⓑ Ⓒ Ⓓ
29 Ⓐ Ⓑ Ⓒ Ⓓ
30 Ⓐ Ⓑ Ⓒ Ⓓ

Section I, Part B

76 Ⓐ Ⓑ Ⓒ Ⓓ
77 Ⓐ Ⓑ Ⓒ Ⓓ
78 Ⓐ Ⓑ Ⓒ Ⓓ
79 Ⓐ Ⓑ Ⓒ Ⓓ
80 Ⓐ Ⓑ Ⓒ Ⓓ
81 Ⓐ Ⓑ Ⓒ Ⓓ
82 Ⓐ Ⓑ Ⓒ Ⓓ
83 Ⓐ Ⓑ Ⓒ Ⓓ
84 Ⓐ Ⓑ Ⓒ Ⓓ
85 Ⓐ Ⓑ Ⓒ Ⓓ
86 Ⓐ Ⓑ Ⓒ Ⓓ
87 Ⓐ Ⓑ Ⓒ Ⓓ
88 Ⓐ Ⓑ Ⓒ Ⓓ
89 Ⓐ Ⓑ Ⓒ Ⓓ
90 Ⓐ Ⓑ Ⓒ Ⓓ

AAAAAAAAAAAAAAAAAAAAAAAAAAAAAAAAAA

CALCULUS BC
SECTION I, Part A
Time—60 minutes
Number of questions—30

A CALCULATOR MAY NOT BE USED ON THIS PART OF THE EXAM.

Directions: Use the space provided to solve each of the following problems. After examining the given choices, decide which of the choices is the best and fill in the corresponding circle on the answer sheet. Credit will only be given for answers appropriately marked on the answer sheet, and not for anything written in the exam book. Do not spend too much time on any one problem.

In this exam:

(1) The domain of a function f should be assumed to be the set of all real numbers x for which $f(x)$ is a real number, unless the question indicates otherwise.

(2) The inverse function notation f^{-1} or the prefix "arc" (e.g., $\sin^{-1} x = \arcsin x$) may be used to indicate the inverse of a trigonometric function.

GO ON TO THE NEXT PAGE.

A A A A A A A A A A A A A A A A A A A A A A A A A A A A A A A A A A A

1. What is the value of $\sum_{n=0}^{\infty}\left(-\frac{3}{4}\right)^{n}$?

   (A) $-\frac{3}{7}$ (B) $\frac{4}{7}$ (C) $\frac{7}{4}$ (D) The series diverges.

---

2. Let $y = f(x)$ be the solution to the differential equation $\frac{dy}{dx} = y - x + 2$ with initial condition $f(2) = -3$. Using Euler's method with 2 steps of equal size, what is the approximation for $f(1.6)$?

   (A) -4.52 (B) -1.96 (C) 0.48 (D) 1.88

---

x	-2	-1	0	1	2	4
$f(x)$	3	4	5	6	5	6

3. Several ordered pairs for the function f are given in the table above. If $\lim_{x \to 1} f(g(x)) = 4$, then $\lim_{x \to 1} g(x) =$

   (A) -1 (B) 3   (C) 4   (D) 6

GO ON TO THE NEXT PAGE.

AAAAAAAAAAAAAAAAAAAAAAAAAAAAAAAAAAAA

x	$f(x)$	$g(x)$	$f'(x)$	$g'(x)$
2	3	4	5	6
3	4	5	6	7
4	5	6	7	8
5	6	7	8	9

4. Two functions, $f(x)$ and $g(x)$, are continuous and differentiable for all real numbers. Some values of the functions and their derivatives are given in the table above. Based on the table, what is the value of $h'(2)$ if $h(x) = f(g(x))$?

(A) 5   (B) 7   (C) 30   (D) 42

5. A function h has a Maclaurin series given by

$$x^2 - \frac{x^4}{2} + \frac{x^6}{3} + \dots + (-1)^n \frac{x^{2n}}{n} + \dots$$

Which of the following is an expression for $h(x)$?

(A) $\cos x$ (B) $\ln(1 + x^2)$ (C) $e^x - \sin x$ (D) $e^x + \sin x$

GO ON TO THE NEXT PAGE.

AAAAAAAAAAAAAAAAAAAAAAAAAAAAAAAAAAA

6. For time $t > 0$, the position of a particle moving in the xy-plane is given parametrically by $x = \dfrac{1}{2t+1}$ and $y = 6t - t^2$. What is the acceleration vector at time $t = 2$?

(A) $\langle -2, 2 \rangle$ (B) $\left\langle \dfrac{-2}{25}, 2 \right\rangle$ (C) $\left\langle \dfrac{8}{125}, -2 \right\rangle$ (D) $\left\langle \dfrac{2}{25}, 2 \right\rangle$

7. $\displaystyle\int \dfrac{4}{x^2 - 1}\, dx =$

(A) $4 \ln\left|x^2 - 1\right| + C$

(B) $2 \ln\left|x - 1\right| + 2 \ln\left|x + 1\right| + C$

(C) $2 \ln\left|\dfrac{x-1}{x+1}\right| + C$

(D) $2 \ln\left|\dfrac{x+1}{x-1}\right| + C$

x	0	4	6	9
$h(x)$	1	4	1	5

8. Some values of the continuous function h are given in the table above. Use the table to approximate the value of $\displaystyle\int_0^9 h(x)\, dx$ using trapezoids with the three subintervals.

(A) 7   (B) 17   (C) 24   (D) 28

GO ON TO THE NEXT PAGE.

AAAAAAAAAAAAAAAAAAAAAAAAAAAAAAAAAA

9. $\lim\limits_{h \to 0} \dfrac{\sqrt[3]{1+h} - \sqrt[3]{1}}{h} =$

    (A) 0   (B) $\dfrac{1}{3}$ (C) 1   (D) Nonexistent

---

10. The radius of convergence for the power series $\sum\limits_{n=1}^{\infty} \dfrac{(x+4)^{2n}}{n}$ is 1. What is the interval of convergence?

    (A) $3 \le x < 5$

    (B) $-1 \le x < 1$

    (C) $-5 < x < -3$

    (D) $-5 \le x < -3$

---

11. If $u = x^2$, which of the following integrals is equivalent to $\int_0^2 xe^{x^2}\, dx$?

    (A) $\int_0^2 e^u\, du$

    (B) $\dfrac{1}{2}\int_0^2 ue^u\, du$

    (C) $\int_0^4 e^u\, du$

    (D) $\dfrac{1}{2}\int_0^4 e^u\, du$

GO ON TO THE NEXT PAGE.

AAAAAAAAAAAAAAAAAAAAAAAAAAAAAAAAAAA

12. Which of the following series converge?

I. $\displaystyle\sum_{n=1}^{\infty} \frac{1}{\sqrt[3]{n}}$

II. $\displaystyle\sum_{n=1}^{\infty} \left(\frac{2}{\pi}\right)^n$

III. $\displaystyle\sum_{n=1}^{\infty} \frac{5(n+1)}{(n+2)!}$

(A) I only

(B) II only

(C) I and II only

(D) II and III only

---

13. A particle travels along the x-axis so that its velocity is given by $v(t) = 2 \cos t$ for $t \geq 0$. The position of the particle at $t = \dfrac{\pi}{2}$ is $x = 5$. The position, $x(t)$, of the particle at any time t is given by

(A) $x(t) = 2 \sin t + 5$

(B) $x(t) = -2 \sin t + 7$

(C) $x(t) = 2 \sin t + 3$

(D) $x(t) = -2 \sin t + 5$

GO ON TO THE NEXT PAGE.

A A A A A A A A A A A A A A A A A A A A A A A A A A A A A A

14. The derivative of $x^3 + xy = 8$ is $\dfrac{dy}{dx} = \dfrac{-3x^2 - y}{x}$. What is the value of $\dfrac{d^2y}{dx^2}$ at $x = 2$?

    (A) -6 (B) -3 (C) 0   (D) 4

x	$g(x)$	$g'(x)$
1	1	3
6	-3	7

15. The function g has a continuous derivative. The table above gives selected values of g and its derivative for $x = 1$ and $x = 6$. If $\displaystyle\int_1^6 g(x)\ dx = 10$, what is the value of $\displaystyle\int_1^6 xg'(x)\ dx$?

    (A) -109 (B) $-36\dfrac{1}{3}$ (C) -29 (D) -19

GO ON TO THE NEXT PAGE.

AAAAAAAAAAAAAAAAAAAAAAAAAAAAAAAA

16. The function f has the property that $f(x), f'(x)$, and $f''(x)$ are positive for all $x > 0$. Which of the following could be part of the graph of f?

(A)

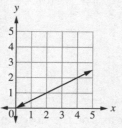

(B)

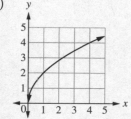

(C)

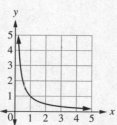

(D)

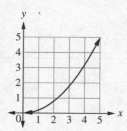

GO ON TO THE NEXT PAGE.

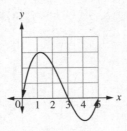

17. The graph of a function f is shown above. The area of the region between f and the x-axis on the closed interval $[0, 3]$ is 9. The area of the region between f and the x-axis on the closed interval $[3, 5]$ is 2. What is the value of $\int_0^5 f(x)\, dx$?

(A) 5   (B) 7   (C) 11   (D) 18

18. The line tangent to the graph of $f(x) = \dfrac{x}{x + 3}$ is parallel to the line $x - 3y = 3$ at which of the following?

(A) $x = 0$ only

(B) $x = -2$ only

(C) $x = -4$ and $x = -2$

(D) $x = 0$ and $x = -6$

GO ON TO THE NEXT PAGE.

A A A A A A A A A A A A A A A A A A A A A A A A A A A A A A A

19. $\dfrac{d}{dx}\displaystyle\int_1^{x^3}\ln\left(5+t^2\right)dt =$

(A) $3x^2\ln\left(5+x^6\right)$

(B) $\dfrac{3x^2}{5+x^6}$

(C) $\ln\left(5+x^6\right)$

(D) $\ln\left(5+x^2\right)$

20. Which of the following functions has the line $y = 3$ as a horizontal asymptote?

   I. $y = \dfrac{\sqrt{9x^2+1}}{x-2}$

   II. $y = \dfrac{3x}{1-x}$

   III. $y = \dfrac{9x^2-2}{2+3x^2}$

(A) I only

(B) I and II only

(C) I and III only

(D) I, II, and III

GO ON TO THE NEXT PAGE.

21. $\int \sec^2 3x \, dx =$

(A) $\frac{1}{3} \tan(3x) + C$

(B) $\frac{\sec^3 3x}{3} + C$

(C) $3 \tan(3x) + C$

(D) $\sec^3 3x + C$

---

22. What is the slope of the line tangent to the polar curve $r = 3\theta$ when $\theta = \frac{\pi}{2}$?

(A) $-\frac{\pi}{2}$ (B) $-\frac{2}{\pi}$ (C) 0   (D) $\frac{\pi}{2}$

---

23. Let f be a function defined by $f(x) = \int_3^{x^3} \sqrt{t^2 + t} \, dt$. $f(2) =$

(A) $2\sqrt{6}$ (B) $4\sqrt{6}$ (C) $2\sqrt{20}$ (D) $4\sqrt{20}$

GO ON TO THE NEXT PAGE.

A A A A A A A A A A A A A A A A A A A A A A A A A A A A A A A A A A A A A A

24. Let $y = (x^5 + \sin x)^4$. $\dfrac{dy}{dx} =$

(A) $4(5x^4 + \cos x)^3$

(B) $4(x^5 + \sin x)^3 (5x^4 - \cos x)$

(C) $4(x^5 + \sin x)^3 (5x^4 + \cos x)$

(D) $4(5x^4 + \cos x)^3$

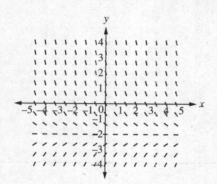

25. A slope field for which of the following differential equations is shown above?

(A) $\dfrac{dy}{dx} = -y - 2$

(B) $\dfrac{dy}{dx} = y + 2$

(C) $\dfrac{dy}{dx} = \dfrac{y + 2}{x}$

(D) $\dfrac{dy}{dx} = \dfrac{y + 2}{x^2}$

GO ON TO THE NEXT PAGE.

AAAAAAAAAAAAAAAAAAAAAAAAAAAAAAAAA

26. Let a function f be defined as $f(x) = x^3 - 2x - 4$ for $x \geq 1$. Let $g(x)$ be the inverse of $f(x)$ and note that $f(2) = 0$. What is the value of $g'(0)$?

    (A) -2 (B) $-\dfrac{1}{2}$ (C) $\dfrac{1}{10}$ (D) 1

27. Suppose the point (x, y) lies on the curve $y = \sqrt{x}$. For what value of x is the distance between (x, y) and the point $(3, 0)$ a minimum?

    (A) $\dfrac{7}{3}$ (B) $\dfrac{5}{2}$ (C) $\dfrac{7}{2}$ (D) $\dfrac{9}{2}$

28. $\displaystyle\lim_{x \to 0} \dfrac{1 - \cos(\pi x)}{\sin(2x)} =$

    (A) $-\dfrac{1}{2}$ (B) 0 (C) $\dfrac{1}{2}$ (D) nonexistent

GO ON TO THE NEXT PAGE.

29. The position of a particle moving along the x-axis is given by $x(t) = e^{2t} - e^t$ for all $t \geq 0$. When the particle is at rest, the acceleration of the particle is

(A) $\frac{1}{2}$　(B) $\frac{1}{4}$　(C) $\ln \frac{1}{2}$　(D) 4

30. $\int \dfrac{2\,dx}{1+16x^2} =$

(A) $\frac{1}{2}\arctan 4x + C$

(B) $\frac{1}{2}\arcsin 4x + C$

(C) $\arctan 4x + C$

(D) $\arcsin 4x + C$

END OF PART A OF SECTION I

**IF YOU FINISH BEFORE TIME IS CALLED, YOU MAY
CHECK YOUR WORK ON PART A ONLY.**

CALCULUS BC
SECTION I, Part B
Time—45 minutes
Number of questions—15

A GRAPHING CALCULATOR IS REQUIRED FOR SOME QUESTIONS ON
THIS PART OF THE EXAM.

Directions: Use the space provided to solve each of the following problems. After examining the given choices, decide which of the choices is the best and fill in the corresponding circle on the answer sheet. Credit will only be given for answers appropriately marked on the answer sheet, and not for anything written in the exam book. Do not spend too much time on any one problem.

In this exam:

(1) When the exact numerical value of the correct answer does not appear among the choices given, select the number that best approximates the numerical value.

(2) The domain of a function f should be assumed to be the set of all real numbers x for which $f(x)$ is a real number, unless the question indicates otherwise.

(3) The inverse function notation f^{-1} or the prefix "arc" (e.g., $\sin^{-1} x = \arcsin x$) may be used to indicate the inverse of a trigonometric function.

GO ON TO THE NEXT PAGE.

B B B B B B B B B

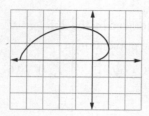

76. The graph above shows the polar curve $r = 3\theta + \sin\theta$ for $0 \le \theta \le \pi$. Let R be the region bounded by the polar curve and the x-axis. What is the area of R ?

(A) 16.804   (B) 20.923   (C) 56.720   (D) 113.439

77. What is the area of the region in the first quadrant bounded by the graphs of $f(x) = \dfrac{5\ln x}{x}$, $x = 1$, and $x = e$?

(A) $\dfrac{5}{3}$ (B) $\dfrac{5}{2}$ (C) $5e$ (D) 15

78. Suppose a function f is continuous for all x. If f has a relative minimum at $(-2, -3)$ and a relative maximum at $(4, 5)$, which of the following statements must be true?

(A) The graph of f has an absolute maximum at $(4, 5)$.

(B) The graph of f is increasing for all x in $-2 < x < 4$.

(C) There exists some c in the interval $-2 < x < 4$ such that $f(c) = 0$.

(D) There exists some c in the interval $-2 < x < 4$ such that $f'(c) = 0$.

GO ON TO THE NEXT PAGE.

79. The growth of a population P is modeled by the differential equation $\frac{dP}{dt} = 0.713P$. If the population is 2 at $t = 0$, what is the population at $t = 5$?

    (A) 5.565   (B) 20.401   (C) 50.810   (D) 70.679

x	0	3	6	9	12	15	18
$g(x)$	−4	−2	3	4	9	5	1

80. Suppose g is a continuous function. A table of selected values of g is shown above. Approximate the value of $\int_0^{18} g(x)\, dx$ using a midpoint Riemann sum with three subintervals of equal length.

    (A) 21   (B) 24   (C) 42   (D) 48

81. The rate of change, $\frac{dP}{dt}$, of the number of people entering an auditorium for a concert is modeled by a logistic differential equation. The capacity of the auditorium is 1,600. At 6:30 pm the number of people in the auditorium is 400 and is increasing at a rate of 600 people per hour. Which of the following differential equations describes the situation?

    (A) $\frac{dP}{dt} = P(1,600 - P) + 400$

    (B) $\frac{dP}{dt} = \frac{1}{600}P(1,600 - P)$

    (C) $\frac{dP}{dt} = \frac{1}{800}P(1,600 - P)$

    (D) $\frac{dP}{dt} = 600P(1,600 - P)$

GO ON TO THE NEXT PAGE.

B B B B B B B B B

82. The first five terms of the Taylor expansion for $f(x)$ about $x = 1$ is given by

$$7 - \frac{3}{2}(x-1) + \frac{7}{2}(x-1)^2 - 6(x-1)^3 + 2(x-1)^4.$$

What is $f'''(1)$?

(A) -36 (B) -18 (C) 0   (D) 6

83. Suppose $f'(x) = \dfrac{-x^3 + 4x - 2}{\sqrt{x^2 + 1}}$. Which of the following statements must be true?

  I. f has a relative minimum at $x = -0.768$.

  II. f has a relative minimum at $x = 0.539$.

  III. f has a relative minimum at $x = 1.675$.

(A) I only

(B) II only

(C) III only

(D) I and II only

GO ON TO THE NEXT PAGE.

84. Suppose R is the region bounded by the graphs of $y = x^3$ and $y = \sqrt{x}$. What is the volume of the solid of revolution that is generated when R is revolved about the y-axis?

    (A) 0.357   (B) 1.122   (C) 1.257   (D) 1.309

85. A spherical balloon is being filled with water so that its volume increases at a rate of 100 cm³/s. How fast is the radius of the balloon increasing when the diameter is 50 cm? The volume of a sphere of radius r is given by the formula $V = \frac{4}{3}\pi r^3$.

    (A) $\frac{1}{100\pi}$ cm/s   (B) $\frac{1}{75\pi}$ cm/s   (C) $\frac{1}{50\pi}$ cm/s   (D) $\frac{1}{25\pi}$ cm/s

86. What is the slope of the curve $y = 3^{\sin x} - 2$ at its first positive x-intercept?

    (A) 0.683   (B) 1.643   (C) 1.705   (D) 1.805

GO ON TO THE NEXT PAGE.

B B B B B B B B B

87. The function f has derivatives of all orders for all real numbers. Let $f^{(4)}(x) = e^{\tan x}$. If the third-degree Taylor polynomial for f centered at $x = 0$ is used to approximate $f(0.5)$, what is the Lagrange Error Bound for the maximum error on the closed interval $[0, 1]$?

(A) 0.002284   (B) 0.004497   (C) 0.042152   (D) 0.677777

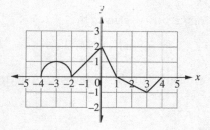

88. Let f be a function defined for $-4 < x < 4$ and consisting of a semi-circle and four line segments, as shown above. Suppose g is the function defined by $g(x) = \int_0^x f(t)\, dt$. The function g is decreasing for which of the following?

(A) $1 \le x \le 4$ only

(B) $1 \le x \le 3$ only

(C) $0 \le x \le 3$ only

(D) $-3 \le x \le -2$ and $0 \le x \le 3$

GO ON TO THE NEXT PAGE.

89. A stream flows into a lake at a rate of $R(t) = \sin\frac{\pi t}{12} + 40t$ gallons per hour. A pump is used to remove water from the lake at a rate of 500 gallons per hour. At the beginning of a certain day, the lake contains 30,000 gallons of water. How much water is in the lake after 24 hours?

    (A) 11,520   (B) 29,520   (C) 41,020   (D) 41,520

---

90. The nth derivative of a function f at $x = 0$ is given by
$$f^{(n)}(0) = (-1)^n \frac{n+2}{(n+1)3^n}$$
    for $n \geq 0$. Which of the following gives the first four nonzero terms for the Maclaurin series for f?

    (A) $2 - \frac{1}{2}x + \frac{4}{27}x^2 - \frac{5}{108}x^3 + \ldots$

    (B) $2 - \frac{1}{2}x + \frac{2}{27}x^2 - \frac{5}{648}x^3 + \ldots$

    (C) $2 + \frac{1}{2}x + \frac{4}{27}x^2 + \frac{5}{108}x^3 + \ldots$

    (D) $2 - 2x + \frac{4}{7}x^2 - \frac{648}{5}x^3 + \ldots$

END OF SECTION I

IF YOU FINISH BEFORE TIME IS CALLED, YOU MAY CHECK YOUR WORK ON PART B ONLY.

CALCULUS BC
SECTION II, Part A
Time—30 minutes
Number of problems—2

A GRAPHING CALCULATOR IS REQUIRED FOR THESE PROBLEMS.

Directions: Write your solution to each part of the following questions in the space provided. Write clearly and legibly. Cross out any errors you make; erased or crossed-out work will not be scored.

You may wish to look over the problems before starting to work on them; on the actual test, it is not expected that everyone will be able to complete all parts of all problems. All problems are given equal weight, but the individual parts of a particular problem are not necessarily given equal weight. You should not spend too much time on any one problem.

- Show all of your work. Clearly label functions, graphs, tables, or anything else that you use to arrive at your final solution. Your work will be scored on the correctness and completeness of your methods as well as your final answers. Answers without supporting work will usually not receive credit.

- Justifications (i.e., the request that you "justify your answer") require that you give mathematical (non-calculator) reasons.

- Do not use calculator syntax. Your work must be expressed in standard mathematical notation.

- Numeric or algebraic answers do not need to be simplified, unless the question indicates otherwise.

- If you use decimal approximations in calculations, your final answers should be accurate to three places after the decimal point, unless the question indicates otherwise.

- The domain of a function f should be assumed to be the set of all real numbers x for which $f(x)$ is a real number, unless the question indicates otherwise.

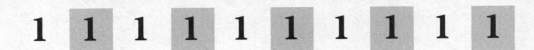

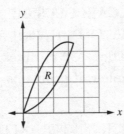

1. Let R be the region in the first quadrant bounded by the graphs of $f(x) = 5x - x^2$ and $g(x) = 2^x - 1$ as shown in the figure above.

   (a) Find the area of region R.

   (b) Let k be the number such that the vertical line $x = k$ divides the area of region R exactly in half, forming two separate regions of equal area. Write, but do not solve, an equation involving an integral that can be used to find the value of k.

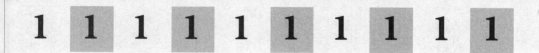

(c) Find the volume of the solid generated when R is rotated about the x-axis.

(d) Suppose region R is the base of a solid whose cross sections cut by a plane perpendicular to the x-axis are squares. Write, but do not evaluate, an integral expression that gives the volume of the solid.

2. A particle moves along the y-axis so that its velocity v at time t is given by $v(t) = t \sin t$, where $t \geq 0$. At time $t = 0$, the particle is at position $y = 2$.

    (a) Find the velocity and the acceleration of the particle at time $t = 4$.

    _____

    (b) Is the speed of the particle increasing or decreasing at time $t = 4$? Justify your answer.

    _____

    (c) Find the total distance traveled by the particle from time $t = 0$ to time $t = 5$.

GO ON TO THE NEXT PAGE.

(d) Find the average velocity of the particle for the time interval from time $t = 0$ to time $t = 5$.

_____

(e) Find the position of the particle at time $t = 5$.

END OF PART A OF SECTION II

**IF YOU FINISH BEFORE TIME IS CALLED, YOU MAY
CHECK YOUR WORK ON PART A ONLY.**

CALCULUS BC
SECTION II, Part B
Time—60 minutes
Number of problems—4

NO CALCULATOR IS ALLOWED FOR THESE PROBLEMS.

Note: If you have extra time, you can go back and work on Part A of Section II, but you cannot use a calculator to complete your work at this time.

3 3 3 3 3 3 3 3 3 3

NO CALCULATOR ALLOWED

3. A 13-foot ladder that is leaning against the wall of a building is pulled away from the wall. At the instant when the base of the ladder is 5 feet from the wall, the base is moving at a rate of 3 feet per second.

(a) At what rate is the top of the ladder sliding down the wall at the instant when the base of the ladder is 5 feet from the wall? Indicate units of measure.

(b) Let θ be the angle between the ladder and the ground. At what rate is this angle changing at the instant when the base of the ladder is 5 feet from the wall? Indicate units of measure.

(c) The wall of the building, the ladder, and the ground form a right triangle. At what rate is the area of the triangle changing at the instant when the base of the ladder is 5 feet from the wall? Indicate units of measure.

GO ON TO THE NEXT PAGE.

NO CALCULATOR ALLOWED

4. Let f be a function satisfying $f'(x) = 4x - 2xf(x)$ for all $x \geq 0$. $\lim\limits_{x \to \infty} f(x) = 4$ and $f(0) = 10$.

(a) Find the value of $\displaystyle\int_0^\infty \left(4x - 2x\,f(x)\right)\,dx$.

(b) Use Euler's method beginning at $x = 0$ to approximate $f(1)$ using 2 intervals of equal width.

(c) Find the unique solution for $y = f(x)$ to the separable differentiable equation

$\dfrac{dy}{dx} = 4x - 2xy$ with the initial condition $f(0) = 10$.

5 5 5 5 5 5 5 5 5 5

NO CALCULATOR ALLOWED

5. Let f be a function with derivative of all orders at $x = 2$. The values of f and its first four derivatives at $x = 2$ are shown below.

$f(2) = 4$

$f'(2) = 0$

$f''(2) = -3$

$f'''(2) = 12$

$f^{(4)}(2) = 3$

(a) Write the third-degree Taylor polynomial for f about $x = 2$ and use it to approximate $f(1.5)$.

(b) Does f have a relative extremum at $x = 2$? If so, determine if the extremum is a relative maximum or relative minimum. If not, explain why not.

GO ON TO THE NEXT PAGE.

5 5 5 5 5 5 5 5 5 5

NO CALCULATOR ALLOWED

(c) It is known that $f^{(4)}(x)$ is strictly increasing on the closed interval $[0, 2]$. Use the Lagrange Error Bound to show that the third-degree Taylor polynomial approximates $f(1.5)$ within 0.01.

(d) Let $h(x) = \int_2^x f(t)\, dt$. Write the third-degree Taylor polynomial for $h(x)$ about $x = 2$.

GO ON TO THE NEXT PAGE.

6 6 6 6 6 6 6 6 6 6

6. The rate of change of the population $P(t)$ of a herd of deer is given by $\dfrac{dP}{dt} = \dfrac{1}{4}(1,200 - P)$, where t is measured in years. When $t = 0$, the population P is 200.

(a) Write an equation of the line tangent to the graph of P at $t = 0$. Use the tangent line to P in order to approximate the population of the herd after 2 years.

(b) Show that $\dfrac{d^2 P}{dt^2} = -\dfrac{1}{16}(1,200 - P)$.

GO ON TO THE NEXT PAGE.

(c) Does the approximation for the population of the herd as found in part (a) overestimate or underestimate the actual population of the herd at time $t = 2$? Support your answer using $\frac{d^2P}{dt^2}$. What does the value of $\frac{d^2P}{dt^2}$ indicate about the shape of the graph?

(d) Using separation of variables, find the particular solution $y = P(t)$ to the differential equation $\frac{dP}{dt} = \frac{1}{4}(1,200 - P)$ with the initial population $P(0) = 200$.

STOP

END OF EXAM

ANSWER KEY

Section I, Part A

1. B	16. C
2. B	17. B
3. A	18. D
4. D	19. A
5. B	20. C
6. C	21. A
7. C	22. B
8. C	23. D
9. B	24. C
10. C	25. A
11. D	26. C
12. D	27. B
13. C	28. B
14. C	29. A
15. C	30. A

Section I, Part B

76. C
77. B
78. C
79. D
80. C
81. D
82. A
83. B
84. C
85. D
86. C
87. B
88. A
89. B
90. B

SCORING

Section II, Parts A & B

Use the scoring rubrics (found after the Section I answers and explanations) to self-score your free-response questions.

Section I, Part A Number Correct: _____

Section I, Part B Number Correct: _____

Section II, Part A Points Earned: _____

Section II, Part B Points Earned: _____

Enter your results to your Practice Exam 3 assignment to see your 1–5 score by logging in at kaptest.com.

Haven't registered your book yet? Go to kaptest.com/booksonline to begin.

BC PRACTICE EXAM 3 ANSWERS AND EXPLANATIONS

SECTION I, Part A

1. B

This is a geometric series with first term of 1.

$$S = \frac{a_0}{1-r} = \frac{1}{1-\frac{-3}{4}} = \frac{4}{7}$$

Therefore, **(B)** is the correct answer.

2. B

$$m = y - x + 2\big|_{(2,-3)}$$

1st iteration:
$$= -3 - 2 + 2 = -3$$
$$T : y = -3(x-2) - 3\big|_{x=1.8}$$
$$= -3(1.8-2) - 3 = -2.4$$

$$m = y - x + 2\big|_{(1.8,-2.4)}$$

2nd iteration:
$$= -2.4 - 1.8 + 2 = -2.2$$
$$T : y = -2.2(x-1.8) - 2.4\big|_{x=1.6}$$
$$= -2.2(1.6-1.8) - 2.4 = -1.96$$

Thus, **(B)** is the correct answer.

3. A

The rule for finding the limit of a composition is:

$$\lim_{x \to a} f(g(x)) = f\left(\lim_{x \to a} g(x)\right) = f(b)$$

Normally, you would start with the inner function, but in this question, you don't have any information about g. You do, however, know that $f(b) = 4$, for some number b. From the table, the value of b is -1 because $f(-1) = 4$. Now, from the rule, this tells you that $\lim_{x \to 1} g(x)$ must be -1, which is **(A)**.

4. D

Apply the chain rule when one function is composed with (inside of) another. To differentiate, take the derivative of the outer function $f(x)$ while leaving $g(x)$ alone inside $f(x)$. Then multiply by the derivative of $g(x)$. That is, $h'(x) = [f'(g(x))] \cdot g'(x)$. Evaluating at $x = 2$, you get:

$$h'(2) = \left[f'(g(2))\right] \cdot g'(2)$$
$$= f'(4) \cdot (6)$$
$$= 7 \cdot 6 = 42$$

That's **(D)**.

5. B

The Maclaurin series for

$$\ln(1 + x) = x - \frac{x^2}{2} + \frac{x^3}{3} + \dots + (-1)^n \frac{x^n}{n} + \dots.$$

Substitute x^2 for x. Hence, **(B)** is the correct answer.

6. C

$$x(t) = (2t + 1)^{-1} \qquad y(t) = 6t - t^2$$
$$x'(t) = -2(2t + 1)^{-2} \qquad y'(t) = 6 - 2t$$
$$x''(t) = 8(2t + 1)^{-3} \qquad y''(t) = -2$$
$$x''(2) = \frac{8}{125} \qquad y''(2) = -2$$
$$a(\vec{t}) = \left\langle \frac{8}{125}, -2 \right\rangle$$

Hence, **(C)** is the correct answer.

7. C

By partial fractions,

$$\frac{4}{x^2 - 1} = \frac{2}{x-1} - \frac{2}{x+1}$$

$$\int \frac{4}{x^2 - 1}\, dx = \int \frac{2}{x-1} - \frac{2}{x+1}\, dx$$
$$= 2 \ln|x - 1| - 2 \ln|x + 1|$$
$$= 2 \ln\left|\frac{x-1}{x+1}\right| + C$$

Therefore, **(C)** is the correct answer.

8. C

Because the x-values in the table are not equally spaced, the area of each trapezoid must be found individually.

$$\int_0^9 h(x)\, dx \approx \frac{1}{2}(1 + 4)(4) + \frac{1}{2}(4 + 1)(2) + \frac{1}{2}(1 + 5)(3)$$
$$= \frac{1}{2}(5)(4) + \frac{1}{2}(5)(2) + \frac{1}{2}(6)(3)$$
$$= (5)(2) + 5(1) + (3)(3) = 24$$

That's **(C)**.

9. B

The definition of the derivative of f at $x = a$ is given by:

$$f'(a) = \lim_{h \to 0} \frac{f(a + h) - f(a)}{h}$$

This means that $\lim\limits_{h \to 0} \dfrac{\sqrt[3]{1 + h} - \sqrt[3]{1}}{h}$ gives $f'(1)$ for $f(x) = \sqrt[3]{x}$.

If $f(x) = \sqrt[3]{x} = x^{\frac{1}{3}}$, then $f'(x) = \frac{1}{3}x^{-\frac{2}{3}}$ and $f'(1) = \frac{1}{3}(1)^{-\frac{2}{3}} = \frac{1}{3}$. That's **(B)**. Note that because the original limit is in the form $\frac{0}{0}$, you could also use L'Hôpital's rule.

10. C

Using the Ratio Test we find that the series converges as follows:

$$\lim_{n \to \infty} \left| \frac{\dfrac{(x + 4)^{2(n+1)}}{n+1}}{\dfrac{(x + 4)^{2n}}{n}} \right| = \lim_{n \to \infty} \left| (x + 4)^2 \right| < 1$$

$$\Rightarrow -1 < x + 4 < 1 \Rightarrow -5 < x < -3$$

The series does not converge at either endpoint because of the p-rule. Thus, **(C)** is the correct answer.

11. D

If $u = x^2$, then $du = 2x\,dx$. You already have the x and the dx, but not the 2, so multiply the integrand by 2, and to offset this, multiply the entire integral by $\frac{1}{2}$. This means either **(B)** or **(D)** is correct.

When looking for the equivalent definite integral, don't forget to convert the limits of integration: $u = x^2$, so $u = 0$ when $x = 0$ and $u = 4$ when $x = 2$. Be sure to replace the u-values where their corresponding x-values were ($x = 2$ was the upper limit so $u = 4$ is the new upper limit).

The integral in terms of x can now be written in terms of u as:

$$\int_0^2 xe^{x^2}\,dx = \frac{1}{2}\int_0^4 e^u\,du$$

That's **(D)**.

12. D

I does not converge by the p-series rule.

II converges because it is a geometric series with a ratio between -1 and 1.

III converges because the factorial in the denominator makes the ratio test equal zero for all values of x.

Hence, **(D)** is the correct answer.

13. C

Velocity is the derivative of position, so integrate the velocity function to find the position function:

$$x(t) = \int v(t)\,dt = \int 2\cos t\,dt = 2\sin t + C$$

The value of C can be found by using the given initial condition $x\left(\frac{\pi}{2}\right) = 5$. Substitute into the position equation to get:

$$x\left(\frac{\pi}{2}\right) = 2\sin\left(\frac{\pi}{2}\right) + C$$
$$5 = 2(1) + C$$
$$3 = C$$

Hence, the position function is $x(t) = 2\sin t + 3$, which is **(C)**.

14. C

You're going to need several values, so start by substituting $x = 2$ into the original equation to find y. The result is $2^3 + 2y = 8$ or $2y = 0$, so $y = 0$.

You will also need the value of $\frac{dy}{dx}$ at $x = 2$, so plug that in as well:

$$\frac{dy}{dx} = \frac{-3 \cdot 2^2 - 0}{2} = \frac{-12}{2} = -6$$

To find $\frac{d^2y}{dx^2}$, use the quotient rule on $\frac{dy}{dx} = \frac{-3x^2 - y}{x}$.

$$\frac{d^2y}{dx^2} = \frac{\left(-6x - \dfrac{dy}{dx}\right)(x) - \left(-3x^2 - y\right)(1)}{x^2}$$

Remember, the derivative of y is $\dfrac{dy}{dx}$. Once you have finished taking your derivative(s), you can then plug in any known values. It is much easier to do the algebra when you have numbers in the equation rather than having to deal with many variables.

Substitute the values of x, y, and $\dfrac{dy}{dx}$ to get:

$$\frac{d^2y}{dx^2} = \frac{2(-6 \cdot 2 - (-6)) - (1)(-3 \cdot 2^2 - 0)}{2^2}$$
$$= \frac{2(-12 + 6) - (-12)}{4}$$
$$= \frac{-12 + 12}{4} = 0$$

That's **(C)**.

15. C

Using integration by parts,

$$\int_1^6 xg'(x)\,dx =$$
$$xg(x) - \int_1^6 g'(x)\,dx$$
$$xg(x) - g(x)\Big|_1^6 = 6(-3) - 1(-1) - 10 = -29$$

Hence, **(C)** is the correct answer.

16. C

First, $f(x) > 0$ means that f is above the x-axis. All of the choices meet this condition (for $x > 0$ as stipulated in the question stem).

Second, $f'(x) > 0$ means that f is increasing. This eliminates (B).

Third, $f''(x) > 0$ means that f is concave up. This eliminates (A), which is linear so it is neither concave up nor down, and (D), which is concave down. Thus, the correct answer is **(C)**, which is above the x-axis, increasing, and concave up.

17. B

The reason that the integral symbol looks like an elongated S is because the definite integral is used to find the sum of an infinite number of really, really small terms. When you are using an integral to find the area under a curve, you are adding the areas of an infinite number of extremely thin rectangles: area = (length)(width) = (y)

(dx). The integral from $x = 0$ to $x = 3$ is positive because the graph is above the x-axis, making the y-values positive. The integral from $x = 3$ to $x = 5$ is negative because the graph is below the x-axis, making the y-values negative.

Thus, **(B)** is correct because:

$$\int_0^5 f(x)\,dx = \int_0^3 f(x)\,dx + \int_3^5 f(x)\,dx$$
$$= 9 - 2 = 7$$

18. D

Parallel lines have the same slope. To find the slope of the given line, rewrite the equation in slope-intercept form:

$$3y = x - 3$$
$$y = \frac{1}{3}x - 1$$

Thus, the slope of both lines is $m = \dfrac{1}{3}$.

To find the slope of f, use the quotient rule to differentiate $f(x) = \dfrac{x}{x + 3}$:

$$f'(x) = \frac{(1)(x + 3) - x(1)}{(x + 3)^2} = \frac{3}{(x + 3)^2}$$

This needs to equal the slope of the other line, so $\dfrac{3}{(x + 3)^2} = \dfrac{1}{3}$.

Now solve for x. Cross-multiply to get $9 = (x + 3)^2$. This becomes $x + 3 = \pm 3$. So $x = -3 \pm 3$, which means $x = -6$ or $x = 0$. That's **(D)**.

19. A

When you see the derivative of an integral and one of the limits of integration is a variable, use the second fundamental theorem of calculus. Don't forget to use the chain rule.

$$\frac{d}{dx} \int_1^{x^3} \ln(5 + t^2)\,dt$$
$$= \ln\left(5 + \left(x^3\right)^2\right) \cdot \left(3x^2\right) - \ln(5 + 1^2)(0)$$
$$= 3x^2 \ln(5 + x^6) - 0$$

That's a perfect match for **(A)**!

20. C

Horizontal asymptotes occur if the function approaches a particular y-value as $x \to \infty$ or $x \to -\infty$. When x approaches positive or negative infinity, the highest powers of x are so much larger than the other xs that you can disregard the smaller powers of x.

For statement I:

$$\lim_{x \to \infty} \frac{\sqrt{9x^2 + 1}}{x - 2} \to \frac{\sqrt{9x^2}}{x} = \frac{3|x|}{x} = \frac{3x}{x} = \frac{3}{1} = 3$$

So, $y = 3$ is a horizontal asymptote. Note, as $x \to -\infty$, $y \to -3$.

For statement II:

$$\lim_{x \to \infty} \frac{3x}{1 - x} \to \frac{3x}{-x} = \frac{3}{-1} = -3$$

and

$$\lim_{x \to -\infty} \frac{3x}{1 - x} \to \frac{3x}{-x} = \frac{-3}{1} = -3,$$

So, $y = -3$ is a horizontal asymptote, but $y = 3$ is not, so eliminate (B) and (D).

For statement III:

$$\lim_{x \to \infty} \frac{9x^2 - 2}{2 + 3x^2} \to \frac{9x^2}{3x^2} = \frac{9}{3} = 3$$

So, $y = 3$ is a horizontal asymptote. You would get the same asymptote if $x \to -\infty$.

Thus, I and III have $y = 3$ as horizontal asymptotes, making **(C)** the correct answer.

21. A

Be sure to memorize the derivatives of the six basic trig functions and their inverses before Test Day. Then write each function and its derivative on the test before you even start (this way you won't forget during the test). The derivative of $\tan u$ is $\sec^2 u$, so:

$$\int \sec^2 u \, du = \tan u + C$$

Here, if $u = 3x$, then $du = 3 \, dx$. You already have the dx, but not the 3, so multiply the integrand by 3, and to offset this, multiply the entire integral by $\frac{1}{3}$. This means **(A)** is correct.

22. B

$$x = 3\theta \cos \theta$$
$$\frac{dx}{d\theta} = 3\theta(-\sin \theta) + \cos \theta \cdot 3$$
$$y = 3\theta \sin \theta$$
$$\frac{dy}{d\theta} = 3\theta \cos \theta + \sin \theta \cdot 3$$

$$\frac{dy}{dx} = \frac{\dfrac{dy}{d\theta}}{\dfrac{dx}{d\theta}} = \frac{3\theta \cos \theta + \sin \theta \cdot 3}{3\theta(-\sin \theta) + \cos \theta \cdot 3}\bigg|_{\theta = \frac{\pi}{2}} = \frac{-1}{\dfrac{\pi}{2}} = \frac{-2}{\pi}$$

Hence, **(B)** is the correct answer.

23. D

$$f(x) = dt \int_1^{x^2} \sqrt{t^2 + t}$$
$$\Rightarrow f(x) = F(t) \big|_1^{x^2} = F(x^2) - F(1)$$
$$f'(x) = F'(x^2)(2x) - 0$$

$$f'(x) = \sqrt{(x^2)^2 + x^2}\,(2x)$$
$$f'(2) = \sqrt{16 + 4}\,(2 \cdot 2) = 4\sqrt{20}$$

Hence, **(D)** is the correct answer.

24. C

Given $y = (x^5 + \sin x)^4$, use the power rule and then the chain rule to find the derivative:

$$\frac{dy}{dx} = 4(x^5 + \sin x)^3 \cdot (5x^4 + \cos x)$$

That's **(C)**.

25. A

Each tangent line has the same slope, $\frac{dy}{dx}$, for any given value of y, so the slope only depends on y (to see this, notice that all the slopes that are horizontal to one another are the same). Thus, the slopes are not affected by the x-values, so the differential equation should only be in terms of the variable y. This eliminates (C) and (D).

Now, you might notice that the slope field has horizontal tangent lines at $y = -2$, so the slope, $\frac{dy}{dx}$, must equal 0 when $y = -2$. Unfortunately, that is true for both (A) and (B), so that doesn't help narrow down the choices.

To decide between (A) and (B), choose simple values (ones that are easy to find on the graph), like $x = 0$ and $y = 0$, to plug into the equation. These are nice values because they lie on the axes. Looking along the x-axis of the graph, the slope is negative when $y = 0$. So **(A)** must be correct.

26. C

Because f and g are inverses of each other, you know that:

$$g'(x) = \frac{1}{f'(g(x))} = \frac{1}{f'(y)}$$

The function $f(x) = x^3 - 2x - 4$ has derivative $f'(x) = 3x^2 - 2$.

Because $(2, 0)$ is a point on f, the corresponding point on g is $(0, 2)$. To find an inverse, remember to swap the x's with y's and vice versa, even in the ordered pairs.

Thus:

$$g'(0) = \frac{1}{f'(2)} = \frac{1}{3(2)^2 - 2} = \frac{1}{10}$$

That's **(C)**.

27. B

The question asks about the distance between two points, so use the distance formula, or the square of the formula to make things simpler:

$$D^2 = (x - 3)^2 + (y - 0)^2 = x^2 - 6x + 9 + y^2$$

Because $y = \sqrt{x}$, this becomes:

$$D^2 = x^2 - 6x + 9 + x = x^2 - 5x + 9$$

The distance is at a minimum at the same value of x where the square of the distance is at a minimum, so there is no need to square root the equation. Instead, call the equation something different (such as L), just to avoid confusion, take the derivative, and set it equal to 0:

$$L = x^2 - 5x + 9$$

$$\frac{dL}{dx} = 2x - 5$$

$$0 = 2x - 5 \rightarrow x = \frac{5}{2}$$

That's **(B)**.

28. B

Substituting 0 into the expression yields:

$$\frac{1 - \cos(0\pi)}{\sin(0)} = \frac{1 - 1}{0} = \frac{0}{0}$$

This means you can use L'Hôpital's rule, which applies to limits of the form $\frac{0}{0}$ or $\frac{\infty}{\infty}$. To use L'Hôpital's rule, take the derivative of the numerator and the derivative of the denominator, and re-evaluate the limit. Don't forget to use the chain rule when necessary:

$$\lim_{x \to 0} \frac{1 - \cos(\pi x)}{\sin(2x)} = \lim_{x \to 0} \frac{0 - (-\sin(\pi x) \cdot \pi)}{\cos(2x) \cdot 2}$$

$$= \frac{\sin(0) \cdot \pi}{\cos(0) \cdot 2}$$

$$= \frac{0 \cdot \pi}{1 \cdot 2} = \frac{0}{2} = 0$$

That's **(B)**.

29. A

The derivative of position is velocity and the derivative of velocity is acceleration. Remember to use the chain rule when taking the derivative of e^{2t}.

Because position is $x(t) = e^{2t} - e^t$, then velocity is $v(t) = 2e^{2t} - e^t$ and acceleration is $a(t) = 4e^{2t} - e^t$.

A particle is at rest when velocity $= 0$, so $v(t) = 2e^{2t} - e^t = 0$. Factor out the GCF to get $e^t(2e^t - 1) = 0$. The first factor will never equal 0. For the second factor, $2e^t = 1$. Solving for t gives $e^t = \frac{1}{2}$ or $t = \ln \frac{1}{2}$. Be careful—the answer isn't (C)! Make sure you answer the question asked.

The acceleration at this time is:

$$a\left(\ln \frac{1}{2}\right) = 4e^{2\ln \frac{1}{2}} - e^{\ln \frac{1}{2}} = 4e^{\ln\left(\frac{1}{2}\right)^2} - e^{\ln \frac{1}{2}}$$

$$= 4\left(\frac{1}{4}\right) - \frac{1}{2} = 1 - \frac{1}{2} = \frac{1}{2}$$

Thus, **(A)** is correct.

30. A

Whenever you see perfect squares in the denominator of an integral, think "inverse trig function." Here, use the fact that $\frac{d}{du}\tan^{-1}u = \frac{1}{1+u^2}$. Let $u = 4x$, which means $du = 4\,dx$. You already have $2\,dx$, but not $4\,dx$, so multiply the integrand by 2 and, to offset that, multiply the entire integral by $\frac{1}{2}$. Now you should be ready to integrate:

$$\int \frac{2\,dx}{1+16x^2} = \frac{1}{2}\int \frac{2\cdot 2\,dx}{1+(4x)^2} = \frac{1}{2}\int \frac{du}{1+u^2}$$

$$= \frac{1}{2}\tan^{-1}(u) + C$$

$$= \frac{1}{2}\tan^{-1}(4x) + C$$

That's **(A)**.

SECTION I, Part B

76. C

The equation for finding a region bounded by a polar curve is

$$\frac{1}{2}\int_a^b r^2\,d\theta$$

$$A = \frac{1}{2}\int_0^\pi (3\theta + \sin\,\theta)^2\,d\theta = 56.7204$$

Hence, **(C)** is correct.

77. B

Graph the equation in your calculator to find that, between 1 and e, the function f is above the x-axis, so the area is given by $\int_1^e \frac{5\,\ln x}{x}\,dx$. Save yourself some time by using the Integrate function on your calculator to arrive at an area of 2.5, which is **(B)**.

To integrate by hand, bring the 5 in front of the integral, use the substitution $u = \ln x$ and $du = \frac{1}{x}dx$, and adjust the limits of integration, which yields:

$$A = 5\int_0^1 u\,du = 5\left[\frac{u^2}{2}\right]_0^1 = 5\left(\frac{1}{2} - 0\right) = \frac{5}{2}$$

78. C

Compare each statement to what you're given in the question. Start with (A): The function f does not have to have an absolute extremum. That is only guaranteed if the function is continuous on a *closed* interval, so eliminate (A).

(B): The graph of f could be increasing for all x in $-2 < x < 4$, but it does not have to be. There could be other extrema and thus other turning points between $x = -2$ and $x = 4$. Eliminate (B).

(C): The statement $f(c) = 0$ means there is a y-value equal to 0 for some x-value equal to c. Because $y = 0$ is between $y = -3$ and $y = 5$, this statement must be true by the intermediate value theorem, so **(C)** is correct.

There is no need to check (D), but in case you're not sure about (C), you can eliminate (D) because the function f could be strictly increasing for x between -2 and 4, in which case $f'(x)$ is always positive (not 0). Note that the only time $f'(c)$ is guaranteed to be 0 is (by Rolle's theorem) when f is continuous and differentiable and the y-values are the same at the endpoints.

79. D

This differential equation is of the form $\frac{dy}{dt} = ky$, where the rate of change is proportional to the amount present. This means that the population's growth can be modeled by a typical exponential equation of the form $P = P_0 e^{kt}$. Using $P_0 = 2$ (the initial population at time $= 0$) and $k = 0.713$, this model becomes $P = 2e^{0.713t}$. At $t = 5$, the population is $P = 2e^{0.713(5)} \approx 70.679$, which is **(D)**. If you didn't happen to recall the exponential model, you could also separate the variables and integrate, find the constant, then evaluate the resulting equation for P at $t = 5$.

80. C

Three uniform subintervals on the closed interval $[0, 18]$ make $\Delta x = \frac{18 - 0}{3} = 6$.

This makes the subintervals $[0, 6]$, $[6, 12]$, and $[12, 18]$ with midpoints at $x = 3$, $x = 9$, and $x = 15$. Using the functional values in the table that correspond to these midpoints gives:

$$\int_0^{18} g(x)\,dx \approx g(3)\Delta x + g(9)\Delta x + g(15)\Delta x$$

$$= \Delta x(-2 + 4 + 5) = 6(7) = 42$$

That's **(C)**.

81. D

A logistic growth curve follows the differential equation $\frac{dP}{dt} = k(P)(M - P)$, where M is the maximum carrying capacity of the system. Only **(D)** meets this condition and the initial condition that the population is increasing at the rate of 600 people per hour.

82. A

$$\frac{f'''(1)}{3!} = -6$$
$$\Rightarrow f'''(1) = -6 \cdot 3! = -36$$

Hence, **(A)** is correct.

83. B

The extrema of f occur at critical points. The critical points of f are the zeros of $f'(x)$. Use your calculator to find that the graph of $f'(x) = \frac{-x^3 + 4x - 2}{\sqrt{x^2 + 1}}$ has zeros at $x = -2.214$, $x = 0.539$, and $x = 1.675$. Eliminate statement I because $x = -0.768$ is not a critical point of f.

Statement II is true. The function f has a relative minimum at $x = 0.539$ because $f'(x)$ goes from negative (decreasing) to positive (increasing). To confirm this, check your calculator again—the graph of $f'(x)$ goes from below the x-axis to above the x-axis.

Statement III is false. Rather than a minimum, f has a relative maximum at $x = 1.675$ because $f'(x)$ goes from positive to negative. (The graph of $f'(x)$ goes from above the x-axis to below the x-axis).

Thus, the only statement that is true is statement II, making **(B)** the correct choice.

84. C

Start by graphing the functions in your calculator. Once you get a general idea of where the region R is located, restrict your viewing window to more reasonable values. Then, sketch in a sample slice—don't forget that you're revolving about the y-axis. Your graph should look something like the one here:

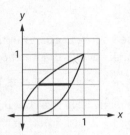

The graphs intersect at the points $(0, 0)$ and $(1, 1)$, so the rectangles (the slices) will range from $[0, 1]$ along the y-axis. This type of rectangle will generate washers with a thickness (or width) of dy. To use the washer method, the outer radius R and the inner radius r are both calculated as $x_{big} - x_{little}$, so you need to solve for x in each of the given equations.

From $y = x^3$, you get $x = \sqrt[3]{y}$. From $y = \sqrt{x}$, you get $x = y^2$. Thus:

$$R = x_{big} - x_{little} = \sqrt[3]{y} - 0 = \sqrt[3]{y}$$
and
$$r = x_{big} - x_{little} = y^2 - 0 = y^2$$

Using the washer method, the volume is:

$$V = \pi \int_0^1 (R^2 - r^2)\, dy$$
$$= \pi \int_0^1 \left(\left(\sqrt[3]{y}\right)^2 - \left(y^2\right)^2 \right) dy$$
$$= 0.400\pi \approx 1.257$$

Thus, **(C)** is correct.

85. D

This is a related rates question. It is given that the volume increases at a rate of 100 cm³/s. The volume's rate can be described by $\frac{dV}{dt}$, so $\frac{dV}{dt} = 100 \, \text{cm}^3/\text{s}$. You are asked to find the rate (how fast) that the radius changes, $\frac{dr}{dt}$. You are also given an equation that relates V and r, which is $V = \frac{4}{3}\pi r^3$. Take the derivative of this equation to relate the rate of the change in the volume to the rate of the change in the radius: $\frac{dV}{dt} = 4\pi r^2 \frac{dr}{dt}$.

Because the diameter you're interested in is 50 cm, the radius is half that, or $r = 25$ cm. Plug the known values into the derivative equation to find that $100 = 4\pi(25)^2 \frac{dr}{dt}$.

Now solve for $\frac{dr}{dt}$ to arrive at the correct answer, which is (D).

$$\frac{dr}{dt} = \frac{100}{4\pi(25)^2} = \frac{\cancel{100}}{\cancel{4}\pi(25)\cancel{(25)}} = \frac{1}{25\pi}\,\text{cm/s}$$

86. C

Use the Calculate Zero function on your calculator to find the first positive x-intercept, which is $x = 0.68275$. To find the slope at this value of x, find the value of the derivative at $x = 0.68275$, using your calculator again. On the $\boxed{Y=}$ screen enter the function and adjust the viewing window as needed. Press $\boxed{\text{CALC}}$ ($\boxed{\text{2nd}}$ $\boxed{\text{TRACE}}$) 6 (for the $\frac{dy}{dx}$ option). This gives y' at $x = 0.68275$ is 1.705, which is (C).

87. B

On the closed interval $[0, 1]$, $e^{\tan x}$ is an increasing function. It will reach its maximum value on the interval $[0, 0.5]$ at $x = 0.5$.

The maximum error

$$E < \left| \frac{f^{(4)}(0.5)}{4!}(0.5 - 0)^4 \right| = 0.004497$$

Thus, (B) is correct.

88. A

You're looking for where g is decreasing, so you want $g'(x) < 0$. Using the second fundamental theorem of calculus on $g(x) = \int_0^x f(t)\,dt$ tells you that $g'(x) = f(x)$. The given graph is the graph of $f(x)$, so it is also the graph of $g'(x)$. The graph is below the x-axis, and thus negative, on the interval $1 < x < 4$, so g is decreasing on $1 \leq x \leq 4$ only, making (A) correct.

89. B

The amount of water that flows into the lake is calculated as:

$$\int_0^{24} R(t)\,dt = \int_0^{24}\left(\sin\frac{\pi t}{12} + 40t\right)dt = 11{,}520 \text{ gallons}$$

Use your calculator to evaluate the integral: Enter the equation for R into $\boxed{Y=}$ and adjust the viewing window as needed. Press $\boxed{\text{CALC}}$ ($\boxed{\text{2nd}}$ $\boxed{\text{TRACE}}$) 7 (for the $\int f(x)\,dx$ option). You'll also need to enter the limits of integration (here, the lower limit is 0 and the upper limit is 24). The result is 11,520, but be careful—don't pick (A), because you're not done yet. The amount of water pumped out of the lake over the 24-hour period is $(500)(24) = 12{,}000$ gallons. Thus, the amount of the water in the lake after 24 hours is computed as $30{,}000 + 11{,}520 - 12{,}000 = 29{,}520$ gallons, which is (B).

90. B

Coefficients are found as follows.

n	$f^{(n)}(0)$	$\dfrac{f^{(n)}(0)}{n!}$
0	2	2
1	$-\dfrac{1}{2}$	$-\dfrac{1}{2}$
2	$\dfrac{4}{27}$	$\dfrac{2}{27}$
3	$-\dfrac{5}{108}$	$-\dfrac{5}{648}$

Therefore, (B) is the correct answer.

SECTION II, Part A

1. Special Note on Decimals: Answers using decimal approximations from the calculator should always be given to at least three places after the decimal, rounded or truncated. However, rounding or truncating should only take place on final answers. Because this value is an intermediate answer that will be used in subsequent computations, store the point of intersection of the two functions in the memory of the calculator in its exact form as $a = 2.8354\ldots$ to avoid approximation errors. At the very least, carry an extra digit or two beyond the necessary three places after the decimal to ensure accuracy in the final answer.

(a)

Solutions	Scoring Rubric
The points of intersection of $f(x)$ and $g(x)$ are $(0, 0)$ and $(2.8354, 6.137)$. Let $a = 2.8354$. The upper boundary of R is $f(x)$ and the lower boundary is $g(x)$, so the typical rectangle has a height given as $h = f(x) - g(x)$. Thus, the area of the region is $$A = \int_0^a f(x) - g(x)\, dx$$ $$= \int_0^a (5x - x^2) - (2^x - 1)\, dx$$ $$= 6.481$$	1: limits 0, 2.8354 1: correctly set up integral 1: $A = 6.481$

(b)

Solutions	Scoring Rubric
To find the area of the half of the region before the vertical line $x = k$, change the upper limit of integration on the integral to k and set it equal to half of the area of region R, $\dfrac{6.481}{2} = 3.240$ or 3.241. This gives the integral: $$A_{half} = \int_0^k f(x) - g(x)\, dx$$ $$\frac{6.481}{2} = \int_0^k (5x - x^2) - (2^x - 1)\, dx$$	1: half $A = \dfrac{6.481}{2}$ or 3.240 or 3.241 1: correctly set up integral with limits 0, k

(c)

Solutions	Scoring Rubric
$R = y_{big} - y_{little} = 5x - x^2 - 0 = 5x - x^2$ and $r = y_{big} - y_{little} = 2^x - 1 - 0 = 2^x - 1$. Using the washer method, the volume is: $$V = \pi \int_0^a R^2 - r^2\, dx$$ $$= \pi \int_0^a (f(x))^2 - (g(x))^2\, dx$$ $$= \pi \int_0^a (5x - x^2)^2 - (2^x - 1)^2\, dx$$ $$= 43.875\pi \text{ or } 43.876\pi$$ $$= (137.840 \text{ or } 137.841)$$	1: correctly set up integral for washer method 1: 43.875π or 43.876π or 137.840 or 137.841

(d)

Solutions	Scoring Rubric
To find the volume of a solid with known cross sections, use $V = \int_0^a A(x)\, dx$. For cross sections that are squares, the area is $A(x) = s^2$. The side of the square is embedded into the region R, so the length of the side is $s = y_{big} - y_{little} = 5x - x^2 - (2^x - 1)$. Thus, $A(x) = (5x - x^2 - (2^x - 1))^2 = (5x - x^2 - 2^x + 1)^2$. The expression that will find the volume of the solid is $$V = \int_0^a A(x)\, dx = \int_0^a (5x - x^2 - 2^x + 1)^2\, dx$$	1: $V = \int_0^a A(x)\, dx$ 1: $A(x) = (5x - x^2 - 2^x + 1)^2$

2. (a)

Solutions	Scoring Rubric
Evaluating the given $v(t)$ function at $t = 4$, $v(4) = 4 \sin(4) = -3.027$. Finding $a(t)$ by differentiating $v(t)$ would require using the product rule. However, we only need to find the value of the derivative at a specific point. This can be accomplished using the numeric derivative feature of a graphing calculator. Using the calculator, $a(4) = v'(4) = -3.371$.	1: $v(4) = -3.027$ 1: $a(4) = -3.371$

(b)

Solutions	Scoring Rubric
As a general rule, when v and a have the same sign, speed is increasing. When v and a have different signs, speed is decreasing. Because $v(4) < 0$ and $a(4) < 0$, the speed is increasing at time $t = 4$.	1: increasing 1: valid reasoning

(c)

Solutions	Scoring Rubric
The total distance traveled on the closed interval [a, b] is found by using $\int_a^b |v(t)|\, dt$. In this case, the total distance traveled is $$\int_0^5 |t \sin t|\, dt = 8.660.$$	1: 8.660

(d)

Solutions	Scoring Rubric
By definition, the average value of f on [a, b] is given by $\dfrac{1}{b-a}\int_a^b f(x)\, dx$. In this case, $$\frac{1}{5-0}\int_0^5 t \sin t\, dt = \frac{1}{5}(-2.377) = -0.475.$$	1: correctly set up integral for average 1: −0.475

(e)

Solutions	Scoring Rubric
The integral $\int_0^5 v(t)\, dt$ finds the displacement of the particle for the time interval [0, 5]. Adding this to the initial position should find the position at time $t = 5$. Thus, $$y(5) = y(0) + \int_0^5 v(t)\, dt = 2 + (-2.377) = -0.377.$$ Another way to calculate the position of the particle at time $t = 5$ is to take the average velocity calculated in step (d), multiply by 5(time), and add the initial position. Make sure to use the exact average velocity and not the rounded number. $$-0.4754 \cdot 5 + y(0) = -2.377 + 2 = -0.377$$	1: −0.377 1: appropriate work shown

SECTION II, Part B

3. (a)

Solutions	Scoring Rubric
The physical situation can be represented as a right triangle, as shown below. It is given that $\frac{dx}{dt} = 3$ when $x = 5$. We want to find $\frac{dy}{dt}$. A basic equation to relate the relevant variables is the Pythagorean theorem, which gives $x^2 + y^2 = 13^2$. Note that when $x = 5$, the value of y is $y = 12$, using the Pythagorean relationship. To get a $\frac{dy}{dt}$ to result, you must differentiate the Pythagorean equation with respect to t. This gives $$2x\frac{dx}{dt} + 2y\frac{dy}{dt} = 0.$$ Substituting the given values, you get $$2(5)(3) + 2(12)\frac{dy}{dt} = 0.$$ Solving for $\frac{dy}{dt}$, you find that $12\frac{dy}{dt} = -15$ or $\frac{dy}{dt} = -\frac{15}{12} = -\frac{5}{4}$ feet/second.	1: differentiated Pythagorean theorem with respect to t 1: $\frac{-5}{4}$ 1: feet/second

(b)

Solutions	Scoring Rubric
In order to find $\frac{d\theta}{dt}$, you need an equation that involves θ. In the given right triangle, choose cosine because this will involve x and the constant hypotenuse. Differentiating the equation $$\cos \theta = \frac{x}{13} \text{ gives } -\sin \theta \cdot \frac{d\theta}{dt} = \frac{1}{13} \cdot \frac{dx}{dt}.$$ From the original triangle scenario, you know that $\sin \theta = \frac{12}{13}$ at the instant when $x = 5$. Now $-\sin \theta \cdot \frac{d\theta}{dt} = \frac{1}{13} \cdot \frac{dx}{dt}$ becomes $-\frac{12}{13} \cdot \frac{d\theta}{dt} = \frac{3}{13}$. This gives $$\frac{d\theta}{dt} = \frac{\frac{3}{13}}{-\frac{12}{13}} = -\frac{1}{4}$$ radians per second.	1: differentiated for $\cos \theta$ with respect to t 1: $\frac{-1}{4}$ 1: radians/second

(c)

Solutions	Scoring Rubric
The area of the right triangle is $A = \frac{1}{2}bh = \frac{1}{2}xy$. To find $\frac{dA}{dt}$, differentiate the area equation with respect to t. Using the product rule, you get $$\frac{dA}{dt} = \frac{1}{2}\left(x\frac{dy}{dt} + y\frac{dx}{dt}\right).$$ Substituting the given values, $$\frac{dA}{dt} = \frac{1}{2}\left(5 \cdot -\frac{5}{4} + 12 \cdot 3\right)$$ square feet per second. On the AP exam, there is no requirement to simplify an answer, so the answer can be left in this form. In any case, you do need to make sure that the answer does include the units.	1: differentiated area of the right triangle with respect to t 1: $$\frac{dA}{dt} = \frac{1}{2}\left(5 \cdot -\frac{5}{4} + 12 \cdot 3\right)$$ 1: square feet/second

4. (a)

Solutions	Scoring Rubric
$$4x - 2x \cdot f(x) = f'(x)$$ $$\Rightarrow \int_0^\infty (4x - 2x \cdot f(x))\, dx = \int_0^\infty f'(x)\, dx$$ $$= \lim_{b \to \infty} \int_0^b f'(x)\, dx = \lim_{b \to \infty} f(x)\Big|_0^b$$ $$= \lim_{b \to \infty} f(b) - f(0) = 4 - 10 = -6$$	1: $\int_0^\infty f'(x)\, dx$ 1: $\lim_{b \to \infty} f(b) - f(0)$ 1: -6

(b)

Orig x	Orig y	dx	$dy = (4x - 2xf(x))dx$	New x	New y	Scoring Rubric
0	10	0.5	$(4(0) - 2(0)(1))(0.5) = 0$	0.5	10	1: $f(0) = 0$ 1: $f(.5) = -4$ 1: $f(1) = 6$
0.5	10	0.5	$(4(0.5) - 2(0.5)(10))(0.5) = -4$	1.0	6	

(c)

Solutions	Scoring Rubric
$\dfrac{dy}{dx} = 4x - 2xy$ $\dfrac{dy}{dx} = 2x(2 - y)$ $\displaystyle\int \dfrac{dy}{2-y} = \int 2x\,dx$ $-\ln\lvert 2 - y\rvert = x^2 + C$ $-\ln\lvert 2 - 10\rvert = 0^2 + C \Rightarrow C = -\ln(8)$ $\ln\lvert 2 - y\rvert = \ln(8) - x^2$ $y = 2 - 8e^{-x^2}$	1: $-\ln\lvert 2 - y\rvert = x^2 + C$ 1: $C = -\ln(8)$ 1: $y = 2 - 8e^{-x^2}$

5. (a)

Solutions	Scoring Rubric
$T_3(x) = f(2) + \dfrac{f'(2)}{1!}(x-2)^1 + \dfrac{f''(2)}{2!}(x-2)^2 + \dfrac{f'''(2)}{3!}(x-2)^3$ $\qquad = 4 - 0(x-2) - \dfrac{3}{2}(x-2)^2 + \dfrac{12}{6}(x-2)^3$ $T_3\!\left(\dfrac{3}{2}\right) = 4 - \dfrac{3}{2}\!\left(\dfrac{3}{2}-2\right)^2 + \dfrac{12}{6}\!\left(\dfrac{3}{2}-2\right)^3 = \dfrac{27}{8}$	1: $4 - 0(x-2) - \dfrac{3}{2}(x-2)^2 + \dfrac{12}{6}(x-3)^3$ 1: $\dfrac{31}{8}$

(b)

Solutions	Scoring Rubric
$f'(x) = 0 + 3(x-2) + 6(x-2)^2 + \ldots$ $\qquad\qquad + \dfrac{f^{(n)}(2)}{n!}(x-2)^n + \ldots$ $f'(2) = 0$ $f''(2) = -3$ Because $f'(2) = 0$ and $f''(2) < 0$, there is a relative maximum at $x = 2$.	1: $f'(x) = 0 + 3(x-2) + 6(x-2)^2 + \ldots$ $\qquad\qquad + \dfrac{f^{(n)}(2)}{n!}(x-2)^n + \ldots$ 1: $f'(2) = 0$ $f''(2) = -3$ 1: explain why this means there is a relative max at $x = 2$

(c)

Solutions	Scoring Rubric
$\text{Error} < \left\| \dfrac{f^4_{\max[1,2]}(z)}{4!}(x-2)^4 \right\|$	1: $\text{Error} < \left\| \dfrac{3}{4!}\left(\dfrac{3}{2}-2\right)^4 \right\| \Rightarrow \text{Error} < \dfrac{1}{128}$
$\text{Error} < \left\| \dfrac{3}{4!}\left(\dfrac{3}{2}-2\right)^4 \right\| \Rightarrow \text{Error} < \dfrac{1}{128}$	1:
which is less than 0.01.	$\dfrac{1}{128} < .01$

(d)

Solutions	Scoring Rubric
$P_3(x) = \displaystyle\int_2^x \left(4 - \dfrac{3}{2}(t-2)^2\right)dt$	1: $P_3(x) = \displaystyle\int_2^x \left(4 - \dfrac{3}{2}(t-2)^2\right)dt$
$\Rightarrow 4t - \dfrac{1}{2}(t-2)^3 \Big\|_2^x = 4x - \dfrac{1}{2}(x-2)^3 - 8$	1:
$\qquad\qquad = 4(x-2) - \dfrac{1}{2}(x-2)^3$	$4(x-2) - \dfrac{1}{2}(x-2)^3$

6. The first step in solving a differential equation must be to separate the variables. If this is not done or at least attempted, you typically will not be eligible for any partial credit on this entire part of the problem.

(a)

Solutions	Scoring Rubric
The point of tangency is $(0, 200)$. At the point $(0, 200)$, the slope is $\dfrac{dP}{dt} = \dfrac{1}{4}(1,200 - 200) = \dfrac{1,000}{4} = 250.$ Using point-slope form, the equation of the tangent line is $y - 200 = 250(t - 0)$ or $y = 250t + 200$. When $t = 2$, the tangent line approximation gives $y = 250(2) + 200 = 700$.	1: $y - 200 = 250(t - 0)$ or $y = 250t + 200$ 1: when $t = 2$, $y = 700$

(b)

Solutions	Scoring Rubric
Starting with $$\frac{dP}{dt} = \frac{1}{4}(1,200 - P),$$ you get $$\frac{d^2P}{dt^2} = 0 - \frac{1}{4}\frac{dP}{dt} = -\frac{1}{4}\frac{dP}{dt}$$ $$= -\frac{1}{4}\left(\frac{1}{4}(1,200 - P)\right)$$ $$= -\frac{1}{16}(1,200 - P)$$	1: showed appropriate work

(c)

Solutions	Scoring Rubric
At $(0, 200)$, the value of $\frac{d^2P}{dt^2}$ is $$\frac{d^2P}{dt^2} = -\frac{1}{16}(1,200 - 200) < 0$$ which indicates that the graph of P is concave down and so the tangent line to the curve at $P = 200$ lies above the curve. Thus, the value of the approximation overestimates the actual value of P.	1: overestimates 1: the value of $\frac{d^2P}{dt^2} < 0$ 1: graph is concave down

(d)

Solutions	Scoring Rubric
For the differential equation $$\frac{dP}{dt} = \frac{1}{4}(1,200 - P),$$ separating variables gives $$\frac{1}{1,200 - P}dP = \frac{1}{4}dt.$$ Integrating both sides is the next step. The anti-derivative of the left side can be determined by inspection or can be handled more thoroughly with a u-substitution where $u = 1200 - P$. $$-\ln|1,200 - P| = \frac{1}{4}t + C$$ Next you isolate P: $$\ln|1,200 - P| = -\frac{1}{4}t - C$$ $$|1,200 - P| = e^{-\frac{1}{4}t - C}$$ $$|1,200 - P| = e^{-\frac{1}{4}t}e^{-C}$$ $$|1,200 - P| = Ce^{-\frac{1}{4}t}$$ $$1,200 - P = Ce^{-\frac{1}{4}t}$$ The absolute values can be dropped, because C can be positive or negative as necessary to handle the issue of the sign. $$-P = Ce^{-\frac{1}{4}t} - 1,200$$ $$P = -Ce^{-\frac{1}{4}t} + 1,200$$ To find the value of the constant of integration, substitute the initial condition $(0, 200)$ $$200 = -Ce^{-\frac{1}{4}(0)} + 1,200$$ $$C = 1,000$$ This gives the final solution as $$P = -1,000e^{-\frac{1}{4}t} + 1,200.$$	1: correctly separated variables and integrated 1: $C = 1,000$ 1: $$P = -1,000e^{-\frac{1}{4}t} + 1,200$$

BC Practice Exam 4

Section I, Part A

1. (A) (B) (C) (D)
2. (A) (B) (C) (D)
3. (A) (B) (C) (D)
4. (A) (B) (C) (D)
5. (A) (B) (C) (D)
6. (A) (B) (C) (D)
7. (A) (B) (C) (D)
8. (A) (B) (C) (D)
9. (A) (B) (C) (D)
10. (A) (B) (C) (D)
11. (A) (B) (C) (D)
12. (A) (B) (C) (D)
13. (A) (B) (C) (D)
14. (A) (B) (C) (D)
15. (A) (B) (C) (D)
16. (A) (B) (C) (D)
17. (A) (B) (C) (D)
18. (A) (B) (C) (D)
19. (A) (B) (C) (D)
20. (A) (B) (C) (D)
21. (A) (B) (C) (D)
22. (A) (B) (C) (D)
23. (A) (B) (C) (D)
24. (A) (B) (C) (D)
25. (A) (B) (C) (D)
26. (A) (B) (C) (D)
27. (A) (B) (C) (D)
28. (A) (B) (C) (D)
29. (A) (B) (C) (D)
30. (A) (B) (C) (D)

Section I, Part B

76. (A) (B) (C) (D)
77. (A) (B) (C) (D)
78. (A) (B) (C) (D)
79. (A) (B) (C) (D)
80. (A) (B) (C) (D)
81. (A) (B) (C) (D)
82. (A) (B) (C) (D)
83. (A) (B) (C) (D)
84. (A) (B) (C) (D)
85. (A) (B) (C) (D)
86. (A) (B) (C) (D)
87. (A) (B) (C) (D)
88. (A) (B) (C) (D)
89. (A) (B) (C) (D)
90. (A) (B) (C) (D)

A A A A A A A A A A A A A A A A A A A A A A A A A A A A A A A A A A

CALCULUS BC
SECTION I, Part A
Time—60 minutes
Number of questions—30

A CALCULATOR MAY NOT BE USED ON THIS PART OF THE EXAM.

Directions: Use the space provided to solve each of the following problems. After examining the given choices, decide which of the choices is the best and fill in the corresponding circle on the answer sheet. Credit will only be given for answers appropriately marked on the answer sheet, and not for anything written in the exam book. Do not spend too much time on any one problem.

In this exam:

(1) The domain of a function f should be assumed to be the set of all real numbers x for which $f(x)$ is a real number, unless the question indicates otherwise.

(2) The inverse function notation f^{-1} or the prefix "arc" (e.g., $\sin^{-1} x = \arcsin x$) may be used to indicate the inverse of a trigonometric function.

GO ON TO THE NEXT PAGE.

A A A A A A A A A A A A A A A A A A A A A A A A A A A A A A A A A

1. The number of bacteria in a petri dish increases at a rate of $R(t)$ bacteria per hour. If there are 1,000 bacteria at time $t = 5$, which of the following expressions gives the total number of bacteria in the petri dish at $t = 7$?

(A) $R(7)$

(B) $1,000 + R(7)$

(C) $1,000 + \int_{5}^{7} R(t)\, dt$

(D) $\int_{5}^{7} R(t)\, dt$

2. Which of the following sequences does not converge?

(A) $a_n = \dfrac{100^{100}}{n^2}$

(B) $a_n = \dfrac{4n(n+3)}{(n-5)(n+1)}$

(C) $a_n = \dfrac{2 \sin n}{3}$

(D) $a_n = \dfrac{(2n-3)(n+1)}{n(2n-1)(n-5)}$

GO ON TO THE NEXT PAGE.

AAAAAAAAAAAAAAAAAAAAAAAAAAAAAAAA

3. The line $y = \frac{3}{2}x + 5$ is tangent to $f(x)$ at $x = 3$. What is the equation of the normal line to $f(x)$ at $x = 2$?

(A) $y = \frac{3}{2}x + \frac{5}{2}$

(B) $y = -\frac{3}{2}x + \frac{5}{2}$

(C) $y = -\frac{3}{2}x + \frac{23}{2}$

(D) $y = \frac{3}{2}x + \frac{23}{2}$

---

4. If $f(x) = \dfrac{2x^4 - 3x^2 + 4}{-7x^4 + 2x^3 - x}$, what is $\lim_{x \to \infty} f(x)$?

(A) $-\infty$ (B) -1 (C) $-\frac{2}{7}$ (D) 0

---

5. If $\lim_{x \to 5} \sqrt{f(x)} = 2$, then $\lim_{x \to 5} (f(x))^2 =$

(A) 2 (B) 4 (C) 8 (D) 16

GO ON TO THE NEXT PAGE.

A A A A A A A A A A A A A A A A A A A A A A A A A A A A A A A A

6. If $x(v) = 3\sin(v)$ and $y(v) = 3\cos(2v)$, find $\dfrac{dy}{dx}$.

    (A) $-4\sin(v)$ (B) $4\sin(v)$ (C) $-2\tan(2v)$ (D) $-4\csc(v)$

7. The graph of $y = x^4 - 3x^3 - 60x^2 + 24x - 36$ is concave up for

    (A) $\left(-\dfrac{5}{2}, 4\right)$

    (B) $\left(-4, \dfrac{5}{2}\right)$

    (C) $\left(-\infty, -\dfrac{5}{2}\right)$ or $(4, \infty)$

    (D) $(-\infty, -4)$ or $\left(\dfrac{5}{2}, \infty\right)$

8. $\displaystyle\int_{-\infty}^{1} 2e^{3x}\, dx =$

    (A) $\dfrac{2}{3}$

    (B) $e - 1$

    (C) $3e$

    (D) $\dfrac{2}{3}e^3$

GO ON TO THE NEXT PAGE.

9. What is the area of the region enclosed by $f(x) = -x^2 + 16$ and the x-axis?

(A) $-\dfrac{256}{3}$ (B) 32   (C) $\dfrac{128}{3}$ (D) $\dfrac{256}{3}$

10. Functions f and g are continuous functions such that $\displaystyle\int_{12}^{-8} g(x)\, dx = -7$, $\displaystyle\int_{-8}^{12} f(x)\, dx = 7$, and $\displaystyle\int_{-8}^{13} f(x)\, dx = 8$. Find $\displaystyle\int_{12}^{13} 3f(x)\, dx + \int_{12}^{12} 4g(x)\, dx$.

(A) -3

(B) 3

(C) 48

(D) 49

11. Find $\displaystyle\lim_{x \to 0} \frac{3\sin(2x)}{2x + 5\sin x}$.

(A) 0

(B) $\dfrac{3}{7}$

(C) $\dfrac{3}{5}$

(D) $\dfrac{6}{7}$

GO ON TO THE NEXT PAGE.

A A A A A A A A A A A A A A A A A A A A A A A A A A A A A A A A A

12. If the geometric series $\displaystyle\sum_{k=0}^{\infty} 7(-0.4)^k$ converges, what is its sum?

(A) $\dfrac{7}{4}$

(B) $\dfrac{14}{5}$

(C) 5

(D) The series does not converge.

13. If the substitution $u = \dfrac{x}{3}$ is made, the integral $\displaystyle\int_{3}^{6} \dfrac{2 - \left(\dfrac{x}{3}\right)^2}{x}\, dx =$

(A) $-\dfrac{3}{2} + 2\ln 2$ (B) $-\dfrac{1}{3}$ (C) $-\dfrac{5}{2} + 2\ln 2$ (D) $-\dfrac{29}{2} + \ln 12$

14. Find $\displaystyle\int_{2}^{6} \dfrac{dx}{6 - 4x}$.

(A) $-\dfrac{80}{81}$

(B) $-\dfrac{\ln 36}{4}$

(C) $-\dfrac{\ln 9}{4}$

(D) $\ln 9$

GO ON TO THE NEXT PAGE.

AAAAAAAAAAAAAAAAAAAAAAAAAAAAAAAAAA

15. If $y = 7^{3x^2+5}$ then $\dfrac{dy}{dx} =$

(A) $\ln(6x)(7^{3x^2+5})$

(B) $(6x)(7^{3x^2+5})$

(C) $(42\ln 7)(x^{3x^2+6})$

(D) $(6x\ln 7)(7^{3x^2+5})$

---

16. $\displaystyle\lim_{x\to\frac{\pi}{6}}\dfrac{\sin^2 x + \cos^2 x}{\tan(2x)} =$

(A) $\dfrac{1}{\sqrt{3}}$ 　　(B) $\sqrt{3}$ 　　(C) $\dfrac{\sqrt{(3)}}{3}$ 　　(D) $\dfrac{3+\sqrt{3}}{2}$

---

17. What is the general solution of the differential equation $\dfrac{dy}{dx} = \dfrac{y\sin x}{1-2y^2}$?

(A) $y - y^2 = -\cos x + C$

(B) $y - y^2 = \cos x + C$

(C) $\ln|y| - y^2 = \cos x + C$

(D) $\ln|y| - y^2 = -\cos x + C$

GO ON TO THE NEXT PAGE.

AAAAAAAAAAAAAAAAAAAAAAAAAAAAAAAAAA

18. $\int \dfrac{7x + 46}{x^2 + 11x + 28}\, dx =$

(A) $\ln\left|\dfrac{x+7}{6(x+4)}\right| + C$

(B) $6\ln|x+7| + \ln|x+4| + C$

(C) $\ln\left|\dfrac{(x+7)}{(x+4)^6}\right| + C$

(D) $\ln|x+7| + 6\ln|x+4| + C$

19. Which of the following is(are) true with regard to polar curves and their derivatives?

I. $\dfrac{dx}{d\theta} = \dfrac{\frac{dy}{d\theta}}{\frac{dy}{dx}}$

II. If $x = 5\cos(\theta)\sin(\theta)$, then $\dfrac{dx}{d\theta} = -5$

III. If $y = -3\sin^2(\theta)$, then $\dfrac{dy}{d\theta} = -6\sin(\theta)\cos(\theta)$

(A) I only

(B) I and II only

(C) I and III only

(D) I, II, and III

GO ON TO THE NEXT PAGE.

AAAAAAAAAAAAAAAAAAAAAAAAAAAAAAAAA

20. If $f(x) = \begin{pmatrix} 2e^x + 3, & x \leq 1 \\ 2\ln(x) + 3, & x > 1 \end{pmatrix}$, then which of the following is true?

(A) $f(x)$ is continuous at $x = 1$

(B) $f(x)$ is differentiable at $x = 1$

(C) The limit as x approaches 1 does not exist

(D) The $\lim\limits_{x \to 1^+} f(x) = 2e + 3$

21. $\int x^2 e^x \, dx =$

(A) $x^2 e^x - 2xe^x + C$

(B) $x^2 e^x + 2xe^x + C$

(C) $x^2 e^x + 2xe^x + 2xe^x + C$

(D) $x^2 e^x - 2xe^x + 2xe^x + C$

22. What is the area of the region R bound by $y = x + 2$, $y = -\frac{5}{3}x + 10$, the y-axis, and the x-axis?

(A) 7.5

(B) 10.5

(C) 18

(D) 30

GO ON TO THE NEXT PAGE.

A A A A A A A A A A A A A A A A A A A A A A A A A A A A A A A A A A A A

23. A region is enclosed by the graph of $y = \sqrt[3]{x}$, the x-axis, and $x = 8$. What is the volume of the solid generated by revolving this region around the line $x = -4$?

(A) $\pi \int_0^2 \left(12^2 - \left(y^3 + 4\right)^2\right) dy$

(B) $\pi \int_0^8 \left(12^2 - \left(\sqrt[3]{x} + 4\right)^2\right) dx$

(C) $\pi \int_0^2 \left(12 - y^3\right)^2 dy$

(D) $\pi \int_0^2 \left(\left(y^3 + 4\right)^2 - 12^2\right) dy$

24. What is the radius of convergence for the power series $\displaystyle\sum_{n=1}^{\infty} \frac{(x-3)^n}{4 \cdot 2^{n+1}}$?

(A) $\dfrac{1}{4}$

(B) 2

(C) 3

(D) 4

GO ON TO THE NEXT PAGE.

AAAAAAAAAAAAAAAAAAAAAAAAAAAAAAAAA

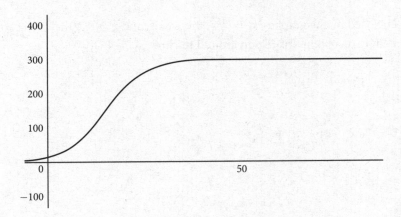

25. The graph shown above is $y = P(t)$, which is the solution to a logistic model differential equation.

    Which of the following could be $\dfrac{dP}{dt}$?

    (A) $\dfrac{dP}{dt} = 60P\left(2 - \dfrac{P}{300}\right)$

    (B) $\dfrac{dP}{dt} = 20P\left(3 - \dfrac{P}{100}\right)$

    (C) $\dfrac{dP}{dt} = 0.2P(200 - P)$

    (D) $\dfrac{dP}{dt} = 0.2P(500 - P)$

26. Which of the following is not equal to the definition of the derivative?

    (A) $\displaystyle\lim_{\Delta x \to 0} \dfrac{f(x + \Delta x) - f(x)}{\Delta x}$

    (B) $\displaystyle\lim_{h \to 0} \dfrac{f(a + h) - f(a)}{h}$

    (C) $\displaystyle\lim_{x \to a} \dfrac{f(x) - f(a)}{x - a}$

    (D) $\displaystyle\lim_{\Delta a \to 0} \dfrac{f(a + h) - f(a)}{h}$

GO ON TO THE NEXT PAGE.

AAAAAAAAAAAAAAAAAAAAAAAAAAAAAAAA

27. $\lim\limits_{x \to 0} \dfrac{x^2 + 5x}{\sin^3 x - \sin x} =$

(A) -5

(B) 0

(C) 5

(D) 6

---

28. Given the function $J(x) = -\sin(x)$, which of the following is equal to the 47th derivative of $J(x)$?

(A) $J^{(47)}(x) = -\cos(x)$

(B) $J^{(47)}(x) = \sin(x)$

(C) $J^{(47)}(x) = \cos(x)$

(D) $J^{(47)}(x) = -\sin(x)$

---

29. $\dfrac{1}{4}\displaystyle\int (e^x - \sqrt{x})\, dx =$

(A) $\dfrac{1}{4}e^x - \dfrac{1}{6}x^{\frac{3}{2}} + C$

(B) $\dfrac{1}{4}e^x - \dfrac{3}{8\sqrt{x}} + C$

(C) $\dfrac{1}{4}e^x - \dfrac{2}{3}x^{\frac{3}{2}} + C$

(D) $\dfrac{1}{8}e^{2x} - \dfrac{1}{6}x^{\frac{3}{2}} + C$

GO ON TO THE NEXT PAGE.

AAAAAAAAAAAAAAAAAAAAAAAAAAAAAAA

30. What is the average rate of change of the function $f(x) = 2\cos(3x)$ on the interval $\left[\dfrac{\pi}{2}, 2\pi\right]$?

(A) $\dfrac{4}{3\pi}$

(B) 2

(C) $\dfrac{3\pi}{2}$

(D) 3π

END OF PART A OF SECTION I

**IF YOU FINISH BEFORE TIME IS CALLED, YOU MAY
CHECK YOUR WORK ON PART B ONLY.**

B B B B B B B B B

CALCULUS BC
SECTION I, Part B
Time—45 minutes
Number of questions—15

A GRAPHING CALCULATOR IS REQUIRED FOR SOME QUESTIONS ON
THIS PART OF THE EXAM.

Directions: Use the space provided to solve each of the following problems. After examining the given choices, decide which of the choices is the best and fill in the corresponding circle on the answer sheet. Credit will only be given for answers appropriately marked on the answer sheet, and not for anything written in the exam book. Do not spend too much time on any one problem.

In this exam:

(1) When the exact numerical value of the correct answer does not appear among the choices given, select the number that best approximates the numerical value.

(2) The domain of a function f should be assumed to be the set of all real numbers x for which $f(x)$ is a real number, unless the question indicates otherwise.

(3) The inverse function notation f^{-1} or the prefix "arc" (e.g., $\sin^{-1} x = \arcsin x$) may be used to indicate the inverse of a trigonometric function.

GO ON TO THE NEXT PAGE.

B B B B B B B B B

76. If $f(x) = 3x^3 - 5x^2 + 8x - 2$, then $(f^{-1})'(4) =$

    (A) $\frac{1}{7}$

    (B) $\frac{1}{4}$

    (C) 1

    (D) 7

77. A particle moves along the x-axis so that at any given time $t \geq 0$ its velocity is given by $v(t) = 6t^2 - 3t + 7$. If the particle is at position $x = -5$ at $t = 0$, what is the position of the particle at time $t = 4$?

    (A) 79.75

    (B) 91

    (C) 127

    (D) 132

78. Which of the following is the equation of the tangent line for the parametric function $x(t) = 1 + 2\sin(t)$ and $y(t) = 3 - 4\cos(t)$ at $t = \frac{\pi}{3}$?

    (A) $y - 0.464 = 1.155(x - 2)$

    (B) $y - 0.866 = 3.464(x - 0.5)$

    (C) $y - 1 = 0.289(x - 2.732)$

    (D) $y - 1 = 3.464(x - 2.732)$

GO ON TO THE NEXT PAGE.

B B B B B B B B B

x	-0.3	0	0.3	0.6
$f'(x)$	0.6	1.7	2.1	3.6

79. The table above gives the values of f', which is the derivative of the function f for selected values of x. If $f(-0.3) = 7$, what is the approximation of $f(0.6)$ obtained by using Euler's method with a step size of 0.3 starting at $x = -0.3$?

(A) 7.18

(B) 7.69

(C) 8.32

(D) 8.8

80. At which x-value does the function $y = x^3 + 3x^2 + 5x - 4 \cos (2x)$ change concavity on the interval $[-2, 1]$?

(A) $x = -1.426$

(B) $x = -1.411$

(C) $x = -0.823$

(D) $x = -0.819$

GO ON TO THE NEXT PAGE.

B B B B B B B B B

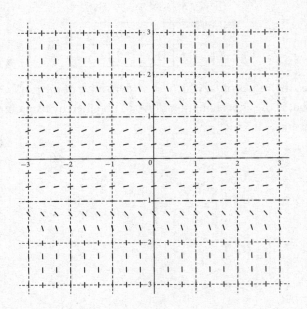

81. The slope field for the differential equation $\dfrac{dy}{dx} = y^2\left(1 - y^2\right)$ is shown above. If $y = g(x)$ is the solution to the differential equation with the initial condition $g(-1) = 3$, then what is $\lim\limits_{x \to \infty} g(x)$?

   (A) -1

   (B) 0

   (C) 1

   (D) 2

82. The base of a solid is the region in the first quadrant bounded by the y-axis, the x-axis, the horizontal line $y = 1$, the vertical line $x = \dfrac{\pi}{2}$, and the graph of $y = \sin x$. For this solid, each cross section perpendicular to the x-axis is a square. What is the volume of the solid?

   (A) 0.178

   (B) 0.356

   (C) 0.571

   (D) 1.571

GO ON TO THE NEXT PAGE.

B B B B B B B B B

83. Given $f(x) = 3x^3 + 2x^2 - 8x + 7$, what is the equation of the tangent line at the point where $f(x)$ changes concavity?

(A) $y - 8.621 = -8.444(x - 0.222)$

(B) $y - 13.135 = -7.444(x - 1.132)$

(C) $y = -7.444x - 10.2753$

(D) $y - 8.843 = -8.444(x + 0.222)$

84. What is the area of the region in the first quadrant enclosed by the graphs of $y = \sin x$, $y = -x + 3$, and the y-axis?

(A) 1.572

(B) 2.180

(C) 2.592

(D) 4.164

GO ON TO THE NEXT PAGE.

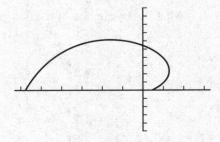

85. The graph above shows the polar curve $r = 3\theta + \sin(\theta)$ for $0 \leq \theta \leq \pi$. What is the area of the region bound by the curve and the x-axis?

(A) 8.188

(B) 47.052

(C) 56.720

(D) 113.439

86. At what point on the graph of $r = 2 + 3\cos(\theta)$ is the tangent line not horizontal?

(A) $(-0.679, 2.675)$

(B) $(3.681, 0.976)$

(C) $(3.681, 5.307)$

(D) $(3.056, 3.608)$

GO ON TO THE NEXT PAGE.

87. Given the polar curve $r = \cos(2\theta)$, find the area of the petals in any two quadrants.

    (A) 0.393

    (B) 0.785

    (C) 1.571

    (D) 2

88. A particle moves along the x-axis so that at any given time $t \geq 0$, its velocity is given by $v(t) = t^3 \ln(t + 6)$. What is the acceleration of the particle at $t = 5$?

    (A) 6.818

    (B) 84.540

    (C) 191.206

    (D) 299.737

89. What is the area of the region in the first quadrant enclosed by the graph $y = \sin(x) + 5$, $y = x$, and the y-axis?

    (A) 4.153

    (B) 8.622

    (C) 13.672

    (D) 22.294

GO ON TO THE NEXT PAGE.

Practice Exams

90. The function f is defined by $f(x) = 4x - 3 \sin (2x + 1)$, and its derivative is $f'(x) = 4 - 6 \cos (2x + 1)$. What are all values of x that satisfy the conclusion of the mean value theorem applied to f on the closed interval $[-1, 2]$?

(A) 0.295 and 1.846

(B) 1.054

(C) –0.025 only

(D) –0.025 and 2.166

END OF SECTION I

**IF YOU FINISH BEFORE TIME IS CALLED, YOU MAY
CHECK YOUR WORK ON PART B ONLY.**

CALCULUS BC
SECTION II, Part A
Time—30 minutes
Number of problems—2

A GRAPHING CALCULATOR IS REQUIRED FOR THESE PROBLEMS.

Directions: Write your solution to each part of the following questions in the space provided. Write clearly and legibly. Cross out any errors you make; erased or crossed-out work will not be scored.

You may wish to look over the problems before starting to work on them; on the actual test, it is not expected that everyone will be able to complete all parts of all problems. All problems are given equal weight, but the individual parts of a particular problem are not necessarily given equal weight. You should not spend too much time on any one problem.

- Show all of your work. Clearly label functions, graphs, tables, or anything else that you use to arrive at your final solution. Your work will be scored on the correctness and completeness of your methods as well as your final answers. Answers without supporting work will usually not receive credit.

- Justifications (i.e., the request that you "justify your answer") require that you give mathematical (non-calculator) reasons.

- Do not use calculator syntax. Your work must be expressed in standard mathematical notation.

- Numeric or algebraic answers do not need to be simplified, unless the question indicates otherwise.

- If you use decimal approximations in calculations, your final answers should be accurate to three places after the decimal point, unless the question indicates otherwise.

- The domain of a function f should be assumed to be the set of all real numbers x for which $f(x)$ is a real number, unless the question indicates otherwise.

GO ON TO THE NEXT PAGE.

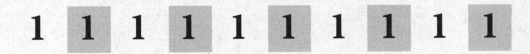

1. Fans entering a stadium for a professional soccer game are measured in people per minute modeled by the function $F(t) = 62 + 5 \sin\left(\dfrac{t}{3}\right)$ for $0 \leq t \leq 30$, where $F(t)$ is the number of people per minute and t is measured in minutes.

   (a) To the nearest whole number how many people enter the stadium over the thirty minute period?

   (b) Is the flow of fans increasing or decreasing at $t = 8$? Give a reason for your answer.

(c) What is the average value of the fans entering the stadium over the time interval $10 \leq t \leq 25$? Indicate units of measure.

(d) What is the average rate of change of the fans entering the stadium over the time interval $10 \leq t \leq 25$? Indicate units of measure.

2 2 2 2 2 2 2 2 2 2

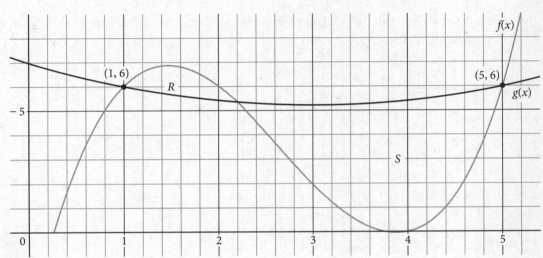

2. Let f and g be the functions defined by $f(x) = x^3 - 8x^2 + 17x - 4$ and $g(x) = \frac{1}{5}x^2 - \frac{6}{5}x + 7$.

    Let R and S be the two regions enclosed by the graphs of f and g shown in the figure above.

    (a) Find the sum of the areas of regions R and S.

(b) Region R is the base of a solid whose cross sections perpendicular to the x-axis are equilateral triangles. Write, but do not evaluate, the definite integral used to find the volume of the solid.

(c) Region S is revolved about the line $y = 7$. Write, but do not evaluate, the definite integral used to find the volume of the resulting solid.

END OF PART A OF SECTION II

**IF YOU FINISH BEFORE TIME IS CALLED, YOU MAY
CHECK YOUR WORK ON PART A ONLY.**

CALCULUS BC
SECTION II, Part B
Time—60 minutes
Number of problems—4

NO CALCULATOR IS ALLOWED FOR THESE PROBLEMS.

Note: If you have extra time, you can go back and work on Part A of Section II, but you cannot use a calculator to complete your work at this time.

3 3 3 3 3 3 3 3 3 3

NO CALCULATOR ALLOWED

3. Let $r = 2\cos\theta - \sqrt{2}$ for $0 < \theta < \pi$. Let S be the region in the first quadrant bounded by the curve and the x-axis.

(a) Write, but do not evaluate, an integral expression for the area of S.

(b) Write expressions for $\frac{dx}{d\theta}$ and $\frac{dy}{d\theta}$ in terms of θ.

(c) Write an equation in terms of x and y for the line tangent to the graph of the polar curve at the point where $\theta = \frac{\pi}{2}$. Show the computations that lead to your answer.

GO ON TO THE NEXT PAGE.

4 4 4 4 4 4 4 4 4 4

NO CALCULATOR ALLOWED

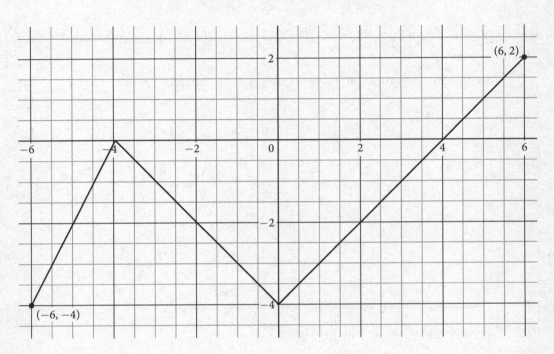

4. The function f, shown in the given graph, is defined on the closed interval $[-6, 6]$ and consists of three line segments. Let g be the function defined by $g(x) = \int_{-4}^{x} f(t)\, dt$.

(a) Find $g(0)$, $g'(0)$ and $g''(0)$. If the value does not exist, explain why.

4 4 4 4 4 4 4 4 4 4

NO CALCULATOR ALLOWED

(b) On what open intervals contained in $-6 < x < 6$ is the graph of g both decreasing and concave up? Give a reason for your answer.

(c) The function h is defined by $h(x) = 3xg(x)$. Find $h'(2)$.

GO ON TO THE NEXT PAGE.

5 5 5 5 5 5 5 5 5 5

NO CALCULATOR ALLOWED

5. Use the function $f(x) = ax^3 + bx^2 + cx$.

Determine the values of a, b, and c so that $f(x)$ will have an inflection point at $(2, 6)$ and the slope of the tangent line to $f(x)$ at $(2, 6)$ will be -1.

(a) Find the first two derivatives and solve for b, in terms of a, given that there is an inflection point at $(2, 6)$.

(b) If the slope of the tangent line to $f(x)$ at $(2, 6)$ is -1 what is c in terms of a?

(c) Solve for a, b, and c.

GO ON TO THE NEXT PAGE.

NO CALCULATOR ALLOWED

6. Let $f(x) = \ln(1 + x^4)$.

   (a) The Maclaurin series for $\ln(1 + x)$ is given by the following:

   $$x - \frac{x^2}{2} + \frac{x^3}{3} - \frac{x^4}{4} + \cdots + (-1)^{n+1}\frac{x^n}{n} + \cdots$$

   Use the series to write the first four nonzero terms and the general term of the Maclaurin Series for f.

   (b) The radius of convergence of the Maclaurin Series for f is 1. Determine the interval of convergence. Show the work that leads to your answer.

GO ON TO THE NEXT PAGE.

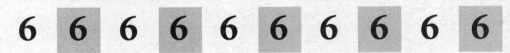

(c) Write the first four nonzero terms of the Maclaurin series for $f'(t^3)$. If $g(x) = \int_0^x f'(t^3)\, dt$, use the first two nonzero terms of the Maclaurin series for g to approximate $g(1)$.

(d) The Maclaurin Series for g, evaluated at $x = 1$, is a convergent alternating series with individual terms that decrease in absolute value to 0. Show that your approximation in part (c) must differ from $g(1)$ by less than $\dfrac{25}{170}$.

STOP

END OF EXAM

ANSWER KEY

Section I, Part A

1.	C	16.	A
2.	C	17.	D
3.	C	18.	D
4.	C	19.	C
5.	D	20.	C
6.	A	21.	D
7.	C	22.	C
8.	D	23.	A
9.	D	24.	B
10.	B	25.	B
11.	D	26.	D
12.	C	27.	A
13.	A	28.	C
14.	C	29.	A
15.	D	30.	A

Section I, Part B

76.	A
77.	C
78.	D
79.	C
80.	D
81.	C
82.	B
83.	D
84.	C
85.	C
86.	D
87.	B
88.	C
89.	C
90.	A

SCORING

Section II, Parts A & B

Use the scoring rubrics (found after the Section I answers and explanations) to self-score your free-response questions.

Section I, Part A Number Correct: _____

Section I, Part B Number Correct: _____

Section II, Part A Points Earned: _____

Section II, Part B Points Earned: _____

Enter your results to your Practice Exam 4 assignment to see your 1–5 score by logging in at kaptest.com.

Haven't registered your book yet? Go to kaptest.com/booksonline to begin.

BC PRACTICE EXAM 4 ANSWERS AND EXPLANATIONS

SECTION I, Part A

1. C

For you to find the total number of bacteria in the petri dish, you must add the number of new bacteria (those added from $t = 5$ to $t = 7$) to the number of bacteria at time $t = 5$. The answer is $1{,}000 + \int_{5}^{7} R(t)\, dt$. The 1,000 represents the starting number of bacteria, and the $\int_{5}^{7} R(t)\, dt$ represents the additional bacteria.

Hence, **(C)** is correct. (A) is the rate at which the number of bacteria is changing at time $t = 7$ and does not represent the total number of bacteria. If the question asked for how many bacteria grow from $t = 5$ to $t = 7$, (D) would have been correct.

2. C

Deciding if a sequence converges as n gets infinitely large is like deciding if a finite limit exists for a function as x approaches infinity. The sequence $a_n = \dfrac{2 \sin n}{3}$ continues to oscillate between positive and negative numbers as n continues to get larger, so it does not converge.

Therefore, **(C)** is correct. Even though the numerator in (A) seems like a large value, it is still a constant. As n gets infinitely large, $a_n = \dfrac{100^{100}}{n^2}$ converges to 0 because the denominator will eventually grow larger than the numerator. (B) can be written as $a_n = \dfrac{4n^2 + 12n}{n^2 - 4n - 5}$. Because the numerator and denominator have the same degree, the sequence converges to the ratio of the leading coefficients, which is 4. (D) can be rewritten as $a_n = \dfrac{2n^2 - n - 3}{2n^3 - 11n^2 - 5n}$. When the sequence is defined by a rational expression and the degree of the denominator is greater than the degree of the numerator, the sequence converges to 0.

3. C

The normal line is perpendicular to the tangent line at the point of tangency, and perpendicular lines have negative reciprocal slopes. The slope of the tangent line is $\dfrac{2}{3}$, so the slope of the normal line must be $-\dfrac{3}{2}$. Because $y = \dfrac{2}{3}x + 5$ is tangent to $f(x)$ at $x = 3$, you can find $f(3)$ by evaluating $y(3)$.

$$y(3) = \frac{2}{3}(3) + 5$$
$$y(3) = 2 + 5$$
$$y(3) = 7$$

The point $(3, 7)$ and the slope $-\dfrac{3}{2}$ can now be put into the point-slope form of a line.

$$y - 7 = -\frac{3}{2}(x - 3)$$
$$y - 7 = -\frac{3}{2}x + \frac{9}{2}$$
$$y = -\frac{3}{2}x + \frac{23}{2}$$

Thus, **(C)** is correct. (A) uses the reciprocal slope of the tangent line instead of the negative reciprocal. (B) uses the correct slope but incorrectly distributes the slope when working with the point-slope form of a line.

4. C

Because the degree of the numerator is the same as the degree of the denominator, the limit as x approaches infinity is the ratio of the leading coefficients. Another way to reach this answer is to divide each term by x^4 and then take the limit of the resulting function as x approaches ∞.

$$\lim_{x\to\infty} \frac{2x^4 - 3x^2 + 4}{-7x^4 + 2x^3 - x}$$

$$= \lim_{x\to\infty} \frac{\frac{2x^4}{x^4} - \frac{3x^2}{x^4} + \frac{4}{x^4}}{\frac{-7x^4}{x^4} + \frac{2x^3}{x^4} - \frac{x}{x^4}}$$

$$= \lim_{x\to\infty} \frac{2 - \frac{3}{x^2} + \frac{4}{x^4}}{-7 + \frac{2}{x} - \frac{1}{x^3}}$$

$$= \frac{2 - 0 + 0}{-7 + 0 - 0}$$

$$= \frac{-2}{7}$$

Therefore, **(C)** is correct. (D) would be the limit if the degree in the numerator was less than the degree in the denominator.

5. D

When taking the limit of a function that is being raised to a power, you can simply raise the limit of the function to the same power. This property needs to be applied twice in this problem. Remember that taking the square root of a function is equivalent to raising the function to the $\frac{1}{2}$ power.

$$\lim_{x\to5} \sqrt{f(x)} = \sqrt{\lim_{x\to5} f(x)} = 2$$

$$\lim_{x\to5} f(x) = 4$$

$$\lim_{x\to5} f(x)^2 = \left[\lim_{x\to5} f(x)\right]^2 = 4^4 = 16$$

Hence, **(D)** is correct. (B) is the limit of $f(x)$. (C) is the product of 4 and 2, which is how you handle exponents within a logarithm but not with a limit.

6. A

In order to find $\frac{dy}{dx}$, you must find the quotient of $\frac{dy}{dv}$ and $\frac{dx}{dv}$ because $\frac{dy}{dx} = \dfrac{\frac{dy}{dv}}{\frac{dx}{dv}}$.

$$\frac{dy}{dx} = \frac{-2(3\sin(2v))}{3\cos(v)}$$

Replace $\sin(2v)$ with $2\sin(v)\cos(v)$ using a double angle identity. Then simplify.

$$\frac{dy}{dx} = \frac{-2(3(2\sin(v)\cos(v)))}{3\cos(v)}$$

$$\frac{dy}{dx} = -4\sin(\partial)$$

Thus, **(A)** is correct. (B) is the result if you use the incorrect derivative for cosine. (C) is found if you incorrectly replace $\frac{-2\sin(2v)}{\cos(v)}$ with $-2\tan(2v)$. (D) occurs if you find $\frac{dx}{dy}$.

7. C

The graph is concave up where the second derivative is positive. Start by calculating the second derivative. Then evaluate the zeros to find the critical values.

$$y = x^4 - 3x^3 - 60x^2 + 24x - 36$$
$$y' = 4x^3 - 9x^2 - 120x + 24$$
$$y'' = 12x^2 - 18x - 120$$
$$0 = 12x^2 - 18x - 120$$
$$0 = 6(2x^2 - 3x - 20)$$
$$0 = 6(2x + 5)(x - 4)$$
$$x = -\frac{5}{2} \text{ or } x = 4$$

Using the factored form $y'' = 6(2x + 5)(x - 4)$, test intervals can be used to determine where the original graph will be concave up.

Test Interval	Sign of y''
$\left(-\infty, -\dfrac{5}{2}\right)$	+
$\left(-\dfrac{5}{2}, 4\right)$	−
$(4, \infty)$	+

From the test intervals, you can determine that the graph is concave up for all x on either the interval $\left(-\infty, -\frac{5}{2}\right)$ or the interval $(4, \infty)$.

Thus, **(C)** is correct. The graph is concave down on the interval in (A). (B) and (D) are the result of a factoring error.

8. D

The function $y = 2e^{3x}$ has a horizontal asymptote at $y = 0$, making this an unbound region. Replace $-\infty$ with the variable a, and evaluate the resulting integral using u-substitution. Let $u = 3x$, which means $du = 3dx$ or $\frac{1}{3} du = dx$. The new integral becomes $\lim\limits_{a \to -\infty} \int_a^1 2e^u \frac{1}{3} \, du$, which can be written as $\lim\limits_{a \to -\infty} \frac{2}{3} \int_a^1 e^u \, du$. Remember that the antiderivative of e^u is e^u.

$$\lim_{a \to -\infty} \frac{2}{3} \int_a^1 e^u \, du$$
$$= \lim_{a \to -\infty} \frac{2}{3} e^u \Big|_{3a}^3$$
$$= \lim_{a \to -\infty} \left[\frac{2}{3}\left(e^3 - e^{3a}\right) \right]$$

The limit of this value can now be taken as a approaches negative infinity.

$$\lim_{a \to -\infty} \left[\frac{2}{3}\left(e^3 - e^{3a}\right) \right]$$
$$= \frac{2}{3} \lim_{a \to -\infty} \left(e^3 - e^{3a}\right)$$
$$= \frac{2}{3}\left(e^3 - e^{-\infty}\right)$$
$$= \frac{2}{3}\left(e^3 - \frac{1}{e^\infty}\right)$$
$$= \frac{2}{3}\left(e^3 - 0\right)$$
$$= \frac{2}{3}e^3$$

Hence, **(D)** is correct.

9. D

To find the area between the function and the x-axis, it is easier to have a picture or sketch of the function in order to properly write the integral that represents the area.

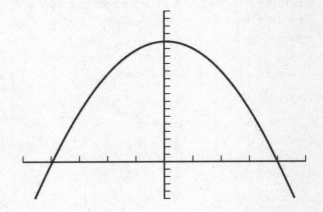

As shown in the sketch, this function is a negative quadratic function translated 16 units up as compared to the parent function. It has x-intercepts of $x = -4$ and $x = 4$ and a y-intercept of $y = 16$. The integral that represents the area is written as $\int_{-4}^4 \left(-x^2 + 16\right) dx$. To evaluate the integral, first take the antiderivative. Then find the difference between the upper and lower bounds.

$$\int_{-4}^4 \left(-x^2 + 16\right) dx$$
$$= -\frac{1}{3}x^3 + 16x \Big|_{-4}^4$$
$$= -\frac{1}{3}(4)^3 + 16(4) - \left[-\frac{1}{3}(-4)^3 + 16(-4)\right]$$
$$= -\frac{64}{3} + 64 - \left[\frac{64}{3} - 64\right]$$
$$= -\frac{128}{3} + 128$$
$$= -\frac{128}{3} + \frac{384}{3}$$
$$= \frac{256}{3}$$

Therefore, **(D)** is correct. (A) is a negative area. Although integrals can be negative, areas can never be. (B) mistakenly never takes the antiderivative and instead subtracts $f(-4)$ from $f(4)$. (C) is the area between the function and the x-axis but also bound in the first quadrant. It is important to evaluate the lower bound as $x = -4$ and not $x = 0$.

10. B

Using the given integrals for $f(x)$, $\int_{-8}^{12} f(x)\,dx = 7$ and $\int_{-8}^{13} f(x)\,dx = 8$, you can deduce that $\int_{12}^{13} f(x)\,dx = 1$.

Evaluating $\int_{12}^{13} 3f(x)\,dx + \int_{12}^{12} 4g(x)\,dx$ requires you to evaluate the integrals separately because the upper and lower bounds are different: $\int_{12}^{13} 3f(x)\,dx = 3$ and $\int_{12}^{12} 4g(x)\,dx = 0$. Notice that on the integral with $g(x)$, the upper and lower bounds are the same. Therefore this integral has a value of zero.

Hence, **(B)** is correct. (A) uses the incorrect sign for $\int_{12}^{13} f(x)\,dx = -1$. (C) and (D) incorrectly use $\int_{12}^{12} 4g(x)\,dx = 48$.

11. D

If you recall from the limits section, you should start by substituting the value x is approaching into the function. If that produces $\frac{0}{0}$ or $\frac{\infty}{\infty}$, it is an indeterminate form and other techniques are required to find the limit. For most limits, some basic algebraic applications are sufficient. With more challenging problems, L'Hospital's rule, can be useful. L'Hospital's rule states if $\lim_{x \to a} \frac{f(x)}{g(x)}$ is in the form of $\frac{0}{0}$ or if $\lim_{x \to a} \frac{f(x)}{g(x)}$ is in the form of $\frac{\pm\infty}{\pm\infty}$, then $\lim_{x \to a} \frac{f(x)}{g(x)}$ is in the form of $\lim_{x \to a} \frac{f'(x)}{g'(x)}$, where a can be any real number, infinity, or negative infinity. In this problem when you substitute 0 for x, you get $\frac{0}{0}$. Basic simplification or rewriting of the original is not obvious. So you can differentiate the numerator and denominator separately and then take the limit of the new quotient. This gives $\lim_{x \to 0} \frac{6\cos(2x)}{2 + 5\cos x}$. Substituting 0 for x results in $\frac{6}{7}$.

Thus, **(D)** is correct. (A) demonstrates a misunderstanding of the meaning behind $\frac{0}{0}$. (C) is the ratio of the coefficients of the sine values. However, using leading coefficients is appropriate only when x is approaching positive or negative infinity and the function is the quotient of two polynomial functions of the same degree. (B) incorrectly differentiates the numerator by not applying the chain rule.

12. C

This geometric series has a starting value of $a = 7$ with a common ratio of $r = -0.4$. A geometric series always converges if $|r| < 1$. To find the value it converges to, use the formula $S = \frac{a}{1-r}$.

$$S = \frac{7}{1 - (-0.4)}$$
$$S = \frac{7}{1 + 0.4}$$
$$S = 5$$

Therefore, **(C)** is correct. (A) mistakenly divides the starting value, a, by the ratio, r. (B) is the product of the starting value and the ratio.

13. A

The u-value is provided, and $du = \frac{1}{3}dx$. Substituting u for $\frac{x}{3}$ in the numerator leaves $\int_3^6 \frac{2 - (u)^2}{x}\,dx$. Solving $u = \frac{x}{3}$ for x allows you to replace the x in the denominator with $3u$, making the new integral $\int_3^6 \frac{2 - (u)^2}{3u}\,dx$. Because $du = \frac{1}{3}dx$, it replaces the dx and the 3 in the denominator when substituted, $\int_3^6 \frac{2 - (u)^2}{u}\,du$. The upper and lower bounds need to be adjusted to make them in terms of u, which means they both need to be divided by 3. The definite integral to be evaluated is now $\int_1^2 \frac{2 - (u)^2}{u}\,du$ or $\int_1^2 \left(\frac{2}{u} - u\right)du$. The antiderivative of $\frac{2}{u}$ is $2\ln|u|$, and the power rule can be applied to u.

$$\int_1^2 \left(\frac{2}{u} - u\right)du$$
$$= 2\ln|u| - \frac{u^2}{2}\Big|_1^2$$
$$= \left[2\ln 2 - \frac{2^2}{2}\right] - \left[\ln 1 - \frac{1^2}{2}\right]$$
$$= [2\ln 2 - 2] - \left[0 - \frac{1}{2}\right]$$
$$= -\frac{3}{2} + 2\ln 2$$

Hence, **(A)** is correct. If you got (B), you ignored the x in the denominator of the original function. C) contains an algebraic error when subtracting the results after substituting the upper and lower bound values. (D) is a result of forgetting to adjust the upper and lower bound values.

14. C

This question requires integrals, u-substitution, and natural logs. The first step is to rewrite the integral with a negative exponent: $\int_{2}^{6}(6-4x)^{-1}dx$. Once you attempt to apply the power rule for antiderivatives by adding 1 to the exponent and dividing by the new power, you should realize there is an issue when you divide by 0. The general rule for the antiderivative raised to the -1 power is that it is the natural log of the absolute value of the integrand. Here you also need to use u-substitution to evaluate the integral correctly.

$$u = (6 - 4x)$$

$$\frac{du}{dx} = -4$$

$$dx = -\frac{1}{4}du$$

$$\int_{2}^{6}(6-4x)^{-1}dx = -\frac{1}{4}\int u^{-1}du$$

$$\int_{2}^{6}(6-4x)^{-1}dx = -\frac{1}{4}\left(\ln|6-4x|\right)\Big|_{2}^{6}$$

$$= -\frac{1}{4}[\ln 18 - \ln 2]$$

$$= -\frac{1}{4}\ln 9$$

$$= -\frac{\ln 9}{4}$$

Thus, **(C)** is correct. Notice the condensing of the natural logs: $\ln 18 - \ln 2 = \ln 9$. Also notice the last step to reach the answer; $-\frac{1}{4}\ln 9 = \frac{-\ln 9}{4}$. (B) is incorrectly condenses the natural logs to get $\ln 18 - \ln 2 = \ln (36)$. (D) ignores the necessary u-substitution and chain rule when you take the antiderivative. (A) takes the derivative instead of the antiderivative of the expression of $(6-4x)^{-1}$, which gives $\frac{dy}{dx} = \frac{4}{(6-4x)^2}$. It also integrates between $x = 2$ and $x = 6$.

15. D

Use the formula $\frac{d}{dx}a^u = a^u \ln a \cdot \frac{du}{dx}$ to find the answer, $(6x \ln 7)\left(7^{3x^2+5}\right)$.

Thus, **(D)** is correct. (C) is incorrect because you cannot multiply $6x$ by the base 7 to get $(42 \ln 7)\left(x^{3x^2+6}\right)$.

16. A

Using the identity $\sin^2 x + \cos^2 x = 1$, evaluate $\lim_{x \to \frac{\pi}{6}}\frac{1}{\tan(2x)}$. So you need to find the reciprocal of $\tan \frac{\pi}{3}$, which is $\sqrt{3}$.

Therefore, the correct answer is $\frac{1}{\sqrt{3}}$, matching **(A)**. (B), although tempting, is the $\lim_{x \to \frac{\pi}{6}}\tan(2x)$. (C) ignores the 2 in the tangent function and the reciprocal; it is the $\lim_{x \to \frac{\pi}{6}}\tan(x)$. (D) ignores the necessary squares on the sine and cosine functions.

17. D

This differential equation is defined in terms of both x and y, so you need to take certain steps to get all the y-terms on one side of the equation and all the x-terms on the other side before integrating. Multiply both sides by dx and $1 - 2y^2$. Then divide both sides by y. Finally, integrate both sides to find the general solution.

$$\frac{dy}{dx} = \frac{y \sin x}{1 - 2y^2}$$

$$(1 - 2y^2)dy = y \sin x \, dx$$

$$\left(\frac{1 - 2y^2}{y}\right)dy = \sin x \, dx$$

$$\left(\frac{1}{y} - 2y\right)dy = \sin x \, dx$$

$$\int \left(\frac{1}{y} - 2y\right)dy = \int \sin x \, dx$$

$$\ln|y| - y^2 = -\cos x + C$$

Therefore, **(D)** is correct. If you got (A), you probably did not divide both sides by the y that is being multiplied by $\sin x$. (C) has the incorrect sign for the antiderivative of the sine function. It should be negative cosine instead of cosine.

18. D

First factor the denominator.

$$\frac{7x + 46}{x^2 + 11x + 28} = \frac{7x + 46}{(x + 7)(x + 4)}$$

Then use constants A and B to represent the numerators.

$$\frac{7x + 46}{x^2 + 11x + 28} = \frac{A}{x + 7} + \frac{B}{x + 4}$$

Distribute the least common denominator $(x + 7)(x + 4)$ to solve for A and B.

$$7x + 46 = A(x + 4) + B(x + 7)$$

If $x = -4$, then $B = 6$. If $x = -7$, then $A = 1$.

$$\int \frac{7x + 46}{x^2 + 11x + 28}\,dx = \int \left(\frac{1}{x + 7} + \frac{6}{x + 4}\right)dx$$

$$= \int \frac{1}{x + 7}\,dx + \int \frac{6}{x + 4}\,dx$$

$$= \ln|x + 7| + 6\ln|x + 4| + C$$

Thus, **(D)** is correct. (B) has the coefficient 6 on the incorrect term.

19. C

Statement I is always true, following standard rules for derivatives with respect to other variables. To confirm, you can divide the given fractions and see that the expression reduces to $\frac{dx}{d\theta}$.

$$\frac{dx}{d\theta} = \frac{\dfrac{dy}{d\theta}}{\dfrac{dy}{dx}}$$

$$\frac{dx}{d\theta} = \frac{dy}{d\theta} \cdot \frac{dx}{dy}$$

$$\frac{dx}{d\theta} = \frac{dx}{d\theta}$$

Statement III is found by applying the chain rule to find the derivative after rewriting the function as $y = -3(\sin \theta)^2$. Remember that the derivative of y with respect to θ is $\frac{dy}{d\theta} = -6 \sin(\theta) \cos(\theta)$.

When finding the derivative of $x = 5 \cos(\theta) \sin(\theta)$, use the product rule.

$$x = 5 \cos(\theta) \sin(\theta)$$

$$\frac{dx}{d\theta} = 5[(\cos \theta)(\cos \theta) + (\sin \theta)(-\sin \theta)]$$

$$\frac{dx}{d\theta} = 5(\cos^2 \theta - \sin^2 \theta)$$

Statement II, $\frac{dx}{d\theta} = -5$, is false because $\cos^2 \theta - \sin^2 \theta \neq -1$.

Hence, **(C)** is correct.

20. C

Because the function limits are not the same from the left and the right, the limit as x approaches 1 does not exist. The $\lim_{x \to 1^-} f(x) = 2e + 3$, and the $\lim_{x \to 1^+} f(x) = 3$.

Therefore, **(C)** is correct. Notice that (D) is the limit from the incorrect side. If you try to find the limit as x approaches 1 from the right, you should use the function $2 \ln(x) + 3$, resulting in $2 \ln(1) + 3$. The natural log of 1 is 0. So the result of $2 \ln(1) + 3 = 3$.

21. D

Integration by Parts allows you to take the antiderivative of the product of two functions. Here the integral represents the product of two functions and requires the use of the Integration by Parts formula: $\int u\,dv = uv - \int v\,du$. In this example, you need to apply the formula twice as shown below. The first step is to identify the most appropriate u and dv for the problem. This is typically the most difficult part of the Integration by Parts process. Your goal is to make the new integral easier to take the antiderivative of or at least have it be a lower power so that Integration by Parts can be used again.

$$u = x^2$$
$$dv = e^x dx$$

Notice the position of how they are written. You should separate what you need to take the derivative of and what you need to take the antiderivative of. The next step is to take the derivative of u to get du and to take the antiderivative of dv to get v.

$$u = x^2 \qquad v = e^x$$
$$du = 2\,x dx \quad dv = e^x dx$$

Then substitute into the formula $\int u\,dv = uv - \int v\,du$.

$$\int x^2 e^x = x^2 e^x - \int 2xe^x\,dx$$

Use Integration by Parts a second time.

$$u = x \qquad v = e^x$$
$$du = 1\,dx \quad dv = e^x\,dx$$

$$\int x^2 e^x = x^2 e^x - 2[xe^x - \int e^x\,dx]$$

$$\int x^2 e^x = x^2 e^x - 2[xe^x - e^x] + C$$

$$\int x^2 e^x = x^2 e^x - 2xe^x + 2e^x + C$$

Thus, **(D)** is correct. (A) is a middle step in the correct process but does not account for the second integral in the formula. (C) has the incorrect signs and uses an addition symbol in the Integration by Parts formula.

22. C

There are two approaches to this question that, when done correctly, produce the same answer. The first approach is to work with integrals. The other is to work with areas between the functions and the x-axis. Either approach requires a basic knowledge of what the functions look like, so a sketch would be appropriate.

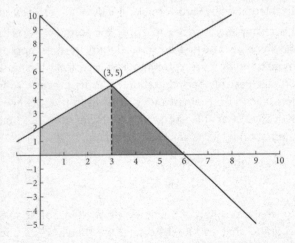

The lines intersect at (3, 5). This can be found from solving the system of equations that were given. The first section (to the left of the intersection point) is the area

of a trapezoid with bases of 2 and 5 and a height of 3. This produces an area of 10.5.

$$\text{Area}_{\text{trapezoid}} = \left(\frac{b_1 + b_1}{2}\right)h$$

$$\text{Area}_{\text{trapezoid}} = \left(\frac{2 + 5}{2}\right)3$$

$$\text{Area}_{\text{trapezoid}} = 10.5$$

The second section (to the right of the intersection point) is a triangle with a base of 3 and a height of 5, which gives an area of 7.5. The total area of the bounded region is $10.5 + 7.5 = 18$ units. To use integrals for this question, you need to write them carefully.

$$\int_0^3 (x + 2)\,dx + \int_3^6 \left(-\frac{5}{3}x + 10\right)dx$$

Two integrals are needed because when the graphs intersect at $x = 3$, the function that creates the boundaries of the region changes. The function changes over the domain [0, 3] to [3, 6]. The upper bound in the second integral is found by setting the expression equal to 0 and solving for x.

$$\int_0^3 (x + 2)\,dx = \frac{1}{2}x^2 + 2x\Big|_0^3$$

$$= 10.5$$

$$\int_3^6 \left(-\frac{5}{3}x + 10\right)dx = -\frac{5}{6}x^2 + 10x\Big|_3^6$$

$$= 7.5$$

Hence, **(C)** is correct. (A) and (B) are the areas of the individual sections, not the area between the functions and the x-axis. (D) is the value of $\int_0^6 (x + 2)\,dx$. This is incorrect because it ignores the second linear function.

23. A

Start by solving the given equation for x because the curve is being revolved around the vertical line $x = -4$. The upper bound is determined by the y-coordinates of the point of intersection of the functions $y = \sqrt[3]{x}$ and $x = 8$, which is $y = 2$. The lower bound is the y-coordinate of the x-axis, which is 0. Integrate the function with respect to y, and remember to multiply by π. By using $\pi \int_{a}^{b} \left(R^2 - r^2\right) dy$, you can find R. It is the distance between $x = -4$ (the center of rotation) and the line $x = 8$. You can find r from the difference between the function and the line $x = -4$. Therefore, the volume of the solid is $\pi \int_{0}^{2} \left(12^2 - \left(y^3 + 4\right)^2\right) dy$.

Therefore, **(A)** is correct. (D) reverses the R with lowercase r.

24. B

If a power series converges when $|x| < r$ and diverges when $|x| > r$, then r is called the radius of convergence. To find the radius of convergence, follow the steps to solve the inequality $\lim_{n \to \infty} \left| \frac{u_{n+1}}{u_n} \right| < 1$ until you get to $|x| < r$. This r-value is considered the radius of convergence.

Use $\lim_{n \to \infty} \left| \frac{u_{n+1}}{u_n} \right| < 1$ on $\sum_{n=1}^{\infty} \frac{(x-3)^n}{4 \cdot 2^{n+1}}$.

$$\lim_{n \to \infty} \left| \frac{(x-3)^{n+1}}{4 \cdot 2^{n+1+1}} \cdot \frac{4 \cdot 2^{n+1}}{(x-3)^n} \right| < 1$$

$$\lim_{n \to \infty} \left| \frac{x-3}{2} \right| < 1$$

$$\left| \frac{x-3}{2} \right| < 1$$

$$\frac{1}{2} |x-3| < 1$$

$$|x-3| < 2$$

The radius of convergence, r, in this question is 2.

Thus, **(B)** is correct. (C) is the c-value where the Taylor series is centered but is not the radius of convergence. (D) is the length between the endpoints of the interval of convergence but is not the radius of convergence. Notice the steps to test the endpoints of $x = 1$ and $x = 5$ are not necessary here because the question is asking for the radius of convergence, not the interval of convergence.

25. B

$\frac{dP}{dt} = 20P\left(3 - \frac{P}{100}\right)$. By analyzing the graph, you should be able to recognize that the capacity of the environment is 300. By putting the choices into the form $\frac{dy}{dt} = ky(A - y)$, where A is the maximum, you can determine which one has a maximum of 300. Factoring $\frac{1}{100}$ from the value inside of the parenthesis of (B) leaves $\frac{dP}{dt} = 0.2P(300 - P)$, which is the correct capacity.

Hence, **(B)** is correct. Factoring $\frac{1}{300}$ from the value inside of the parenthesis of (A) leaves $\frac{dP}{dt} = 0.2P(600 - P)$, which has a capacity of 600. (C) and (D) are already in the appropriate form to identify the capacity of the environment, but neither has 300 as its capacity.

26. D

$\lim_{\Delta a \to 0} \frac{f(a+h) - f(a)}{h}$ is the only answer choice that does not truly represent the instantaneous rate of change as the distance between two points gets smaller and smaller. (A), (B), and (C) are all standard definitions of the derivative. You should understand and memorize these for the exam. Close investigation of **(D)** should lead you to realize that the change in a approaching zero does not produce a derivative.

27. A

Factor the numerator and denominator. This limit simplifies to $\lim_{x \to 0} \frac{x(x + 5)}{\sin x(\sin^2 x - 1)}$. When you apply the rule for the $\lim_{x \to 0} \frac{x}{\sin x} = 1$, the result is $\lim_{x \to 0} \frac{(x + 5)}{(\sin^2 x - 1)}$. Use the trigonometric identity $\sin^2 x + \cos^2 x = 1$ to replace denominator with $-\cos^2 x$. Then evaluate the limit as x approaches 0. You are left with $\frac{(0 + 5)}{-\cos^2(0)}$, which equals -5. Be careful to follow the correct order of operations. After you square the cosine of 0, multiply the result by -1 to become -1.

Thus, **(A)** is correct. If you incorrectly use $-\cos^2 0 = 1$, you would get (C).

28. C

Do not find 47 derivatives. Basic sine and cosine derivatives always repeat in patterns of 4:

$$J(x) = -\sin(x)$$
$$J'(x) = -\cos(x)$$
$$J''(x) = \sin(x)$$
$$J^{(3)}(x) = \cos(x)$$
$$J^{(4)}(x) = -\sin(x)$$

The pattern just repeats in this cyclical manner. To find the 47th derivative, identify which part of the repeating pattern is the same. Because the remainder is 3 when 47 is divided by 4, the 47th derivative is the same as the third derivative, $J^{(47)}(x) = \cos(x)$, **(C)**.

29. A

Rewriting $\sqrt{x}$ as $x^{\frac{1}{2}}$ may help you identify this as a problem that requires the power rule. The antiderivative of e^x is e^x. If e was being raised to a more complex exponent, u-substitution would be necessary. When taking an indefinite integral, don't forget to include the constant of integration, C.

$$\frac{1}{4} \int (e^x - \sqrt{x})\, dx$$
$$= \frac{1}{4} \int \left(e^x - x^{\frac{1}{2}}\right) dx$$
$$= \frac{1}{4} \left(e^x - \frac{2}{3} x^{\frac{3}{2}} + C\right)$$
$$= \frac{1}{4} e^x - \frac{1}{6} x^{\frac{3}{2}} + C$$

Hence, **(A)** is correct. (B) incorrectly differentiates the x-term. (C) does not distribute the $\frac{1}{4}$ to the middle term. (D) incorrectly handles the e-term by attempting to increase the exponent by x and then dividing the coefficient by 2.

30. A

The average rate of change of a function on a closed interval can be determined by calculating the slope of the secant line. You are given two x-values. Determine the corresponding y-values by evaluating the function. You can either find the y-values separately or substitute the function directly into the slope formula to represent y_1 and y_2.

$$\text{Average Rate of Change} = \frac{y_2 - y_1}{x_2 - x_1}$$

$$= \frac{2\cos(3 \cdot 2\pi) - 2\cos\left(3 \cdot \frac{\pi}{2}\right)}{2\pi - \frac{\pi}{2}}$$

$$= \frac{2\cos(6\pi) - 2\cos\left(\frac{3\pi}{2}\right)}{\frac{3\pi}{2}}$$

$$= \frac{2 \cdot 1 - 2 \cdot 0}{\frac{3\pi}{2}}$$

$$= \frac{2}{\frac{3\pi}{2}}$$

$$= \frac{4}{3\pi}$$

Thus, **(A)** is correct. (B) is the difference of the resulting function values. (C) is the difference of the interval's boundaries. (D) incorrectly simplifies the complex fraction in the final step of the solution.

SECTION I, Part B

76. A

Notice the notation shown in the question; this is asking for the derivative of the inverse. This question requires you to use the formula $g'(x) = \dfrac{1}{f'(g(x))}$ but with a slightly different notation: $\left(f^{-1}\right)'(x) = \dfrac{1}{f'(f^{-1}(x))}$. At first you may not realize you need your calculator to solve this. However, as stated in the AP guidelines, calculators may be needed to solve an equation numerically. When a question like this appears in the calculator use section, it is a good indicator that using your calculator may be helpful. After you set this equation equal to 4, you can solve for the x in the calculator using a variety of methods, all resulting in $x = 1$. Once you have established you need the derivative at $x = 1$, it also important to find $f'(x) = 9x^2 - 10x + 8$ and $f'(1) = 7$. Then substitute 4 in the formula.

$$\left(f^{-1}\right)'(x) = \frac{1}{f'(f^{-1}(x))}$$

$$\left(f^{-1}\right)'(4) = \frac{1}{f'(f^{-1}(4))}$$

$$\left(f^{-1}\right)'(4) = \frac{1}{f'(1)}$$

$$\left(f^{-1}\right)'(4) = \frac{1}{7}$$

Hence, **(A)** is correct. You could also use the calculator to evaluate the derivative of the function at a specific point, provided you remember to take the reciprocal to get the final answer, as the formula requires. (B) uses $f(1)$ in the denominator instead of $f'(1)$. (C) is simply the x-coordinate where you are solving for the derivative. (D) ignores the numerator in the formula.

77. C

The new position can be found by adding the integral of the velocity function from $t = 0$ to $t = 4$ to the initial starting position of $x = -5$. Let $x(t)$ be the position function.

$$x(t_2) = x(t_1) + \int_{t_1}^{t_2} v(t)\, dt$$

$$x(4) = x(0) + \int_0^4 6t^2 - 3t + 7\, dt$$

Using your calculator, $x(4) = 127$.

Thus, **(C)** is correct. If you mistakenly take the derivative of the velocity function and integrate the acceleration function, $x(4) = x(0) + \int_0^4 a(t)\, dt$, you get (A). (B) is $v(4)$, which is not the new position at $t = 4$. If you do not add the -5 to the integral, you get (D), which is the distance traveled but not the new position at $t = 4$.

78. D

To write the equation of the tangent line, you need the x- and y-coordinates as well as the slope. Because this is calculator active, do not forget you can compute all of these decimals on the calculator, which may be more efficient than doing so by hand. In order to find the x- and y-coordinates simply put $t = \dfrac{\pi}{3}$ into $x(t) = 1 + 2\sin(t)$ and $y(t) = 3 - 4\cos(t)$. This produces the coordinate pair (2.732, 1). The slope or $\dfrac{dy}{dx}$ is found by dividing $\dfrac{dy}{dt}$ by $\dfrac{dx}{dt}$, which is $\dfrac{dy}{dx} = \dfrac{3.464}{1}$. Substituting into the point-slope formula for the equation of a line yields $y - 1 = 3.464(x - 2.732)$.

Therefore, **(D)** is correct. Note that your calculator must be set in radian mode to calculate the values correctly. If left in degree mode, you will get $y + 0.999 = 0.037(x - 1.037)$. (A) uses $t = \dfrac{\pi}{6}$ or occurs if you incorrectly switch the coordinates for $t = \dfrac{\pi}{3}$. (B) is the result if you ignore the constants in $x(t) = 1 + 2\sin(t)$ and $y(t) = 3 - 4\cos(t)$ and therefore mistakenly get (0.866, 0.5) as a coordinate. (C) has the correct coordinates but the incorrect slope calculated from $\dfrac{dx}{dy}$ instead of from $\dfrac{dy}{dx}$.

79. C

Euler's method requires the use of the slope formula to find the change in y at $x = -0.3$.

$$\Delta y = \Delta x \frac{dy}{dx}$$
$$\Delta y = 0.3\,(0.6)$$
$$\Delta y = 0.18$$

This $\Delta y = 0.18$ is added to the y-coordinate to get the new point, (0, 7.18). The next coordinate is reached by calculating the Δy at $x = 0$.

$$\Delta y = 0.3\,(1.7)$$
$$\Delta y = 0.51$$

Then add $\Delta y = 0.51$ to (0, 7.18) to get the new point, (0.3, 7.69). The next and final coordinate is reached by calculating the Δy at $x = 0.3$.

$$\Delta y = 0.3\,(2.1)$$
$$\Delta y = 0.63$$

Then add $\Delta y = 0.63$ to (0.3, 7.69) to get the new point, (0.6, 8.32). So the approximation to $f(0.6)$ using Euler's method is 8.32.

Hence, **(C)** is correct. (A) is $f(0)$, and (B) is $f(0.3)$. (D) generalizes the three steps. It assumes that Δy would be 1.8 and incorrectly adds that value to 7 to get 8.8.

80. D

When given the option, you should use the calculator. First, though, consider what needs to be graphed. If you graph the given function you can approximate where concavity changes. However, when two options are very close, you will not know which one to pick. You have a couple of options for this question. You could graph the derivative of the given function on the interval and find the x-value of the turning point using the calculator. Another option would be to graph the second derivative on the interval and find the x-intercept using the calculator. Both options will give you the correct answer, $x = -0.819$. Don't forget to use the chain rule when taking the derivative of the trigonometric term.

$$y = x^3 + 3x^2 + 5x - 4\cos(2x)$$
$$y' = 3x^2 + 6x + 5 + 8\sin(2x)$$
$$y'' = 6x + 6 + 16\cos(2x)$$

Hence, **(D)** is correct. (B) is the zero of the first derivative, but that represents a maximum on the original function.

81. C

Instead of attempting to separate the variables and integrating both sides, you can follow the pattern of the slope field starting at the initial condition. The slope field shows a horizontal asymptote at both $y = 1$ and $y = -1$. Because the initial condition has a y-value greater than 1, the particular solution must approach the horizontal asymptote $y = 1$ as x approaches infinity. If the initial condition had a y-value less than -1, then the limit would be -1 as x approaches infinity.

82. B

Because the cross sections are squares, the definite integral is $\int_a^b (\text{base})^2\, dx$. The lower and upper limits are defined as $x = 0$ and $x = \frac{\pi}{2}$. Because the function $y = \sin x$ is never greater than 1, you know that $y = 1$ is the top of the region. This makes the base of each square $1 - \sin x$. Substitute this along with the limits into the definite integral, and evaluate using a calculator.

$$V = \int_0^{\frac{\pi}{2}} (1 - \sin x)^2\, dx = 0.356$$

Therefore, **(B)** is correct. You get (C) if you forget to square the base when entering the integral on your calculator.

83. D

In order to answer this question, you need to find the coordinates of the point of inflection. Points of inflection occur when the second derivative changes signs. The point of inflection occurs at (−0.222, 8.843). You can find this using either by using your calculator or by setting the second derivative equal to 0 and confirming a sign change in that value's vicinity. Because $f'(-0.2220) = -8.444$ plugging into $y - y_1 = m(x - x_1)$ produces $y - 8.843 = -8.444(x + 0.222)$.

Thus, **(D)** is correct. If you solve for y incorrectly, you get (C), which has the incorrect sign in the parentheses before placing the equation into slope-intercept form.

84. C

Graph the given functions to visualize which function is on top and which is on the bottom. The graph also allows you to locate the x-value where $y = \sin x$ and $y = -x + 3$ intersect, which is the upper bound of the definite integral. Because $\sin x = -x + 3$ at $x = 2.180$ and the linear function is the top function, the definite integral is

$$\int_0^{2.180} (-x + 3 - \sin x)\, dx.$$ Once again, there is no reason to take the antiderivative because the calculator can evaluate this definite integral for you. The area is 2.592, which is **(C)**.

(A) is just the area under the sine function. (B) is the x-coordinate of the point of intersection, which needs to be used as a bound. (D) is the area under the linear function on the same interval.

85. C

To find the area between the curve and the x-axis, use the formula Area $= \dfrac{1}{2}\int_a^b (r)^2\, d\theta$, where a and b are values of θ.

Here the area is found using Area $= \dfrac{1}{2}\int_0^\pi (3\theta + \sin(\theta))^2\, d\theta$.

The upper and lower bounds are found by the portion of the graph shown. Use the calculator to evaluate the equation.

$$\text{Area} = \frac{1}{2}\int_\theta^\pi (3\theta + \sin(\theta))^2\, d\theta$$

$$\text{Area} = 56.720$$

Hence, **(C)** is correct. Another method of using the calculator is to type the function into r and then on the home screen type.

$$.5* \int_0^\pi (r1^2)\, d\theta$$

$$56.71959114$$

Be sure your calculator is in polar and radian mode.

(A) is obtained if you place the squared sign within the integral on only the sine function, for example Area $= \dfrac{1}{2}\int_0^\pi (3\theta + (\sin(\theta))^2)\, d\theta = 8.188$. (B) is the result when your calculator is in degree mode instead of radian mode. (D) neglects the necessary $\dfrac{1}{2}$ in the formula.

86. D

In order for the tangent line to the graph to be horizontal, its derivative must equal 0. For $\dfrac{dy}{dx} = 0$, $\dfrac{dy}{d\theta} = 0$. There are several approaches, especially because the calculator is active for this particular question. This relation is a limacon with an inner loop because the constant is less than the coefficient of the cosine function.

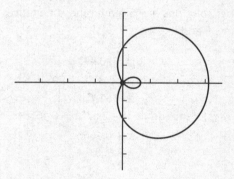

There are 4 points where the tangent is horizontal. Because $y = r \sin \theta$, replace r with the given equation, $y = (2 + 3 \cos \theta)(\sin \theta)$. To find $\dfrac{dy}{d\theta}$, use the product rule.

$$\frac{dy}{d\theta} = (2 + 3\cos \theta)(\cos \theta) + (\sin \theta)(-3\sin \theta)$$

$$\frac{dy}{d\theta} = 3\cos^2 \theta + 2\cos \theta - 3\sin^2 \theta$$

Replace $\sin^2 \theta$ with $1 - \cos^2 \theta$ and simplify.

$$\frac{dy}{d\theta} = 3\cos^2 \theta + 2\cos \theta - 3(1 - \cos^2 \theta)$$

$$\frac{dy}{d\theta} = 6\cos^2 \theta + 2\cos \theta - 3$$

Set $\dfrac{dy}{d\theta} = 0$ and use the quadratic formula to solve.

$$6\cos^2\theta + 2\cos\theta - 3 = 0$$

$$\cos\theta = \dfrac{-2 \pm \sqrt{(2)^2 - 4(6)(-3)}}{2(6)}$$

$$\cos\theta = \dfrac{-2 \pm \sqrt{76}}{12}$$

When cosine is positive, θ is in quadrants I and IV.

$$\cos(\theta) = 0.5598$$

$$\cos^{-1}(\cos(\theta)) = \cos^{-1}(0.5598)$$

$$\theta = 0.9766 \text{ QI}$$

$$QIV \ 2\pi - 0.9766$$

$$\theta = 5.3066$$

The other solutions are when cosine is negative; θ is in quadrants II and III.

$$\cos(\theta) = -0.8931$$

$$\cos^{-1}(\cos(\theta)) = \cos^{-1}(-0.8931)$$

$$\theta = 2.6751 \text{ QII}$$

$$\text{reference angle } \pi - 2.6751 = 0.4666$$

$$QIII \ \pi + 0.4666$$

$$\theta = 3.6081$$

It is important to use inverse cosine with the solutions to find the reference angle and then find θ in the appropriate quadrants as shown. Lastly, to find the coordinates in polar form, (r, θ) simply plug the θ value into the original equation $r = 2 + 3\cos(\theta)$. Thus, **(D)** is correct.

(A), (B), and (C) are all points at which the tangent line is horizontal. The horizontal tangent line associated with the θ value of (D) occurs at the polar coordinate $(-0.680, 3.608)$.

87. B

A picture or sketch may be helpful, so graph in polar mode with an appropriate window. This curve is a rose with four petals as shown below.

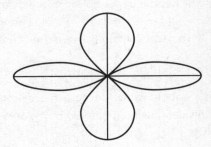

Due to the symmetry of the graph, the area of the petals in any two quadrants can be found in multiple ways. One way is to find the area of the first half of a petal in the first quadrant and then multiply that area by 4. As shown in the picture, any two quadrants have a total of two petals. The endpoints of the first half of a petal are $\theta = 0$ and $\dfrac{\pi}{4}$. Calculate the area of that first half.

$$\text{Area} = \dfrac{1}{2}\int_0^{\frac{\pi}{4}} (\cos(2\theta))^2 \, d\theta$$

$$\text{Area} = 0.196$$

Multiply this area by 4 to get the area of the petals in two quadrants: $(4)(0.196) = 0.785$.

Therefore, **(B)** is correct. (A) is the area of one complete petal, but not the sum in two quadrants. (C) is found if you multiply the half petal by 8, and (D) occurs if you forget the exponent in the area formula.

88. C

Acceleration is the first derivative of the velocity function. Remember to use your calculator to find the numerical derivative at $t = 5$. The result is 191.206. Do not attempt to calculate this derivative by hand. This is a calculator question, and you should use the calculator to your advantage.

Hence, **(C)** is correct. (A) is the result if you take the derivative incorrectly by hand, ignoring the product rule and substituting $t = 5$. (B) is the second derivative of the velocity function. If you were given the position function, then this would be the correct answer looking for acceleration. However, you were given the velocity function, so only one derivative is necessary. (D) is the velocity at $t = 5$.

89. C

This area is found by the integral $\int_{0}^{4.153} (\sin(x) + 5 - x) \, dx$.

To find the upper bound, you must determine the point of intersection of the two functions. To find the enclosed area, use 0 for the lower bound. Also use $\sin(x) + 5$ as the top function and x as the bottom function. The correct solution is found by subtracting the two areas formed by these functions. Be sure your calculator is set to radian mode to answer this question.

$$\int_{0}^{4.153} (\sin(x) + 5 - x) \, dx = 13.672$$

Thus, **(C)** is correct. (A) is the x-coordinate of the intersection point, not the area enclosed. (B) is the value of $\int_{0}^{4.153} x \, dx$, and (D) is the value of $\int_{0}^{4.153} (\sin(x) + 5) \, dx$.

90. A

You must use your calculator to solve this correctly. First, determine the y-coordinates of the function. Then find the slope, m, of the secant line.

$$f(-1) = -1.476$$
$$f(2) = 10.877$$

$$m = \frac{10.877 - (-1.476)}{2 - (-1)}$$

$$m = 4.118$$

To find the value(s) of c that satisfy the mean value theorem, graph the derivative and the function $y = 4.118$. Then find the intersection points. These graphs intersect at 0.295 and 1.846. Both of these values are within the domain and satisfy the conclusion of the mean value theorem.

Therefore, **(A)** is correct. (B) is the result if you find the intersection of 4.118 with the original function. This is not the conclusion of the mean value theorem. (C) and (D) occur if you set the second derivative equal to 4.118. You should notice the second solution of (D) is not even in the interval $[-1, 2]$ and therefore should not be considered.

SECTION II, Part A

1. (a)

Solutions	Scoring Rubric
To find the total number of people that enter the stadium in the given time interval, integrate the rate from $t = 0$ to $t = 30$. $$\int_0^{30} 62 + 5\sin\left(\frac{t}{3}\right) dt = 1{,}887.586073$$ $$\int_0^{30} 62 + 5\sin\left(\frac{t}{3}\right) dt \approx 1{,}888 \text{ people}$$	1: set up integral correctly 1: 1,888 people

(b)

Solutions	Scoring Rubric
$F'(8) = -1.482$ Because $F'(8) < 0$, flow of fans is decreasing at $t = 8$.	1: $F'(8) = -1.482$ 1: $t = 8$

(c)

Solutions	Scoring Rubric
$$\frac{1}{25-10}\int_{10}^{25} 62 + 5\sin\left(\frac{t}{3}\right) dt = 61.47953003$$ $$\frac{1}{15}\int_{10}^{25} 62 + 5\sin\left(\frac{t}{3}\right) dt \approx 61.480 \text{ people per minute}$$	1: set up integral correctly 1: 61.480 1: correct units of measure

(d)

Solutions	Scoring Rubric
$$\frac{F(25) - F(10)}{25 - 10} = 0.3593 \text{ people per min}^2$$	1: 0.3593 1: correct units of measure

2. (a)

Solutions	Scoring Rubric
Before you can find the area of region R or region S, you need to know the missing point of intersection. This is found by graphing each of the functions and using the intersect feature of the calculator. The intersection point is (2.2, 5.328). For region R, $f(x)$ is the top function and $g(x)$ is the bottom function. For region S, $g(x)$ is the top function and $f(x)$ is the bottom function. The sum of the areas of the two regions can be found with the following: $$\int_{1}^{2.2}\left(f(x)-g(x)\right)dx + \int_{2.2}^{5}\left(g(x)-f(x)\right)dx$$ Use the calculator to evaluate these definite integrals to get 10.492.	1: intersection (2.2, 5.328) 1: correct integral set up 1: 10.492

(b)

Solutions	Scoring Rubric
Each side of the equilateral triangle has a length of $f(x) - g(x)$. You just need to determine the appropriate area formula to integrate. Sketch an equilateral triangle. 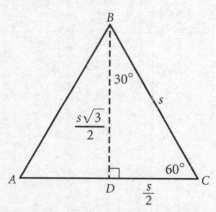 When a height is drawn, it cuts the base in half and creates two 30-60-90 triangles. This should be a familiar triangle from trigonometry. It makes the height of the equilateral triangle equal to $\frac{3}{2}$ times the base. From the diagram above, you can see the appropriate area formula is $A = \frac{s}{2}\left(\frac{\sqrt{3}s}{2}\right) = \frac{\sqrt{3}s^2}{4}$. To find the volume of the solid, evaluate the definite integral of the area formula. $$V = \frac{\sqrt{3}}{4}\int_{1}^{2.2}\left(f(x)-g(x)\right)^2 dx$$	1: correct work shown/ drawing figure 1: correct Area formula $$V = \frac{\sqrt{3}s^2}{4}$$ 1: $$V = \frac{\sqrt{3}}{4}\int_{1}^{2.2}\left(f(x)-g(x)\right)^2 dx$$

(c)

Solutions	Scoring Rubric
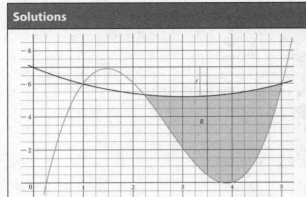	1: correct work shown/drawing figure 1: use the correct integral formula $$\pi \int_a^b \left(R^2 - r^2\right) dx$$ 1: $$V = \pi \int_{2.2}^5 \left[\left(7 - f(x)\right)^2 - \left(7 - g(x)\right)^2\right] dx$$

The cross sections of the solid are washers, so the integral is in the form $\pi \int_a^b \left(R^2 - r^2\right) dx$. R is defined by $7 - f(x)$, and r is defined by $7 - g(x)$. Recall that the radius values for the disk or washer method depend on the distance from the axis of rotation to the function value. Visuals can be very helpful in determining the appropriate relationship to use. Plugging in this information along with the limits found in the previous questions gives the definite integral.

$$V = \pi \int_{2.2}^5 \left[\left(7 - f(x)\right)^2 - \left(7 - g(x)\right)^2\right] dx$$

SECTION II, Part B

3. (a)

Solutions	Scoring Rubric
To find the area of a polar curve, use the formula $A = \frac{1}{2}\int_a^b (r)^2 \, d\theta$. The enclosed section in the first quadrant for the lower and upper bounds are found when $\theta = 0$ and when $r = 0$. $$2\cos\theta - \sqrt{2} = 0$$ $$2\cos\theta = \sqrt{2}$$ $$\cos\theta = \frac{\sqrt{2}}{2}$$ $$\theta = \frac{\pi}{4}$$ Thus, $A = \frac{1}{2}\int_0^{\frac{\pi}{2}}\left(2\cos\theta - \sqrt{2}\right)^2 \, d\theta$	1: use correct formula, $A = \frac{1}{2}\int_a^b (r)^2 \, d\theta$ 1: $\theta = \frac{\tau}{4}$ 1: $A = \frac{1}{2}\int_0^{\frac{\pi}{2}}\left(2\cos\theta - \sqrt{2}\right)^2 \, d\theta$

(b)

Solutions	Scoring Rubric
Use the fact that $$x = r\cos\theta$$ $$r = 2\cos\theta - \sqrt{2}$$ to solve for $\frac{dx}{d\theta}$. $\frac{dr}{d\theta} = -2\sin\theta$ $\frac{dx}{d\theta} = r\sin\theta + \cos\theta\frac{dr}{d\theta}$ $\frac{dx}{d\theta} = \left(2\cos\theta - \sqrt{2}\right)\sin\theta + \cos\theta(-2\sin\theta)$ $\frac{dx}{d\theta} = 2\cos\theta\sin\theta - \sqrt{2}\sin\theta - 2\cos\theta\sin\theta$ $\frac{dx}{d\theta} = -\sqrt{2}\sin\theta$ Use the fact that $$y = r\sin\theta$$ $$r = 2\cos\theta - \sqrt{2}$$ to solve for $\frac{dy}{d\theta}$. $\frac{dr}{d\theta} = -2\sin\theta$ $\frac{dy}{d\theta} = r\cos\theta + \sin\theta\frac{dr}{d\theta}$ $\frac{dy}{d\theta} = \left(2\cos\theta - \sqrt{2}\right)\cos\theta + \sin\theta(-2\sin\theta)$ $\frac{dy}{d\theta} = 2\cos^2\theta - \sqrt{2}\cos\theta - 2\sin^2\theta$	1: $\frac{dr}{d\theta} = -2\sin\theta$ 1: $\frac{dx}{d\theta} = -\sqrt{2}\sin\theta$ 1: $\frac{dy}{d\theta} = 2\cos^2\theta - \sqrt{2}\cos\theta - 2\sin^2\theta$

(c)

Solutions	Scoring Rubric
In order to write the equation of the tangent line, find the x-coordinate and the y-coordinate using $x = r \cos \theta$ and $y = r \sin \theta$, and find the slope using $\frac{dy}{dx} = \frac{\frac{dy}{d\theta}}{\frac{dx}{d\theta}}$.	1: $\left(0, -\sqrt{2}\right)$ 1: $m = \dfrac{2}{\sqrt{2}}$ 1: $y + \sqrt{2} = \dfrac{2}{\sqrt{2}} x$

$$x = r \cos \theta \qquad\qquad y = r \sin \theta$$
$$x = \left(2 \cos \theta - \sqrt{2}\right)(\cos \theta) \qquad y = \left(2 \cos \theta - \sqrt{2}\right)(\sin \theta)$$
$$x = \left(2 \cos\left(\frac{\pi}{2}\right) - \sqrt{2}\right)\left(\cos\left(\frac{\pi}{2}\right)\right) \qquad y = \left(2 \cos\left(\frac{\pi}{2}\right) - \sqrt{2}\right)\left(\sin\left(\frac{\pi}{2}\right)\right)$$
$$x = 0 \qquad\qquad y = -\sqrt{2}$$

Therefore:

$$\frac{dy}{dx} = \frac{\frac{dy}{d\theta}}{\frac{dx}{d\theta}}$$

$$\frac{dy}{dx} = \frac{2 \cos^2 \theta - \sqrt{2} \cos \theta - 2 \sin^2 \theta}{-\sqrt{2} \sin \theta}$$

$$\left.\frac{dy}{dx}\right|_{\theta = \frac{\pi}{2}} = \frac{2 \cos^2\left(\frac{\pi}{2}\right) - \sqrt{2} \cos\left(\frac{\pi}{2}\right) - 2 \sin^2\left(\frac{\pi}{2}\right)}{-\sqrt{2} \sin\left(\frac{\pi}{2}\right)}$$

$$\left.\frac{dy}{dx}\right|_{\theta = \frac{\pi}{2}} = \frac{2}{\sqrt{2}}$$

Then plug into the point slope formula for the equation of a line.

$$y - y_1 = m(x - x_1)$$
$$y - \left(-\sqrt{2}\right) = \frac{2}{\sqrt{2}}(x - 0)$$
$$y + \sqrt{2} = \frac{2}{\sqrt{2}} x$$

4. (a)

Solutions	Scoring Rubric
$g(0)$ is found by evaluating the integral between $x = -4$ and $x = 0$. Because the shape between the function and the x-axis makes a triangle, simply find the area of the triangle using the area formula, $A = \frac{1}{2}bh$. The area is 8 units, but because the function is below the x-axis, the value of integral is -8. To find $g'(0)$, $g(x) = f(t)$. Because $f(0) = -4$, $g'(0) = -4$. To find $g''(0)$, $g''(x) = f'(t)$. Because $f(t)$ is not differentiable at $x = 0$, $g''(0)$ does not exist.	1: $g(0) = -8$ 1: $g'(0) = -4$ 1: $g''(0)$ does not exist.

(b)

Solutions	Scoring Rubric
On those intervals, $g'(x) < 0$ and $g''(x) < 0$. Using the second fundamental theorem of calculus, $g'(x) = f(x)$. That means the function g is decreasing whenever $f(x) < 0$. From the graph, that means g is decreasing on the intervals $-6 < x < -4$ and $-4 < x < 4$. The second derivative is necessary for determining concavity, $g''(x) = f'(x)$. This tells you that g is concave up whenever $f'(x) > 0$ or when $f(x)$ is increasing. From the graph, that occurs on the intervals $-6 < x < -4$ and $0 < x < 6$. That means the intervals where g is both decreasing and concave up must be $-6 < x < -4$ and $0 < x < 4$.	1: $-6 < x < -4$ 1: $0 < x < 4$ 1: justification

(c)

Solutions	Scoring Rubric
Using the product rule, $h'(x) = 3g(x) + 3xg'(x)$ and $h'(2) = 3g(2) + 3(2)g'(2)$. To evaluate $g(2)$, calculate the area under f on the closed interval $[-4, 2]$. This can be done by breaking the region into a triangle on the interval $[-4, 0]$ and into a trapezoid on the interval $[0, 2]$ and then adding the areas together. Alternatively, you can find the area of the triangle on the interval $[-4, 4]$ and subtract the area of the triangle on the interval $[2, 4]$. Either technique gives you $g(2) = \int_{-4}^{2} f(t)\, dt = -14$. Use the second fundamental theorem of calculus $g'(x) = f(x)$. From the graph, $g'(2) = f(2) = -2$. Substituting these values makes the derivative $h'(2) = 3(-14) + 3(2)(-2)$, which simplifies to $h'(2) = -54$.	1: $g(2) = -14$ 1: $g'(2) = -2$ 1: $h'(2) = -54$

5.　(a)

Solutions	Scoring Rubric
This question provides an inflection point, which comes from the second derivative. It also provides a slope, which comes from the first derivative. Start by finding those derivative functions. $$f(x) = ax^3 + bx^2 + cx$$ $$f'(x) = 3ax^2 + 2bx + c$$ $$f''(x) = 6ax + 2b$$ At the point (2, 6), there is an inflection point. This means the second derivative must equal 0 when 2 is substituted for x. This information allows you to solve for b in terms of a. $$f''(x) = 6ax + 2b$$ $$0 = 6a(2) + 2b$$ $$0 = 12a + 2b$$ $$b = -6a$$	1: $f'(x) = 3ax^2 + 2bx + c$ 1: $f''(x) = 6ax + 2b$ 1: reasoning on how to use inflection point (2, 6) 1: $b = -6a$

(b)

Solutions	Scoring Rubric
The slope of the tangent line at (2, 6) is -1. This means the first derivative must equal -1 when 2 is substituted for x. You can also substitute $-6a$ for b, allowing you to solve for c in terms of a. $$f'(x) = 3ax^2 + 2bx + c$$ $$-1 = 3a(2)^2 + 2(-6a)(2) + c$$ $$-1 = 12a - 24a + c$$ $$-1 = -12a + c$$ $$c = -1 + 12a$$	1: reasoning on how to use slope of tangent line 1: $c = -1 + 12a$

(c)

Solutions	Scoring Rubric
Now that b and c are both solved in terms of a, the original function can be evaluated using the point $(2, 6)$ to solve for a. $$f(x) = ax^3 + bx^2 + cx$$ $$6 = a(2)^3 + (-6a)(2)^2 + (-1+12a)(2)$$ $$6 = 8a - 24a - 2 + 24a$$ $$6 = 8a - 2$$ $$8 = 8a$$ $$a = 1$$ Once you have a, you can solve for both b and c. $$b = -6a$$ $$b = -6(1)$$ $$b = -6$$ $$c = -1 + 12a$$ $$c = -1 + 12(1)$$ $$c = 11$$	1: $a = 1$ 1: $b = -6$ 1: $c = 11$

6. (a)

Solutions	Scoring Rubric
To find the first four non zero terms, replace the x's in the Maclaurin series for $\ln(1 + x)$ with x^4 and simplify $$x - \frac{x^2}{2} + \frac{x^3}{3} - \frac{x^4}{4} + \cdots + (-1)^{n+1}\frac{x^n}{n} + \cdots$$ $$\left(x^4\right) - \frac{\left(x^4\right)^2}{2} + \frac{\left(x^4\right)^3}{3} - \frac{\left(x^4\right)^4}{4} + \cdots + (-1)^{n+1}\frac{\left(x^4\right)^n}{n} + \cdots$$ $$x^4 - \frac{x^8}{2} + \frac{x^{12}}{3} - \frac{x^{16}}{4} + \cdots + (-1)^{n+1}\frac{x^{4n}}{n} + \cdots$$	1: substitute in x^4 correctly 1: $x^4 - \frac{x^8}{2} + \frac{x^{12}}{3} - \frac{x^{16}}{4} + \cdots$ $+ (-1)^{n+1}\frac{x^{4n}}{n} + \cdots$

(b)

Solutions	Scoring Rubric
Test each of these points, $x = -1$ and $x = 1$, for convergence. Test at $x = -1$. $$\sum (-1)^{n+1} \frac{x^{4n}}{n}$$ $$\sum (-1)^{n+1} \frac{(-1)^{4n}}{n}$$ $$\sum (-1)^{n+1} \frac{\left((-1)^4\right)^n}{n}$$ $$\sum (-1)^{n+1} \frac{1}{n}$$ This is an alternating harmonic series that converges. Now test at $x = 1$. $$\sum (-1)^{n+1} \frac{x^{4n}}{n}$$ $$\sum (-1)^{n+1} \frac{(1)^{4n}}{n}$$ $$\sum (-1)^{n+1} \frac{\left((1)^4\right)^n}{n}$$ $$\sum (-1)^{n+1} \frac{1}{n}$$ This is an alternating harmonic series that converges. The interval of convergence is therefore $-1 \leq x \leq 1$.	1: work shown for testing at point $x = -1$ 1: work shown for testing at point $x = 1$ 1: the interval of convergence is therefore $-1 \leq x \leq 1$.

(c)

Solutions	Scoring Rubric
The Maclaurin series for $f'(x)$, $f'(t^3)$, and $g(x)$ are shown below: $$f'(x): \sum_{n=1}^{\infty}(-1)^{n+1}4 \cdot x^{4n-1} = 4x^3 - 4x^7 + 4x^{11} - 4x^{15} - \cdots$$ $$f'(t^3): \sum_{n=1}^{\infty}(-1)^{n+1}4 \cdot t^{12n-3} = 4t^9 - 4t^{21} + 4t^{33} - 4t^{45} + \cdots$$ $$g(x): \sum_{n=1}^{\infty}(-1)^{n+1}\frac{4}{12n-2} \cdot t^{12n-2} = \frac{4}{10}t^{10} - \frac{4}{22}t^{22} + \frac{4}{34}t^{34} - \frac{4}{46}t^{46} + \cdots$$ Then $$g(1) \approx \frac{4}{10}(1)^{10} - \frac{4}{22}(1)^{22}$$ $$g(1) \approx \frac{4}{10} - \frac{4}{22}$$ $$g(1) \approx \frac{88}{220} - \frac{40}{220}$$ $$g(1) \approx \frac{48}{220}$$	1: correctly solve for $f'(t^3)$ 1: $g(1) \approx \dfrac{48}{220}$ or something equivalent

(d)

Solutions	Scoring Rubric
The Maclaurin Series for g, evaluated at $x = 1$ is an alternating series with individual terms that decrease in absolute value to 0. The error will always be less than the first omitted term, in this case the third term. $$\left| g(1) - \frac{48}{220} \right| < \frac{4}{34}(1)^{34}$$ $$\left| g(1) - \frac{48}{220} \right| < \frac{4}{34}$$ $$\frac{4}{34} < \frac{25}{170}$$ $$\left| g(1) - \frac{48}{220} \right| < \frac{25}{170}$$	1: deduce that the first omitted term is the third term 1: $\left| g(1) - \dfrac{48}{220} \right| < \dfrac{25}{170}$

BC Practice Exam 5

Section I, Part A

1 Ⓐ Ⓑ Ⓒ Ⓓ
2 Ⓐ Ⓑ Ⓒ Ⓓ
3 Ⓐ Ⓑ Ⓒ Ⓓ
4 Ⓐ Ⓑ Ⓒ Ⓓ
5 Ⓐ Ⓑ Ⓒ Ⓓ
6 Ⓐ Ⓑ Ⓒ Ⓓ
7 Ⓐ Ⓑ Ⓒ Ⓓ
8 Ⓐ Ⓑ Ⓒ Ⓓ
9 Ⓐ Ⓑ Ⓒ Ⓓ
10 Ⓐ Ⓑ Ⓒ Ⓓ
11 Ⓐ Ⓑ Ⓒ Ⓓ
12 Ⓐ Ⓑ Ⓒ Ⓓ
13 Ⓐ Ⓑ Ⓒ Ⓓ
14 Ⓐ Ⓑ Ⓒ Ⓓ
15 Ⓐ Ⓑ Ⓒ Ⓓ
16 Ⓐ Ⓑ Ⓒ Ⓓ
17 Ⓐ Ⓑ Ⓒ Ⓓ
18 Ⓐ Ⓑ Ⓒ Ⓓ
19 Ⓐ Ⓑ Ⓒ Ⓓ
20 Ⓐ Ⓑ Ⓒ Ⓓ
21 Ⓐ Ⓑ Ⓒ Ⓓ
22 Ⓐ Ⓑ Ⓒ Ⓓ
23 Ⓐ Ⓑ Ⓒ Ⓓ
24 Ⓐ Ⓑ Ⓒ Ⓓ
25 Ⓐ Ⓑ Ⓒ Ⓓ
26 Ⓐ Ⓑ Ⓒ Ⓓ
27 Ⓐ Ⓑ Ⓒ Ⓓ
28 Ⓐ Ⓑ Ⓒ Ⓓ
29 Ⓐ Ⓑ Ⓒ Ⓓ
30 Ⓐ Ⓑ Ⓒ Ⓓ

Section I, Part B

76 Ⓐ Ⓑ Ⓒ Ⓓ
77 Ⓐ Ⓑ Ⓒ Ⓓ
78 Ⓐ Ⓑ Ⓒ Ⓓ
79 Ⓐ Ⓑ Ⓒ Ⓓ
80 Ⓐ Ⓑ Ⓒ Ⓓ
81 Ⓐ Ⓑ Ⓒ Ⓓ
82 Ⓐ Ⓑ Ⓒ Ⓓ
83 Ⓐ Ⓑ Ⓒ Ⓓ
84 Ⓐ Ⓑ Ⓒ Ⓓ
85 Ⓐ Ⓑ Ⓒ Ⓓ
86 Ⓐ Ⓑ Ⓒ Ⓓ
87 Ⓐ Ⓑ Ⓒ Ⓓ
88 Ⓐ Ⓑ Ⓒ Ⓓ
89 Ⓐ Ⓑ Ⓒ Ⓓ
90 Ⓐ Ⓑ Ⓒ Ⓓ

A A A A A A A A A A A A A A A A A A A A A A A A A A A A A A A A A A A A A

CALCULUS BC
SECTION I, Part A
Time—60 minutes
Number of questions—30

A CALCULATOR MAY NOT BE USED ON THIS PART OF THE EXAM.

Directions: Use the space provided to solve each of the following problems. After examining the given choices, decide which of the choices is the best and fill in the corresponding circle on the answer sheet. Credit will only be given for answers appropriately marked on the answer sheet, and not for anything written in the exam book. Do not spend too much time on any one problem.

In this exam:

(1) The domain of a function f should be assumed to be the set of all real numbers x for which $f(x)$ is a real number, unless the question indicates otherwise.

(2) The inverse function notation f^{-1} or the prefix "arc" (e.g., $\sin^{-1} x = \arcsin x$) may be used to indicate the inverse of a trigonometric function.

GO ON TO THE NEXT PAGE.

A A A A A A A A A A A A A A A A A A A A A A A A A A A A A A A A A

1. Given $f(x) = \begin{cases} \sin(7x) & x \le \dfrac{\pi}{6} \\ \cos(ax) & x > \dfrac{\pi}{6} \end{cases}$

   For which value of a does $\lim\limits_{x \to \frac{x}{6}} f(x)$ exist?

   (A) 2

   (B) 4

   (C) 5

   (D) 10

2. The infinite series $\sum\limits_{k=3}^{\infty} a_k$ has an nth partial sum $S_n = \dfrac{n^3 + 7}{n^2 - 4}$ for $n \ge 3$. What is the sum of the first 5 terms of a_k?

   (A) $-\dfrac{7}{4}$

   (B) $\dfrac{255}{441}$

   (C) 1

   (D) $\dfrac{132}{21}$

GO ON TO THE NEXT PAGE.

AAAAAAAAAAAAAAAAAAAAAAAAAAAAAAAAAAAAAA

3. The Taylor series for a function f about $x = 0$ is given by $\sum_{n=1}^{\infty} \frac{(-1)^n}{(3n-1)!} x^{2n+1}$ and converges for all real numbers of x. If the 5th-degree Taylor polynomial for f about $x = 0$ is used to approximate $f\left(\frac{1}{2}\right)$, what is the alternating series error bound?

(A) $\dfrac{-1}{2^7 \cdot 8!}$

(B) $\dfrac{1}{2^{13} \cdot 17!}$

(C) $\dfrac{1}{2^7 \cdot 8!}$

(D) $\dfrac{1}{2 \cdot 8!}$

---

4. $\displaystyle\lim_{x \to -\infty} \frac{x^5 - 4x^4 - 2x + 10}{-2x^2 - x - 6} =$

(A) $-\dfrac{5}{3}$

(B) $-\dfrac{1}{2}$

(C) 0

(D) nonexistent

GO ON TO THE NEXT PAGE.

AAAAAAAAAAAAAAAAAAAAAAAAAAAAAAAAAAAA

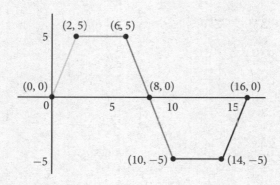

5. The function $g(x)$ is shown above. Put these in order from least to greatest.

I. $\displaystyle\int_{8}^{16} g(x)\, dx$

II. $\displaystyle\int_{0}^{16} g(x)\, dx$

III. the area between $g(x)$ and the x-axis bound by $x = 10$ and $x = 16$

IV. the area between $g(x)$ and the x-axis bound by $x = 8$ and $x = 16$

(A) I, II, III, IV

(B) IV, III, II, I

(C) II, III, I, IV

(D) not enough information

GO ON TO THE NEXT PAGE.

A A A A A A A A A A A A A A A A A A A A A A A A A A A A A A A A A A A A A A

6. Which of the following converge(s)?

   I. $\sum_{n=1}^{\infty} \dfrac{|\cos(n)|}{n^3}$

   II. $\sum_{n=1}^{\infty} 8e^{-n}$

   III. $\sum_{n=1}^{\infty} \dfrac{(3n+7)(n-1)}{4n+3n^2-7}$

   (A) I only

   (B) II only

   (C) I and II only

   (D) I, II, and III

7. A 12-foot ladder slides down the side of a building at a rate of 3 feet per second. At what rate is the angle formed by the ground and ladder changing when the top of the ladder is 8 feet from the ground?

   (A) $-\dfrac{\sqrt{5}}{12}$

   (B) $-\dfrac{3}{4\sqrt{5}}$

   (C) $-\dfrac{3}{8}$

   (D) $-\dfrac{2}{3}$

GO ON TO THE NEXT PAGE.

A A A A A A A A A A A A A A A A A A A A A A A A A A A A A A A A A A A A

8. At time $t \geq 0$, a particle moving along the xy-plane has a velocity vector given by $v(t) = \langle \sin(3t), e^{4t} \rangle$. What is the acceleration vector of the particle?

(A) $\langle 3\cos(3t), 4e^{4t} \rangle$

(B) $\langle -3\cos(3t), 4e^{4t} \rangle$

(C) $\langle 3\cos(3t), e^{4t} \rangle$

(D) $\langle -3\cos(3t), e^{4t} \rangle$

9. $\displaystyle\int_0^{\frac{1}{2}} \frac{dx}{\sqrt{1-x^2}}\, dx =$

(A) $-\dfrac{\pi}{6}$

(B) 0

(C) $\dfrac{\pi}{6}$

(D) $\dfrac{\pi}{3}$

GO ON TO THE NEXT PAGE.

A A A A A A A A A A A A A A A A A A A A A A A A A A A A A A A A

10. If f is the quadratic function defined by $f(x) = x^2 - 2$ and if $0 < a < b$, then $\int_a^b f''(x)\, dx =$

(A) $2b - 2a$

(B) 0

(C) $b - a$

(D) 2

---

11. Let $f(x) = \begin{cases} x^2 + 2x & x < -1 \\ x & -1 \le x \le 1 \\ -5x + 4 & x > 1 \end{cases}$

Where is $f(x)$ not continuous?

(A) $x = -1$

(B) $x = 1$

(C) both $x = -1$ and $x = 1$

(D) neither $x = -1$ nor $x = 1$

GO ON TO THE NEXT PAGE.

A A A A A A A A A A A A A A A A A A A A A A A A A A A A A A A A A A A A A A A

12. If $\dfrac{dy}{dx} = \dfrac{xe^{3x^2}}{y}$ and $y(0) = 2$, what is the particular solution $y(x)$?

(A) $y = -\sqrt{12e^{3x^2} - 8}$

(B) $y = -\sqrt{\dfrac{1}{3}e^{3x^2} + \dfrac{11}{3}}$

(C) $y = \sqrt{\dfrac{1}{3}e^{3x^2} + \dfrac{11}{3}}$

(D) $y = \sqrt{12e^{3x^2} - 8}$

13. Which of the following is an equation of the line tangent to the graph of $2x^2 - 4xy = 2x$ at the point $(3, -1)$?

(A) $y - 1 = -\dfrac{1}{2}(x - 3)$

(B) $y - 1 = \dfrac{5}{2}(x - 3)$

(C) $y - 1 = \dfrac{1}{2}(x - 3)$

(D) $y - 1 = 3(x - 3)$

GO ON TO THE NEXT PAGE.

AAAAAAAAAAAAAAAAAAAAAAAAAAAAAAAAAAA

14. The length of the curve $y = \sin(4x)$ from $x = 0$ to $x = \frac{\pi}{6}$ is given by

(A) $\int_0^{\frac{\pi}{6}} \left(1 + 16\cos^2(4x)\right) dx$

(B) $\int_0^{\frac{\pi}{6}} \sqrt{1 + 16\cos^2(4x)} \, dx$

(C) $\int_0^{\frac{\pi}{6}} \left(1 + \sin^2(4x)\right) dx$

(D) $\int_0^{\frac{\pi}{6}} \sqrt{\left(1 + 4\cos^2(4x)\right)} \, dx$

15. What is the area of the region enclosed by the graphs of $f(x) = -2x^2 + 9$ and $g(x) = -6x + 13$?

(A) $-\frac{1}{3}$

(B) $\frac{1}{3}$

(C) $\frac{5}{3}$

(D) $\frac{13}{3}$

GO ON TO THE NEXT PAGE.

AAAAAAAAAAAAAAAAAAAAAAAAAAAAAAAAA

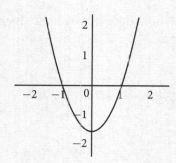

16. The graph of $y = f'(x)$ is shown above. Which of the following could be the graph of $f(x)$?

(A)

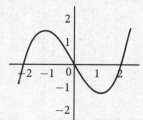

(B)

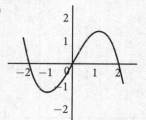

(C)

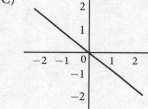

(D)

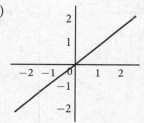

17. What radius value will minimize the surface area of a cylindrical can with a lid if the can must have a volume of 9π cubic units?

(A) 0

(B) $\sqrt[3]{2}$

(C) $\dfrac{4}{\sqrt[3]{4}}$

(D) $\dfrac{4}{\sqrt[3]{2}}$

---

18. The Taylor polynomial of degree 50 for the function about $x = 2$ is given by the following.

$$P(x) = (x - 2) - \frac{(x-2)^4}{2!} + \frac{(x-2)^6}{3!} + \cdots (-1)^{n+1}\frac{(x-2)^{2n}}{n!} + \cdots - \frac{(x-2)^{50}}{25!}$$

What is the value of $f^{(20)}(2)$?

(A) $\dfrac{-20!}{10!}$

(B) -2

(C) $\dfrac{-1}{10!}$

(D) $\dfrac{20!}{10!}$

GO ON TO THE NEXT PAGE.

A A A A A A A A A A A A A A A A A A A A A A A A A A A A A A

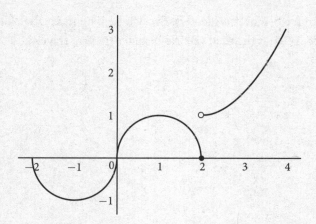

19. The graph of $f(x)$ above has a vertical tangent at the point $(0, 0)$ and horizontal tangents at the points $(-1, -1)$ and $(1, 1)$. For what values of x, $-2 < x < 4$, is $f(x)$ not differentiable?

(A) 2 only

(B) -1 and 1 only

(C) 0 and 2 only

(D) $-1, 0, 1,$ and 2

20. The radius of convergence for the power series $\sum_{n=1}^{\infty} \frac{(x-4)^n}{4^n}$ is equal to 4. What is the interval of convergence?

(A) $-4 < x < 4$

(B) $0 < x < 8$

(C) $0 < x \leq 8$

(D) $0 \leq x < 8$

GO ON TO THE NEXT PAGE.

21. A ball is dropped from a height of 12 meters. Every time it bounces, the ball goes half as high as the previous bounce. What is the total vertical distance the ball travels?

    (A) 24 meters

    (B) 36 meters

    (C) 48 meters

    (D) The series does not converge.

---

22. Consider the curve $y = -x^3 + 5x + 1$. Find an equation for the tangent to the curve at the point $(2, 3)$.

    (A) $y = 7y - 19$

    (B) $y = -7x + 17$

    (C) $y = -7x + 23$

    (D) $y = 7y - 11$

---

23. Let f and g be differentiable functions with $f(x) < 0$ for all of x and $g(0) = 2$.

    If $h(x) = \dfrac{f(x)}{g(x)}$ and $h'(x) = \dfrac{f'(x)g(x)}{g(x)^2}$, then $g(x) =$

    (A) $g'(x)$ (B) $f(x)$ (C) 2   (D) 0

GO ON TO THE NEXT PAGE.

A A A A A A A A A A A A A A A A A A A A A A A A A A A A A A A A

24. What is the average value of $\frac{1}{x^2}$ on the closed interval $[1, 4]$?

(A) $-\frac{5}{16}$

(B) $\frac{1}{4}$

(C) $\frac{2}{3}$

(D) $\frac{3}{4}$

---

25. Which of the following is the average rate of change of the function $f(x) = e^{x^2 + 5x}$ over the closed interval $[0, 1]$?

(A) 1   (B) 5   (C) $e^6 - 1$ (D) $7e^6$

---

26. Find $\int_0^\infty e^{-x}\, dx$

(A) -1

(B) 1

(C) e

(D) The improper integral diverges.

GO ON TO THE NEXT PAGE.

A A A A A A A A A A A A A A A A A A A A A A A A A A A A A A A A

27. The nth derivative of a function at $x = 0$ is given by $f^{(n)}(0) = (-1)^n \dfrac{n+1}{(n+3)3^n}$ for all $n \geq 0$. Which of the following is a possible 3rd-degree Maclaurin series for f?

(A) $\dfrac{1}{3} - \dfrac{1}{6}x + \dfrac{1}{30}x^2 - \dfrac{1}{243}x^3$

(B) $\dfrac{1}{3} - \dfrac{1}{6}x + \dfrac{1}{15}x^2 - \dfrac{2}{81}x^3$

(C) $\dfrac{1}{3} - \dfrac{1}{6}x - \dfrac{1}{30}x^2 + \dfrac{2}{81}x^3$

(D) $\dfrac{1}{3} + \dfrac{1}{6}x + \dfrac{1}{30}x^2 + \dfrac{1}{243}x^3$

---

28. What does the Ratio Test tell you about the series $\displaystyle\sum_{n=1}^{\infty} \dfrac{(n+2)!}{4n}$?

(A) $\displaystyle\lim_{n \to \infty} \dfrac{a_{n+1}}{a_n} < 1$, so $\displaystyle\sum_{n=1}^{\infty} \dfrac{(n+2)!}{4n}$ converges

(B) $\displaystyle\lim_{n \to \infty} \dfrac{a_{n+1}}{a_n} > 1$, so $\displaystyle\sum_{n=1}^{\infty} \dfrac{(n+2)!}{4n}$ diverges

(C) $\displaystyle\lim_{n \to \infty} \dfrac{a_{n+1}}{a_n} > 1$, so $\displaystyle\sum_{n=1}^{\infty} \dfrac{(n+2)!}{4n}$ converges

(D) $\displaystyle\lim_{n \to \infty} \dfrac{a_{n+1}}{a_n} = 1$, so the test is inconclusive

GO ON TO THE NEXT PAGE.

A A A A A A A A A A A A A A A A A A A A A A A A A A A A A A A A A A A

29. Using the line tangent to the function $f(x) = 4x^3 - 5x^2 - x$ at $x = 2$, approximate $f(2.1)$.

    (A) 7.3

    (B) 10

    (C) 12.7

    (D) 27

30. $\int_1^e \dfrac{dx}{1-3x} =$

    (A) $-\dfrac{1}{3}\ln|3-3e|$

    (B) $-\dfrac{1}{3}\ln\left|\dfrac{1-3e}{2}\right|$

    (C) $\dfrac{1}{3}\ln\left|\dfrac{1-3e}{-2}\right|$

    (D) $\dfrac{1}{3}\ln|3-3e|$

END OF PART A OF SECTION I

**IF YOU FINISH BEFORE TIME IS CALLED, YOU MAY
CHECK YOUR WORK ON PART B ONLY.**

CALCULUS BC
SECTION I, Part B
Time—45 minutes
Number of questions—15

A GRAPHING CALCULATOR IS REQUIRED FOR SOME QUESTIONS ON THIS PART OF THE EXAM.

Directions: Use the space provided to solve each of the following problems. After examining the given choices, decide which of the choices is the best and fill in the corresponding circle on the answer sheet. Credit will only be given for answers appropriately marked on the answer sheet, and not for anything written in the exam book. Do not spend too much time on any one problem.

In this exam:

(1) When the exact numerical value of the correct answer does not appear among the choices given, select the number that best approximates the numerical value.

(2) The domain of a function f should be assumed to be the set of all real numbers x for which $f(x)$ is a real number, unless the question indicates otherwise.

(3) The inverse function notation f^{-1} or the prefix "arc" (e.g., $\sin^{-1} x = \arcsin x$) may be used to indicate the inverse of a trigonometric function.

B B B B B B B B B

76. The function $f(x)$ has its first derivative given by $f'(x) = \dfrac{\sqrt{2x}}{x^3 + 2x + 1}$. What is the x-coordinate of the inflection point of the graph of $f(x)$?

    (A) 0

    (B) 0.372

    (C) 0.431

    (D) 0.480

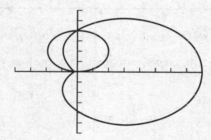

77. Find the area common to the curves $r = 2\sin(\theta)$ and $r = 2 + 2\cos(\theta)$ as shown in the figure above.

    (A) 0.283

    (B) 2.283

    (C) 4.566

    (D) 5.425

GO ON TO THE NEXT PAGE.

78. Let $y = g(x)$ be the solution to the differential equation $\dfrac{dy}{dx} = x + y + 2$ with the initial condition $g(1) = 5$. What is the approximation for $g(1.6)$ if Euler's method is used starting at $x = 1$ and using two steps of equal size?

(A) 8.27

(B) 10.61

(C) 15.15

(D) 18.029

79. Let R be the region enclosed by the graph of $y = 4 + \ln(\cos^2 x)$, the x-axis, and the lines $x = -\dfrac{4}{5}$ and $x = \dfrac{4}{5}$. What is the closest integer approximation of the area of R?

(A) 0

(B) 3

(C) 4

(D) 6

GO ON TO THE NEXT PAGE.

80. The position of a particle moving in the xy-plane is given by the parametric functions $x(t)$ and $y(t)$ for which $x'(t) = t \cos(t)$ and $y'(t) = 6e^{-2t} + 3$. What is the slope of the line tangent to the path of the particle at the point where $t = 2$?

    (A) −3.737

    (B) −0.268

    (C) 0.984

    (D) 10.168

81. A particle moves along a straight line with its position given by $x(t) = 7 + e^{\frac{t}{5}}$ for time $t \geq 0$. What is the velocity of the particle at $t = 8$?

    (A) 0.198

    (B) 0.494

    (C) 0.991

    (D) 11.953

82. What is the average value of $y = \dfrac{3\sin x + 2}{x^2 - 2x + 3}$ on the closed interval $[-2, 3]$?

    (A) 0.440

    (B) 0.926

    (C) 1.544

    (D) 4.632

GO ON TO THE NEXT PAGE.

83. A cup of hot chocolate is cooling in a room that has a constant temperature of 72°F. The initial temperature of the hot chocolate is 197°F, and the temperature of the hot chocolate changes at a rate of $R(t) = -7.2e^{-0.047t}$ °F per minute for the first 8 minutes. What is the temperature of the hot chocolate to the nearest degree after 4 minutes?

(A) 26°F

(B) 171°F

(C) 184°F

(D) 223°F

---

84. A home uses energy at the rate of $r(t) = 180 + 280\sin\left(\dfrac{t}{50}\right)$ kilowatts per month, where t is the number of days from the beginning of the cooling season. To the nearest kilowatt, what is the amount of energy used from $t = 0$ to $t = 4$ days?

(A) 22

(B) 76

(C) 720

(D) 765

GO ON TO THE NEXT PAGE.

85. The value of a tablet is decreasing at a rate that is proportional at any time to the value of the tablet at that time. The tablet was initially worth \$750, and it is worth \$480 after 2 years. How much will the tablet be worth after 5 years to the nearest dollar?

(A) \$157

(B) \$246

(C) \$384

(D) \$416

---

86. Let $f(x)$ be the function with its derivative given by $f'(x) = \cos(x^2 - 1)$. How many relative extrema does $f(x)$ have on the interval $1 \le x \le 3$?

(A) 1

(B) 2

(C) 3

(D) 4

---

87. What is the length of the arc of $y = \frac{4}{5}x^{\frac{5}{4}}$ from $x = 0$ to $x = 4$?

(A) 5.830

(B) 6.076

(C) 7.258

(D) 9.333

GO ON TO THE NEXT PAGE.

88. If $P(x)$ is a 4th-degree Taylor polynomial for $f(x) = \ln(x)$ centered at $x = 1$, then $P(1.4) =$

    (A) 0.334

    (B) 0.335

    (C) 0.336

    (D) 0.341

89. The population of a country is growing at a rate proportional to its population. If the grown rate per year is 3% of the current population, how many years will it take for the population to double in size?

    (A) 2.310

    (B) 5.673

    (C) 18.951

    (D) 23.105

90. A particle moves along the x-axis with a velocity given by $v(t) = t^2 - 7t + 10$ for time $t > 0$. If the particle is at $x = 12$ at $t = 1$, what is its position at $t = 5$?

    (A) −2.667

    (B) 9.333

    (C) 16.167

    (D) 18.333

END OF SECTION I

**IF YOU FINISH BEFORE TIME IS CALLED, YOU MAY
CHECK YOUR WORK ON PART B ONLY.**

CALCULUS BC
SECTION II, Part A
Time—30 minutes
Number of problems—2

A GRAPHING CALCULATOR IS REQUIRED FOR THESE PROBLEMS.

Directions: Write your solution to each part of the following questions in the space provided. Write clearly and legibly. Cross out any errors you make erased or crossed-out work will not be scored.

You may wish to look over the problems before starting to work on them; on the actual test, it is not expected that everyone will be able to complete all parts of all problems. All problems are given equal weight, but the individual parts of a particular problem are not necessarily given equal weight. You should not spend too much time on any one problem.

- Show all of your work. Clearly label functions, graphs, tables, or anything else that you use to arrive at your final solution. Your work will be scored on the correctness and completeness of your methods as well as your final answers. Answers without supporting work will usually not receive credit.

- Justifications (i.e., the request that you "justify your answer") require that you give mathematical (non-calculator) reasons.

- Do not use calculator syntax. Your work must be expressed in standard mathematical notation.

- Numeric or algebraic answers do not need to be simplified, unless the question indicates otherwise.

- If you use decimal approximations in calculations, your final answers should be accurate to three places after the decimal point, unless the question indicates otherwise.

- The domain of a function f should be assumed to be the set of all real numbers x for which $f(x)$ is a real number, unless the question indicates otherwise.

GO ON TO THE NEXT PAGE.

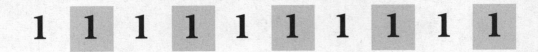

1. Let R be the region enclosed by the graph of $y = 4\sqrt{x-5} - 12$, the vertical line $x = 5$, and the x-axis.

   (a) Find the area of R.

   (b) Find the volume of the solid generated when R is revolved about the horizontal line $y = 2$.

   (c) Find the volume of the solid generated when R is revolved around the y-axis.

GO ON TO THE NEXT PAGE.

2 2 2 2 2 2 2 2 2 2

2. For $t \geq 0$, a particle moves along the x-axis. The velocity at time t is given by $v(t) = 2 + 3\cos\left(\dfrac{t^2}{2}\right)$. The particle is at position $x = 5$ at time $t = 3$.

(a) At time $t = 3$, is the particle speeding up or slowing down?

(b) Find all the times in the interval $0 < t < 4$ when the particle changes direction. Justify your answer.

GO ON TO THE NEXT PAGE.

2 2 2 2 2 2 2 2 2 2

(c) Find the particle's position at $t = 0$.

(d) Find the total distance traveled from $t = 0$ to $t = 3$.

END OF PART A OF SECTION II

**IF YOU FINISH BEFORE TIME IS CALLED, YOU MAY
CHECK YOUR WORK ON PART A ONLY.**

CALCULUS BC
SECTION II, Part B
Time—60 minutes
Number of problems—4

NO CALCULATOR IS ALLOWED FOR THESE PROBLEMS.

Note: If you have extra time, you can go back and work on Part A of Section II, but you cannot use a calculator to complete your work at this time.

3 3 3 3 3 3 3 3 3 3

NO CALCULATOR ALLOWED

3. The function g has derivatives of all orders, and the Maclaurin series for g is

$$\sum_{n=0}^{\infty}(-1)^n \frac{x^{2n+3}}{2n+4} = \frac{x^3}{4} - \frac{x^5}{6} + \frac{x^7}{8} - \cdots$$

(a) Using the ratio test, determine the interval of convergence for the Maclaurin series for g.

(b) The Maclaurin series for g evaluated at $x = \frac{1}{2}$ is an alternating series whose terms decrease in absolute value to 0. The approximation for $g\left(\frac{1}{2}\right)$ using the first two nonzero terms of this series is $\frac{5}{192}$. Show that this approximation differs from $g\left(\frac{1}{2}\right)$ by less than $\frac{1}{1,000}$.

(c) Write the first three nonzero terms and the general term of the Maclaurin series for $g'(x)$.

GO ON TO THE NEXT PAGE.

4 4 4 4 4 4 4 4 4 4

NO CALCULATOR ALLOWED

4. The function f is defined by $f(x) = x^3 - 6x^2 + 9x - 3$.

(a) On what interval(s) is the function increasing?

(b) On what interval(s) is the function decreasing and concave down?

(c) Identify all local extrema.

GO ON TO THE NEXT PAGE.

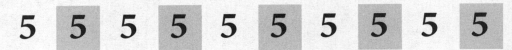

NO CALCULATOR ALLOWED

5. The Maclaurin series for $\ln(1 + x)$ is given by the following:

$x - \dfrac{x^2}{2} + \dfrac{x^3}{3} - \dfrac{x^4}{4} + \cdots + (-1)^{n+1}\dfrac{x^n}{n} + \cdots$ On its interval of convergence, the series converges to

$\ln(1 + x)$. Let f be the function defined by $f(x) = x^2 \ln\left(1 + \dfrac{x}{2}\right)$.

(a) Write the first four nonzero terms and the general term of the Maclaurin series for f.

(b) Define the interval of convergence. Show the work that leads to your answer.

(c) Let $P_5(x)$ be the 5th-degree Taylor polynomial for f about $x = 0$. Use the alternating series
     error bound to find the upper bound for $\left|P_5\left(\dfrac{1}{3}\right) - f\left(\dfrac{1}{3}\right)\right|$.

GO ON TO THE NEXT PAGE.

6 6 6 6 6 6 6 6

NO CALCULATOR ALLOWED

6. The functions f and g are given by $f(x) = \int_0^{2x} \sqrt{4 + t^3} \, dt$ and $g(x) = f(\cos x)$.

   (a) Find $f'(x)$ and $g'(x)$.

   _____

   (b) Write an equation for the line tangent to the graph of $y = g(x)$ at $x = \dfrac{\pi}{2}$.

   _____

   (c) Use the equation of the tangent line found in part b to approximate the value of $g(0)$. Is this approximation an under or over estimation? Explain why.

STOP

END OF EXAM

_____

ANSWER KEY

Section I, Part A

1.	B	16.	A
2.	D	17.	B
3.	C	18.	A
4.	D	19.	C
5.	A	20.	B
6.	C	21.	B
7.	B	22.	B
8.	A	23.	C
9.	C	24.	B
10.	A	25.	C
11.	B	26.	B
12.	C	27.	A
13.	C	28.	B
14.	B	29.	C
15.	B	30.	B

Section I, Part B

76.	B
77.	B
78.	B
79.	D
80.	A
81.	C
82.	B
83.	B
84.	D
85.	B
86.	C
87.	B
88.	B
89.	D
90.	B

SCORING

Section II, Parts A & B

Use the scoring rubrics (found after the Section I answers and explanations) to self-score your free-response questions.

Section I, Part A Number Correct: _____

Section I, Part B Number Correct: _____

Section II, Part A Points Earned: _____

Section II, Part B Points Earned: _____

Enter your results to your Practice Exam 5 assignment to see your 1–5 score by logging in at kaptest.com.

Haven't registered your book yet? Go to kaptest.com/booksonline to begin.

BC PRACTICE EXAM 5 ANSWERS AND EXPLANATIONS

SECTION I, Part A

1. B

The limit as x approaches $\frac{\pi}{6}$ from the left is $\sin \frac{7\pi}{6}$, or $-\frac{1}{2}$. In order for the limit to exist, the limit as x approaches $\frac{\pi}{6}$ from the right must also equal $-\frac{1}{2}$. You are looking for a value of a that will make $\cos \frac{a\pi}{6} = -\frac{1}{2}$. Simple substitution to make $\cos u = -\frac{1}{2}$ and solving results in $u = \frac{2\pi}{3}$. Setting $\frac{a\pi}{6} = \frac{2\pi}{3}$ and solving for a gives $a = 4$.

Therefore, **(B)** is correct. (A) would be correct if the left limit evaluated to $\frac{1}{2}$ instead of $-\frac{1}{2}$.

2. D

The partial sum formula is given to use as the partial sum of the series. In order to find the partial sum of the series, replace 5 for n in the partial sum formula and evaluate. When you are given the partial sum formula, use it to calculate the sum for any number of n-terms.

$$S_5 = \frac{(5)^3 + 7}{(5)^2 - 4}$$

$$S_5 = \frac{132}{21}$$

Thus, **(D)** is correct. (A) is just the quotient of the constants, ignoring the n in the numerator and the n in the denominator. (B) is the derivative of the partial sum formula evaluated at $n = 5$.

3. C

The alternating series error bound formula can be used to find the amount of error in the approximation of a power series versus the actual function in an alternating series by taking the absolute values of the next omitted term. The formula for the alternating series error bound is $|R_n| < a_{n+1}$, where the error is less than the absolute value of the next term. Since the polynomial was a 5th degree, as shown in the series formula, a fifth degree polynomial would be created when $n = 2$. To find the alternating series error bound, you take the absolute value of the next term, in this case $n = 3$. For $n = 3$ and $x = \frac{1}{2}$, use the alternating series error bound formula.

$$\sum_{n=1}^{\infty} \frac{(-1)^n}{(3n-1)!} x^{2n+1}$$

$$\left| \frac{(-1)^3}{(3(3)-1)!} \left(\frac{1}{2}\right)^{2(3)+1} \right|$$

$$\left| \frac{-1}{2^7 \cdot 8!} \right|$$

$$\frac{1}{2^7 \cdot 8!}$$

Hence, **(C)** is correct. (A) fails to use the absolute value at the end and keeps the difference negative. This is not a correct usage of the alternating series error bound formula. (B) uses $n = 6$, which would form a 13-degree polynomial. (D) is missing the exponent on the base 2 in the denominator.

4. D

Use long division to divide $x^5 - 4x^4 - 2x + 10$ by $-2x^2 - x - 6$. The result is $-\frac{1}{2}x^3 + \cdots$ This means the function takes on the properties of a negative cubic function as shown below.

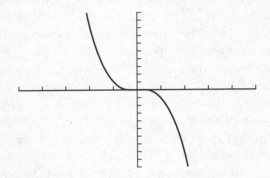

If you take the limit as x approaches $-\infty$, the function approaches $+\infty$. Therefore "nonexistent" is the best choice.

Therefore, **(D)** is correct. (A) incorrectly divides the constants at the end of the polynomial. (B) is the quotient of the leading coefficients. This approach should be used only if the degree of the numerator equals the degree of the denominator. (C) would be correct if there was a horizontal asymptote at $y = 0$, which happens when the degree of the denominator is more than the degree of the numerator.

5. A

There are several approaches to this question. The simplest is to use the symmetry of the function around $(8, 0)$. The integral from $x = 8$ to $x = 16$ (statement I) is negative since the function is below the x-axis. The integral from $x = 0$ to $x = 16$ (statement II) is 0 since the equal areas cancel each other out. The area between $x = 8$ and $x = 16$ (statement III) is slightly bigger than the area between $x = 10$ and $x = 16$ (statement IV). Therefore, I, II, III, IV are the statements in order from least to greatest, **(A)**.

Statements I and IV do not have the same values. Statement I is asking for the integral, which is negative. Statement IV is asking for the area. Since area can never be negative, the answer is the absolute value of the integral, which is positive. It is important to recognize that you can get a negative result when evaluating an integral, but area is always positive.

6. C

In statement I, $\sum_{n=1}^{\infty} \frac{|\cos(n)|}{n^3}$, the denominator is much larger than the numerator. Therefore, this series converges. The numerator oscillates between 0 and 1; however, the denominator grows very rapidly, causing the series to converge. This can also be compared to a p-series with $p = 3$, which converges. By rewriting statement II, $\sum_{n=1}^{\infty} 8e^{-n}$, with a positive exponent, $\sum_{n=1}^{\infty} \frac{8}{e^n}$, you can see that this series also converges. If you continue to rewrite it as $8\sum_{n=1}^{\infty} \frac{1}{e^n}$ and then as $8\sum_{n=1}^{\infty} \left(\frac{1}{e}\right)^n$, you can see that this is a geometric series with a constant ratio less than 1, causing the series to converge. Statement III, $\sum_{n=1}^{\infty} \frac{(3n+7)(n-1)}{4n+3n^2-7}$, should be rewritten as $\sum_{n=1}^{\infty} \frac{(3n+7)(n-1)}{3n^2+4n-7}$, factored into $\sum_{n=1}^{\infty} \frac{(3n+7)(n-1)}{(3n+7)(n-1)}$, and simplified to $\sum_{n=1}^{\infty} 1$. From the nth Term Divergence Test since $\lim_{n\to\infty} \frac{(3n+7)(n-1)}{4n+3n^2-7} \neq 0$, this series diverges.

Hence, **(C)** is correct.

7. B

In this problem you are looking for the rate at which the angle θ is changing with respect to time as the ladder slides down a building. The hypotenuse (ladder) of the triangle is constant, and a height value is provided along with the rate at which the height is changing with respect to time.

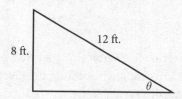

With this information, the sine ratio provides a useful formula.

$$\sin\theta = \frac{\text{opposite}}{\text{hypotenuse}}$$

Since the hypotenuse is constant, it can be substituted into the ratio now. Then the ratio can be differentiated with respect to time.

$$\sin\theta = \frac{a}{12}$$
$$\cos\theta \cdot \frac{d\theta}{dt} = \frac{1}{12}\frac{dh}{dt}$$

You are given $\frac{dh}{dt} = -3$. By using the Pythagorean theorem, you can find the missing side of the right triangle.

$$8^2 + b^2 = 12^2$$
$$64 + b^2 = 144$$
$$b^2 = 80$$
$$b = 4\sqrt{5}$$

This means $\cos\theta = \frac{4\sqrt{5}}{12} = \frac{\sqrt{5}}{3}$. Now you have all the necessary information to find $\frac{d\theta}{dt}$.

$$\cos\theta \cdot \frac{d\theta}{dt} = \frac{1}{12}\frac{dh}{dt}$$
$$\frac{\sqrt{5}}{3}\frac{d\theta}{dt} = \frac{1}{12}\cdot(-3)$$
$$\frac{\sqrt{5}}{3}\frac{d\theta}{dt} = -\frac{1}{4}$$
$$\frac{d\theta}{dt} = -\frac{3}{4\sqrt{5}}$$

Therefore, **(B)** is correct. (A) multiplies both sides of the equation by $\frac{\sqrt{5}}{3}$ instead of by $\frac{3}{\sqrt{5}}$. (C) is a result of substituting the sine ratio of $\frac{2}{3}$ instead of the cosine ration of $\frac{\sqrt{5}}{3}$.

8. A

The acceleration vector is found by taking the derivative of the velocity functions. **(A)**, $\langle 3\cos(3t), 4e^{4t}\rangle$, is the only choice that correctly uses the chain rule on both the x and y velocity functions.

(B) and (D) have the incorrect sign for the derivative of the sine function, and (C) neglects the chain rule on the y velocity function.

9. C

This is a special case antiderivative. Remember that $\frac{1}{\sqrt{1-x^2}}$ is the derivative of arcsin x, making arcsin x the antiderivative of $\frac{1}{\sqrt{1-x^2}}$. The inverse trigonometric functions typically show up once or twice on any given AP Calculus exam, so they should be memorized.

$$\int_0^{\frac{1}{2}}\frac{dx}{\sqrt{1-x^2}}dx$$
$$= \arcsin x\Big|_0^{\frac{1}{2}}$$
$$= \arcsin\frac{1}{2} - \arcsin 0$$
$$= \frac{\pi}{6} - 0$$
$$= \frac{\pi}{6}$$

Thus, **(C)** is correct. (A) would be correct if the given expression was $-\frac{1}{\sqrt{1-x^2}}$, which would make arccos x the antiderivative.

10. A

The antiderivative of $f''(x)$ is $f'(x)$. The derivative of $f(x)$ using the power rule is $f'(x) = 2x$. This is the function that should be evaluated at the upper and lower limits.

$$\int_a^b f''(x)\,dx$$
$$= f' \big|_a^b$$
$$= 2x \big|_a^b$$
$$= 2b - 2a$$

Hence, **(A)** is correct. If you use the second derivative instead of the first derivative, you get (B). (D) is the average value of the integral.

11. B

By using the test for continuity, we can show that the function is continuous at $x = -1$.

$$\lim_{x \to -1^-} f(x) = f(-1) = \lim_{x \to -1^+} f(x) = -1$$

The test for continuity also shows that the function is not continuous at $x = 1$.

$$\lim_{x \to 1^-} f(x) = f(x) = 1$$
$$\text{but}$$
$$\lim_{x \to 1^+} f(x) = -1$$

Therefore, $f(x)$ is not continuous at $x = 1$, which is **(B)**.

12. C

This differential equation is defined by both x and y. When this happens, you should start by separating the variables. Multiplying both sides by y and by dx gives the equation $y\,dy = xe^{3x^2}\,dx$. You can now take the antiderivative of both sides using the power rule for the left side and u-substitution for the right side.

$$y\,dy = xe^{3x^2}\,dx$$
$$\int y\,dy = \int xe^{3x^2}\,dx$$
$$\frac{y^2}{2} + C_1 = \frac{1}{6}e^{3x^2} + C_2$$

When you subtract C_1 from C_2, you just get another constant, C. You can now substitute the given condition $y(0) = 2$ and solve for C.

$$\frac{y^2}{2} = \frac{1}{6}e^{3x^2} + C$$
$$\frac{(2)^2}{2} = \frac{1}{6}e^{3(0)^2} + C$$
$$2 = \frac{1}{6} + C$$
$$C = \frac{11}{6}$$

Substitute this value into $\frac{y^2}{2} = \frac{1}{6}e^{3x^2} + C$ and solve for y.

$$\frac{y^2}{2} = \frac{1}{6}e^{3x^2} + \frac{11}{6}$$
$$y^2 = \frac{1}{3}e^{3x^2} + \frac{11}{3}$$
$$y = \pm\sqrt{\frac{1}{3}e^{3x^2} + \frac{11}{3}}$$

There are two options for y. You know from the original condition, though, that the y-value at $x = 0$ is positive 2, so the principal square root is the correct answer.

Thus, **(C)** is correct. (A) and (D) are what you get if the integral was multiplied by 6 during u-substitution instead of by $\frac{1}{6}$. (B) would be correct if the given condition had a negative y-value.

13. C

When calculating a line tangent to a graph, you need to know a point and the instantaneous rate of change at that point. In this example, the point (3, 1) is provided. The derivative of the function is used to find slope, but y is not explicitly defined in terms of x. It is possible to solve for y, but implicit differentiation is a more effective method in this situation.

$$2x^2 - 4xy = 2x$$
$$\frac{d}{dx}(2x^2 - 4xy) = \frac{d}{dx}(2x)$$
$$\frac{d}{dx}(2x^2) - \frac{d}{dx}(4xy) = \frac{d}{dx}(2x)$$

Using the power rule, $\frac{d}{dx}(2x^2) = 4x$. When taking this derivative, you really get $4x\frac{dx}{dx}$. However, $\frac{dx}{dx} = 1$, so it is not expected that you write that part. It is only being shown here to help clear up what happens when taking the derivative of a y-term with respect to the variable x. Using the product rule on the second term gives the following. Remember that $\frac{dx}{dx} = 1$.

$$\frac{d}{dx}(4xy)$$
$$= \left(y \cdot \frac{d}{dx}4x\right) + \left(4x \cdot \frac{d}{dx}y\right)$$
$$= \left(y \cdot 4\frac{dx}{dx}\right) + \left(4x \cdot \frac{dy}{dx}\right)$$
$$= 4y + 4x\frac{dy}{dx}$$

On the right side of the equal sign, use the power rule again: $\frac{d}{dx}(2x) = 2$. That leaves you with $4x - \left(4y + 4x\frac{dy}{dx}\right) = 2$. Remember that $\frac{dy}{dx}$ represents the slope. To find it at the given point, replace x with 3 and y with 1 to get $\frac{dy}{dx} = \frac{1}{2}$. Plugging this information into the point-slope form of a line gives $y - 1 = \frac{1}{2}(x - 3)$.

Therefore, **(C)** is correct. (A) mistakenly does not distribute the subtraction sign through the product rule derivative. You may get (B) if you do not use the product rule for the derivative of $4xy$ and use $4\frac{dy}{dx}$ instead.

14. B

The arc length formula is $s = \int_a^b \sqrt{1 + \left(\frac{dy}{dx}\right)^2}\, dx$. Make sure to memorize this formula before taking the AP exam.

$$y = \sin(4x)$$
$$\frac{dy}{dx} = 4\cos(4x)$$
$$\left(\frac{dy}{dx}\right)^2 = 16\cos^2(4x)$$
$$s = \int_a^b \sqrt{1 + \left(\frac{dy}{dx}\right)^2}\, dx$$
$$s = \int_0^{\frac{\pi}{6}} \sqrt{1 + 16\cos^2(4x)}\, dx$$

Hence, **(B)** is correct. Be sure to square the derivative correctly. The coefficient should be 16, not 4, like in (D). (A) and (C) both use an incorrect arc length formula.

15. B

Calculating area requires a definite integral, so the first step is to determine appropriate upper and lower bounds. These interval values are determined by finding the x-values where $f(x) = g(x)$.

$$f(x) = g(x)$$
$$-2x^2 + 9 = -6x + 13$$
$$2x^2 - 6x + 4 = 0$$
$$2(x - 1)(x - 2) = 0$$
$$x = 1, x = 2$$

Along with the bounds, it is also important to identify which function forms the top of the region and which one forms the bottom of the region. In this problem, the parabola is the top function and the line is the bottom function. The integral $\int_a^b (\text{top} - \text{bottom})\, dx$ produces the area between the curves.

$$\int_1^2 \left[(-2x^2 + 9) - (-6x + 13)\right] dx$$

$$= \int_1^2 (-2x^2 + 6x - 4)\, dx$$

$$= -\frac{2}{3}x^3 + 3x^2 - 4x \Big|_1^2$$

$$= \left(-\frac{2}{3}(2)^3 + 3(2)^2 - 4(2)\right) - \left(-\frac{2}{3}(1)^3 + 3(1)^2 - 4(1)\right)$$

$$= \left(-\frac{4}{3}\right) - \left(-\frac{5}{3}\right)$$

$$= \frac{1}{3}$$

Thus, **(B)** is correct. Another technique would be to take the definite integral of each function using the lower bound of $x = 1$ and the upper bound of $x = 2$. The results of the definite integrals can then be subtracted. Although this method works, it can be more time consuming. (A) is the result of subtracting the functions in the wrong order. Although definite integrals can give negative results, the area of a region between two curves is always positive.

16. A

The derivative is a quadratic function, meaning the original function must be a cubic function and narrows down the options. Since the graph of the derivative is provided, the y-coordinates represent the instantaneous rate of change of $f(x)$. The y-coordinates of $f'(x)$ transition from positive to negative left of $x = -1$. This means $f(x)$ must transition from increasing to decreasing at the same x-value (local maximum). The y-coordinates of $f'(x)$ transition from negative to positive right of $x = 1$, which means $f(x)$ must transition from increasing to decreasing at the same x-value (local minimum). **(A)** is correct because it meets these criteria.

(B) would be the appropriate function if the given graph was reflected over the x-axis. (D) is the derivative of the given graph, or the second derivative of $f(x)$.

17. B

The constraint on this problem is that the volume must be 4π. Using the formula $V = \pi r^2 h$ will allow you to write the surface area formula in terms of a single variable. Since the question is concerned with the radius, it is a good idea to solve for h in terms of r.

$$V = \pi r^2 h$$
$$4\pi = \pi r^2 h$$
$$h = \frac{4}{r^2}$$

The goal is to optimize the surface area, $A = 2\pi rh + 2\pi r^2$, which can be rewritten using the constraint information above.

$$A = 2\pi rh + 2\pi r^2$$
$$A = 2\pi r \frac{4}{r^2} + 2\pi r^2$$
$$A = \frac{8\pi}{r} + 2\pi r^2$$
$$A = 2\pi\left(r^2 + 4r^{-1}\right)$$

This is the equation that needs to be optimized, which means it needs to be differentiated to identify critical values.

$$A = 2\pi\left(r^2 + 4r^{-1}\right)$$
$$A' = 2\pi\left(2r - 4r^{-2}\right)$$
$$0 = 2\pi\left(2r - 4r^{-2}\right)$$
$$0 = 4\pi r^{-2}\left(r^3 - 2\right)$$
$$0 = 4\pi r^{-2} \text{ or } 0 = (r^3 - 2)$$
$$r = 0 \qquad\qquad r = \sqrt[3]{2}$$

$A' < 0$ on the interval $0 < r < \sqrt[3]{2}$ and $A' > 0$ for $r > \sqrt[3]{2}$. This transition from negative to positive translates to a minimum surface area. The critical value $r = 0$ does not need to be analyzed because it is not appropriate for the context of the problem.

Therefore, **(B)** is correct. (C) represents the height of the can when the surface area is minimized.

18. A

This question is asking about the 20th derivative. After closely inspecting the Taylor polynomial formula for the 20th derivative, you should realize that $n = 10$. Use the inner part of the formula with $n = 10$ to find the term.

$$P(x) = \cdots (-1)^{n+1} \frac{(x-2)^{2n}}{n!} + \cdots$$

$$= (-1)^{10+1} \frac{(x-2)^{2(10)}}{10!}$$

$$= \frac{-1(x-2)^{20}}{10!}$$

Remember that the term includes the division, which is by 20! in this case. This is based on the following formula.

$$f(x) \approx f(c) + f'(c)(x-c) + \frac{f''(c)}{2}(x-c)^2$$

$$+ \frac{f^{(3)}(c)}{3!}(x-c)^3 + \cdots + \frac{f^{(n)}(c)}{n!}(x-c)^n$$

To find only the derivative, you need to multiply the coefficient by 20!

$$f^{(20)}(2) = \frac{-1}{10!} \cdot 20!$$

$$f^{(20)}(2) = \frac{-20!}{10!}$$

Hence, **(A)** is correct. (B) incorrectly divides the factorials. (C) is the coefficient but not the derivative. (D) neglects the negative sign.

19. C

There are a few key indicators for identifying when a function is not differentiable. The first indicator is continuity. If a function is not continuous at a point, then it is not differentiable. For this graph, a discontinuity occurs at $x = 2$. Another indicator is the presence of a vertical tangent line. The derivative describes the slope of $f(x)$, and a vertical line has an undefined slope. The information provided states that there is a vertical tangent at $(0, 0)$. These two indicators make **(C)**, 0 and 2 only, the correct answer. The function has turning points at $x = -1$ and $x = 1$, which means the derivative will be 0 at these points.

20. B

You should use the information in the question to help you identify the endpoints of the interval of convergence. Since the Taylor polynomial was written centered at $x = 4$ with a radius of convergence of 4, the endpoints of the interval must be $x = 0$ and $x = 8$. You must test the endpoints $x = 0$ and $x = 8$ to determine if the series converges or diverges at these particular endpoints. They are independent of each other and should be tested separately. Use the tests for convergence and divergence to test at $x = 0$:

$$\sum_{n=1}^{\infty} \frac{(x-4)^n}{4^n}$$

$$\sum_{n=1}^{\infty} \frac{(0-4)^n}{4^n}$$

$$\sum_{n=1}^{\infty} \frac{(-4)^n}{4^n}$$

$$\sum_{n=1}^{\infty} \left(\frac{-4}{4}\right)^n$$

$$\sum_{n=1}^{\infty} (-1)^n$$

This series diverges because it alternates between -1 and $+1$ without bound, so the series has no sum and diverges. The value of $x = 0$ should not be included in the interval of convergence. Now test at $x = 8$.

$$\sum_{n=1}^{\infty} \frac{(x-4)^n}{4^n}$$

$$\sum_{n=1}^{\infty} \frac{(8-4)^n}{4^n}$$

$$\sum_{n=1}^{\infty} \frac{(4)^n}{4^n}$$

$$\sum_{n=1}^{\infty} \left(\frac{4}{4}\right)^n$$

$$\sum_{n=1}^{\infty} (1)^n$$

This series diverges because it continually adds the number 1 without bounds. The value of $x = 8$ should not be included in the interval of convergence. The interval of convergence is $0 < x < 8$.

Thus, **(B)** is correct. You should not waste time by finding the radius of convergence using $\lim\limits_{n\to\infty}\left|\dfrac{u_{n+1}}{u_n}\right| < 1$ since it was already given in the question. (A) is the correct radius of convergence but centered at $x = 0$ and not correctly at $x = 4$.

21. B

On the initial drop, the ball falls 12 meters and bounces back up 6 meters. On the second bounce, the ball falls 6 meters and bounces back up 3 meters. On the third bounce, the ball falls 3 meters and bounces back up $\dfrac{3}{2}$ meters. Each vertical distance traveled by the ball is defined by $12\left(\dfrac{1}{2}\right)^n$. The scenario can be represented by the series

$$12 + 12\left(\frac{1}{2}\right) + 12\left(\frac{1}{2}\right) + 12\left(\frac{1}{2}\right)^2 + 12\left(\frac{1}{2}\right)^2 + 12\left(\frac{1}{2}\right)^3 + 12\left(\frac{1}{2}\right)^3 + \cdots$$

which can be simplified to

$12 + 24\left(\dfrac{1}{2}\right) + 24\left(\dfrac{1}{2}\right)^2 + 24\left(\dfrac{1}{2}\right)^3 + \cdots$ If the initial 12 is rewritten as $-12 + 24$, the series becomes

$-12 + 24 + 24\left(\dfrac{1}{2}\right) + 24\left(\dfrac{1}{2}\right)^2 + 24\left(\dfrac{1}{2}\right)^3 + \cdots$ and can be written in the familiar geometric form $-12 + \sum\limits_{k=0}^{\infty} 24\left(\dfrac{1}{2}\right)^k$. The sum of a geometric series with $|r| < 1$ is $S = \dfrac{a}{1-r}$. In this situation, that becomes $S = \dfrac{24}{1 - \dfrac{1}{2}} = 48$. Don't forget to add the -12 to this sum, giving you a total vertical distance of 36 meters. Hence, **(B)** is correct. (A) is $\sum\limits_{k=0}^{\infty} 12\left(\dfrac{1}{2}\right)^n$, which does not account for the ball traveling the same vertical distance twice after the initial drop. (C) is just $\sum\limits_{k=0}^{\infty} 24\left(\dfrac{1}{2}\right)^k$.

22. B

To write the equation of a line, you must know a point on the line and the slope of the line. The point (2, 3) is given. You can find the slope of the tangent by taking the derivative of the given function and evaluating it at $x = 2$.

$$y = -x^3 + 5x + 1$$
$$y' = -3x^2 + 5$$
$$y'(2) = -3(2)^2 + 5$$
$$y'(2) = -7$$

You can now plug the slope and the given point into the point-slope form of a line. All the multiple-choice options are in slope-intercept form, so you need to solve for y.

$$y - 3 = -7(x - 2)$$
$$y - 3 = -7x + 14$$
$$y = -7x + 17$$

Therefore, **(B)** is correct. (A) uses 7 instead of -7 in the point-slope form. (C) is the result of switching x and y when substituting values into point-slope form.

23. C

The function $h(x)$ is defined as the quotient of two differentiable functions, and the derivative of $h(x)$ is provided. Using the quotient rule, you can compare the actual derivative to what you get when following the rule.

$$h(x) = \frac{f(x)}{g(x)}$$
$$h'(x) = \frac{f'(x)g(x) - g'(x)f(x)}{g(x)^2} = \frac{f'(x)g(x)}{g(x)^2}$$

You can see the given numerator is missing the second term from the quotient rule, meaning the second term must be equal to 0. For $g'(x)\,f(x)$ to equal 0, either $f(x) = 0$ or $g'(x) = 0$. You know from the given information (i) that $f(x)$ must be negative for all x-values, so that means $g'(x)$ must equal 0. The derivative of a constant is 0, making $g(x)$ a constant. The second piece of given information (ii) tells you the value of constant $g(x)$. This makes **(C)** correct.

24. B

Using the average value theorem, you get the following.

$$f(c) = \frac{1}{4-1}\int_1^4 x^{-2}\, dx$$

$$f(c) = \frac{1}{3}\left[-x^{-1}\right]\bigg|_1^4$$

$$f(c) = \frac{1}{3}\left[-\frac{1}{4}+1\right]$$

$$f(c) = \frac{1}{3}\left[\frac{3}{4}\right]$$

$$f(c) = \frac{1}{4}$$

Hence, **(B)** is correct. (A) is the mean value of the function on the same interval. Be careful not to mix up the two theorems. The mean value theorem is for average *rate*, and the average value theorem is for the average *value*. (D) is missing the $\frac{1}{4-1}$ portion of the theorem.

25. C

The average rate of change is the same as the slope of the secant line between $x = 0$ and $x = 1$. The notation [0, 1] represents the domain over which you are finding the average rate of change. The first step to solve this problem is to find the y-coordinates that correspond to the x-values. When you plug in 0 and then 1 into the original function, it produces (0, 1) and (1, e^6). The average rate of change is the slope of these two points: $\frac{e^6-1}{1-0}$ or simply $e^6 -1$.

Thus, **(C)** is correct. (A) is $f(0)$, (B) is $f'(0)$, and (D) is $f'(1)$, none of which is the average rate of change.

26. B

Having ∞ as one of the limits is a clear indicator of an improper integral. After replacing ∞ with a variable, b, evaluate the integral.

$$\int_0^\infty e^{-x}\, dx$$

$$= \lim_{b\to\infty}\int_0^b e^{-x}\, dx$$

$$= \lim_{b\to\infty} -e^{-x}\big|_0^b$$

$$= \lim_{b\to\infty}\left(-e^{-b}+e^0\right)$$

$$= \lim_{b\to\infty}\left(-\frac{1}{e^b}+1\right)$$

Now you can evaluate the limit of this value as b approaches infinity. Remember that if a constant is divided by an infinitely large number, the quotient approaches 0.

$$\lim_{b\to\infty}\left(-\frac{1}{e^b}+1\right)$$

$$= 0+1$$

$$= 1$$

Hence, **(B)** is correct. Be careful when integrating e^{-x}. The antiderivative is $-e^{-x}$, but if you forget to include the negative sign, you will get (A).

27. A

The Maclaurin series is based on the Taylor polynomial written at $x = 0$. The Taylor polynomial formula is the following.

$$f(x) \approx f(c) + f'(c)(x-c) + \frac{f''(c)}{2}(x-c)^2$$

$$+ \frac{f^{(3)}(c)}{3!}(x-c)^3 + \cdots + \frac{f^{(n)}(c)}{n!}$$

When $c = 0$, the Maclaurin series formula is the following.

$$f(x) \approx f(0) + f'(0)x + \frac{f''(0)}{2}x^2 + \frac{f^{(3)}(0)}{3!}x^3 + \cdots + \frac{f^{(n)}(0)}{n!}x^n$$

In order to answer this question, you need to evaluate $f(0)$ and the first three derivatives when $x = 0$.

$$f(0) = \frac{1}{3}$$

$$f'(0) = -\frac{1}{6}$$

$$f''(0) = \frac{1}{15}$$

$$f^{(3)}(0) = -\frac{2}{81}$$

When you plug these into the formula, remember to multiply the denominators by the appropriate factorials and simplify.

$$f(x) \approx f(0) + \frac{-1}{6}x + \frac{1}{15} \cdot \frac{1}{2}x^2 + \frac{-2}{81} \cdot \frac{1}{3!}x^3$$

$$f(x) \approx f(0) - \frac{1}{6}x + \frac{1}{30}x^2 - \frac{1}{243}x^3$$

$$f(x) \approx \frac{1}{3} - \frac{1}{6}x + \frac{1}{30}x^2 - \frac{1}{243}x^3$$

Therefore, **(A)** is correct. Notice that the correct choice is raised to the 3rd degree since the question asked for a 3rd-degree polynomial. Be careful that the number of terms is different than the degree. (B) neglects to divide by the correct factorials in the formula. (D) is incorrect because it ignores the alternating signs from the alternator in the derivative.

28. B

This question requires you to know the details of the Ratio Test. The best way to remember the details about the Ratio Test is to divide any two consecutive terms. Say you divide a_{n+1} by a_n. If the terms are getting smaller, that quotient should be less than 1; the series converges. If the quotient is more than 1, the terms are getting bigger; the series diverges. If the quotient is equal to 1, the test is inconclusive. (C) is always false, so that is not an option for this question. When you evaluate the ratio of the terms, dividing by a_n is the same as multiplying by the reciprocal. Notice how the factorials reduce from the numerator and the denominator as well as how the limit is evaluated.

$$\lim_{x \to \infty} \frac{a_{n+1}}{a_n} = \frac{(n+1+1)!}{4(n+1)} \cdot \frac{4n}{(n+1)!}$$

$$\lim_{x \to \infty} \frac{a_{n+1}}{a_n} = \frac{(n+2)!}{4(n+1)} \cdot \frac{4n}{(n+1)!}$$

$$\lim_{x \to \infty} \frac{a_{n+1}}{a_n} = \frac{(n+2)(n+1)(n)(n-1)\cdots}{4(n+1)} \cdot \frac{4n}{(n+1)(n)(n-1)\cdots}$$

$$\lim_{x \to \infty} \frac{a_{n+1}}{a_n} = \frac{n+2}{n+1} \cdot \frac{n}{1}$$

$$\lim_{x \to \infty} \frac{a_{n+1}}{a_n} = \frac{n(n+2)}{n+1}$$

$$\lim_{x \to \infty} \frac{a_{n+1}}{a_n} > 1$$

Remember that (A), (C), and (D) are all true in general. In this particular series since $\lim_{n \to \infty} \frac{a_{n+1}}{a_n} > 1$, the series diverges, making **(B)** correct.

29. C

After determining the equation of the line tangent to the function $f(x)$ at $x = 2$, you should then substitute 2.1 for x and solve for y.

$$f(x) = 4x^3 - 5x^2 - x$$
$$f(2) = 4(2)^3 - 5(2)^2 - (2)$$
$$f(2) = 10$$
$$f'(x) = 12x^2 - 10x - 1$$
$$f'(2) = 27$$
$$y - 10 = 27(x - 2)$$
$$y - 10 = 27(2.1 - 2)$$
$$y - 10 = 2.7$$
$$y = 12.7$$

Therefore, **(C)** is correct. (A) contains an algebraic error after substituting the correct values into the tangent line equation. (B) is the function value at $x = 2$. (D) is the instantaneous rate of change at $x = 2$.

30. B

By letting $u = 1 - 3x$, $du = -3dx$. This means the integral must be multiplied by $-\dfrac{1}{3}$ when replacing the variable x with the variable u. The bounds must also be adjusted when rewriting the integral.

$$-\frac{1}{3}\int_{-2}^{1-3e} \frac{1}{u}\, du$$
$$= -\frac{1}{3}\ln u \Big|_{-2}^{1-3e}$$
$$= -\frac{1}{3}\big[\ln|1 - 3e| - \ln|-2|\big]$$
$$= -\frac{1}{3}\ln\left|\frac{1 - 3e}{2}\right|$$

Hence, **(B)** is correct. (A) subtracts the expressions after the natural logs instead of applying the properties of logarithms and dividing those values. (C) forgets to include the absolute value symbols with the natural log.

SECTION I, Part B

76. B

The inflection point occurs when the second derivative changes sign. Since the first derivative is provided, locating the turning point corresponds to a sign change in $f''(x)$. Just like $f'(x)$ describes the rate of change of $f(x)$, $f''(x)$ describes the rate of change of $f'(x)$. The turning point is at $x = 0.372$.

Therefore, **(B)** is correct. Your calculator can also graph a derivative for you with different steps based on which calculator you are using. You may be able to enter $\dfrac{d}{dx}\left(\dfrac{\sqrt{2x}}{x^3 + 2x + 1}\right)\Big|_{x=x}$ directly into the graphing part of your calculator. Now you can find the x-intercept by using the calculator. You still get the same result: $x = 0.372$. (D) is the y-coordinate of the turning point of $f'(x)$.

77. B

One way to approach this question is to separate the region into two part. The first is $r = 2\sin(\theta)$ from $\theta = 0$ to $\theta = \dfrac{\pi}{2}$.

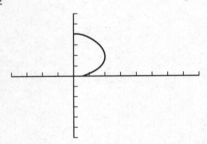

The second is $r = 2 + 2\cos(\theta)$ from $\theta = \dfrac{\pi}{2}$ to $\theta = \pi$.

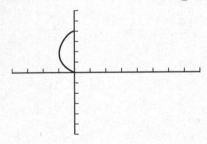

To find the areas, use the formula and substitute. Be sure the calculator is in radian mode. Since the lower and upper bounds are different, you must calculate the areas separately and add them for the final answer. Separating into different parts and using the symmetry of the graphs are both important approaches when calculating area in polar graph questions.

$$\text{Area} = \frac{1}{2}\int_0^{\frac{\pi}{2}}(2\sin(\theta))^2\, d\theta + \frac{1}{2}\int_{\frac{\pi}{2}}^{\pi}(2 + 2\cos(\theta))^2\, d\theta$$

$$\text{Area} = 1.571 + 0.712$$

$$\text{Area} = 2.283$$

Thus, **(B)** is correct. (C) forgets the necessary $\dfrac{1}{2}$. The other choices are combinations of incorrect integrals that do not represent the area common to both graphs.

78. B

In this question, $\Delta x = 0.3$ since the change in x is 0.6 and there are two equal-size steps. Since $\dfrac{dy}{dx} = x + y + 2$, use (1, 5) to get $\dfrac{dy}{dx}_{(1, 5)} = 8$ for the first approximation.

Use the formula $\Delta y = \Delta x \dfrac{dy}{dx}$ to find the next step.

$$\Delta y = 0.3(8)$$
$$\Delta y = 2.4$$

This makes the next coordinate (1.3, 7.4). Repeat this process using the same Δx. Just remember to calculate $\dfrac{dy}{dx}$ at (1.3, 7.4)

$$\frac{dy}{dx}_{(1.3, 7.4)} = 10.7$$
$$\Delta y = 0.3(10.7)$$
$$\Delta y = 3.21$$

When Δx and Δy are added to the coordinates (1.3, 7.4), the result is (1.6, 10.61). This question is asking for an approximation of $g(1.6)$ so only the y-coordinate is necessary, 10.61.

Hence, **(B)** is correct. (C) uses a step size of 0.5, and (D) uses a step size of $\dfrac{1}{3}$ as opposed to 0.3.

79. D

This is a calculator-allowed question, so there is no need to take any antiderivatives. The upper and lower limits are also both provided. Use the calculator to evaluate the integral $\int_{-\frac{4}{5}}^{\frac{4}{5}} \left(4 + \ln(\cos^2 x)\right) dx$.

$$\int_{-\frac{4}{5}}^{\frac{4}{5}} \left(4 + \ln\left(\cos^2 x\right)\right) dx = 6.03367$$

The question asks for the closest integer approximation, which is 6, **(D)**. It is critical that you can quickly evaluate definite integrals and derivatives using the calculator.

80. A

The slope of the tangent line to the path of the particle is equal to the derivative $\dfrac{dy}{dx}$. In parametric motion problems, $\dfrac{dy}{dx} = \dfrac{\frac{dy}{dt}}{\frac{dx}{dt}}$. The answer can be found by calculating $\dfrac{dy}{dx} = \dfrac{y'(2)}{x'(2)}$. These functions were given, so carefully evaluate them using the calculator in radian mode. The result is -3.737.

Thus, **(A)** is correct. (B) occurs if you mistakenly divide $\dfrac{dx}{dt}$ by $\dfrac{dy}{dt}$. (C) and (D) are incorrect because they both use the second derivative.

81. C

Velocity is the first derivative of the position function. Remember to use your calculator to find the numerical derivative at $t = 8$. The result is 0.991.

Therefore, **(C)** is correct. (A) is the acceleration at time $t = 8$. (B) is the average value of the position function between times $t = 0$ and $t = 8$. (D) is the position function at $t = 8$.

82. B

Use your calculator to evaluate the definite integral $\int_{-2}^{3} \left(\dfrac{3\sin x + 2}{x^2 - 2x + 3}\right) dx$ and then divide by $3 - (-2) = 5$. Be careful when entering complex functions into the calculator by making sure to include parentheses when necessary. In this case, the parentheses should be used to separate the numerator and the denominator clearly.

Hence, **(B)** is correct. (A) is the result if you do not include parentheses around the numerator when entering the rational expression into the calculator. (D) is the solution to the definite integral without dividing by 5.

83. B

If $H(t)$ represents the temperature of the hot chocolate after t minutes, then use the calculator to evaluate $H(4)$.

$$H(4) = H(0) + dt \int_0^4 R(t)$$

$$H(4) = 197 + dt \int_0^4 R(t)$$

$$H(4) = 197 - 26.255$$

$$H(4) = 170.745$$

To the nearest degree, the temperature of the hot chocolate after 4 minutes is 171°F.

Therefore, **(B)** is correct. (A) represents the net change, which in this case is 26°F, and does not account for the initial temperature. (C) is the result if you input the rate in the calculator without including the 0 in the tenths place of the exponent: $R(t) = -7.2e^{-0.47t}$. This is not the given function for the rate of change. (D) occurs if you mistakenly subtract the integral. Notice the integral of the rate is already negative, representing a negative net change of the temperature. If you subtract the integral, you increase the temperature of the hot chocolate. Also note that if the hot chocolate is cooling, there would be no reason that the temperature increased after the first 4 minutes.

84. D

To calculate total usage, find the integral of the rate from $t = 0$ to $t = 4$: $\int_0^4 r(t)\,dt = 764.776$. To the nearest kilowatt, this is 765.

Thus, **(D)** is correct. **(B)** is incorrect. If you mistakenly add 180 and 280 to get $r(t) = 460 \sin\left(\dfrac{t}{50}\right)$ in the original function, then you would get 76 as the amount of energy. **(C)** is incorrect because your calculator must be set in radian mode and not in degrees.

85. B

When you are told the value is decreasing at a proportional rate, you should use the exponential model. Since $\dfrac{dV}{dt} = kV$, you know that $V = V_0 e^{kt}$. Substituting the initial value $V_0 = 750$ makes the equation $V = 750 e^{kt}$. You are also given the value of the tablet after 2 years. Substituting that information allows you to solve for k.

$$480 = 750 e^{2k}$$
$$0.64 = e^{2k}$$
$$\ln 0.64 = 2k$$
$$k = \frac{\ln 0.64}{2}$$

Using the initial value, $k = \dfrac{\ln 0.64}{2}$, and $t = 5$ will now allow you to find the value of the tablet after 5 years.

$$V = 750 e^{\frac{5 \ln 0.64}{2}}$$
$$V = 245.76$$
$$V \approx 246$$

Hence, **(B)** is correct. You could also use $V = 480 e^{\frac{3 \ln 0.64}{2}}$ to solve for V. **(A)** is the answer to $V = 480 e^{\frac{5 \ln 0.64}{2}}$, which is the value of the tablet after 7 years. **(C)** is the answer to $V = 750 e^{\frac{3 \ln 0.64}{2}}$. If you are going to use a time of 3 years, you must use the value at year 2 as the initial value.

86. C

Relative or local extrema on $f(x)$ correspond with sign changes in $f'(x)$, which is provided. Use the calculator to graph $f'(x)$. Make sure to create a window that is representative of the interval provided in the question. The graph of $f'(x)$ has 3 sign changes on the given interval, making **(C)** the correct answer.

(A) is the number of inflection points on $f'(x)$. **(B)** is the number of relative extrema for $f'(x)$ on the same interval.

87. B

Differentiate $y = \dfrac{4}{5} x^{\frac{5}{4}}$ using the power rule to get $\dfrac{dy}{dx} = x^{\frac{1}{4}}$. Substitute this into the arc length formula, $s = \int_a^b \sqrt{1 + \left(\dfrac{dy}{dx}\right)^2}\,dx$, which gives you

$$s = \int_0^4 \sqrt{1 + \left(x^{\frac{1}{4}}\right)^2}\,dx \quad \text{or} \quad s = \int_0^4 \sqrt{1 + \left(x^{\frac{1}{2}}\right)}\,dx.$$ Use the calculator to evaluate this definite integral. The answer is 6.076.

Therefore, **(B)** is correct. If you forgot to square $\dfrac{dy}{dx}$, you would get **(A)**. **(D)** is the result if you mistakenly leave the radical out of the arc length formula.

88. B

There are very few calculator Taylor polynomial questions on the exam; however, they appear often enough that one is usually included as a multiple-choice question. Resist the urge to simply type ln (1.4) into the calculator, or you will incorrectly wind up with (C). To write the 4th-degree Taylor polynomial, you can use the ln (x) formula. Alternatively, you can accomplish this by hand by taking the derivatives, evaluating the first four derivatives at $x = 1$, and plugging them into the power series formula.

$$\ln(x) = (x-1) - \frac{1}{2}(x-1)^2 + \frac{1}{3}(x-1)^3 - \frac{1}{4}(x-1)^4 + \cdots$$

$$+ (-1)^{n+1}\frac{1}{n}(x-1)^n$$

Since this is a power series approximation, you will denote it as $P(x)$.

$$P(x) \approx (x-1) - \frac{1}{2}(x-1)^2 + \frac{1}{3}(x-1)^3 - \frac{1}{4}(x-1)^4$$

Using your calculator to evaluate $x = 1.4$ results in 0.335.

Thus, **(B)** is correct. If you mistakenly used a 3rd-degree Taylor polynomial, $P(x) \approx (x-1) - \frac{1}{2}(x-1)^2 + \frac{1}{3}(x-1)^3$, to evaluate $P(1.4)$, you would get 0.341, which is (D) and is incorrect.

89. D

The rate is given as $\frac{dP}{dt} = 0.03P$, which means the solution fits the exponential model $P = P_0 e^{0.03t}$, where P_0 represents the starting population. The goal is for the population to double, so you could solve $2P_0 = P_0 e^{0.03t}$. You will notice, though, that the first step is to divide both sides by P_0, leaving $2 = e^{0.03t}$. Having a firm understanding of the exponential model will allow you to skip that first step. The calculator should be used to solve for t. If $2 = e^{0.03t}$, then $t = 23.105$.

Hence, **(D)** is correct. If you got (A), you used 0.3 as 3% instead of 0.03.

90. B

If the position is represented by $x(t)$, then

$$x(5) = x(1) + \int_1^5 v(t)\, dt.$$ It is important to remember the initial starting position to obtain the new position at $t = 5$. (A) ignores the starting position. (C) is if you mistakenly put the lower bound in the integral as $t = 0$ instead of correctly using $t = 1$. (D) adds the total distance traveled to the original starting position by using the absolute value of the velocity function. However, this question is asking for the new position so displacement is more important; do not calculate total distance traveled.

SECTION II, Part A

1. (a)

Solutions	Scoring Rubric
To find the area, integrate the function between $x = 5$ and $x = 14$. The function intersects the x-axis at $x = 14$. If you set the function equal to zero and solve for x, you can identify the x-intercept. $$A = \int_5^{14} 4\sqrt{x-5} - 12 \ dx$$ $$A = 36$$	1: correct bounds on integral 1: rest of integral set up correctly 1: $A = 36$

(b)

Solutions	Scoring Rubric
Use the formula: $V = \pi \int_a^b (R)^2 - (r)^2 \ dx$ $$V = \pi \int_5^{14} \left(2 - \left(4\sqrt{x-5} - 12\right)\right)^2 - (2)^2 \ dx$$ Enter into the calculator to get $V = 1{,}130.973$	1: use correct volume formula $V = \pi \int_a^b (R)^2 - (r)^2 \ dx$ 1: enter info into formula correctly $V = \pi \int_5^{14} \left(2 - \left(4\sqrt{x-5} - 12\right)\right)^2 - (2)^2 \ dx$ 1: $V = 1{,}130.973$

(c)

Solutions	Scoring Rubric
Since the axis of revolution is the y-axis, the upper and lower bounds as well as the integral must be written in terms of y. Solving $y = 4\sqrt{x-5} - 12$ for x yields $x = \left(\dfrac{y+12}{4}\right)^2 + 5$. The volume is therefore $$V = \pi \int_{-12}^{0} \left(\left(\dfrac{y+12}{4}\right)^2 + 5\right)^2 \ dy$$ Enter into the calculator to get $V = 2{,}684.177$	1: $x = \left(\dfrac{y+12}{4}\right)^2 + 5$ 1: set up integral correctly $V = \pi \int_{-12}^{0} \left(\left(\dfrac{y+12}{4}\right)^2 + 5\right)^2 \ dy$ 1: $V = 2{,}684.177$

Enter feedback here.

2. (a)

Solutions	Scoring Rubric
You can determine whether the speed is increasing or decreasing based on the signs of the velocity and acceleration functions at a particular time t. If the signs are the same, the particle is speeding up. If they are different, the particle is slowing down. Acceleration, $a(t)$, is the derivative of velocity, $v(t)$. The signs are the same at $t = 3$: $v(3) = 1.367$ and $a(3) = 8.798$. So the particle is speeding up.	1: $v(3) = 1.367$ or $a(3) = 8.798$ 1: speeding up

(b)

Solutions	Scoring Rubric
Use the calculator to graph the velocity function over the interval $[0, 4]$, and find the zeros of the function. At $t = 2.14$, the velocity function changes from positive to negative. At $t = 2.822$, the velocity changes from negative to positive.	1: $t = 2.14$ 1: $t = 2.822$ 1: justification

(c)

Solutions	Scoring Rubric
$$x(3) = x(0) + \int_0^3 v(t)\, dt$$ $$5 = x(0) + \int_0^3 v(t)\, dt$$ $$x(0) = 5 - \int_0^3 v(t)\, dt$$ $$x(0) = 5 - 7.7294$$ $$x(0) = -2.729$$ At time $t = 0$, the particle is at position $x = -2.729$	1: set up integral correctly 1: $x(0) = -2.729$

(d)

Solutions	Scoring Rubric
Total distance is found by integrating the absolute value of the velocity function from time $t = 0$ to $t = 3$. $$TD = \int_{t_1}^{t_2} |v(t)|\, dt$$ $$TD = \int_0^3 \left| 2 + 3\cos\left(\frac{t^2}{2}\right) \right|\, dt$$ $$TD = 8.618$$	1: set up integral correctly 1: total distance = 8.618

SECTION II, Part B

3. (a)

Solutions	Scoring Rubric
Using the ratio test,	1: determine the inequality to test is $-1 < x < 1$
$$\lim_{n \to \infty} \left| \frac{(-1)^{n+1} x^{2(n+1)+3}}{2(n+1)+4} \cdot \frac{2n+4}{(-1)^n x^{2n+3}} \right| < 1$$	1: test the $x = -1$ endpoint
$$\lim_{n \to \infty} \left| \frac{(-1)^{n+1} x^{2n+5}}{2n+6} \cdot \frac{2n+4}{(-1)^n x^{2n+3}} \right| < 1$$	1: test the $x = 1$ endpoint
$$\lim_{n \to \infty} \left| (-1) x^2 \frac{2n+4}{2n+6} \right| < 1$$	1: the interval of convergence is $-1 \leq x \leq 1$
$$\lim_{n \to \infty} x^2 \frac{2n+4}{2n+6} < 1$$	
Then	
$$\lim_{n \to \infty} x^2 \frac{2n+4}{2n+6} < 1$$	
$$x^2 < 1$$	
$$-1 < x < 1$$	
Test each of these points, $x = -1$ and $x = 1$, for convergence. Test at $x = -1$.	
$$\sum (-1)^n \frac{x^{2n+3}}{2n+4}$$	
$$\sum (-1)^{n+1} \frac{(-1)^{2n+3}}{2n+4}$$	
$$\sum \frac{(-1)^{3n+4}}{2n+4}$$	
This is an alternating harmonic series that converges. Now test at x	
$$\sum (-1)^n \frac{x^{2n+3}}{2n+4}$$	
$$\sum (-1)^{n+1} \frac{(1)^{2n+3}}{2n+4}$$	
$$\sum (-1)^{n+1} \frac{1}{2n+4}$$	
This is also an alternating harmonic series that converges. The interval of convergence is therefore $-1 \leq x \leq 1$.	

(b)

Solutions	Scoring Rubric
Since the Maclaurin series for g evaluated at $x = \frac{1}{2}$ is an alternating series whose terms decrease in absolute value to 0, the approximation for $g\left(\frac{1}{2}\right)$ using the first two nonzero terms will always be less than the first omitted term. In this case that would be the third term. $$\left|g\left(\frac{1}{2}\right) - \frac{5}{192}\right| < \frac{x^7}{8}$$ $$\left|g\left(\frac{1}{2}\right) - \frac{5}{192}\right| < \left(\frac{1}{2}\right)^7 \frac{1}{8}$$ $$\left|g\left(\frac{1}{2}\right) - \frac{5}{192}\right| < \left(\frac{1}{128}\right)\frac{1}{8}$$ $$\left|g\left(\frac{1}{2}\right) - \frac{5}{192}\right| < \frac{1}{1,024}$$ $$\frac{1}{1,024} < \frac{1}{1,000}$$	1: determine that the first omitted term is the third term 1: correct work shown 1: $\dfrac{1}{1,024} < \dfrac{1}{1,000}$

(c)

Solutions	Scoring Rubric
To find the first three nonzero terms, take the derivative of the first three given terms for the series g. To find the general term, use the power rule on the general term of the series for g and simplify. $$g(x) = \frac{x^3}{4} - \frac{x^5}{6} + \frac{x^7}{8} - \cdots$$ $$g'(x) = \frac{3x^{(3-1)}}{4} - \frac{5x^{(5-1)}}{6} + \frac{7x^{(7-1)}}{8} - \cdots$$ $$g'(x) = \frac{3x^2}{4} - \frac{5x^4}{6} + \frac{7x^6}{8} - \cdots$$	1: correct work shown 1: $g'(x) = \dfrac{3}{4}x^2 - \dfrac{5}{6}x^4 + \dfrac{7}{8}x^6 - \cdots (-1)^n \dfrac{2n+3}{2n+4}x^{2n+2} + \cdots$

4. (a)

Solutions	Scoring Rubric
The derivative should be used to determine when a function is increasing or decreasing. Differentiate the function and set the derivative equal to 0. The solutions will represent critical values, which define the test intervals for determining if the function is increasing or decreasing. Values that make the derivative undefined should also be used as critical values, but that will not apply in this problem.	1: $f'(x) = 3x^2 - 12x + 9$ 1: critical values are $x = 1, 3$ 1: increasing intervals $(-\infty, 1)$ and $(3, \infty)$

$$f(x) = x^3 - 6x^2 + 9x - 3$$
$$f'(x) = 3x^2 - 12x + 9$$
$$0 = 3x^2 - 12x + 9$$
$$0 = 3(x^2 - 4x + 3)$$
$$0 = 3(x - 1)(x - 3)$$
$$x = 1, 3$$

The test intervals are $(-\infty, 1)$, $(1, 3)$, and $(3, \infty)$. It may be helpful to organize this information on a number line or in a table to determine where the function is increasing or decreasing. If the derivative is positive on the test interval, then $f(x)$ is increasing on that interval. If the derivative is negative on the test interval, then $f(x)$ is decreasing on that interval.

Interval	Sign of $f'(x)$
$(-\infty, 1)$	$+$
$(1, 3)$	$-$
$(3, \infty)$	$+$

The function is increasing on the intervals $(-\infty, 1)$ and $(3, \infty)$.

(b)

Solutions	Scoring Rubric
To analyze the concavity of the function, set the second derivative equal to 0 and solve to identify the critical values. Like in part (a), x-values that make the second derivative undefined should also be included as critical values, but that will not apply for this problem. The critical values will create the test intervals for the second derivative. If the second derivative is positive on the test interval, then $f(x)$ is concave up on that interval. If the second derivative is negative on the test interval, then $f(x)$ is concave down on that interval.	1: $f''(x) = 6x - 12$ 1: test intervals are $(-\infty, 2)$ and $(2, \infty)$ 1: decreasing and concave down on the interval $(1, 2)$

$$f'(x) = 3x^2 - 12x + 9$$
$$f''(x) = 6x - 12$$
$$0 = 6x - 12$$
$$0 = 6(x - 2)$$
$$x = 2$$

The test intervals are $(-\infty, 2)$ and $(2, \infty)$.

Interval	Sign of $f''(x)$
$(-\infty, 2)$	−
$(2, \infty)$	+

The function is concave down on the interval $(-\infty, 2)$. Combining this with the information obtained from the first derivative allows you to conclude that the function is decreasing and concave down on the interval $(1, 2)$.

(c)

Solutions	Scoring Rubric
There are a few ways to identify local extrema. One is to look for sign changes in the first derivative. Since the sign of the first derivative changes from positive to negative at $x = 1$, there must be a local maximum when $x = 1$. Since the sign of the first derivative changes from negative to positive at $x = 3$, there must be a local minimum when $x = 3$. Another way to determine these values is to use the second derivative. Since $x = 1$ is a critical value in the first derivative test and based on the second derivative test the function is concave down when $x = 1$, there must be a maximum there. Since $x = 3$ is a critical value in the first derivative test and based on the second derivative test the function is concave up when $x = 3$, there must be a minimum there. To find the corresponding y-values, you must substitute 1 and 3 for x in the original function and evaluate.	1: properly use the first and/or second derivative to locate where the local extremas ocur 1: $x = 1$ and $x = 3$ are where the local extrema are located 1: local extremas $(1, 1)$ and $(3, -3)$

$$f(1) = (1)^3 - 6(1)^2 + 9(1) - 3$$
$$f(1) = 1$$
$$f(3) = (3)^3 - 6(3)^2 + 9(3) - 3$$
$$f(3) = -3$$

Therefore, the local extrema are located at $(1, 1)$ and $(3, -3)$.

5. (a)

Solutions	Scoring Rubric
The series for $\ln(1 + x)$ was given by the following: $x - \dfrac{x^2}{2} + \dfrac{x^3}{3} - \dfrac{x^4}{4} + \cdots + (-1)^{n+1} \dfrac{x^n}{n} + \cdots$ In order to find $f(x) = x^2 \ln\left(1 + \dfrac{x}{2}\right)$, first replace each x in the formula with $\dfrac{x}{2}$. Then simplify and multiply by x^2.	1: substitute $\dfrac{x}{2}$ in for all x 1: correct work shown 1: $\dfrac{x^3}{2} - \dfrac{x^4}{8} + \dfrac{x^5}{24} - \dfrac{x^6}{64} + \cdots + (-1)^{n+1} \dfrac{x^{n+2}}{2^n \cdot n}$

$$\left(x^2\right)\left[\left[\frac{x}{2} - \frac{\left(\frac{x}{2}\right)^2}{2} + \frac{\left(\frac{x}{2}\right)^3}{3} - \frac{\left(\frac{x}{2}\right)^4}{4} + \cdots + (-1)^{n+1} \frac{\left(\frac{x}{2}\right)^n}{n} + \cdots\right]\right]$$

$$= \frac{x^3}{2} - \frac{x^4}{8} + \frac{x^5}{24} - \frac{x^6}{64} + \cdots + (-1)^{n+1} \frac{x^{n+2}}{2^n \cdot n}$$

(b)

Solutions	Scoring Rubric
To find an interval of convergence, first determine the radius of convergence and then test the endpoints for convergence. Use the Ratio Test and solve the equation $\lim\limits_{n \to \infty}\left\|\dfrac{u_{n+1}}{u_n}\right\| < 1$ on $(-1)^{n+1}\dfrac{x^{n+1}}{2^n \cdot n}$	1: determine that the interval $-2 \le x \le 2$ needs to be tested 1: test both endpoints 1: the interval of convergence is therefore $-2 < x \le 2$.

$$\lim_{n \to \infty}\left|\frac{u_{n+1}}{u_n}\right| < 1$$

$$\lim_{n \to \infty}\left|\frac{(-1)^{n+1+1}x^{n+2+1}}{2^{n+1}\cdot(n+1)}\cdot\frac{2^n\cdot 2}{(-1)^{n+1}x^{n+2}}\right| < 1$$

$$\lim_{n \to \infty}\left|-\frac{1}{2}x\cdot\frac{n}{n-1}\right| < 1$$

$$\left|-\frac{1}{2}x\right| < 1$$

$$2|x| < 1$$

$$|x| < 2$$

Test each of these points, $x = 2$ and $x = -2$, for convergence. Test at $x = 2$.

$$\sum(-1)^{n+1}\frac{x^{n+2}}{2^n\cdot 2}$$

$$\sum(-1)^{n+1}\frac{(2)^{n+2}}{2^n\cdot 2}$$

$$\sum(-1)^{n+1}\frac{4}{n}$$

$$4\sum(-1)^{n+1}\frac{1}{n}$$

This is an alternating harmonic series that converges. Now test at $x = -2$.

$$\sum(-1)^{n+1}\frac{(-2)^{n+2}}{2^n\cdot 2}$$

$$\sum(-1)^{n+1}\frac{(-1)^{n+2}(2)^{n+2}}{2^n\cdot 2}$$

$$\sum(-1)^{2n+3}\frac{2^n\cdot 2^2}{2^n\cdot n}$$

$$\sum(-1)^{2n+3}\frac{4(1)^n}{n}$$

$$-4\sum\frac{1}{n}$$

This is a harmonic series that diverges. The series does not alternate and is strictly negative. The interval of convergence is therefore $-2 < x \le 2$.

(c)

Solutions	Scoring Rubric
Since $P_5(x)$ is a 5th-degree polynomial, take the absolute value of the term raised to the 6th power in order to use the alternating series error bound formula.	1: determine that the 6th-degree term needs to be used
$$\frac{x^3}{2} - \frac{x^4}{8} + \frac{x^5}{24} - \frac{x^6}{64} + \cdots + (-1)^{n+1}\frac{x^{n+2}}{2^n \cdot n} + \cdots$$	1: correct work shown
The term raised to the 6th-degree is $-\frac{x^6}{64}$. Using the x-coordinate of $\frac{1}{3}$ and the absolute value, the error is less than $\frac{1}{3^6 \cdot 64}$.	1: $\dfrac{1}{3^6 \cdot 64}$
$$\left| -\frac{x^6}{64} \right|$$ $$= \left| -\frac{\left(\frac{1}{3}\right)^6}{64} \right|$$ $$= \left| -\frac{1}{3^6 \cdot 64} \right|$$ $$= \frac{1}{3^6 \cdot 64}$$	

6. (a)

Solutions	Scoring Rubric
The derivative of $f(x)$ is $$f'(x) = 2\sqrt{4 + (2x)^3}$$ The derivative of $g(x)$ is $$g'(x) = f'(\cos x)(-\sin x)$$ $$f'(\cos x) = 2\sqrt{4 + (2\cos x)^3}$$ $$g'(x) = 2\sqrt{4 + (2\cos x)^3}\,(-\sin x)$$ $$g'(x) = (-2\sin x)\sqrt{4 + (2\cos x)^3}$$	1: $f'(x) = 2\sqrt{4 + (2x)^3}$ 1: $g'(x) = (-2\sin x)\sqrt{4 + (2\cos x)^3}$

(b)

Solutions	Scoring Rubric
In order to write the equation of the tangent line, first find the y-coordinate of the g function for $x = \dfrac{\pi}{2}$ $$g\left(\frac{\pi}{2}\right) = f\left(\cos\left(\frac{\pi}{2}\right)\right)$$ $$g\left(\frac{\pi}{2}\right) = f(0)$$ $$g\left(\frac{\pi}{2}\right) = 0$$ Next, find the slope using $$g'(x) = (-2\sin x)\sqrt{4 + (2\cos x)^3} \text{ at } x = \frac{\pi}{2}$$ $$g'\left(\frac{\pi}{2}\right) = \left(-2\sin\left(\frac{\pi}{2}\right)\right)\sqrt{4 + \left(2\cos\left(\frac{\pi}{2}\right)\right)^3}$$ $$g'\left(\frac{\pi}{2}\right) = (-2)(\sqrt{4})$$ $$g'\left(\frac{\pi}{2}\right) = -4$$ The tangent line is written using $y - y_1 = m(x - x_1)$, $$y - 0 = -4\left(x - \frac{\pi}{2}\right)$$ $$y = -4\left(x - \frac{\pi}{2}\right)$$	1: $g\left(\dfrac{\pi}{2}\right) = 0$ 1: $g'\left(\dfrac{\pi}{2}\right) = -4$ 1: $y = -4\left(x - \dfrac{\pi}{2}\right)$

(c)

Solutions	Scoring Rubric
In order to approximate the value of $g(0)$ using the equation of the tangent line, plug $x = 0$ in and simplify. $g(0) \approx 2\pi$ To determine if this is an over or under approximation, you must look at the concavity at $x = 0$ and that is determined by the second derivative. $g'(x) = (-2 \sin x)\sqrt{4 + (2 \cos x)^3}$ $g''(x) = (-2 \sin x)\frac{1}{2}\left(4 + (2 \cos x)^3\right)^{-\frac{1}{2}}(24 \cos^2 x)(-\sin x) + (-2 \cos x)\sqrt{4 + (2 \cos x)^3}$ $g''(0) = (-2 \sin(0))\frac{1}{2}\left(4 + (2 \cos(0))^3\right)^{-\frac{1}{2}}(24 \cos^2 (0))(-\sin(0)) + (-2 \cos(0))\sqrt{4 + (2 \cos(0))^3}$ $g''(0) = (-2 \cos(0))\sqrt{4 + (2 \cos(0))^3}$ $g''(0) < 0$ Since the second derivative is less than 0, the function must be concave down, making the tangent line approximation an over approximation.	1: $y \approx 2\pi$ 1: correct work for $g''(x)$ 1: over approximation 1: justification

BC Practice Exam 6

Section I, Part A

1. (A) (B) (C) (D)
2. (A) (B) (C) (D)
3. (A) (B) (C) (D)
4. (A) (B) (C) (D)
5. (A) (B) (C) (D)
6. (A) (B) (C) (D)
7. (A) (B) (C) (D)
8. (A) (B) (C) (D)
9. (A) (B) (C) (D)
10. (A) (B) (C) (D)
11. (A) (B) (C) (D)
12. (A) (B) (C) (D)
13. (A) (B) (C) (D)
14. (A) (B) (C) (D)
15. (A) (B) (C) (D)
16. (A) (B) (C) (D)
17. (A) (B) (C) (D)
18. (A) (B) (C) (D)
19. (A) (B) (C) (D)
20. (A) (B) (C) (D)
21. (A) (B) (C) (D)
22. (A) (B) (C) (D)
23. (A) (B) (C) (D)
24. (A) (B) (C) (D)
25. (A) (B) (C) (D)
26. (A) (B) (C) (D)
27. (A) (B) (C) (D)
28. (A) (B) (C) (D)
29. (A) (B) (C) (D)
30. (A) (B) (C) (D)

Section I, Part B

76. (A) (B) (C) (D)
77. (A) (B) (C) (D)
78. (A) (B) (C) (D)
79. (A) (B) (C) (D)
80. (A) (B) (C) (D)
81. (A) (B) (C) (D)
82. (A) (B) (C) (D)
83. (A) (B) (C) (D)
84. (A) (B) (C) (D)
85. (A) (B) (C) (D)
86. (A) (B) (C) (D)
87. (A) (B) (C) (D)
88. (A) (B) (C) (D)
89. (A) (B) (C) (D)
90. (A) (B) (C) (D)

A A A A A A A A A A A A A A A A A A A A A A A A A A A A A A A A A A A

CALCULUS BC
SECTION I, Part A
Time—60 minutes
Number of questions—30

A CALCULATOR MAY NOT BE USED ON THIS PART OF THE EXAM.

Directions: Use the space provided to solve each of the following problems. After examining the given choices, decide which of the choices is the best and fill in the corresponding circle on the answer sheet. Credit will only be given for answers appropriately marked on the answer sheet, and not for anything written in the exam book. Do not spend too much time on any one problem.

In this exam:

(1) The domain of a function f should be assumed to be the set of all real numbers x for which $f(x)$ is a real number, unless the question indicates otherwise.

(2) The inverse function notation f^{-1} or the prefix "arc" (e.g., $\sin^{-1} x = \arcsin x$) may be used to indicate the inverse of a trigonometric function.

GO ON TO THE NEXT PAGE.

AAAAAAAAAAAAAAAAAAAAAAAAAAAAAAAA

1. A rectangle in the first quadrant is bound by the x-axis and y-axis with the upper right-hand vertex on the curve $y = -\frac{1}{5}x^2 + 6$. For which value of x will the rectangle have the largest area?

   (A) $x = \sqrt{10}$

   (B) $x = \sqrt{30}$

   (C) $x = 4\sqrt{10}$

   (D) $x = 15$

2. Which of the following statements is true for $\displaystyle\sum_{k=1}^{\infty} \frac{6k^2}{k^3+1}$?

   (A) The series $\displaystyle\sum_{k=1}^{\infty} \frac{6k^2}{k^3+1}$ converges from the p-Series Test.

   (B) The series $\displaystyle\sum_{k=1}^{\infty} \frac{6k^2}{k^3+1}$ diverges from the Geometric Series Test.

   (C) The series $\displaystyle\sum_{k=1}^{\infty} \frac{6k^2}{k^3+1}$ diverges from the Integral Test.

   (D) The series $\displaystyle\sum_{k=1}^{\infty} \frac{6k^2}{k^3+1}$ converges from the Integral Test.

3. $\displaystyle\lim_{x \to 2} \frac{4x^2-4}{x^2-x-2} =$

   (A) -3

   (B) 0

   (C) 4

   (D) nonexistent

GO ON TO THE NEXT PAGE.

AAAAAAAAAAAAAAAAAAAAAAAAAAAAAAAAAAA

4. $\displaystyle\int u\csc^2(u)\,du =$

(A) $-u\cot(u) + \ln|\sin(u)| + C$

(B) $-u\cot(u) + \ln|\cos(u)| + C$

(C) $-u\tan(u) - \ln|\sin(u)| + C$

(D) $u\cot(u) + \ln|\sin(u)| + C$

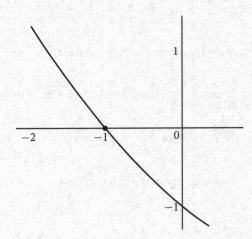

5. The graph of a twice differentiable function $f(x)$ is shown above. Which of the following is true?

(A) $f'(-1) < f(-1) < f''(-1)$

(B) $f''(-1) < f'(-1) < f(-1)$

(C) $f''(-1) < f(-1) < f'(-1)$

(D) $f'(-1) < f''(-1) < f(-1)$

GO ON TO THE NEXT PAGE.

A A A A A A A A A A A A A A A A A A A A A A A A A A A A A A A A A A

6. If $\int_0^1 f(x)\, dx = 3$ and $\int_0^5 f(x)\, dx = -5$, then $\int_1^5 (3f(x) + 2)\, dx =$

    (A) -32

    (B) -16

    (C) -7

    (D) 32

7. Which of the following series can be used with the Limit Comparison Test to determine whether the series $\sum_{n=1}^{\infty} \dfrac{n^2 + 2}{n^3 + 3}$ converges or diverges?

    (A) $\sum_{n=1}^{\infty} \dfrac{1}{n}$

    (B) $\sum_{n=1}^{\infty} \dfrac{1}{n^3 + 3}$

    (C) $\sum_{n=1}^{\infty} \dfrac{n + 2}{n + 3}$

    (D) $\sum_{n=1}^{\infty} \dfrac{\frac{2}{3}}{n^2 + 1}$

GO ON TO THE NEXT PAGE.

AAAAAAAAAAAAAAAAAAAAAAAAAAAAAAAA

8. $\int_{-2}^{3} \frac{|x|}{x} dx$ is

   (A) -1

   (B) 0

   (C) 1

   (D) 5

---

9. If $f'(x) = (x+2)^2(x-1)(x-5)$, where does $f(x)$ have a local maximum?

   (A) $x = -2$, $x = 1$, and $x = 5$

   (B) $x = -2$ and $x = 5$ only

   (C) $x = 5$ only

   (D) $x = 1$ only

---

10. A particle moves along the x-axis so that at any time $t \geq 0$, its position is given by $x(t) = (\sin(3t))^2$. What is the speed of the particle at $t = \frac{\pi}{4}$?

   (A) -3

   (B) -1

   (C) 1

   (D) 3

GO ON TO THE NEXT PAGE.

AAAAAAAAAAAAAAAAAAAAAAAAAAAAAAAA

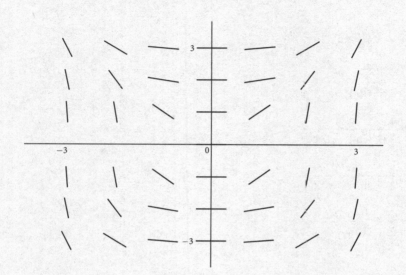

11. The slope field shown above corresponds with which differential equation?

(A) $\dfrac{dy}{dx} = \dfrac{x^3}{y^2}$ (B) $\dfrac{dy}{dx} = \dfrac{x^2}{y^3}$ (C) $\dfrac{dy}{dx} = \dfrac{x^3}{y}$ (D) $\dfrac{dy}{dx} = \dfrac{x}{y}$

12. The Alternating Series Test can be used to show convergence for which of the following alternating series?

I. $4 - \dfrac{1}{4} + \dfrac{12}{27} - \dfrac{1}{16} + a_n + \cdots$ where $a_n = \begin{cases} \dfrac{12}{3^n} & \text{if } n \text{ is odd} \\[2mm] -\dfrac{1}{2^n} & \text{if } n \text{ is even} \end{cases}$

II. $\dfrac{1}{2} - \dfrac{1}{4} + \dfrac{1}{6} - \dfrac{1}{8} + a_n + \cdots$ where $a_n = \dfrac{(-1)^{n+1}}{2n}$

III. $-\dfrac{2}{5} + \dfrac{3}{8} - \dfrac{4}{11} + \dfrac{5}{14} + a_n + \cdots$ where $a_n = (-1)^n \left(\dfrac{n+1}{3n+2} \right)$

(A) I only

(B) II only

(C) III only

(D) II and III only

GO ON TO THE NEXT PAGE.

AAAAAAAAAAAAAAAAAAAAAAAAAAAAAAAA

13. Given $\cos y - x^3 y^2 = 2y$, find $\dfrac{dy}{dx}$.

(A) $\dfrac{dy}{dx} = \dfrac{3x^2 y^2}{2 + \sin y + 2x^3 y}$

(B) $\dfrac{dy}{dx} = \dfrac{-3x^2 y^2}{2 + \sin y - 2x^3 y}$

(C) $\dfrac{dy}{dx} = \dfrac{-3x^2 y^2}{2 - \sin y + 2x^3 y}$

(D) $\dfrac{dy}{dx} = \dfrac{-3x^2 y^2}{2 + \sin y + 2x^3 y}$

---

14. Using the substitution $u = \sqrt{2x}$, $\int_2^8 \dfrac{e^{\sqrt{2x}}}{\sqrt{2x}}\, dx$ is equal to which of the following?

I. $\displaystyle\int_2^8 e^u\, du$

II. $\displaystyle\int_2^4 e^u\, du$

III. $e^{2x}\big|_2^8$

(A) I only

(B) II only

(C) I and II only

(D) II and III only

GO ON TO THE NEXT PAGE.

15. Which of the following is a power series expansion of $2(e^x + e^{-x})$?

   (A) $1 + \dfrac{1}{2!}x^2 + \dfrac{1}{4!}x^4 + \ldots + \dfrac{1}{(2n)!}x^{2n}$

   (B) $2 + \dfrac{2}{2!}x^2 + \dfrac{2}{4!}x^4 + \ldots + \dfrac{2}{(2n)!}x^{2n}$

   (C) $4 + \dfrac{4}{2!}x^2 + \dfrac{4}{4!}x^4 + \ldots + \dfrac{4}{(2n)!}x^{2n}$

   (D) $4 + 4x + \dfrac{4}{2!}x^2 + \dfrac{4}{3!}x^3 + \ldots + \dfrac{4}{n!}x^n$

16. Which of the following is(are) true?

   I. $\displaystyle\lim_{h \to 0} \dfrac{\left((x+h)^3 - 5\right) - \left(x^3 - 5\right)}{h} = 3x^2$

   II. $\displaystyle\lim_{h \to 0} \dfrac{\left((4+h)^2\right) - 16}{h} = 8$

   III. $\displaystyle\lim_{h \to 0} \dfrac{\left((x+h)^2 + 8x + 8h\right) - x^2 - 8x}{2h} = x + 4$

   (A) I only

   (B) II only

   (C) I and II only

   (D) I, II, and III

GO ON TO THE NEXT PAGE.

AAAAAAAAAAAAAAAAAAAAAAAAAAAAAAAA

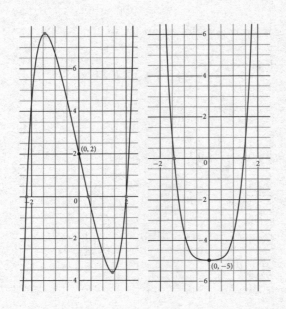

17. Given the graphs of $f(x)$ and $f'(x)$ above, what is the best approximation for $f(-0.2)$ using the line tangent to f at $x = 0$?

(A) -4 (B) 1   (C) 2   (D) 3

18. Which of the following series is(are) conditionally convergent?

I. $\displaystyle\sum_{n=1}^{\infty} \frac{(-2)^n}{n!}$

II. $\displaystyle\sum_{n=1}^{\infty} \frac{(-1)^{2n}}{3n}$

III. $\displaystyle\sum_{n=1}^{\infty} \frac{(-1)^{n+3}(5)^n}{(7)^{n+1}}$

(A) II only

(B) I and II only

(C) III only

(D) none

GO ON TO THE NEXT PAGE.

AAAAAAAAAAAAAAAAAAAAAAAAAAAAAAAAAA

19. What are the values of x for which the function $f(x) = x^3 - 6x^2 - 36x + 12$ is decreasing?

   (A) $-2 < x < 6$

   (B) $x < -2$ or $x > 6$

   (C) $x < -6$ or $x > 2$

   (D) $-6 < x < 2$

---

20. What is the slope of the tangent line to the graph of $y = \dfrac{e^x}{x-3}$ at $x = 1$?

   (A) $-\dfrac{3}{4}e$ (B) $-\dfrac{1}{2}e$ (C) $-\dfrac{1}{4}e$ (D) e

---

21. $\displaystyle\int_0^x \cos(t^6)\, dt =$

   (A) $x - \dfrac{x^{13}}{13\cdot 2!} + \dfrac{x^{25}}{25\cdot 4!} - \dfrac{x^{37}}{37\cdot 6!} + \ldots + (-1)^{n-1}\dfrac{\left(x^6\right)^{2n-2+1}}{(6(2n-2)+1)(2n-2)!}$

   (B) $x - \dfrac{x^{12}}{2!} + \dfrac{x^{24}}{4!} - \dfrac{x^{36}}{6!} + \ldots + (-1)^{n-1}\dfrac{\left(x^6\right)^{2n-2}}{(2n-2)!}$

   (C) $t + \dfrac{t^{13}}{13\cdot 2!} + \dfrac{t^{25}}{25\cdot 4!} + \dfrac{t^{37}}{37\cdot 6!} + \ldots + (-1)^{n-1}\dfrac{\left(t^6\right)^{2n-2+1}}{(6(2n-2)+1)(2n-2)!}$

   (D) $-x + \dfrac{x^{13}}{13\cdot 2!} - \dfrac{x^{25}}{25\cdot 4!} + \dfrac{x^{37}}{37\cdot 6!} + \ldots - (-1)^{n-1}\dfrac{\left(x^6\right)^{2n-2+1}}{(6(2n-2)+1)(2n-2)!}$

GO ON TO THE NEXT PAGE.

AAAAAAAAAAAAAAAAAAAAAAAAAAAAAAAA

22. Let f be the function given by $f(x) = x^3 - 6x^2$. What are all values of c that satisfy the conclusion of the mean value theorem of differential calculus on the closed interval $[0, 6]$?

    (A) 0 only   (B) 4 only   (C) 0 and 4   (D) 4 and 6

23. Let $f(x)$ be the function defined by $f(x) = \int_0^x \left(2t^3 - 9t^2 + 12t\right) dt$. On which of the following intervals is the graph of $y = f(x)$ concave up?

    (A) $(1, 2)$ only

    (B) $(-\infty, 1)$ and $(2, \infty)$

    (C) $(0, \infty)$ only

    (D) $(2, \infty)$ only

24. $\int \left(\dfrac{1}{4x + 5} + \dfrac{1}{x + 7} + \dfrac{5x - 7}{x^2 - 49}\right) dx =$

    (A) $\frac{1}{4}\ln|4x + 5| + \ln\left[(x - 7)^2 \cdot (x + 7)^4\right] + C$

    (B) $\frac{1}{4}\ln|4x + 5| + \ln\left[(x - 7)^2 \cdot (x + 7)^3\right] + C$

    (C) $\ln\left[(4x + 5)^4 \cdot (x - 7)^2 \cdot (x + 7)^4\right] + C$

    (D) $\ln\left[(4x + 5) \cdot (x - 7) \cdot (x + 7)\right] + C$

GO ON TO THE NEXT PAGE.

A A A A A A A A A A A A A A A A A A A A A A A A A A A A A A A A A

25. A particle moves along the x-axis so that at time $t \geq 0$, the acceleration of the particle is $a(t) = 30\sqrt{t}$. The position of the particle is 16 at $t = 0$, and the position of the particle is 9 at $t = 1$. What is the velocity at $t = 0$?

(A) -15 (B) 9   (C) 15   (D) 30

x	0	1	2	3	4	5	6
$f'(x)$	5	6	4	7	6	8	3

26. The function f satisfies $f(0) = 15$. The first derivative of f satisfies the inequality $0 \leq f'(x) \leq 8$ for all x in the interval $[0, 6]$. Selected values of $f'(x)$ are shown in the table above. The function f has a continuous second derivative for all real numbers. Use a midpoint rectangular approximation sum with 3 subintervals of equal length as indicated by the table to approximate $f(6)$.

(A) 41   (B) 42   (C) 45   (D) 57

27. Which of the following statements is false?

(A) If $\lim\limits_{n \to \infty} a_n = 0$, then the series $\sum\limits_{n=1}^{\infty} a_n$ converges.

(B) If $\sum\limits_{n=1}^{\infty} a_n$ converges, then $\lim\limits_{n \to \infty} a_n = 0$.

(C) If $0 < a_n < b_n$ and if $\sum\limits_{n=1}^{\infty} b_n$ converges, then $\sum\limits_{n=1}^{\infty} a_n$ converges.

(D) If $\lim\limits_{n \to \infty} a_n \neq 0$, then $\sum\limits_{n=1}^{\infty} a_n$ diverges.

GO ON TO THE NEXT PAGE.

A A A A A A A A A A A A A A A A A A A A A A A A A A A A A A A A A

28. What are all values of x for which the series $\sum_{n=1}^{\infty} \frac{(x-2)^n}{n}$ converges?

(A) $-2 \leq x < 2$ (B) $1 \leq x < 3$ (C) $1 < x < 3$ (D) $x > 3$

29. Which of the following is not a higher-order derivative of $f(x) = \left(\frac{1}{2}x + 1\right)^5$?

(A) $\frac{5}{2}\left(\frac{1}{2}x + 1\right)^4$

(B) $5\left(\frac{1}{2}x + 1\right)^3$

(C) $15\left(\frac{1}{2}x + 1\right)^2$

(D) $\frac{15}{2}\left(\frac{1}{2}x + 1\right)$

30. If $f''(x) = \frac{|x^2 - 9|}{3 - x}$, then $f(x)$ is concave down on the interval

(A) $(-\infty, \infty)$

(B) $(-\infty, 3)$

(C) $(-3, \infty)$

(D) $(3, \infty)$

END OF PART A OF SECTION I

**IF YOU FINISH BEFORE TIME IS CALLED, YOU MAY
CHECK YOUR WORK ON PART B ONLY.**

CALCULUS BC
SECTION I, Part B
Time—45 minutes
Number of questions—15

A GRAPHING CALCULATOR IS REQUIRED FOR SOME QUESTIONS ON THIS PART OF THE EXAM.

Directions: Use the space provided to solve each of the following problems. After examining the given choices, decide which of the choices is the best and fill in the corresponding circle on the answer sheet. Credit will only be given for answers appropriately marked on the answer sheet, and not for anything written in the exam book. Do not spend too much time on any one problem.

In this exam:

(1) When the exact numerical value of the correct answer does not appear among the choices given, select the number that best approximates the numerical value.

(2) The domain of a function f should be assumed to be the set of all real numbers x for which $f(x)$ is a real number, unless the question indicates otherwise.

(3) The inverse function notation f^{-1} or the prefix "arc" (e.g., $\sin^{-1} x = \arcsin x$) may be used to indicate the inverse of a trigonometric function.

GO ON TO THE NEXT PAGE.

76. What is the area between the curves $r = 7\sin(5\theta)$ and $r = 4\sin(5\theta)$?

    (A) 1.279

    (B) 7.069

    (C) 25.918

    (D) 51.836

77. The population P of a city grows according to the differential equation $\frac{dP}{dt} = kP$, where k is a constant and t is time measured in years. If the population of the city triples every 8 years, what is the value of k ?

    (A) 0.137

    (B) 0.173

    (C) 0.214

    (D) 0.260

78. The base of a solid is a region in the first quadrant bounded by the x-axis, the y-axis, and the line $2x + 3y = 12$. If cross sections of the solid perpendicular to the x-axis are right isosceles triangles with a leg on the region, what is the volume of the solid?

    (A) 16

    (B) 32

    (C) 16π

    (D) 32π

GO ON TO THE NEXT PAGE.

79. An experimental population of fruit flies increases according to the law of exponential growth. There were 200 flies after the second day of the experiment and 600 flies after the fourth day. Approximately how many flies were in the original population?

(A) 22

(B) 67

(C) 115

(D) 172

80. The first derivative of the function $f(x)$ is given by $f'(x) = -x + 6e^{\cos x}$. How many points of inflection does the graph of f have on the interval $0 \leq x \leq 5\pi$?

(A) 2

(B) 3

(C) 4

(D) 5

81. A particle moves along the x-axis so that its velocity at $t \geq 0$ is given by $v(t) = \dfrac{t^2 - 3}{3t^2 + 2}$. What is the displacement from $t = 0$ to $t = 2$?

(A) 0.393

(B) 0.786

(C) 1.104

(D) 1.125

GO ON TO THE NEXT PAGE.

82. If $f(x) = \sum_{k=1}^{\infty} \left(\cos^2 x\right)^k$ then $f(1) =$

    (A) 0.412

    (B) 0.642

    (C) 1.557

    (D) 2.423

---

83. A particle moves along the x-axis so that at $t > 0$, its position is given by $x(t) = \sin\left(\sqrt{t}\right)$. What is the velocity of the particle the first time the particle is at the origin?

    (A) −0.159

    (B) 0

    (C) 0.008

    (D) 9.870

---

84. If $P(x)$ is the Maclaurin series approximation for $\sin(x)$ centered at $x = 2$, a polynomial of which degree guarantees that $\left|P(2) - f(2)\right| \leq \frac{1}{50}$?

    (A) 3

    (B) 4

    (C) 5

    (D) 7

GO ON TO THE NEXT PAGE.

85. Consider the curve $1.4x^{3.2} + 2.5y^{3.2} - 27.6xy = 0$. At which x-value(s) does the curve have a horizontal tangent?

    (A) $x = -3.84$, $x = 0$, and $x = 3.84$

    (B) $x = -2.84$, $x = 0$, and $x = 2.84$

    (C) $x = -1.84$, $x = 0$, and $x = 1.84$

    (D) only $x = 0$

86. Given the function $g(x) = \sqrt{x}$ for $x > 0$, which statement is true about the approximation for $g(14)$ using Euler's method with step sizes of 5 starting at $x = 4$?

    (A) The approximation is greater than $g(14)$.

    (B) The approximation is less than $g(14)$.

    (C) Using the coordinates $(4, 2)$, the first $\Delta y = \frac{1}{4}$.

    (D) For approximating $g(x)$, the y-coordinate is used to find Δy.

87. The position of a particle moving in the xy-plane is given by the parametric equations $x(t) = \sin(3^t)$ and $y(t) = \cos(3^t)$ for time $t \geq 0$. What is the speed of the particle at $t = 2.7$?

    (A) 0.640

    (B) 1

    (C) 2.542

    (D) 21.332

GO ON TO THE NEXT PAGE.

88. A polynomial $p(x)$ has a relative minimum at $(-3, 1)$, a relative maximum at $(1, 4)$, a relative minimum at $(4, -2)$, and no other critical points. How many zeros does $p(x)$ have?

    (A) 1

    (B) 2

    (C) 3

    (D) 4

---

89. What is the x-intercept of the tangent line to the function $f(x) = e^x + 3$ at $x = 1$?

    (A) -1.104

    (B) 1.104

    (C) 3

    (D) 5.350

---

90. If the derivative of function f is defined as $f'(x) = 4x^3 - 6x^2 + 6$ and if $f(1) = 7.158$, which of the following is $f(3.5)$?

    (A) 2.185

    (B) 31

    (C) 85.313

    (D) 87.471

END OF SECTION I

**IF YOU FINISH BEFORE TIME IS CALLED, YOU MAY
CHECK YOUR WORK ON PART B ONLY.**

CALCULUS BC
SECTION II, Part A
Time—30 minutes
Number of problems—2

A GRAPHING CALCULATOR IS REQUIRED FOR THESE PROBLEMS.

Directions: Write your solution to each part of the following questions in the space provided. Write clearly and legibly. Cross out any errors you make; erased or crossed-out work will not be scored.

You may wish to look over the problems before starting to work on them; on the actual test, it is not expected that everyone will be able to complete all parts of all problems. All problems are given equal weight, but the individual parts of a particular problem are not necessarily given equal weight. You should not spend too much time on any one problem.

- Show all of your work. Clearly label functions, graphs, tables, or anything else that you use to arrive at your final solution. Your work will be scored on the correctness and completeness of your methods as well as your final answers. Answers without supporting work will usually not receive credit.

- Justifications (i.e., the request that you "justify your answer") require that you give mathematical (non-calculator) reasons.

- Do not use calculator syntax. Your work must be expressed in standard mathematical notation.

- Numeric or algebraic answers do not need to be simplified, unless the question indicates otherwise.

- If you use decimal approximations in calculations, your final answers should be accurate to three places after the decimal point, unless the question indicates otherwise.

- The domain of a function f should be assumed to be the set of all real numbers x for which $f(x)$ is a real number, unless the question indicates otherwise.

GO ON TO THE NEXT PAGE.

1 1 1 1 1 1 1 1 1 1

x	1	2	3	4	5	6
f(x)	2	3	5	2	1	1.5

1. A large factory receives online orders for a special anniversary sale. They process these orders in one of two ways, either manually using factory workers or from a computer-generated robot. Orders come in according to the table above, where x represents the time in hours since 8 a.m. and $f(x)$ is the number of orders *in hundreds* per hour. Orders are processed and filled by either factory workers or a computer-generated robot. Both work an adjusted schedule and start work at 2 p.m. The factory workers' rate of processing orders is $W(t) = 132t - 38$ and the robot's rate is $R(t) = 25.8t^2 + 8.7t + 40$. In both functions, t represents the number of hours since 2 p.m., which is when the orders start to get processed.

   (a) Estimate the total number of online orders received during the first 6 hours using a right rectangular Riemann sum.

   (b) Find $\int_0^4 W(t)\, dt$. Using the correct units, explain the meaning of $\int_0^4 W(t)\, dt$ in the context of this problem.

GO ON TO THE NEXT PAGE.

1 1 1 1 1 1 1 1 1 1

(c) If only the factory workers work to fill the online orders, how long will they take to complete this work?

(d) Using the right rectangular Riemann sum, which is the more efficient way to fill the orders, with only the robot or with only the factory workers? Explain.

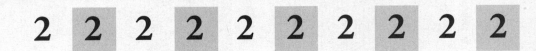

2. A metallurgist is making a memorial statue made of beryllium. The base of the statue is the region R in the first quadrant under the graph of $y = f(x)$ for $0 \le x \le 40$, where $f(x) = 10 \cos\left(\frac{\pi}{40}x\right) + 10$.

Both x and y are measured in feet. The derivative of f is $f'(x) = -\frac{\pi}{4} \sin\left(\frac{\pi}{40}x\right)$.

(a) The region R is cut out of a 20 by 40 feet rectangular sheet of beryllium and the remaining beryllium is discarded. Find the area of the discarded beryllium.

(b) The statue is a solid with base R. Cross sections of the statue perpendicular to the x-axis are squares. If the beryllium weighs 113.7 pounds per cubic foot, find the weight of the statue.

(c) Find the perimeter of the base of the statue.

END OF PART A OF SECTION II

**IF YOU FINISH BEFORE TIME IS CALLED, YOU MAY
CHECK YOUR WORK ON PART A ONLY.**

CALCULUS BC
SECTION II, Part B
Time—60 minutes
Number of problems—4

NO CALCULATOR IS ALLOWED FOR THESE PROBLEMS.

Note: If you have extra time, you can go back and work on Part A of Section II, but you cannot use a calculator to complete your work at this time.

NO CALCULATOR ALLOWED

3. Two particles are moving along the x-axis. For $0 \leq t \leq 10$, the position of particle D at time t is given by $x_D(t) = \ln(t^2 - 4t + 11)$ while the velocity of particle J at time t is given by $v_J(t) = t^2 - 6t + 8$. Particle J is at position $x = 5$ at time $t = 1$.

   (a) For $0 \leq t \leq 10$, when is particle D moving left?

   (b) For $0 \leq t \leq 10$, find all times t when the particles are traveling in the same direction.

GO ON TO THE NEXT PAGE.

NO CALCULATOR ALLOWED

(c) Find the acceleration of particle J at time $t = 3$. Is the speed increasing, decreasing, or neither?

(d) Find the position of particle J the first time it changes direction.

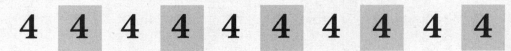

4. Functions $f(x)$, $g(x)$, $r(x)$, and $h(x)$ are defined in the following questions.

(a) Let f be the function defined by $f(x) = 4x^2e^x$. Write the first 4 nonzero terms and the general term of the Maclaurin series for f.

(b) Let $g(x) = \displaystyle\sum_{n=1}^{\infty} \frac{3x^{n+3}}{(n+2)!}$. Find the first 4 nonzero terms of $g'(x)$ and the general term of $g'(x)$.

GO ON TO THE NEXT PAGE.

4 4 4 4 4 4 4 4 4 4

NO CALCULATOR ALLOWED

(c) Let $r(x) = f(x) + 2g'(x)$. Find the first 4 nonzero terms of $r(x)$ and the coefficient of the 10th-degree term for $r(x)$.

(d) Write the first four nonzero terms and the general term of the Maclaurin series for $h(x) = \int_0^x f(t)\, dt$.

NO CALCULATOR ALLOWED

5. The function f is differentiable for all real numbers. The graph of $y = f(x)$ contains the point $\left(2, \frac{1}{2}\right)$, and the slope at each point (x, y) on $f(x)$ is given by $\frac{dy}{dx} = y^2\left(2 + x^3\right)$.

   (a) Find $\frac{d^2y}{dx^2}$ and evaluate it at $\left(2, \frac{1}{2}\right)$.

   _____

   (b) Find $y = f(x)$ by solving the differential equation $\frac{dy}{dx} = y^2\left(2 + x^3\right)$ using the initial condition $f(2) = \frac{1}{2}$.

   _____

   (c) Is $f(x)$ increasing or decreasing at $(-1, 2)$?

6. A function has derivatives of all orders at $x = 0$. Let $P_n(x)$ denote the nth-degree Taylor polynomial for f about $x = 0$.

(a) It is known that $f(0) = 8$, $f''(0) = 2$, and $P_2\left(\dfrac{1}{2}\right) = 9\dfrac{3}{4}$. Find $f'(0)$, and show the work that leads to your answer.

(b) It is known that $f^{(3)}(0) = \dfrac{1}{3}f'(0)$, $f^{(4)}(0) = \dfrac{1}{2}f^{(3)}(0)$, and $4f^{(5)}(0) = 2f^{(4)}(0)$. Find $P_5(x)$.

NO CALCULATOR ALLOWED

(c) The function h has the first derivative given by $h'(x) = f(2x)$. If $h(0) = 6$, find a 3rd-degree Taylor polynomial for h about $x = 0$.

(d) Drew thinks that $\left| P_4\left(\frac{1}{2}\right) - h\left(\frac{1}{2}\right) \right| < \frac{1}{950}$. Determine whether he is correct and explain why.

STOP

END OF EXAM

ANSWER KEY

Section I, Part A

1. A	16. D
2. C	17. D
3. D	18. D
4. A	19. A
5. A	20. A
6. B	21. A
7. A	22. B
8. C	23. B
9. D	24. A
10. D	25. A
11. A	26. D
12. B	27. A
13. D	28. B
14. B	29. C
15. C	30. D

Section I, Part B

76. C
77. A
78. A
79. B
80. C
81. C
82. A
83. A
84. D
85. B
86. A
87. D
88. B
89. A
90. D

SCORING

Section II, Parts A & B

Use the scoring rubrics (found after the Section I answers and explanations) to self-score your free-response questions.

Section I, Part A Number Correct: _____

Section I, Part B Number Correct: _____

Section II, Part A Points Earned: _____

Section II, Part B Points Earned: _____

Enter your results to your Practice Exam 6 assignment to see your 1–5 score by logging in at kaptest.com.

Haven't registered your book yet? Go to kaptest.com/booksonline to begin.

BC PRACTICE EXAM 6 ANSWERS AND EXPLANATIONS

SECTION I, Part A

1. A

The area of a rectangle is calculated by multiplying the base, x, by the height, $y = -\frac{1}{5}x^2 + 6$. This gives $A = x\left(-\frac{1}{5}x^2 + 6\right)$, which is the formula you are trying to optimize. Start by differentiating and identifying all critical values.

$$A = x\left(-\frac{1}{5}x^2 + 6\right)$$
$$A = -\frac{1}{5}x^3 + 6x$$
$$A' = -\frac{3}{5}x^2 + 6$$
$$0 = -\frac{3}{5}x^2 + 6$$
$$-6 = -\frac{3}{5}x^2$$
$$10 = x^2$$
$$x = \pm\sqrt{10}$$

Having a negative x-value is unreasonable for the context of the problem, making $x = \sqrt{10}$ the only critical value. Using test intervals will show that A' transitions from positive to negative at $x = \sqrt{10}$, which means there is a maximum at that x-value.

Thus, **(A)** is correct. (B) is an x-intercept of the area function, but that value would leave a height of 0 and an area of 0. (C) is the maximum area, but this question is focused on the x-value.

2. C

This series is not geometric or in the form for the p-Series Test. Therefore, (A) and (B) can be discarded. Use the Integral Test to determine whether (C) or (D) is correct.

Let $f(x) = \frac{6x^2}{x^3 + 1}$, and evaluate $\lim\limits_{b \to \infty} \int_1^b f(x)\, dx$.

$$\lim_{b \to \infty} dx \int_1^b \frac{6x^2}{x^3 + 1}$$
$$u = x^3 + 1$$
$$\frac{du}{dx} = 3x^2$$
$$dx = \frac{1}{3x^2}\, du$$
$$\lim_{b \to \infty} \int 6x^2 u^{-1} \frac{1}{3x^2}\, du$$
$$\lim_{b \to \infty} 2 \int u^{-1}\, du$$
$$\lim_{b \to \infty} 2\left[\ln|u|\right]$$
$$\lim_{b \to \infty} 2\left[\ln|x^3 + 1|\right]\Big|_1^b$$
$$\lim_{b \to \infty} 2\left[\ln|b^3 + 1| - \ln|1^3 + 1|\right]$$
$$2\left[\ln|\infty^3 + 1| - \ln|2|\right] = \infty$$

Because the integral diverges, then the series diverges as well, making **(C)** correct.

3. D

The limit at $x = 2$ does not exist. You must understand the type of discontinuity that is occurring at $x = 2$. Once the function is properly factored, $\lim\limits_{x \to 2} \dfrac{4(x-1)(x+1)}{(x-2)(x+1)}$, you need to reduce the $(x + 1)$ from the numerator and denominator and keep the $(x - 2)$ in the denominator. Because the $(x + 1)$ was simplified from both the numerator and the denominator, this is a deleted point. Because the binomial $(x - 2)$ remains in the denominator, there is a vertical asymptote at $x = 2$. Further investigation shows that the limit does not exist because the limit from the left is negative infinity $\lim\limits_{x \to 2^-} \dfrac{4(x-1)}{(x-2)} = \dfrac{+}{-}$ and the limit from the right is positive infinity $\lim\limits_{x \to 2^+} \dfrac{4(x-1)}{(x-2)} = \dfrac{+}{+}$. In order for the limit to exist, the value must be the same from both sides.

Hence, **(D)** is correct. (C) is the limit as x approaches infinity.

4. A

This question requires some rarely used integrals. However, you still need to know these integrals for the exam. The first two listed below are used in this example.

$$\int \csc^2(x)\, dx = -\cot(x) + C$$

$$\int \cot(x) = \ln|\sin(x)| + C$$

$$\int \tan(x) = -\ln|\cos(x)| + C$$

The Integration by Parts formula $\int u\, dv = uv - \int v\, du$ using $\begin{array}{ll} u = u & v = -\cot(u) \\ du = 1 & dv = \csc^2(u)\, du \end{array}$ yields the following.

$$\int u\csc^2(u) = -u\cot(u) - \int \left(-\cot(u)\right) du$$

$$\int u\csc^2(u) = -u\cot(u) + \int \cot(u)\, du$$

$$\int u\csc^2(u) = -u\cot(u) + \ln|\sin(u)| + C$$

Hence, **(A)** is correct. The other choices are all variations of incorrect antiderivatives.

5. A

For questions involving a graph, it is important to identify which function is shown. For this problem, $f(x)$ is provided. Using that graph, you must interpret $f'(x)$ and $f''(x)$ at $x = -1$. From the graph, $f(-1) = 0$. At $x = -1$, $f(x)$ is decreasing, which means the derivative, $f'(-1)$, must be negative. At $x = -1$, $f(x)$ is concave up. This means the second derivative, $f''(-1)$, must be positive. Using this information, negative $< 0 <$ positive makes $f'(-1) < f(-1) < f''(-1)$, **(A)**, the correct answer.

6. B

To evaluate this integral, first find $\int_1^5 f(x)\, dx$. Using the given integrals, you can use subtraction to realize that $\int_1^5 f(x)\, dx = -8$. To evaluate $\int_1^5 \left(3f(x) + 2\right) dx$, separate the integrals and move the constant.

$$\int_1^5 \left(3f(x) + 2\right) dx$$

$$= \int_1^5 \left(3f(x)\right) dx + dx\int_1^5 2$$

$$= 3\int_1^5 \left(f(x)\right) dx + dx\int_1^5 2$$

$$= 3(-8) + 2(4)$$

$$= -16$$

Therefore, **(B)** is correct. (A) occurs if you subtract $\int_1^5 \left(3f(x) - 2\right) dx$. (D) is the result of calculating $\int_1^5 f(x)\, dx = 8$.

7. A

For the Limit Comparison Test, you want to choose a function that meets two requirements. First, you know if it will converge or diverge. Second, it has similar properties to the series you are investigating. In this case, because the question is asking about $\sum_{n=1}^{\infty} \frac{n^2+2}{n^3+3}$, the best function to compare it to would be $\sum_{n=1}^{\infty} \frac{1}{n}$, which is **(A)**. Notice that if you looked at the expression $\frac{n^2+2}{n^3+3}$, it would take on the characteristics of $\frac{1}{n}$. You do not need to determine convergence. This question is only asking what the best series to compare it to would be.

8. C

When x is less than 0, this function simplifies to -1. When x is greater than 0, the function simplifies to 1. For that reason, this integral can be split into two separate integrals that can be evaluated individually. Then combine the results.

$$\int_{-2}^{3} \frac{|x|}{x}\,dx$$
$$= \int_{-2}^{0} -1\,dx + \int_{0}^{3} 1\,dx$$
$$= -x\big|_{-2}^{0} + x\big|_{0}^{3}$$
$$= [0-2] + [3-0]$$
$$= -2 + 3$$
$$= 1$$

Hence, **(C)** is correct. (B) would be correct if the lower bound was -3. (D) just simplifies the original to 1 and neglects to consider the consequences of x being a negative number.

9. D

When trying to identify local maximum or minimum values, apply the first derivative test. For a local maximum to exist, the rate of change of the function, $f(x)$, must change from increasing to decreasing. This means the derivative transitions from a positive value to a negative value. This transition can take place only when the derivative is equal to 0 or is undefined. The values where the derivative is equal to 0 or undefined are called critical values. In this problem, the derivative is provided. By setting it equal to 0 you get critical values of $x = -2, x = 1$, and $x = 5$. There are no x-values that will make the derivative undefined. To determine if any of these critical values are local maximums, you must consider the surrounding intervals.

Test Interval	Sign of $f'(x)$
$(-\infty, -2)$	$+$
$(-2, 1)$	$+$
$(1, 5)$	$-$
$(5, \infty)$	$+$

The derivative changes from positive to negative at $x = 1$, making it the only local maximum on $f(x)$.

Therefore, **(D)** is correct. There is no sign change at $x = -2$. The change at $x = 5$ is from negative to positive, which means $f(x)$ has a local minimum at this point. (A) lists the critical values. (C) would be correct if the question was asking for a local minimum on $f(x)$.

10. D

Speed is the absolute value of the velocity. Velocity is the first derivative of the position function. When taking the derivative, remember to use the chain rule.

$$x(t) = (\sin(3t))^2$$
$$v(t) = 6(\sin(3t))(\cos(3t))$$

Plug in $t = \dfrac{\pi}{4}$.

$$v\left(\frac{\pi}{4}\right) = 6\left(\sin\frac{3\pi}{4}\right)\left(\cos\frac{3\pi}{4}\right)$$
$$v\left(\frac{\pi}{4}\right) = 6\left(\frac{\sqrt{2}}{2}\right)\left(\frac{-\sqrt{2}}{2}\right)$$
$$v\left(\frac{\pi}{4}\right) = -3$$

The result is 3, not −3, because speed is always positive. So be careful not to choose (A).

Thus, **(D)** is correct. (B) and (C) are a result of not using the chain rule correctly. If you found $v(t) = 2(\sin(3t))(\cos(3t))$, you would be working with an incorrect equation. The result would be a velocity of −1 and a speed of 1.

11. A

The undefined slopes when $y = 0$ do not help because that would be true for all the given differential equations. The same can be said about the zero slopes when $x = 0$. All the slopes left of the y-axis are negative, and all of the slopes right of the y-axis are positive. For this to happen, x must have an odd degree attached to it and y must have an even degree attached to it, which happens in $\dfrac{dy}{dx} = \dfrac{x^3}{y^2}$.

Hence, **(A)** is correct. For (B), a transition from positive to negative slopes would take place over the x-axis. The slopes given in (C) would be positive in quadrants I and III and would be negative in quadrants II and IV. On the line $y = x$, (D) would have a consistent slope of 1.

12. B

The Alternating Series Test shows convergence for an alternating series that meets the following two requirements. First, $|a_{n+1}| < |a_n|$ for all n. Second, $\lim\limits_{n\to\infty} a_n = 0$. When checking for these conditions, it is usually easier to ignore or separate the alternator and evaluate the remaining expression. For example, start with statement II, $a_n = \dfrac{(-1)^{n+1}}{2n}$. By just using the expression without the alternator, $a_n = \dfrac{1}{2n}$, you can see that $|a_{n+1}| < |a_n|$ for all n and $\lim\limits_{n\to\infty} a_n = 0$. This is the basis for the Alternating Series Test and confirms that the series converges. In statement III, $a_n = (-1)^n\left(\dfrac{n+1}{3n+2}\right)$, the $\lim\limits_{n\to\infty} a_n = \dfrac{1}{3}$. So the Alternating Series Test does not confirm convergence. In statement I,

$$a_n = \begin{cases} \dfrac{12}{3^n} & \text{if } n \text{ is odd} \\[2mm] \dfrac{-1}{2^n} & \text{if } n \text{ is even} \end{cases}$$
. The terms are not getting smaller,

so the first requirement $|a_{n+1}| < |a_n|$ is not met. Thus, **(B)** is correct.

13. D

Start by differentiating each term with respect to x. Remember to use the product rule when differentiating x^3y^2.

$$\cos y - x^3y^2 = 2y$$

$$\frac{d}{dx}(\cos y) - \frac{d}{dx}(x^3y^2) = \frac{d}{dx}(2y)$$

$$\frac{d}{dx}(\cos y) = -\sin y\left(\frac{dy}{dx}\right)$$

$$\frac{d}{dx}(x^3y^2) = 3x^2y^2 + 2x^3y\left(\frac{dy}{dx}\right)$$

$$\frac{d}{dx}(2y) = 2\left(\frac{dy}{dx}\right)$$

$$-\sin y\left(\frac{dy}{dx}\right) - \left(3x^2y^2 + 2x^3y\left(\frac{dy}{dx}\right)\right) = 2\left(\frac{dy}{dx}\right)$$

Now that you have the derivative, solve for $\frac{dy}{dx}$. This requires you to put all terms containing $\frac{dy}{dx}$ on the same side of the equal sign so it can be factored out as a greatest common factor and then isolated. Be aware that if you move all of the $\frac{dy}{dx}$ terms to the left side instead of the right, your answer will appear as $\frac{dy}{dx} = \frac{3x^2y^2}{-2 - \sin y - 2x^3y}$, which is equivalent to $\frac{dy}{dx} = \frac{-3x^2y^2}{2 + \sin y + 2x^3y}$.

Therefore, **(D)** is correct. (A) is missing the negative sign in the numerator. (B) does not distribute the subtraction sign through the product rule derivative. (C) incorrectly uses sine as the derivative of cosine instead of negative sine.

14. B

If $u = \sqrt{2x}$, by the chain rule $du = \frac{1}{\sqrt{2x}}dx$ after substitution. Adjusting the bounds gives the new integral: $\int_2^4 e^u\, du$. So statement II is equivalent to the original integral. Statement I does not have the correct bounds for the variable u. Statement III incorrectly uses e^{2x} as the antiderivative of e^{2x} when it should be $\frac{1}{2}e^{2x}$.

Thus, statement II is the only correct answer, **(B)**.

15. C

This question also uses the Maclaurin polynomial formula for e^x.

$$e^x = 1 + x + \frac{1}{2!}x^2 + \frac{1}{3!}x^3 + \frac{1}{4!}x^4 + \cdots + \frac{1}{n!}x^n$$

If necessary, you should be able to reproduce the formula. However, it would be in your best interest to have it memorized. Other Taylor and Maclaurin series formulas that are used frequently on the BC exam include the following.

$$\ln(x) = (x-1) - \frac{1}{2}(x-1)^2 + \frac{1}{3}(x-1)^3 - \frac{1}{4}(x-1)^4 + \cdots$$
$$+ (-1)^{n+1}\frac{1}{n}(x-1)^n$$

$$\sin(x) = x - \frac{1}{3!}x^3 + \frac{1}{5!}x^5 - \frac{1}{7!}x^7 + \cdots$$
$$+ (-1)^{n-1}\frac{1}{(2n-1)!}x^{2n-1}$$

$$\cos(x) = 1 - \frac{1}{2!}x^2 + \frac{1}{4!}x^4 - \frac{1}{6!}x^6 + \cdots$$
$$+ (-1)^{n-1}\frac{1}{(2n-2)!}x^{2n-2}$$

Notice the natural log formula cannot be centered at $x = 0$ because the natural log function has a limited domain. You can save valuable time by having these formulas memorized. It is a common question to require you to write a new Taylor and/or Maclaurin series that is already known. For this question, $2(e^x + e^{-x})$, you must find e^x, add e^{-x}, and double the sum.

$$e^x = 1 + x + \frac{1}{2!}x^2 + \frac{1}{3!}x^3 + \frac{1}{4!}x^4 + \cdots$$

$$e^{-x} = 1 - x + \frac{1}{2!}(-x)^2 + \frac{1}{3!}(-x)^3 + \frac{1}{4!}(-x)^4 + \cdots$$

Add $e^x + e^{-x}$.

$$e^x + e^{-x} = 1 + 1 + x - x + \frac{1}{2!}x^2 + \frac{1}{2!}(x)^2$$
$$+ \frac{1}{3!}x^3 - \frac{1}{3!}(x)^3 + \frac{1}{4!}x^4 + \frac{1}{4!}(x)^4 \cdots$$

$$e^x + e^{-x} = 2 + \frac{2}{2!}x^2 + \frac{2}{4!}x^4 + \cdots$$

Then double that sum.

$$2\left(e^x + e^{-x}\right) = 4 + \frac{4}{2!}x^2 + \frac{4}{4!}x^4 + \cdots \frac{4}{(2n)!}x^{2n}$$

Hence, **(C)** is correct. (A) mistakenly multiplies all the terms by $\frac{1}{2}$ instead of by 2. (B) does not multiply by the constant of 2. (D) is the result of $2(e^x + e^x)$; notice that none of the odd-degree terms canceled out.

16. D

This question requires true and complete understanding of the idea that a derivative can be expressed as a limit. In statement I, $\lim\limits_{h \to 0} \dfrac{\left((x+h^3) - 5\right) - \left(x^3 - 5\right)}{h} = 3x^2$, the original function is $f(x) = x^3 - 5$. So its derivative is, in fact, $3x^2$. In statement II, $\lim\limits_{h \to 0} \dfrac{\left((4+h)^2\right) - 16}{h} = 8$, you could prove this is true in one of two ways. If you simplified the numerator to $\lim\limits_{h \to 0} \dfrac{\left((16 + 8h + h)^2 - 16\right)}{h}$, you would get $\lim\limits_{h \to 0} \dfrac{h^2 + 8h}{h}$. Then you could actually find the limit: $\lim\limits_{h \to 0} h + 8 = 8$. Proving that statement III,

$$\lim\limits_{h \to 0} \frac{\left((x+h)^2 + 8x + 8h\right) - x^2 - 8x}{2h} = x + 4,$$ is true

requires you to manipulate the expression in several ways to make the question appear more as a derivative question. For example in the numerator, you can factor an 8 out of the $8x + 8h$ and a -1 out of the $-x^2 - 8x$ to produce $\lim\limits_{h \to 0} \dfrac{\left(\left((x+h)^2 + 8(x+h)\right) - \left(x^2 + 8x\right)\right)}{2h} = x + 4$.
The original function is $f(x) = x^2 + 8x$, and its derivative is $f'(x) = 2x + 8$. If you look carefully at the denominator, you will notice a $2h$ as opposed to the normal h. You could multiply both sides by 2 or simply take $\frac{1}{2}$ of the derivative to prove statement III true as well.

Because all statements are true, **(D)** is correct.

17. D

From the graph of $f(x)$, you can obtain the point $f(0) = 2$. From the graph of $f'(x)$, you know the instantaneous rate of change at $x = 0$ is -5. This is all the information needed to create the tangent line equation of $y - 2 = -5(x - 0)$. The approximate y-value can now be found by substituting -0.2 for x and solving for y.

$$y - 2 = -5(x - 0)$$
$$y = (-5)(-0.2) + 2$$
$$y = 1 + 2 = 3$$

Therefore, **(D)** is correct. If you mistakenly switch the slope with the coordinate, it will result in (A). (B) uses 0.2 instead of -0.2 when substituting into the tangent line.

18. D

Being conditionally convergent means that the series diverges without the alternator. Statement I, $\sum\limits_{n=1}^{\infty} \dfrac{(-2)^n}{n!}$, converges with or without the alternator, so it is not conditionally convergent. You may wish to rewrite the series as $\sum\limits_{n=1}^{\infty} \dfrac{(-1)^n (2)^n}{n!}$. However, further investigation shows it converges with or without the alternator. In statement II, $\sum\limits_{n=1}^{\infty} \dfrac{(-1)^{2n}}{3n}$, pay special attention to the exponent of 2. This series does not alternate to begin with, so it cannot be conditionally convergent. It does diverge based on the harmonic series, but it is not conditionally convergent. Statement III, $\sum\limits_{n=1}^{\infty} \dfrac{(-1)^{n+3}(5)^n}{(7)^{n+1}}$, can be rewritten as $\sum\limits_{n=1}^{\infty} \dfrac{(-1)^{n+3}(5)^n}{(7)(7)^n}$ and again as $\sum\limits_{n=1}^{\infty} \dfrac{(-1)^{n+3}}{7} \cdot \left(\dfrac{5}{7}\right)^n$ so it will more clearly resemble a geometric series. Because $|r| < 1$, the series converges. This series also converges with or without the alternator.

Because all 3 statements are not conditionally convergent, **(D)** is correct.

19. A

A function, $f(x)$, is decreasing when $f'(x) < 0$. Start by taking the derivative of $f(x)$, and then find the zeros.

$$f(x) = x^3 - 6x^2 - 36x + 12$$
$$f'(x) = 3x^2 - 12x - 36$$
$$0 = 3x^2 - 12x - 36$$
$$0 = 3(x^2 - 4x - 12)$$
$$0 = 3(x - 6)(x + 2)$$
$$x = -2 \text{ or } x = 6$$

The test intervals $(-\infty, -2)$, $(-2, 6)$, and $(6, \infty)$ can now be used to determine where the derivative is negative. Having an understanding of the graph of $f'(x)$ may be a helpful time saver. Because $f'(x)$ is a parabola that opens upward, the derivative must be negative between the zeros, making **(A)**, $-2 < x < 6$, the correct answer.

(B) represents the intervals on which $f(x)$ is increasing. (C) and (D) are the result of a factoring error.

20. A

Use the quotient rule and remember that the derivative of e^x is e^x. The quotient rule uses the following formula to find the derivative of rational functions that cannot be easily simplified using more efficient rules like the power rule.

$$\frac{d}{dx}\left[\frac{f(x)}{g(x)}\right] = \frac{g(x)f'(x) - f(x)g'(x)}{(g(x))^2}$$

$$f(x) = \frac{e^x}{(x-3)}$$

$$f'(x) = \frac{(x-3)e^x - e^x(1)}{(x-3)^2}$$

$$f'(x) = \frac{e^x(x-3-1)}{(x-3)^2}$$

$$f'(x) = \frac{e^x(x-4)}{(x-3)^2}$$

Plug in $x = 1$.

$$f'(x) = \frac{e^x(x-4)}{(x-3)^2}$$

$$f'(1) = \frac{e(-3)}{4}$$

$$f'(1) = -\frac{3e}{4}$$

Thus, **(A)** is correct. (C) is incorrect because the quotient rule has subtraction, not addition, in the numerator.

21. A

This is a difficult, multistep problem but one that you should be able to accomplish because you have made it this far in your calculus journey. The first step in this problem is to write the expansion for the polynomial for $\cos(x)$, replacing x with t^6. Use the Maclaurin series formula for $\cos(x)$.

$$\cos(x) = 1 - \frac{1}{2!}x^2 + \frac{1}{4!}x^4 - \frac{1}{6!}x^6 + \cdots$$
$$+ (-1)^{n-1}\frac{1}{(2n-2)!}x^{2n-2}$$

$$\cos(t)^6 = 1 - \frac{1}{2!}(t^6)^2 + \frac{1}{4!}(t^6)^4 - \frac{1}{6!}(t^6)^6 + \cdots$$
$$+ (-1)^{n-1}\frac{1}{(2n-2)!}(t^6)^{2n-2}$$

Then simplify.

$$\cos(t^6) = 1 - \frac{1}{2!}t^{12} + \frac{1}{4!}t^{24} - \frac{1}{6!}t^{36} + \cdots$$
$$+ (-1)^{n-1}\frac{1}{(2n-2)!}(t^6)^{2n-2}$$

The next step is to integrate by taking the antiderivative.

$$\int \cos(t^6) = t - \frac{1}{13 \cdot 2!}t^{13} + \frac{1}{25 \cdot 4!}t^{25} - \frac{1}{37 \cdot 6!}t^{37} + \cdots$$
$$+ (-1)^{n-1}\frac{1}{(6(2n-2)+1)(2n-2)!}(t^6)^{2n-2+1}$$

Evaluate between $t = x$ and $t = 0$.

$$x - \frac{1}{13 \cdot 2!}x^{13} + \frac{1}{25 \cdot 4!}x^{25} - \frac{1}{37 \cdot 6!}x^{37} + \cdots$$
$$+ (-1)^{n-1}\frac{1}{(6(2n-2)+1)(2n-2)!}(x^6)^{2n-2+1}$$

To get (A), rewrite with the *x*-terms in the numerator.

$$x - \frac{x^{13}}{13 \cdot 2!} + \frac{x^{25}}{25 \cdot 4!} - \frac{x^{37}}{37 \cdot 6!} + \cdots$$

$$+ (-1)^{n-1} \frac{\left(x^6\right)^{2n-2+1}}{(6(2n-2)+1)(2n-2)!}$$

Hence, **(A)** is correct. (B) neglects to take the antiderivative. (C) has all *t*-variables instead of *x*-variables. (D) has the incorrect signs as though someone did the lower bound minus the upper bound in $\int\limits_0^x \cos\left(t^6\right) dt$ or evaluated $\int\limits_x^0 \cos\left(t^6\right) dt$ accidentally.

22. B

The mean value theorem requires you to find the corresponding *y*-coordinates of the given endpoints of the domain in question: $f(0) = 0$ and $f(6) = 0$. Setting the derivative equal to the slope of these points and solving for *x* produces the *c* value that is considered the conclusion of the mean value theorem. Because the coordinates are (0, 0) and (6, 0), the slope is 0. Setting the derivative equal to zero (applying Rolle's theorem) and solving for *x* produces the following.

$$f(x) = 3x^2 - 12x$$

$$0 = 3x^2 - 12x$$

$$0 = 3x(x - 4)$$

$$x = 0 \text{ and } x = 4$$

(C) looks very good. However, the correct answer is **(B)**, 4 only, because the *c*-value must be inside the given domain, also known as "between the endpoints". In this case, $x = 0$ does not constitute a *c*-value because it is an endpoint of the given domain.

23. B

To analyze concavity, take the second derivative of *y*. The first derivative relies on the fundamental theorem, and the second derivative uses the power rule.

$$y = f(x) = \int_0^x \left(2t^3 - 9t^2 + 12t\right) dt$$

$$y' = f'(x) = 2x^3 - 9x^2 + 12$$

$$y'' = f''(x) = 6x^2 - 18x + 12$$

For *y* to be concave up, the second derivative must be positive. Set the second derivative equal to 0, and solve to locate the critical values.

$$0 = 6x^2 - 18x + 12$$

$$0 = 6(x - 3x + 2)$$

$$0 = 6(x - 1)(x - 2)$$

$$x = 1 \text{ or } x = 2$$

You could use test intervals of $(-\infty, 1)$, $(1, 2)$, and $(2, \infty)$ to determine where the second derivative is positive. Alternatively, you could recognize that this is an upward-opening parabola, which tells you it must be above the axis on the intervals in (B). Therefore, $y = f(x)$ is concave up on the intervals $(-\infty, 1)$ and $(2, \infty)$.

Hence, **(B)** is correct. (A) is the interval where *y* is concave down. (D) is one of the correct intervals, but the other interval is missing.

24. A

This integral needs to be split into the sum of different integrals in order to evaluate it correctly. To evaluate $\int \frac{1}{4x+5} dx$, you have to use the chain rule and u-substitution.

$$\int \left(\frac{1}{4x+5} + \frac{1}{x+7} + \frac{5x-7}{x^2-49} \right) dx = \int \frac{1}{4x+5} dx$$
$$+ \int \frac{1}{x+7} dx + \int \frac{5x-7}{x^2-49} dx$$

$$\int \frac{1}{4x+5} dx = \frac{1}{4} \ln|4x+5| + C_1$$

$$\int \frac{1}{x+7} dx = \ln|x+7| + C_2$$

To evaluate $\int \frac{5x-7}{x^2-49} dx$, use partial fractions.

$$\int \frac{5x-7}{x^2-49} dx = \frac{A}{x-7} + \frac{B}{x+7}$$

$$5x - 7 = A(x+7) + B(x-7)$$

If $x = -7$, then $-42 = -14B \rightarrow 3 = B$. If $x = 7$, then $28 = 14A \rightarrow 2 = A$.

$$\int \frac{5x-7}{x^2-49} dx = \int \left(\frac{2}{x-7} + \frac{3}{x+7} \right) dx$$

$$= 2 \ln|x-7| + 3 \ln|x+7| + C_3$$

Add this to the results from the other integrals you split from the original. Then simplify. Remember that C stands for an unknown constant in each expression. So you can add $C_1 + C_2 + C_3$ to get C.

$$\left(\frac{1}{4} \ln|4x+5| + C_1 \right) + \left(\ln|x+7| + C_2 \right)$$
$$+ \left(2 \ln|x-7| + 3 \ln|x+7| + C_3 \right)$$
$$= \frac{1}{4} \ln|4x+5| + 2 \ln|x-7| + 4 \ln|x+7| + C$$
$$= \frac{1}{4} \ln|4x+5| + \ln\left[(x-7)^2 \cdot (x+7)^4 \right] + C$$

Thus, **(A)** is correct. (B) is not the answer because it incorrectly has the exponent of 3 on the $(x+7)$ binomial. (C) and (D) neglect to use the chain rule and u-substitution when evaluating $\int \frac{1}{4x+5} dx$. Remember that $\int \frac{1}{4x+5} dx \neq \ln|4x+5|$.

25. A

Because the acceleration function is given, you must first take the antiderivative to get the velocity function. Once you have the velocity function, take another antiderivative to get the position function. Remember to add a constant "$+ C$" whenever you take an antiderivative. Because $a(t) = 30\sqrt{t}$, rewrite it as $a(t) = 30t^{\frac{1}{2}}$. Integrating both sides produces $v(t) = 20t^{\frac{3}{2}} + C$. The given information was with regard to position, so you must integrate again. Do not use the same constant "$+ C$." Integrating both sides using a different constant produces $x(t) = 8t^{\frac{5}{2}} + Ct + B$. The position at $t = 0$ was given as 16. Substituting $t = 0$ and $x(t) = 16$ into the position function gives the constant B, which must equal to 16. Using $x(t) = 8t^{\frac{5}{2}} + Ct + 16$ and the position of the particle at $t = 1$ allows you to solve for C. The position at $t = 1$ was given as 9. When you substitute $t = 1$ and $x(t) = 9$ into the position function, the constant C equals -15. The question is asking for $v(0)$. Because $v(t) = 20t^{\frac{3}{2}} - 15$, $v(0) = -15$.

Therefore, **(A)** is correct. (B) is $x(1)$. (C) is the result when you solve for C incorrectly. (D) is $a(1)$.

26. D

Remember that $f(6) = f(0) + \int_0^6 f'(x)\,dx$ and that $f(0)$ is given as 15. Because the interval is from 0 to 6 and there are 3 equal subintervals, each subinterval must be 2 wide. To find the integral, use the midpoints of rectangles, specifically $f'(1) = 6$, $f'(3) = 7$, and $f'(5) = 8$. Use the area formula to find the net change.

$$(2)(6) + (2)(7) + (2)(8) = 42$$

To find $f(6)$, add this net change to $f(0)$.

$$15 + 42 = 57$$

Hence, **(D)** is correct. B) is the net change ignoring the original $f(0)$. (A) and (C) use the left and right rectangular approximations, respectively, added to the original to approximate $f(6)$.

27. A

Although $\lim\limits_{n \to \infty} a_n = 0$ is a requirement for convergence, it is not a guarantee of convergence.

Thus, **(A)** is correct. (C) describes the comparison test for convergence. (D) is true according to the Nth Term Divergence Test, if the limit does not equal 0, then the series diverges.

28. B

If $x = 1$, the series is $\sum\limits_{n=1}^{\infty} \dfrac{(-1)^n}{n}$. Because $a_n = \dfrac{1}{n}$ is positive and decreasing, this is an alternating series that converges. If $x = 3$, the series becomes $\sum\limits_{n=1}^{\infty} \dfrac{1}{n}$, which is the divergent harmonic series. If $x < 1$ or $x > 3$, the exponential numerator grows at a faster rate than the linear denominator and the series diverges.

Therefore, the series converges when $1 \le x < 3$, **(B)**.

29. C

Using the chain rule, you get the following derivatives.

$$f(x) = \left(\frac{1}{2}x + 1\right)^5$$

$$f'(x) = 5\left(\frac{1}{2}x + 1\right)^4 \cdot \frac{1}{2} = \frac{5}{2}\left(\frac{1}{2}x + 1\right)^4$$

$$f''(x) = 10\left(\frac{1}{2}x + 1\right)^3 \cdot \frac{1}{2} = 5\left(\frac{1}{2}x + 1\right)^3$$

$$f^{(3)}(x) = 15\left(\frac{1}{2}x + 1\right)^2 \cdot \frac{1}{2} = \frac{15}{2}\left(\frac{1}{2}x + 1\right)^2$$

$$f^{(4)}(x) = 15\left(\frac{1}{2}x + 1\right) \cdot \frac{1}{2} = \frac{15}{2}\left(\frac{1}{2}x + 1\right)$$

Hence, **(C)** is correct.

30. D

A function, $f(x)$, is concave down when the second derivative is negative. Because the second derivative is given, you must determine where it is below the x-axis. The entire numerator is enclosed by absolute value symbols, which means it will always be nonnegative. To end up with a negative result, the denominator must be negative. That will occur only when x is greater than 3. Therefore, $f(x)$ is concave down on the interval $(3, \infty)$, **(D)**.

(B) represents the interval where $f(x)$ is concave up.

SECTION I, Part B

76. C

Creating a picture or sketch will make this question easier.

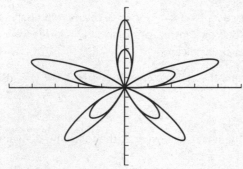

It is important to realize that one cycle through $0 < \theta < \pi$ completes each of the curves. If extended to 2π, there would be two complete cycles of the curves. The area between the curves is found by subtracting the area of $r = 4 \sin (5\theta)$ from $r = 7 \sin (5\theta)$. This can be done either individually or together. The following shows individual integrals being used.

$$\text{Area} = \frac{1}{2} \int_0^\pi \left(7 \sin(5\theta)\right)^2 d\theta - \frac{1}{2} \int_0^\pi \left(4 \sin(5\theta)\right)^2 d\theta$$

$$\text{Area} = 25.918$$

This equation shows one integral being used.

$$\text{Area} = \frac{1}{2} \int_0^\pi \left(\left(7 \sin(5\theta)\right)^2 - \left(4 \sin(5\theta)\right)^2\right) d\theta$$

$$\text{Area} = 25.918$$

Hence, **(C)** is correct. (A) is the result of your calculator being in degree mode instead of radian mode. It is important that you do not square the difference of the functions; For example, $\text{Area} = \frac{1}{2} \int_0^\pi \left(7 \sin(5\theta) - 4 \sin(5\theta)\right)^2 d\theta$ produces the wrong answer, (B). (D) does not use the necessary $\frac{1}{2}$. (D) can also be found by incorrectly integrating from $0 \le \theta \le 2\pi$.

77. A

The solution to the differential equation is the exponential model $P(t) = Ae^{kt}$, which in this problem is $P(t) = Ae^{8t}$. The initial amount, A, is irrelevant in this situation because for any starting amount to triple in 8 years, e^{8k} must equal 3. The calculator should be used to solve for k. When $e^{8k} = 3$, $k = 0.137$.

Hence, **(A)** is correct

78. A

The cross sections are perpendicular to the x-axis, so start by solving the function for y: $y = -\frac{2}{3}x + 4$. This line has an x-intercept of 6, making that the upper limit for the integral. The y-axis is provided as a bound, which means $x = 0$ is the lower limit. The next step is to determine the appropriate area formula to integrate. Because a leg of the triangle is on the region, the other leg makes up the height of the triangle. Legs of an isosceles triangle are equal in length, so the necessary integral is $\int_a^b \left(\frac{1}{2} (\text{leg})^2\right) dx$. You know the limits, and the length of the leg is defined by the function $y = -\frac{2}{3}x + 4$, so the integral is $\frac{1}{2} \int_0^6 \left(-\frac{2}{3}x + 4\right)^2 dx$. Now use the calculator to evaluate this integral.

$$\frac{1}{2} \int_0^6 \left(-\frac{2}{3}x + 4\right)^2 dx = 16$$

Therefore, **(A)** is correct. Be sure to multiply by $\frac{1}{2}$, or you will get (B).

79. B

The population changes according to the law of exponential growth, $P = P_0 e^{kt}$. Two conditions are given, and by substituting that information, you get a system of equations: $200 = P_0 e^{2k}$ and $600 = P_0 e^{4k}$. You can divide these equations to get $3 = e^{2k}$ and then solve for k to get $k = \frac{1}{2}\ln 3$. Using this value along with either of the given conditions will allow you to solve for P_0.

$$200 = P_0 e^{2\left(\frac{1}{2}\ln 3\right)}$$
$$200 = P_0 e^{\ln 3}$$
$$200 = 3P_0$$
$$P_0 \approx 67$$

Thus, **(B)** is correct. If you got (A), you used $k = \ln 3$ instead of $k = \frac{1}{2}\ln 3$. If you got (C), you forgot to include the k-value when substituting to solve for P_0.

80. C

A change in sign for $f''(x)$ translates into a point of inflection for $f(x)$. The second derivative $f''(x)$ is the rate of change of $f'(x)$. When $f'(x)$ is given, you should focus on when the rate of change of the given function has a sign change (turning points in $f'(x)$). Once you have graphed the function on an appropriate interval ($0 \le x \le 5\pi$), you can easily count the number of turning points.

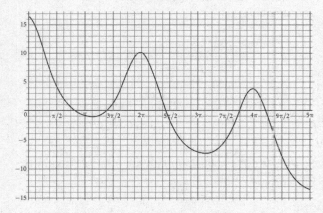

The derivative $f'(x)$ has 4 turning points in the given interval. This means $f''(x)$ changes sign 4 times in that interval, resulting in 4 points of inflection for $f(x)$. If the graph of $f''(x)$ had been given, the focus would have been on how many times the graph passes through 0.

Hence, **(C)** is correct.

81. C

Displacement is represented by the net change in position of the particle. To find displacement, integrate the velocity function from each of the time values: displacement $= \int_0^2 \frac{t^2 - 3}{3t^2 + 2}\, dt$. When evaluated on the calculator, the answer is -1.104. The negative sign represents a motion to the left, but displacement is a net change, so 1.104 is correct.

Therefore, **(C)** is correct. (A) is the average value of the velocity function over the same interval: $\frac{v(2) - v(0)}{2 - 0}$. (B) is the average rate of change from the velocity function $\frac{v(2) - v(0)}{2 - 0}$. Both of (A) and (B) are different from the displacement. (D) is the total distance traveled. It is calculated by using the same integral and the absolute value of the velocity function: total distance $= \int_0^2 \left| \frac{t^2 - 3}{3t^2 + 2} \right| dt$. Including the absolute value sign accounts for the particle moving both left and right. So this function determines the sum of the distance traveled as opposed to the displacement, which is a net change.

82. A

This is a geometric series with an initial value of $\cos^2 x$ and $r = \cos^2 x$. Because $\cos^2 x < 1$ whenever $\cos x \ne 1$ or -1, the series converges according to the formula $S = \frac{a}{1 - r}$.

$$S = \frac{\cos^2 x}{1 - \cos^2 x}$$
$$S = \frac{\cos^2 x}{\sin^2 x}$$
$$S = \cot^2 x$$
$$\cot^2(1) = 0.412$$

Therefore, **(A)** is correct. (B) is cot(1), which is an incorrect simplification of $\frac{\cos^2 x}{\sin^2 x}$. (D) is $\tan^2(1)$.

83. A

It is very important that you find the time referred to in the question. It says the first time the particle is at the origin, which means the particle's position equals 0. Setting $x(t) = 0$ and solving for t gives $t = 9.870$. Notice the domain for the position function does not include $t = 0$. So you cannot include the particle being at the origin; do not use $t = 0$. The velocity is the first derivative of the position function and can easily be calculated on the calculator: $x'(9.870) = -0.159$.

Hence, **(A)** is correct. (B) is the position at time $t = 0$, and (C) is the acceleration or second derivative. Both of these are incorrect. (D) is the result of misreading the position function as $x(t) = \sqrt{\sin(t)}$, which is not the same as $x(t) = \sin(\sqrt{t})$.

84. D

This question requires you to remember the formula for $\sin(x)$.

$$\sin(x) = x - \frac{1}{3!}x^3 + \frac{1}{5!}x^5 - \frac{1}{7!}x^7 + \cdots$$
$$+ (-1)^{n-1}\frac{1}{(2n-1)!}x^{2n-1}$$

You should also realize that the use of the alternating series error bound will help determine when the error is less than $\frac{1}{50}$ (or 0.02). Now view polynomials of different degrees.

$$P_9 = x - \frac{1}{3!}x^3 + \frac{1}{5!}x^5 - \frac{1}{7!}x^7 + \frac{1}{9!}x^9$$
$$P_7 = x - \frac{1}{3!}x^3 + \frac{1}{5!}x^5 - \frac{1}{7!}x^7$$
$$P_5 = x - \frac{1}{3!}x^3 + \frac{1}{5!}x^5$$
$$P_3 = x - \frac{1}{3!}x^3$$

In order for the error to be less than $\frac{1}{50}$, evaluate each of the last terms at $x = 2$ and see which is less than $\frac{1}{50}$.

Using your calculator, you will see that $\frac{\left(\frac{1}{2}\right)^9}{9!} < \frac{1}{50}$. Be careful because the 9th-degree term is less than the given

amount of $\frac{1}{50}$. So the polynomial that guarantees $\frac{1}{50}$ is a polynomial of degree 7, which is the previous term in the alternating series.

Therefore, **(D)** is correct. Remember the alternating series error bound says that the error is less than the absolute value of the first omitted term. (A) guarantees that the difference will be less than 0.266 by using the 5th-degree term. (C) guarantees that the difference will be less than 0.025 by using the seventh-degree term. Neither choice provides a guarantee that the difference will be less than $\frac{1}{50}$. There is no 4th-degree term, so (B) is not an option

85. B

First differentiate each side of the equation with respect to x. Remember to use the product rule when differentiating $-27.6xy$.

$$1.4x^{3.2} + 2.5y^{3.2} - 27.6xy = 0$$
$$\frac{d}{dx}1.4x^{3.2} + \frac{d}{dx}2.5y^{3.2} - \frac{d}{dx}27.6xy = \frac{d}{dx}0$$
$$4.48x^{2.2} + 8y^{2.2}\frac{dy}{dx} - \left(27.6y + 27.6x\frac{dy}{dx}\right) = 0$$
$$\frac{dy}{dx} = \frac{-4.48x^{2.2} + 27.6y}{8y^{2.2} - 27.6x}$$

Recognize that to have a horizontal tangent line, the derivative must equal 0. Solve for y by cross-multiplying and simplifying.

$$\frac{-4.48x^{2.2} + 27.6y}{8y^{2.2} - 27.6x} = 0$$
$$-4.48x^{2.2} + 27.6y = 0$$
$$27.6y = 4.48x^{2.2}$$
$$y \approx 0.162x^{2.2}$$

Substitute this y-value back into the original equation to find the corresponding x-values.

$$1.4x^{3.2} + 2.5\left(0.162x^{2.2}\right)^{3.2} - 27.6x\left(0.162x^{2.2}\right) = 0$$

Enter this into your calculator. Be extremely careful about the parenthesis. Use the Calculate Zero function to find when the graph crosses the x-axis. Therefore, **(B)** is correct.

86. A

To answer this question, you must realize that Euler's method is based on small steps in the tangent line approximations to reach an approximation for a particular x-value. The function is given as $g(x) = \sqrt{x}$, so the derivative is $g'(x) = \dfrac{1}{2\sqrt{x}}$. At the first coordinate of $x = 4$ the point $(4, 2)$ is on the curve. Find its derivative.

$$g'(4) = \frac{1}{2\sqrt{4}}$$
$$g'(4) = \frac{1}{4}$$
$$\Delta y = \Delta x \frac{dy}{dx}$$
$$\Delta y = 5\left(\frac{1}{4}\right)$$
$$\Delta y = \frac{5}{4}$$

So the approximation for $x = 9$ is $\left(9, 3\frac{1}{4}\right)$. Calculate the next value.

$$g'(9) = \frac{1}{2\sqrt{9}}$$
$$g'(9) = \frac{1}{6}$$
$$\Delta y = \Delta x \frac{dy}{dx}$$
$$\Delta y = 5\left(\frac{1}{6}\right)$$
$$\Delta y = \frac{5}{6}$$

So the approximation for $x = 14$ is $\left(14, 4\frac{1}{12}\right)$. Because $\sqrt{14} \approx 3.742$, the approximation for $g(14)$ is greater than $g(14)$. Because the function is strictly concave down (its second derivative is less than 0) tangent line approximation or, in this case, the result from Euler's method is greater than the actual y-coordinate of the function.

Hence, **(A)** is correct. (C) is incorrect as $\Delta y = \dfrac{\varepsilon}{2}$. (D) is incorrect because the derivative is strictly defined by the x-coordinates in this case, not the y-coordinates

87. D

The speed of a parametric function is the absolute value of the velocity. However, because the velocity is a vector having a magnitude and direction, you must use the following formula.

$$\text{Speed} = \left| \sqrt{\left(\frac{dy}{dt}\right)^2 + \left(\frac{dx}{dt}\right)^2} \right|$$

You should definitely use your calculator. Be sure to type the functions correctly and have your calculator in radian mode. The derivatives should be found in the calculator as well, using the numerical derivative button for an equation when $x = x$.

Plot1 Plot2 Plot3	Plot1 Plot2 Plot3	$\sqrt{(Y_2(2.7))^2 + (Y_4($
$\backslash Y_3 = \sin(3^x)$	$\backslash Y_1 = \cos(3^x)$	21.33236388
$\backslash Y_4 = \frac{d}{dx}(Y_3)\vert_{x=x}$	$\backslash Y_2 = \frac{d}{dx}(Y_1)\vert_{x=x}$	■
$\backslash Y_5 =$	$\backslash Y_3 = $ ■	
$\backslash Y_6 =$	$\backslash Y_4 =$	
$\backslash Y_7 =$	$\backslash Y_5 =$	
$\backslash Y_8 =$		

The speed of the particle is 21.332.

Thus, **(D)** is correct. (A) is the absolute value of the velocity quotients but is not the speed in parametric motion. (B) evaluates the position functions incorrectly in the speed formula. (C) neglects to square the derivatives in the formula.

88. B

The fact that $p(x)$ is a polynomial tells you that its graph is continuous everywhere. Because the first local minimum is above the x-axis, that point does not cause zeros. The local maximum point also does not cause any zeros due to its location. The second minimum is located below the axis, which means the graph must go through zero to reach the minimum. Because that point is a local minimum and there are no other critical points, the graph turns back up after $(4, -2)$. This forces the graph to go back through the x-axis, creating a second zero.

You could plot these points in your calculator and connect the dots to see that there will be two zeros.

Thus, **(B)** is correct. (A) gets you to the second minimum but does not account for the graph turning back up. (C) is the number of zeros for $f'(x)$.

89. A

First find the equation of the tangent line, $y - (e + 3) = e(x - 1)$. Solve for y and enter this into your calculator. Use the Calculate Zero function to find the solution $x = -\dfrac{3}{e}$, **(A)**.

90. D

Because $f'(x)$ is given, the antiderivative must be taken using the power rule to find $f(x)$. The power rule involves increasing the exponent by 1 and then dividing the coefficient by the new exponent. Don't forget to include a constant of integration whenever taking the antiderivative of a function.

$$f'(x) = 4x^3 - 6x^2 + 6$$
$$f(x) = x^4 - 2x^3 + 6x + C$$

Using the given value $f(1) = 7.158$, you can determine the constant, C.

$$7.158 = (1)^4 - 2(1)^3 + 6(1) + C$$
$$7.158 = 5 + C$$
$$C = 2.158$$

Now substitute 3.5 into the antiderivative. Use the calculator's Calculate Value function for fast calculations. $f(x) = (x)^4 - 2(x)^3 + 6(x) + 2.158$, to get $f(3.5) = 87.471$.

Hence, **(D)** is correct. (B) is $f'(2.5)$, and (C) is the result if you forget to include the constant, (A) in your antiderivative constant.

SECTION II, Part A

1. (a)

Solutions	Scoring Rubric
The right rectangular Riemann sum is found using the following: $(1)[300 + 500 + 200 + 100 + 150] = 1{,}250$. The given data make it appropriate to use five rectangles, one for each hour. Notice the answer is in hundreds; 12.5 is not correct unless written as 12.5 hundreds.	1: Use correct Riemann sum formula 1: 1,250 or 12.5 hundreds

(b)

Solutions	Scoring Rubric
Make sure to select the correct function, $W(t) = 132t - 38$, to use as the integrand. Then use your calculator to find the value. $$W(t) = 132t - 38$$ $$\int_0^4 W(t)\, dt = (132t - 38)\, dt$$ $$\int_0^4 W(t)\, dt = 904$$ The number 904 represents the total number of orders the factory workers can process in the first 4 hours they work, specifically from 2 p.m. to 6 p.m.	1: $\int_0^4 W(t)\, dt = 904$ or 9.04 hundreds 1: use correct units 1: explain the meaning of $\int_0^4 W(t)\, dt$ correctly

(c)

Solutions	Scoring Rubric
Notice the number of orders is given in terms of hundreds of orders. For the factory workers to fill all the online orders, solve using the following: $$\int_0^t W(t)\, dt = 1{,}450$$ $$\int_0^t (132t - 38)\, dt = 66t^2 - 38t \Big|_0^t$$ $$66t^2 - 38t \Big|_0^t = 66t^2 - 38t$$ $$66t^2 - 38t = 1{,}450$$ $$66t^2 - 38t - 1{,}450 = 0$$ The equation $66t^2 - 38t - 1{,}450 = 0$ can be solved any number of ways, including graphically or using the quadratic formula. The result is $t = 4.984$ hours.	1: $66t^2 - 38t - 1450 = 0$ 1: $t = 4.984$ hours

(d)

Solutions	Scoring Rubric
There are two ways to solve this question. You can determine the time for the robot to process the 1,250 orders. Alternatively, you can compare the number of orders processed by the robot to the number processed by the factory workers at time $t = 4.984$. Because $\int_0^{4.984} R(t)\,dt = 1{,}372.128$, the robots are more efficient. The other option is to calculate the following: $$\int_0^t R(t)\,dt = 1{,}250$$ $$\int_0^t \left(25.8t^2 + 8.7t + 40\right) dt = \frac{25.8t^3}{3} + \frac{8.7}{2}t^2 + 40t \Big|_0^t$$ $$\frac{25.8t^3}{3} + \frac{8.7}{2}t^2 + 40t \Big|_0^t = \frac{25.8t^3}{3} + \frac{8.7}{2}t^2 + 40t$$ $$\frac{25.8t^3}{3} + \frac{8.7}{2}t^2 + 40t = 1{,}250$$ $$\frac{25.8t^3}{3} + \frac{8.7}{2}t^2 + 40t - 1{,}250 = 0$$ The equation above can be solved any number of ways, including graphically. The result is $t = 4.810$ hours, which makes the robots more efficient.	1: robots are more efficient 1: justification

2. (a)

Solutions	Scoring Rubric
To find the area of the discarded beryllium, subtract the area of the region R from the rectangular region it is being cut from. The 20 by 40 feet rectangular sheet has an area of 800 feet^2. The area of region R is found by the integral of the function between $x = 0$ and $x = 40$. $$A = \int_0^{40} 10\cos\left(\frac{\pi}{40}x\right) + 10 \; dx$$ $$A = 400 \text{ ft}^2$$	1: set up integral correctly 1: $A = 400$ ft^2

(b)

Solutions	Scoring Rubric
To find the weight of the statue, multiply the density of beryllium by the volume of the solid. In order to find the volume, integrate the areas of the cross sections between $x = 0$ and $x = 40$. Each cross section would have an area $\left[10\cos\left(\dfrac{\pi}{40}x\right) + 10\right]^2$ so the volume would be: $$V = \int_0^{40} \left(10\cos\left(\frac{\pi}{40}x\right) + 10\right)^2 dx$$ $$V = 6{,}000 \text{ ft}^3$$	1: set up integral correctly 1: $V = 6{,}000$ 1: correct units, ft^3

(c)

Solutions	Scoring Rubric
To find the perimeter of the base of the statue add the lengths along the x- and y-axis to the length of the curve. The length along the y-axis is 20 feet, the length along the x-axis is 40 feet, and the length along the curve is found by using the arc length formula. $$\text{Arc Length} = \int_a^b \sqrt{1 + \left(\frac{dx}{dy}\right)^2}\, dx$$ $$\text{Arc Length} = \int_0^{40} \sqrt{1 + \left(-\frac{\pi}{4}\sin\left(\frac{\pi}{40}x\right)\right)^2}\, dx$$ $$\text{Arc Length} = 45.594 \text{ feet}$$	1: use correct arc length formula 1: set up integral correctly 1: 45.594 1: correct units, feet

SECTION II, Part B

3. (a)

Solutions	Scoring Rubric
Particles and objects move left when the velocity function is less than 0. The position function is given for particle D, so take the derivative to find the velocity and set equal to 0. $$v_D(t) = \frac{2t-4}{t^2-4t+11} = 0$$ $$2t-4 = 0$$ $$t = 2$$ Testing values less than $t = 2$ and greater than $t = 2$ prove that at time $t = 2$, the velocity changes from negative to positive, so the particle moves left $0 < t < 2$.	1: critical number at $t = 2$ 1: particle is moving left at $0 < t < 2$

(b)

Solutions	Scoring Rubric
Compare the velocities of the two particles. In the previous question, you found that particle D changes its velocity from negative to positive at $t = 2$. Determine when particle J changes velocity by setting its velocity function equal to 0. $$v_J(t) = t^2 - 6t + 8$$ $$t^2 - 6t + 8 = 0$$ $$(t-4)(t-2) = 0$$ $$t = 2, 4$$ At times $t = 2$ and $t = 4$, particle J changes direction, because its velocity is changing signs in the surrounding intervals. See the table below to determine the velocity and direction of particles D and J at different times	1: $t = 2, 4$ 1: correct table of directions 1: particles D and J are both traveling in the same direction, to the right, when $4 < t < 10$

Time	$v_D(t) = \dfrac{2t-4}{t^2-4t+11}$	Direction	$v_J(t) = t^2 - 6t + 8$	Direction
$0 < t < 2$	$v_D(t) \leq 0$	Left	$v_J(t) \geq 0$	Right
$2 < t < 4$	$v_D(t) \geq 0$	Right	$v_J(t) \leq 0$	Left
$4 < t < 10$	$v_D(t) \geq 0$	Right	$v_J(t) \geq 0$	Right

Particles D and J are both traveling in the same direction, to the right, when $4 < t < 10$

(c)

Solutions	Scoring Rubric
To find the acceleration of particle J, take the derivative of the velocity function. Then find the acceleration at $t = 3$. $$v_J(t) = t^2 - 6t + 8$$ $$a_J(t) = 2t - 6$$ $$a_J(3) = 2(3) - 6$$ $$a_J(3) = 0$$ Although $v(3) = -1$, the speed is neither increasing nor decreasing because the velocity and acceleration cannot be either the same sign or different signs when acceleration equals 0.	1: $a_J(3) = 0$ 1: neither

(d)

Solutions	Scoring Rubric
As determined in the answer to part b, the first time particle J changes direction is at time $t = 2$. $$x(2) = x(1) + \int_1^2 v(t)\, dt$$ $$x(2) = 5 + \int_1^2 \left(t^2 - 6t + 8\right) dt$$ $$\int_1^2 \left(t^2 - 6t + 8\right) dt = \frac{1}{3}t^3 - 3t^2 + 8t \Big|_1^2$$ $$= \frac{1}{3}(2)^3 - 3(2)^2 + 8(2) - \left[\frac{1}{3}(1)^3 - 3(1)^2 + 8(1)\right]$$ $$= \frac{8}{3} - 12 + 16 - \left[\frac{1}{3} - 3 + 8\right]$$ $$= \frac{8}{3} - 12 + 16 - \frac{1}{3} + 3 - 8$$ $$\int_1^2 \left(t^2 - 6t + 8\right) dt = \frac{4}{3}$$ $$x(2) = 5 + \frac{4}{3}$$ $$x(2) = 6\frac{1}{3}$$	1: use correct formula $x(2) = x(1) + \int_1^2 v(t)\, dt$ 1: $x(2) = 6\frac{1}{3}$ or something equivalent

4. (a)

Solutions	Scoring Rubric
This question requires you either to remember the formula for e^x or to be able to recreate it from the Taylor polynomial formula. Once established, multiply each term by $4x^2$ and determine the general form.	1: use the formula for e^x
	1: $4x^2 e^x = 4x^2 + 4x^3 + \dfrac{4}{2!}x^4 +$
$$e^x = 1 + x + \frac{1}{2!}x^2 + \frac{1}{3!}x^3 + \frac{1}{4!}x^4 + \cdots + \frac{1}{n!}x^n$$	$\dfrac{4}{3!}x^5 + \cdots + \dfrac{4x^{n+2}}{n!}$
$$4x^2 e^x = 4x^2\left[1 + x + \frac{1}{2!}x^2 + \frac{1}{3!}x^3 + \frac{1}{4!}x^4 + \cdots + \frac{1}{n!}x^n\right]$$	
$$4x^2 e^x = 4x^2 + 4x^3 + \frac{4}{2!}x^4 + \frac{4}{3!}x^5 + \cdots + \frac{4x^{n+2}}{n!}$$	

(b)

Solutions	Scoring Rubric
One approach to this question is to identify some terms and then take the derivative of each of the terms individually.	1: $$g'(x) = \frac{12}{3!}x^3 + \frac{15}{4!}x^4 + \frac{18}{5!}x^5 + \frac{21}{6!}x^6 + \cdots$$
$$g(x) = \sum_{n=1}^{\infty} \frac{3x^{n+3}}{(n+2)!}$$	1: $g'(x) = \dfrac{3(n+3)x^{n+2}}{(n+2)!}$
$$g(x) = \frac{3}{3!}x^4 + \frac{3}{4!}x^5 + \frac{3}{5!}x^6 + \frac{3}{6!}x^7 + \cdots$$	
$$g'(x) = \frac{12}{3!}x^3 + \frac{15}{4!}x^4 + \frac{18}{5!}x^5 + \frac{21}{6!}x^6 + \cdots$$	
As far as the general term, you can apply the power rule to the general term to find $g'(x)$.	
$$g(x) = \sum_{n=1}^{\infty} \frac{3x^{n+3}}{(n+2)!}$$	
$$g'(x) = \frac{3(n+3)\left(x^{n+3-1}\right)}{(n+2)!}$$	
$$g'(x) = \frac{3(n+3)x^{n+2}}{(n+2)!}$$	

(c)

Solutions	Scoring Rubric
Because $r(x) = f(x) + 2g'(x)$, simply add $f(x)$ to twice $g'(x)$. Because the question is asking for the first four terms, you will need the first four terms from $f(x)$ but only the first three terms from $g'(x)$.	1: $r(x) = 4x^2 + 8x^3 + \dfrac{13}{4}x^4 + \dfrac{49}{30}x^5$ 1: correct work shown for calculating coefficients 1: $\dfrac{4}{8!} + \dfrac{66}{10!}$

$$4x^2 + 4x^3 + \frac{4}{2!}x^4 + \frac{4}{3!}x^5 + 2\left[\frac{12}{3!}x^3 + \frac{15}{4!}x^4 + \frac{18}{5!}x^5\right]$$

$$= 4x^2 + 4x^3 + \frac{4}{2!}x^4 + \frac{4}{3!}x^5 + \frac{24}{3!}x^3 + \frac{30}{4!}x^4 + \frac{36}{5!}x^5$$

$$= 4x^2 + 4x^3 + \frac{24}{3!}x^3 + \frac{4}{2!}x^4 + \frac{30}{4}x^4 + \frac{4}{3!}x^5 + \frac{36}{5!}x^5$$

$$= 4x^2 + 4x^3 + 4x^3 + 2x^4 + \frac{5}{4}x^4 + \frac{2}{3}x^5 - \frac{3}{10}x^5$$

$$= 4x^2 + 8x^3 + \frac{13}{4}x^4 + \frac{49}{30}x^5$$

For the coefficient of the x^{10}-term for $r(x)$, $n = 8$ for both the $f(x)$ and $g'(x)$ functions.

$$f(x) = \frac{4x^{n+2}}{n!}$$

$$f(x) = \cdots + \frac{4x^{(8)+2}}{(8)!} + \cdots$$

$$f(x) = \cdots + \frac{4x^{10}}{8!} + \cdots$$

Then

$$g'(x) = \frac{3(n+3)x^{n+2}}{(n+2)!}$$

$$g'(x) = \cdots + \frac{3((8)+3)x^{(8)+2}}{((8)+2)!} + \cdots$$

$$g'(x) = \cdots + \frac{3(11)x^{10}}{10!} + \cdots$$

$$g'(x) = \cdots + \frac{33x^{10}}{10!} + \cdots$$

Do not forget to double $g'(x)$. So the coefficient for x^{10} is $\dfrac{4}{8!} + \dfrac{66}{10!}$.

(d)

Solutions	Scoring Rubric
To evaluate this integral, take the antiderivative of $f(t)$ and evaluate between the upper bound x and the lower bound 0. $$h(x) = \int_0^x f(t)\,dt$$ $$h(x) = \int_0^x 4t^2 + 4t^3 + \frac{4t^4}{2!} + \frac{4t^5}{3!} + \cdots + \frac{4t^{n+2}}{n!}\,dt$$ $$= \left[\frac{4t^3}{3} + \frac{4t^4}{4} + \frac{4t^5}{5\cdot 2!} + \frac{4t^5}{6\cdot 3!} + \cdots + \frac{4t^{n+2+1}}{(n+2+1)(n!)}\right]_0^x$$ $$= \frac{4x^3}{3} + \frac{4x^4}{4} + \frac{4x^5}{5\cdot 2!} + \frac{4x^6}{6\cdot 3!} + \cdots + \frac{4x^{n+3}}{(n+3)(n!)}$$ $$= \frac{4}{3}x^3 + x^4 + \frac{2}{5}x^5 + \frac{x^6}{9} + \cdots + \frac{4x^{n+3}}{(n+3)(n!)}$$	1: correct integration $$\frac{4t^3}{3} + \frac{4t^4}{4} + \frac{4t^5}{5\cdot 2!} + \frac{4t^6}{6\cdot 3!} + \cdots$$ $$+ \frac{4t^{n+2+1}}{(n+2+1)(n!)}$$ 1: $\frac{4}{3}x^3 + x^4 + \frac{2}{5}x^5 + \frac{x^6}{9} + \cdots$ $$+ \frac{4x^{n+3}}{(n+3)(n!)}$$

5. (a)

Solutions	Scoring Rubric
Differentiate using the product rule to get $\frac{d^2y}{dx^2} = 2y\frac{dy}{dx}(2+x^3) + 3x^2y^2$. Substitute 2 for x and $\frac{1}{2}$ for y in the original differential equation to find $\frac{dy}{dx}$. $$\frac{dy}{dx} = y^2(2+x^3)$$ $$\frac{dy}{dx} = \left(\frac{1}{2}\right)^2(2+(2)^3)$$ $$\frac{dy}{dx} = \frac{5}{2}$$ This can be substituted along with 2 for x and $\frac{1}{2}$ for y to find $\frac{d^2y}{dx^2}$. $$\frac{d^2y}{dx^2} = 2y\frac{dy}{dx}(2+x^3) + 3x^2y^2$$ $$\frac{d^2y}{dx^2} = 2\left(\frac{1}{2}\right)\left(\frac{5}{2}\right)(2+(2)^3) + 3(2)^2\left(\frac{1}{2}\right)^2$$ $$\frac{d^2y}{dx^2} = 28$$	1: $\frac{d^2y}{dx^2} = 2y(2+x^3) + 3x^2y^2\frac{dy}{dx}$ 1: $\frac{dy}{dx} = \frac{5}{2}$ 1: $\frac{d^2y}{dx^2} = 28$

(b)

Solutions	Scoring Rubric
To separate the variables in the given differential equation, multiply both sides of the equation by dx and divide both sides by y^2. $$\frac{dy}{dx} = y^2\left(2 + x^3\right)$$ $$\frac{dy}{y^2} = \left(2 + x^3\right) dx$$ Now integrate both sides and solve for y. $$\int \frac{dy}{y^2} = \int \left(2 + x^3\right) dx$$ $$-y = 2x + \frac{x^4}{4} + C$$ $$y = -2x - \frac{x^4}{4} + C$$ Substitute the given condition to find the constant of integration. $$y = -2x - \frac{x^4}{4} + C$$ $$\left(\frac{1}{2}\right) = -2(2) - \frac{(2)^4}{4} + C$$ $$\left(\frac{1}{2}\right) = -4 - 4 + C$$ $$\frac{17}{2} = C$$ Hence, $y = -2x - \frac{x^4}{4} + \frac{17}{2}$	1: separate variables correctly $$\frac{dy}{y^2} = \left(2 + x^3\right) dx$$ 1: $y = -2x - \frac{x^4}{4} + C$ 1: $C = \frac{17}{2}$ 1: $y = -2x - \frac{x^4}{4} + \frac{17}{2}$

(c)

Solutions	Scoring Rubric
Use $\frac{dy}{dx} = y^2\left(2 + x^3\right)$ and the point $(-1, 2)$ to determine whether the function is increasing or decreasing. $$\frac{dy}{dx} = y^2\left(2 + x^3\right)$$ $$\frac{dy}{dx} = (2)^2\left(2 + (-1)^3\right)$$ $$\frac{dy}{dx} = 4(2 - 1)$$ $$\frac{dy}{dx} = 4$$ Because the first derivative is positive, $f(x)$ is increasing at $(-1, 2)$.	1: $\frac{dy}{dx} = 4$ 1: $f(x)$ is increasing at $(-1, 2)$.

6. (a)

Solutions	Scoring Rubric
To answer this question, write the Maclaurin polynomial using the following formula. $$f(x) \approx f(0) + f'(0)(x) + \frac{f''(0)}{2}(x)^2 + \cdots$$ Because you are given $P_2\left(\frac{1}{2}\right) = 9\frac{3}{4}$, write the 2nd-degree polynomial. $$P_2(x) = 8 + f'(0)(x) + \frac{2}{2}x^2$$ $$9\frac{3}{4} = 8 + f'(0)\left(\frac{1}{2}\right) + \left(\frac{1}{2}\right)^2$$ $$9\frac{3}{4} = 8 + f'(0)\left(\frac{1}{2}\right) + \frac{1}{4}$$ $$9\frac{3}{4} = 8\frac{1}{4} + f'(0)\left(\frac{1}{2}\right)$$ $$\frac{3}{2} = f'(0)\left(\frac{1}{2}\right)$$ $$3 = f'(0)$$	1: correct work shown 1: $f'(0) = 3$

(b)

Solutions	Scoring Rubric
To answer this question, make sure you have all the derivatives you need before you substitute into the polynomial formula: $$P_5(x) = f(0) + f'(0)(x) + \frac{f''(0)}{2}(x)^2 + \frac{f^{(3)}(0)}{3!}(x)^3 + \frac{f^{(4)}(0)}{4!}(x)^4 + \frac{f^{(5)}(0)}{5!}(x)^5$$ $f(0) = 3$ and $f''(0) = 2$. $f^{(3)}(0) = \frac{1}{3}f'(0)$, so $f^{(3)}(0) = 1$. $f^{(4)}(0) = \frac{1}{2}f^{(3)}(0)$, so $f^{(4)}(0) = \frac{1}{2}$. $f^{(5)}(0) = 2f^{(4)}(0)$, so $f^{(5)}(0) = \frac{1}{4}$. $$P_5(x) = 8 + 3x + x^2 + \frac{x^3}{3!} + \frac{\frac{1}{2}x^4}{4!} + \frac{\frac{1}{4}x^5}{5!}$$ $$P_5(x) = 8 + 3x + x^2 + \frac{x^3}{6} + \frac{x^4}{48} + \frac{x^5}{480}$$	1: deduce that $f^{(4)}(0) = \frac{1}{2}$ 1: deduce that $f^{(5)}(0) = \frac{1}{4}$ 1: $P_5(x) = 8 + 3x + x^2$ $+ \frac{x^3}{6} + \frac{x^4}{48} + \frac{x^5}{480}$

(c)

Solutions	Scoring Rubric
Be sure to calculate the necessary derivatives before plugging them into the polynomial formula. You need to use the chain rule when taking these derivatives. $$h'(x) = f(2x) = f(0) = 8$$ $$h''(x) = 2f'(2x) = 2f'(0) = 2(3) = 6$$ $$h^{(3)}(x) = 4f''(2x) = 4f''(0) = 4(2) = 8$$ $$P_3(x) = h(0) + h'(0)(x) + \frac{h''(0)x^2}{2} + \frac{h^{(3)}(0)x^3}{3!}$$ $$P_3(x) = 6 + 8x + \frac{6x^2}{2} + \frac{8x^2}{3!}$$	1: calculate $h'(x)$, $h''(x)$, and $h(3)(x)$ 1: $P_3(x) = 6 + 8x + \frac{6x^2}{2} + \frac{8x^2}{3!}$

(d)

Solutions	Scoring Rubric
To find the error for the 4th-degree polynomial, use the 5th-degree term. Start by finding the 5th derivative. $$h^{(3)}(x) = 4f''(2x) = 4f''(0) = 4(2) = 8$$ $$h^{(4)}(x) = 8f^{(3)}(2x) = 8f^{(3)}(0) = 8(1) = 8$$ $$h^{(5)}(x) = 8f^{(4)}(2x) = 8f^{(4)}(0) = 8\left(\frac{1}{2}\right) = 4$$ The 5th-degree term is $P(x) = \cdots + \frac{h^{(5)}(0)x^5}{5!} + \cdots$ $$\text{Error} < \left| \frac{h^{(5)}(0)x^5}{5!} \right|$$ $$\text{Error} < \left| \frac{4\left(\frac{1}{2}\right)^5}{5!} \right|$$ $$\text{Error} < \left| \frac{4\left(\frac{1}{32}\right)}{5!} \right|$$ $$\text{Error} < \left| \frac{\frac{1}{8}}{5!} \right|$$ $$\text{Error} < \left| \frac{1}{8 \cdot 5 \cdot 4 \cdot 3 \cdot 2 \cdot 1} \right|$$ $$\text{Error} < \left| \frac{1}{960} \right|$$ Because $\frac{1}{960} < \frac{1}{950}$, Drew is correct.	1: because $\frac{1}{960} < \frac{1}{950}$, Drew is correct 1: explanation/work shown

Derivative Rules

TABLE OF DERIVATIVE FORMULAS

Name	Formula
Derivative of a constant	$\frac{d}{dx}\big(f(x)\big) = \frac{d}{dx}(c) = 0$
Power rule	$\frac{d}{dx}\big(x^n\big) = nx^{n-1}$, n an integer
Derivative of $\sqrt{x}$	$\frac{d}{dx}\big(\sqrt{x}\big) = \frac{1}{2\sqrt{x}}$
Constant multiple rule	$\frac{d}{dx}\big(cf(x)\big) = c \cdot \frac{d}{dx}\big(f(x)\big)$, c a real number
Sum and difference rule	$\frac{d}{dx}\big(f(x) \pm g(x)\big) = \frac{d}{dx}\big(f(x)\big) \pm \frac{d}{dx}\big(g(x)\big)$ for f and g differentiable at x
Product rule	$\frac{d}{dx}\big(f(x) \cdot g(x)\big) = \frac{d}{dx}\big(f(x) \cdot g(x)\big) + f(x) \cdot \frac{d}{dx}\big(g(x)\big)$ for f and g differentiable at x
Quotient rule	$\frac{d}{dx}\left(\frac{f(x)}{g(x)}\right) = \dfrac{\frac{d}{dx}\big(f(x)\big) \cdot g(x) - f(x) \cdot \frac{d}{dx}\big(g(x)\big)}{\big(g(x)\big)^2}$ for f and g differentiable at x, $g(x) \neq 0$
Chain rule	$\frac{d}{dx}h\big(g(x)\big) = \frac{dh}{dx}\big(g(x)\big) \cdot \frac{d}{dx}g(x)$ for g differentiable at x and h differentiable at $g(x)$
Derivatives of trigonometric functions	$\frac{d}{dx}(\sin x) = \cos x$ $\frac{d}{dx}(\tan x) = \sec^2 x$ $\frac{d}{dx}(\sec x) = \sec x \tan x$ $\frac{d}{dx}(\cos x) = -\sin x$ $\frac{d}{dx}(\cot x) = -\csc^2 x$ $\frac{d}{dx}(\csc x) = -\csc x \cot x$
Derivative of e^x	$\frac{d}{dx}\big(e^x\big) = e^x$
Derivative of e^u	$\frac{d}{dx}\big(e^u\big) = e^u \frac{du}{dx}$
Generalized power rule	$\frac{d}{dx}\big(x^r\big) = rx^{r-1}$ for r a real number

Appendices

Derivative of $\ln x$	$\frac{d}{dx}(\ln x) = \frac{1}{x}$
Derivative of $\ln u$	$\frac{d}{dx}(\ln u) = \left(\frac{1}{u}\right)\frac{du}{dx}$
Derivative of f^{-1}	$\frac{d}{dx}\left(f^{-1}(x)\right) = \dfrac{1}{\frac{d}{dx}f\left(f^{-1}(x)\right)}$ for all x where f is differentiable and $f'(f^{-1}(x)) \neq 0$
Derivative of b^x	$\frac{d}{dx}\left(b^x\right) = \ln b \cdot b^x$ for $b > 0$
Derivative of $\log_b x$	$\frac{d}{dx}(\log_b x) = \frac{1}{\ln b \cdot x}, b > 0$
Derivatives of inverse trigonometric functions	$\frac{d}{dx}(\arctan x) = \frac{1}{1+x^2}$ $\frac{d}{dx}(\arcsin x) = \frac{1}{\sqrt{1-x^2}}$ $\frac{d}{dx}(\text{arcsec } x) = \frac{1}{|x|\sqrt{x^2-1}}$ $\frac{d}{dx}(\text{arccot } x) = \frac{-1}{1+x^2}$ $\frac{d}{dx}(\arccos x) = \frac{-1}{\sqrt{1-x^2}}$ $\frac{d}{dx}(\text{arccsc } x) = \frac{-1}{|x|\sqrt{x^2-1}}$

APPENDIX B

Integration Rules

TABLE OF INTEGRATION FORMULAS

Name	Formula
Integration of one	$\int 1 \, dx = x + C$
Integration of powers	$\int x^r \, dx = \dfrac{x^{r+1}}{r+1} + C$
Intergration of trigo-nometric functions	$\int \cos x \, dx = \sin x + C$ $\int \sec^2 x \, dx = \tan x + C$ $\int \sec x \, \tan x \, dx = \sec x + C$ $\int \sin x \, dx = -\cos x + C$ $\int \csc^2 x \, dx = -\cot x + C$ $\int \csc x \, \cot x \, dx = -\csc x + C$
Integration of e^x	$\int e^x \, dx = e^x + C$
Integration of $^1/x$	$\int \dfrac{1}{x} \, dx = \ln|x| + C$
Integration of b^x	$\int b^x \, dx = \dfrac{b^x}{\ln b} + C$
Integration leading to inverse trigono-metric functions	$\int \dfrac{1}{1+x^2} \, dx = \arctan x + C$ $\int \dfrac{1}{\sqrt{1-x^2}} \, dx = \arcsin x + C$ $\int \dfrac{1}{x\sqrt{x^2-1}} \, dx = \operatorname{arcsec}|x| + C$
First fundamental theorem of calculus	If f is continuous on $[a, b]$ and F is any antiderivative of f then $\int_a^b f(x) \, dx = F(b) - F(a)$.
Second fundamental theorem of calculus	The function $G(x) = \int_a^x f(t) \, dt = F(x) - F(a)$ is differentiable and $\dfrac{d}{dx}\big(G(x)\big) = f(x)$, that is, $\dfrac{d}{dx} \int_a^x f(t) \, dt = f(x)$.
Left Riemann sum	$\underset{\text{left endpoint}}{R_n} \approx \displaystyle\sum_{i=1}^{n} f(x_{i-1}) \, \Delta x$ $= \big(f(x_0) + f(x_1) + \ldots + f(x_{i-1})\big)\dfrac{b-a}{n}$

Right Riemann sum	$$R_n \underset{\text{right endpoint}}{\approx} \sum_{i=1}^{n} f(x_i)\, \Delta x$$ $$= \left(f(x_1) + f(x_2) + \ldots + f(x_n)\right)\frac{b-a}{n}$$
Midpoint Riemann sum	$$R_n \underset{\text{midpoint}}{\approx} \sum_{i=1}^{n} f\left(x_{i-1} + \frac{b-a}{2n}\right) \Delta x$$ $$= \left(f\left(x_0 + \frac{b-a}{2n}\right) + f\left(x_1 + \frac{b-a}{2n}\right) + \ldots + f\left(x_{i-1} + \frac{b-a}{2n}\right)\right)\frac{b-a}{n}$$
Trapezoidal approximation	$$T_n = \left(\frac{1}{2}\left(f(x_0) + f(x_1)\right)\cdot \Delta x\right) + \left(\frac{1}{2}\left(f(x_1) + f(x_2)\right)\cdot \Delta x\right) + \ldots + \left(\frac{1}{2}\left(f(x_{n-1}) + f(x_n)\right)\cdot \Delta x\right)$$ $$= \left(\frac{1}{2}f(x_0) + f(x_1) + f(x_2) + \ldots + f(x_{n-1}) + \frac{1}{2}f(x_n)\right)\Delta x$$
Properties of definite integrals	$$\int_a^b f(x) \pm g(x)\, dx = \int_a^b f(x)\, dx \pm \int_a^b g(x)\, dx$$ $$\int_a^b c \cdot f(x)\, dx = c\, dx \int_a^b f(x)$$ $$\int_a^b f(x)\, dx = -\int_b^a f(x)\, dx$$ $$\int_a^b f(x)\, dx = \int_a^c f(x)\, dx + \int_c^b f(x)\, dx$$
Area between two curves	$$\int_a^b (\text{function on top} - \text{function on bottom})\, dx$$
Area bounded by two curves	$$\int_c^d (\text{function on right} - \text{function on left})\, dy$$
Volume of a solid	$$\int_a^b A(x)\, dx \text{ or } \int_a^b A(y)\, dy$$
Volume of a solid of revolution	$$V = \int_a^b \pi r^2\, dx \text{ or } V = \int_a^b \tau R^2 - \pi r^2\, dx$$
Average value integral	$$\frac{1}{b-a}\int_a^b f(x)\, dx$$
Accumulation function	$$G(x) = \int_0^x f(t)\, dt$$
Motion along a line	$$v(t) = \int_a^t a(t)\, dt$$ $$x(t) = \int_a^t v(t)\, dt$$
Rate of change	$$\frac{dy}{dx} = f(x)$$
Exponential growth	Differential equation of the form $\frac{dy}{dx} = ky$ or $\frac{dy}{y} = f(x)\, dx$, solution of the form $y = Ae^{g(x)}$